Biotechnology: Advanced Principles and Applications

Edited by Joy Adam

SYRAWOOD
PUBLISHING HOUSE

New York

Published by Syrawood Publishing House,
750 Third Avenue, 9th Floor,
New York, NY 10017, USA
www.syrawoodpublishinghouse.com

Biotechnology: Advanced Principles and Applications
Edited by Joy Adam

© 2019 Syrawood Publishing House

International Standard Book Number: 978-1-68286-750-1 (Hardback)

Cataloging-in-Publication Data

Biotechnology : advanced principles and applications / edited by Joy Adam.
    p. cm.
Includes bibliographical references and index.
ISBN 978-1-68286-750-1
1. Biotechnology. 2. Genetic engineering. I. Adam, Joy.
TP248.2 .B56 2019
660.6--dc23

TABLE OF CONTENTS

PREFACE

This book has been an outcome of determined endeavour from a group of educationists in the field. The primary objective was to involve a broad spectrum of professionals from diverse cultural background involved in the field for developing new researches. The book not only targets students but also scholars pursuing higher research for further enhancement of the theoretical and practical applications of the subject.

The vast field of biotechnology encompasses a number of procedures that involve the modification of living organisms for specific purposes. The applications of biotechnology are found in diverse areas of agriculture, medicine, industry, marine and aquatic environment, etc. Depending on such specializations, this field is categorized into multiple sub-fields. Green, white and yellow biotechnology are its most significant sub-fields. Green biotechnology is involved with the production of genetically modified seeds and the development of environmentally friendly fertilizers and biopesticides. Yellow biotechnology is involved in food production, like in the fermentation of beer, wine and cheese as well as in the control of harmful insects. The applications of bio-catalysis for industrial purposes falls under the domain of white biotechnology. The objective of this book is to give a general view of the different areas of biotechnology and their applications. The topics included in this book are of the utmost significance and bound to provide incredible insights to readers. As this field is emerging at a rapid pace, the contents of this book will help the readers understand the modern concepts and applications of the subject.

It was an honour to edit such a profound book and also a challenging task to compile and examine all the relevant data for accuracy and originality. I wish to acknowledge the efforts of the contributors for submitting such brilliant and diverse chapters in the field and for endlessly working for the completion of the book. Last, but not the least; I thank my family for being a constant source of support in all my research endeavours.

Editor

Recent updates on different methods of pretreatment of lignocellulosic feedstocks

Adepu Kiran Kumar[*] and Shaishav Sharma

Abstract

Lignocellulosic feedstock materials are the most abundant renewable bioresource material available on earth. It is primarily composed of cellulose, hemicellulose, and lignin, which are strongly associated with each other. Pretreatment processes are mainly involved in effective separation of these complex interlinked fractions and increase the accessibility of each individual component, thereby becoming an essential step in a broad range of applications particularly for biomass valorization. However, a major hurdle is the removal of sturdy and rugged lignin component which is highly resistant to solubilization and is also a major inhibitor for hydrolysis of cellulose and hemicellulose. Moreover, other factors such as lignin content, crystalline, and rigid nature of cellulose, production of post-pretreatment inhibitory products and size of feed stock particle limit the digestibility of lignocellulosic biomass. This has led to extensive research in the development of various pretreatment processes. The major pretreatment methods include physical, chemical, and biological approaches. The selection of pretreatment process depends exclusively on the application. As compared to the conventional single pretreatment process, integrated processes combining two or more pretreatment techniques is beneficial in reducing the number of process operational steps besides minimizing the production of undesirable inhibitors. However, an extensive research is still required for the development of new and more efficient pretreatment processes for lignocellulosic feedstocks yielding promising results.

Keywords: Pretreatment, Lignocellulosic biomass, Cellulose, Lignin, Reducing sugars

Background

Lignocellulosic feedstock represents an extraordinarily large amount of renewable bioresource available in surplus on earth and is a suitable raw material for vast number of applications for human sustainability. The main composition of lignocellulosic feedstocks is cellulose, hemicellulose, and lignin (Table 1). However, many obstacles are associated with effective utilization of lignocellulosic materials. Some of the major factors are the recalcitrance of the plant cell wall due to integral structural complexity of lignocellulosic fractions and strong hindrance from the inhibitors and byproducts that are generated during pretreatment. In addition, few more challenges still remain, like understanding the

physicochemical architecture of feedstock cell walls, suitable pretreatment method and extent of cell wall deconstruction for generation of value-added products etc.

There are several criteria for the selection of a suitable pretreatment method: (a) the selected method should avoid the size reduction of biomass particles, (b) hemicellulose fraction must be preserved, (c) minimize the formation of degradation products, (d) minimize the energy demands and lastly, (e) should involve a low-cost pretreatment catalyst and/or inexpensive catalyst recycle and regeneration of high-value lignin co-product (Wyman 1999). The result of the pretreatment must not only defend but also justify its impact on the cost of downstream processing steps and the tradeoff between operating costs, capital costs, biomass costs, etc. (Lynd et al. 1996).

The pretreatment techniques for overcoming biomass recalcitrance are broadly divided into two classes:

*Correspondence: kiranbio@gmail.com
Bioconversion Technology Division, Sardar Patel Renewable Energy
Research Institute, Vallabh Vidyanagar, Anand 388 120, Gujarat, India

Table 1 Cellulose, hemicellulose, and lignin content in common lignocellulosic feedstocks

Lignocellulosic feedstocks	Cellulose (%)	Hemicellulose (%)	Lignin (%)
Sugar cane bagasse	42	25	20
Sweet sorghum	45	27	21
Hardwood	40–55	24–40	18–25
Softwood	45–50	25–35	25–35
Corn cobs	45	35	15
Corn stover	38	26	19
Rice Straw	32	24	18
Nut shells	25–30	25–30	30–40
Newspaper	40–55	25–40	18–30
Grasses	25–40	25–50	10–30
Wheat straw	29–35	26–32	16–21
Banana waste	13.2	14.8	14
Bagasse	54.87	16.52	23.33
Sponge gourd fibers	66.59	17.44	15.46
Agricultural residues	5–15	37–50	25–50
Hardwood	20–25	45–47	25–40
Softwood	30–60	40–45	25–29
Grasses	0	25–40	35–50
Waste papers from chemical pulps	6–10	50–70	12–20
Newspaper	12	40–55	25–40
Sorted refuse	60	20	20
Leaves	15–20	80–85	0
Cotton seed hairs	80–95	5–20	0
Paper	85–99	0	0–15
Switch grass	45	31.4	12

biochemical and thermochemical (Laser et al. 2009). Based on the operating temperatures, thermochemical pretreatment is again of two types: pyrolysis and gasification. The advantage of thermochemical conversion is that it is a fast process with low residence time and is able to handle a broad range of feedstock in a continuous manner, but major drawback is its non-specific nature of biomass deconstruction. On the other hand, biochemical pretreatment is highly selective in biomass deconstruction to their desired product formation. However, biochemical conversion first uses low-severity thermochemical pretreatment to partially break down the cell wall and expose the cellulose and hemicellulose fractions for improving enzyme accessibility. Elucidating the physicochemical effects of the possible pretreatments upon subsequent hydrolysis and fermentation of biomass has been a significant challenge.

Although several reviews have been present which describe the various categories of pretreatment processes individually, however, a comprehensive review covering different types of pretreatment processes along with their advantages and disadvantages was the need of the hour. Therefore, this review covers all the techniques that have been developed and used for pretreatment of lignocellulosic biomass, recent advancements in pretreatment technology, their mechanism of action, and effect on various lignocellulosic feedstocks.

Methods of pretreatment

The pretreatment of lignocellulosic feedstocks is an essential step and is required to alter the structure of biomass residues and expose the lignocellulosic fractions for easy access to enzymes during enzymatic hydrolysis and enhance the rate and yield of reducing sugars (Alvira et al. 2010). Basically, the pretreatment processes are classified into two major regimes viz. non-biological and biological. A list of promising and most commonly used pretreatment methods are listed in Fig. 1. Based on the type of the treatment process involved, lignocellulosic biomass pretreatment methods are broadly classified into two groups: Non-biological and biological. Non-biological pretreatment methods do not involve any microbial treatments and are roughly divided into different categories: physical, chemical, and physico-chemical methods. Here, we have reviewed the advances in few selective treatment methods that are most commonly employed in pretreatment process of a broad range of lignocellulosic feedstocks.

Physical pretreatment
Mechanical extrusion
It is the most conventional method of biomass pretreatment where the feedstock materials are subjected to heating process (>300 °C) under shear mixing. This pretreatment process results mainly in production of gaseous products and char from the pretreated lignocellulosic biomass residues (Shafizadeh and Bradbury 1979). Due to the combined effects of high temperatures that are maintained in the barrel and the shearing force generated by the rotating screw blades, the amorphous and crystalline cellulose matrix in the biomass residues is disrupted. However, this method requires significant amount of high energy making it a cost intensive method and difficult to scale up for industrial purposes (Zhu and Pan 2010). Karunanithy et al. (2008) studied on the defibrillation and shortening of the biomass fibers and concomitant increase in the overall content of the carbohydrates and its availability for enzymatic hydrolysis process.

Zheng and Rehmann (2014) studied different process parameters of mechanical extraction process and found that the type of the screw design, compression ratio, screw speed, and barrel temperature affected the biomass pretreatment. Similarly, Karunanithy and Muthukumarappan (2010) also studied the effect of temperature

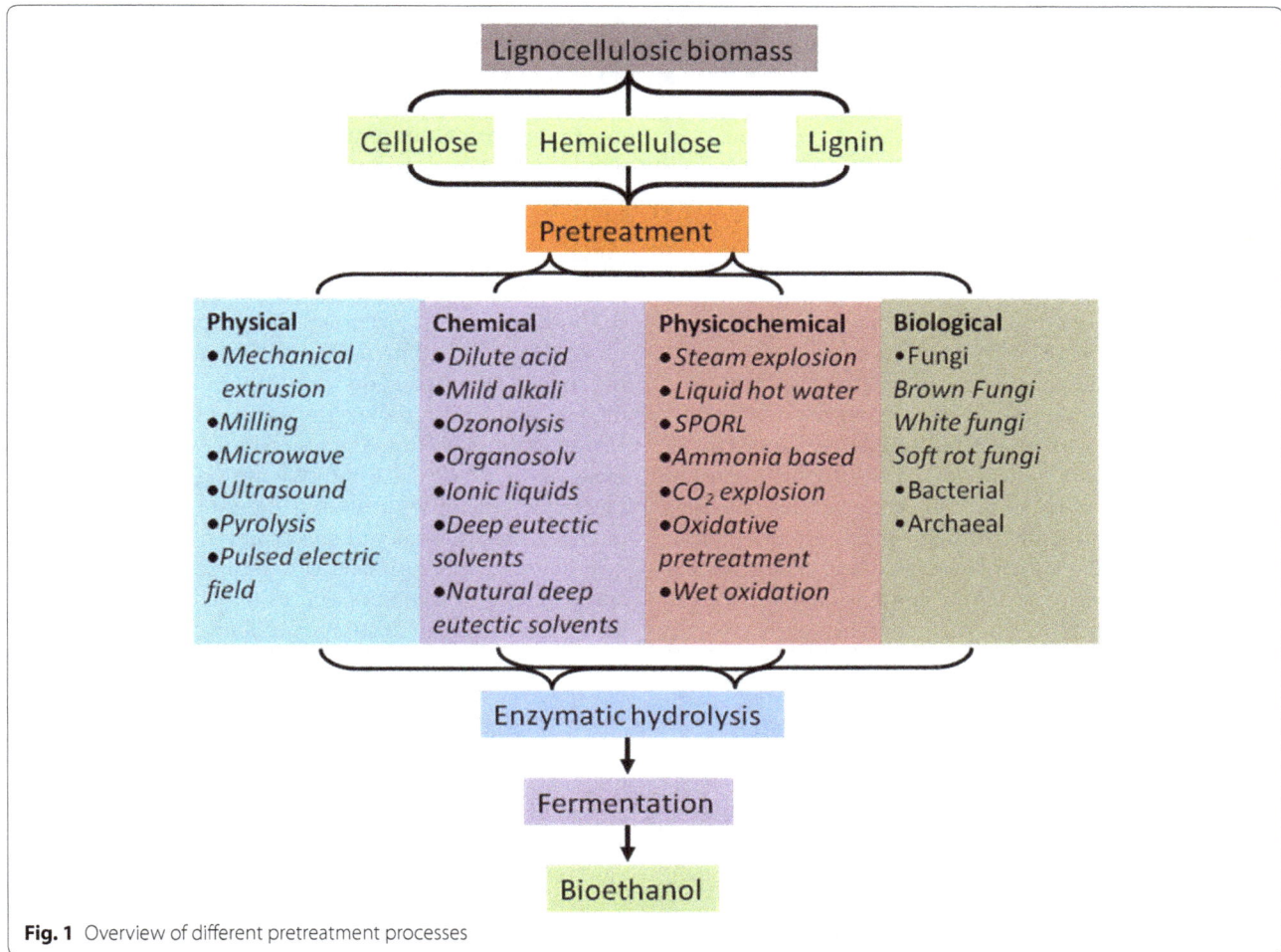

Fig. 1 Overview of different pretreatment processes

and screw speed on pretreatment of corn cobs with different cellulose degrading enzymes and their ratios. When pretreatment was carried out at different temperatures (25, 50, 75, 100, and 125 °C) and different screw speeds (25, 50, 75, 100, and 125 rpm), maximum concentrations of glucose (75%), xylose (49%), and combined sugars (61%) were obtained at 75 rpm and 125 °C using cellulase and β-glucosidase in the ratio of 1:4, which were nearly 2.0, 1.7, and 2.0 times higher than the controls. These clearly indicated that optimization of the pretreatment process conditions and enzyme concentrations had a synergetic effect on the overall yields of reducing sugars.

Moreover, in another study, Karunanithy et al. (2013) selected different varieties of warm season grasses viz. switch grass, big bluestem, and prairie cord grass and studied the effect of different screw speeds (100, 150, and 200 rpm), barrel temperatures (50, 75, 100, 150, and 200 °C) and different concentrations of cellulase with β-glucosidase (1:1 to 1:4). In all the experiments,

maximum reducing sugars were obtained when the ratio of cellulase and β-glucosidase was maintained at 1:4. The reducing sugar yields from the switchgrass pretreated at screw speed of 200 rpm and barrel temperature of 75 °C produced 28.2%, while big bluestem pretreated at screw speed of 200 rpm and 150 °C barrel temperature produced 66.2% and with prairie cord grass pretreated at 150 rpm and 100 °C produced 49.2%. Although the sugar yields are high, mechanical extrusion cannot alone suffice pretreatment of a range of lignocellulosic feedstocks with varied cellulose, hemicellulose, and lignin contents. Thus, it needs better pretreatment methods for higher sugar yields. Besides, sugar recovery is also significantly influenced by the properties of the biomass.

Karunanithy and Muthukumarappan (2010) studied the effect of varying moisture contents (15, 25, 35, and 45% wb) on the sugar recovery from switch grass and prairie cord grass at compression ratio (2:1 and 3:1), screw speed (50, 100 and 150 rpm), and barrel temperature (50, 100, and 150 °C). After enzymatic hydrolysis of the pretreated biomass, maximum 45.2% sugar was recovered from switch grass with 15% moisture content at screw speed of 50 rpm and barrel temperature of 150 °C, whereas a maximum of 65.8% sugar was recovered from prairie cord grass with 25% moisture content at screw speed of 50 rpm and barrel temperature of 50 °C. Alongside, low concentrations of glycerol and acetic acid in the range of 0.02–0.18 g/L were also produced. It is well known that glycerol and acetic acid are the byproducts that are formed during the pretreatment of lignocellulosic feedstocks. However, in this report, unlike hot compressed hot water and acid hydrolysis, the byproduct formation was significantly lower because in mechanical extraction only physical interactions were observed between the feedstock and the barrel blades. Similarly, Lamsal et al. (2010) also compared effects of grinding with extrusion on wheat bran and soybean hull. Better sugar yield was obtained in wheat bran through extrusion but not in soybean hulls. The most plausible reason could be due to the difference in the lignin contents between these biomass residues. Soybean hulls contain nearly twofold higher lignin content than the wheat bran. The residual high-lignin bound to the pretreated biomass could have shown a direct impact on the enzymatic hydrolysis. It is well known that the cellulose degrading enzymes avidly and irreversibly bind to lignin and thus not readily available for effective cellulose disruption. The combination of screw speed and barrel temperature maintained was 7 Hz/150 °C and 3.7 Hz/110 °C where highest sugar yield was produced.

Moreover, particle size of biomass plays an important role on the overall sugar recovery. Studies performed by Karunanithy and Muthukumarappan (2011) showed maximum sugar recovery from big blue stem obtained with 8-mm particle size, 20% wb moisture content at a barrel temperature of 180 °C with screw speed of 150 rpm, where 71.3% glucose, 78.5% xylose, and 56.9% combined sugars were obtained. While with switch grass, at similar particle size and moisture contents, but at a barrel temperature of 176 °C and screw speed of 155 rpm, maximum sugars of 41.4% glucose, 62.2% xylose, and 47.4% combined sugars were obtained. In another study, Zhang et al. (2012a, b) used a twin screw extruder for sugar recovery from corn stover. At 27.5% moisture content with a screw speed of 80 rpm and enzyme dose of 0.028 g enzyme/g dry biomass, glucose, xylose, and combined sugar recovery were 48.79, 24.98, and 40.07%, respectively. These were 2.2, 6.6, and 2.6 times more than that of untreated corn stover. Yoo (2011) compared a thermo-mechanical pretreatment process on soybean hulls. Under optimum processing conditions at screw speed of 350 rpm, barrel temperature of 80 °C and 40% moisture content, 95% cellulose was converted glucose.

These above studies clearly demonstrate that mechanical extrusion treatment had a significant effect on breakdown of cellulose and hemicelluloses fractions from a wide variety of lignocellulosic feedstocks; however, when combined with other pretreatment methods, mechanical extrusion performs better and might enhance the overall yields of the reducing sugars.

Milling

Mechanical grinding (milling) is used for reducing the crystallinity of cellulose. It mostly includes chipping, grinding, and/or milling techniques. Chipping can reduce the biomass size to 10–30 mm only while grinding and milling can reduce the particle size up to 0.2 mm. However, studies found that further reduction of biomass particle below 0.4 mm has no significant effect on rate and yield of hydrolysis (Chang et al. 1997). Chipping reduces the heat and mass transfer limitations while grinding and milling effectively reduce the particle size and cellulose crystallinity due to the shear forces generated during milling. The type and duration of milling and also the kind of biomass determine the increase in specific surface area, final degree of polymerization, and the net reduction in cellulose crystallinity. Different milling methods viz. two-roll milling, hammer milling, colloid milling, and vibratory milling are used to improve the digestibility of the lignocellulosic materials (Taherzadeh and Karimi 2008). Compared to ordinary milling process, vibratory ball milling is found to be more effective in reducing cellulose crystallinity and improving the digestibility of spruce and aspen chips. Also, wet disk milling has been a popular mechanical pretreatment because of its low energy consumption. Disk milling enhances cellulose hydrolysis by producing fibers and is more effective as compared to hammer milling which produces finer bundles (Zhua et al. 2009). Hideno et al. (2009) compared the effect of wet disk milling and conventional ball milling pretreatment method over rice straw. The optimal conditions obtained were 60 min of milling in case of dry ball milling while 10 repeated milling operations were required in case of wet disk milling. Maximum glucose (89.4%) and xylose (54.3%) were obtained with conventional ball milling method as compared to 78.5% glucose and 41.5% xylose with wet disk milling method. However, wet disk milling had lower energy requirement, high effectiveness for enzymatic hydrolysis, and did not produce inhibitors. Lin et al. (2010) found wet milling better than dry milling for the pretreatment of corn stover. The optimum parameters for milling were particle size 0.5 mm, solid/liquid ratio of 1:10, 20 number of steel balls of 10 mm dia each, ball speed of 350 rpm/min grounded for 30 min. Better results were obtained when milling was combined with alkaline pretreatment method. As compared to wet milling process, alkaline milling treatment increased the enzymatic hydrolysis efficiency of corn stover by 110%. Sant Ana da Silva et al. (2010) performed a comparative analysis on effects of ball milling and wet disk milling on treating sugarcane bagasse and straw and found ball milling better pretreatment method than wet disk milling in terms of glucose and xylose hydrolysis yields. Ball milling-treated bagasse and straw produced 78.7 and 72.1 and 77.6 and 56.8%, glucose and xylose, respectively. Kim et al. (2013) compared three different milling methods i.e., ball, attrition, and planetary milling. Attrition and planetary mills were found more effective in reducing the size of biomass as compared to ball milling. Planetary mill produced highest amount of glucose and galactose than other milling methods tested. It is to be noted that all the mill pretreatment methods do not produce any toxic compounds like hydroxymethylfurfuraldehyde (HMF) and levulinic acid. This makes milling pretreatment a good choice of preliminary pretreatment method for a wide variety of lignocellulosic feed stocks. In another study, oil palm frond fiber when pretreated through ball mill produced glucose and xylose yields of 87 and 81.6%, respectively, while empty fruit bunch produced glucose and xylose yields of 70 and 82.3%, respectively (Zakaria et al. 2014).

Microwave

Microwave irradiation is a widely used method for lignocellulosic feedstock pretreatment because of various reasons such as (1) easy operation, (2) low energy requirement, (3) high heating capacity in short duration of time, (4) minimum generation of inhibitors, and (5) degrades structural organization of cellulose fraction. Moreover, addition of mild-alkali reagents is preferred for more effective breakdown. A study on microwave-based alkali pretreatment of switch grass yielded nearly 70–90% sugars (Hu and Wen 2008). Microwave-based alkali treatment of switchgrass and coastal bermudagrass using different alkalis found sodium hydroxide as the most suitable alkali. Under optimum conditions, switchgrass produced 82% glucose and 63% xylose while coastal bermudagrass produced 87% glucose and 59% xylose (Keshwani and Cheng 2010). Although not significant, the authors have correlated the differences in reducing sugars with the difference in the lignin content (19% in bermudagrass vs 22% in switchgrass) in these lignocellulosic feedstocks. Lu et al. (2011) studied microwave pretreatment of rape straw at different powers for different time durations. The higher power of microwave resulted in higher glucose production but treatment time did not have a significant effect at a specific power setting. Chen et al. (2011a, b) optimized the microwave heating at 190 °C for 5 min for bagasse pretreatment in terms of

lignocellulosic structural disruption. In another investigation, Zhu et al. (2015a, b, 2016) have extensively studied the effects of microwave on chemically pretreated *Miscanthus*. Where, microwave treatment was applied to NaOH- and H_2SO_4-pretreated *Miscanthus* and found 12-times high sugar yield in half the time as compared to conventional heating NaOH and H_2SO_4 pretreatment. This was mainly due to the pre-disruption of crystalline cellulose and lignin solubilization with the chemical pretreatment. The maximum sugar yield obtained was 75.3% and glucose yield was 46.7% when pretreated with 0.2 M H_2SO_4 for 20 min at 180 °C. Similarly, Xu et al. (2011) developed an orthogonal design to optimize the microwave pretreatment of wheat straw and increased the ethanol yield from 2.678 to 14.8%. Bonmanumsin et al. (2012) reported substantial increase in yield of monomeric sugars from *Miscanthus sinensis* with microwave-assisted ammonium hydroxide treatment. Microwave pretreatment of oil palm empty fruit bunch fiber in the presence of alkaline conditions showed 74% reduction in lignin (Nomanbhay et al. 2013).

Ultrasound

Sonication is relatively a new technique used for the pretreatment of lignocellulosic biomass. However, studies in the laboratory have found sonication a feasible pretreatment option. Ultrasound waves produce both physical and chemical effects which alter the morphology of lignocellulosic biomass. Ultrasound treatment leads to formation of small cavitation bubbles which rupture the cellulose and hemicellulose fractions thereby increasing the accessibility to cellulose degrading enzymes for effective breakdown into simpler reducing sugars. Yachmenev et al. (2009) reported that the maximum cavitation was formed at 50 °C which is also the optimum temperature for many cellulose degrading enzymes. The ultrasonic field is primarily influenced by ultrasonic frequency and duration, reactor geometry and its type and solvent used. Furthermore, biomass characteristics, reactor configuration, and kinetics also influence the pretreatment through sonication (Bussemaker and Zhang 2013). Duration of sonication has maximum effect on pretreatment of biomass. However, prolonging sonication beyond a certain limit has no additional effect in terms of delignification and sugar release (Rehman et al. 2013). Sonication of corn starch slurry for 40 s increased the sugar yield by 5–6 times as compared to control (Montalbo et al. 2010). Sonication of alkaline pretreated wheat straw for 15–35 min increased delignification by 7.6–8.4% as compared to control (Sun and Tomkinson 2002). Besides duration, the frequency of sonication directly determines the power of sonication, which is also an important factor affecting the lignocellulosic feedstock pretreatment.

Most of the researchers have used ultrasound frequency of 10–100 kHz for the pretreatment process which has been enough for cell breakage and polymer degradation (Gogate et al. 2011). However, higher sonication power level is reported to adversely affect the pretreatment process. High power leads to formation of bubbles near tip of ultrasound transducer which hinders the transfer of energy to the liquid medium (Gogate et al. 2011). Increased oxidation of cellulose has been observed in when the sonication power was increased to 400 W in 200 mL of slurry (Aimin et al. 2005). Similarly, poplar wood cellulose powder suspension turned viscous when treated with high power of 1200 W sonication (Chen et al. 2011a, b). Therefore, power and duration of sonication should be optimized based on the biomass and slurry characteristics to meet the desired pretreatment objectives.

Pyrolysis

Pyrolysis has also been employed for the pretreatment of lignocellulosic biomass, however, in biorefinery processes. Unlike bioethanol applications, pyrolysis treatment is used for production of bio-oil from lignocellulosic feedstocks. Although limited studies have been reported on use of pyrolysis for reducing sugars production, there are few reports on use of pyrolysis in pretreatment of chemically pretreated biomass. Hence, we have included a brief section on pyrolysis pretreatment in this review. Fan et al. (1987) applied mild acid hydrolysis (1 N sulfuric acid, at 97 °C for 2.5 h) on the pyrolysis-pretreated biomass and found ~85% conversion of cellulose to reducing sugars and >5% glucose. In brief, pyrolysis is a thermal degradation process where biomass was subjected to high-temperature treatment, generally operated at 500–800 °C in the absence of oxidizing agent. At this temperature, cellulose rapidly decomposes leading to formation of end products such as gaseous substances, pyrolysis oil, and charcoal (Kilzer and Broido 1965). Pyrolysis is divided into slow and fast pyrolysis based on the heating rate. The amount of each end product varies depending on the type of pyrolysis, biomass characteristics, and reaction parameters. Besides production of high value energy-rich products, pyrolysis is adapted by thermal industries due to easy transport management, storage, combustion, and retrofitting and is flexible in production and marketing. Pyrolysis is found to be more efficient when carried out in the presence of oxygen at lower temperatures (Shafizadeh and Bradbury 1979; Kumar et al. 2009). Shafizadeh and Bradbury carried out the pyrolysis in the presence of oxygen as well as nitrogen and found that a large number of bonds were broken in the presence of oxygen as compared to nitrogen. It was estimated that at

25 °C, 7.8×10^9 bonds/min/g cellulose is cleaved in the presence of oxygen as compared to 1.7×10^8 bonds with nitrogen under similar conditions.

Biomass to liquid (BtL) route is used for the production of transportation of fuels from biomass which includes conversion of biomass to syngas to high-quality Fischer–Tropsch (FT) fuels. Zwart et al. (2006) compared alternative BtL routes comprising chipping, torrefaction, pelletization, and pyrolysis. The most efficient and commercially feasible route was found to be based on torrefaction followed by pyrolysis and pelletization. The study also clearly demonstrated the advantage of pretreatment at the front end of BtL production route by decreasing the cost of FT product by ~3 Euro/GJ.

Pulsed-electric field

Pulsed-electric field (PEF) pretreatment exposes the cellulose present in the biomass by creating the pores in the cell membrane thereby allowing the entry of agents that will break the cellulose into constituent sugars. In PEF pretreatment, the biomass is subjected to a sudden burst of high voltage between 5.0–20.0 kV/cm for short durations (nano to milliseconds). The advantages of PEF are low energy requirement due to very short duration (100 μs) of pulse time and the treatment can be carried out at ambient conditions. Also, the PEF instrument is simple in design due to lack of moving parts (Kumar et al. 2009). Salerno et al. (2009) applied PEF to waste activated sludge and pig manure for increasing the production of methane during anaerobic digestion. Methane production increased twofold from sludge and 80% from pig manure as compared to untreated sludge and manure. Kumar et al. (2011) designed and developed a PEF system for the pretreatment of wood chip and switchgrass. They studied the effect of PEF on untreated and treated samples through the uptake of neutral red dye. Both switch grass and woodchip were found resistant to structural change at low field strengths. Switchgrass showed higher neutral red uptake at field strength ≥ 8 kV/cm while woodchip showed similar results at 10 kV/cm. Electric field strength and pulse duration are the two interdependent processing parameters affecting electroporation through PEF. Two different durations in the range of milliseconds and microseconds were applied to *Chlorella vulgaris* and found irreversible electroporation at >4 kV/cm in the millisecond range and at ≥ 10 kV/cm in the microseconds range (Luengo et al. 2015). Yu et al. (2016) optimized pressure, electric field strength, and pulse number on the juice expression yield, total polyphenols, and total proteins content in the expressed juices of rapeseed stem biomass. The optimum conditions of electric field strength $E = 8$ kV/cm, pressure $P = 10$ bar and pulse number tPEF $= 2$ ms increased juice yield from

34 to 81%. Total polyphenols and total proteins content increased significantly after PEF pretreatment.

Chemical pretreatment

Dilute acid

Although acid treatment is the most commonly used conventional pretreatment method of lignocellulosic feedstocks, it is less attractive due to the generation of high amount of inhibitory products such as furfurals, 5-hydroxymethylfurfural, phenolic acids, and aldehydes. The corrosive and toxic nature of most acids requires a suitable material for building the reactor which can sustain the required experimental conditions and corrosive nature of acids (Saha et al. 2005). Still it is the most widely employed pretreatment method on industrial scale. Based on the type of end application, two types of acid pretreatments are developed; high temperature (above 180 °C) for short duration (1–5 min) and low temperature (<120 °C) for long duration (30–90 min), respectively. In some cases, enzymatic hydrolysis step could easily be avoided as acid itself hydrolyses the biomass into fermentable sugars. However, extensive washing is necessary to remove acid before fermentation of sugars (Sassner et al. 2008). Different types of reactors such as percolation, plug flow, shrinking-bed, batch, flow-through, and counter current reactors have been developed. However acid treatment generates inhibitors which need to be removed before further processing. Also, the concentrated acid must be recovered after hydrolysis in order to make the process economically feasible. Different acids have been used for the pretreatment of a variety of biomass. Some of the commonly used acids are discussed here:

Sulfuric acid The most common commercially used acid is dilute sulphuric acid (H_2SO_4). It has been widely used to pretreat switchgrass (Digman et al. 2010), corn stover (Xu et al. 2009), spruce (Shuai et al. 2010), and poplar (Kumar and Wyman 2009). Pretreatment of bermuda grass and rye straw with 1.5% sulphuric acid followed by enzymatic hydrolysis yielded 19.71 and 22.93% reducing sugars from bermuda grass and rye straw, respectively (Sun and Cheng 2005). Kim et al. (2011a, b) carried pretreatment of rice straw in two-stage process using aqueous ammonia and dilute H_2SO_4 in percolation mode. The yield of reducing sugars was observed to be 96.9 and 90.8%, respectively, indicating that combination of these two processes resulted in better removal of lignin and hemicelluloses. Pretreatment liquor of *Eulaliopsis binata* (a perennial grass commonly found in India and China) with diluted H_2SO_4 at optimum conditions resulted in 21.02% total sugars, 3.22% lignin, and 3.34% acetic acid with the generation of low levels of inhibitors (Tang et al.

2013). Acid pretreatment of wheat and rice straw gave maximum sugar yield of 565 and 287 mg/g, respectively, with no furfural and hydroxymethyl furfural formation (Saha et al. 2005).

Due to its low cost, pretreatment of lignocellulosic biomass through sulfuric acid is a conventional method. However, it has certain disadvantages such as production of inhibitory compounds and corrosion of reaction vessel (Lee and Jeffries 2011). Therefore researchers have carried out the pretreatment of lignocellulosic biomass through various other acids such as oxalic acid and maleic acid which are discussed later in the review (Kootstra et al. 2009; Lu and Mosier 2007; Lee et al. 2009).

Dicarboxylic acids: oxalic and maleic acid As described earlier, other class of acids called as dicarboxylic acids are being tested by researchers in order to overcome the drawbacks associated with sulfuric acid. Such acids have higher pKa values than sulfuric acid and therefore have a higher solution pH as compared to sulfuric acid which is a type of mineral acid. Dicarboxylic organic acids exhibit two pKa values which make them more efficient for carrying out the hydrolysis of the substrate over a range of temperature and pH values (Lee and Jeffries 2011).

Apart from above mentioned advantages, oxalic acid is less toxic to yeasts and other microorganisms than sulfuric and acetic acids, does not hamper glycolysis and does not produce odor. Lee and coworkers (2011) used oxalic acid for the pretreatment of corn cobs. Corn cob was heated to 168 °C and kept for 26 min. A total sugar yield of 13% was obtained through oxalic acid pretreatment. Also, it produced very less amount of inhibitors.

Maleic acid is another common dicarboxylic acid used for the pretreatment purpose. Along with the advantages mentioned above, maleic acid in particular has k_{hyd}/k_{deg} which favors cellulose hydrolysis to glucose over glucose degradation (Mosier et al. 2002). Lee and Jeffries (2011) investigated the effects of oxalic, maleic, and sulfuric acid on hydrolysis and degradation of lignocellulosic biomass at same combined severity factor (CSF) during hydrolysis. At low CSF values, xylose and glucose concentrations were found to be highest in maleic acid followed by oxalic acid and sulfuric acid. The subsequent fermentation with pretreated biomass yielded maximum ethanol (19.2 g/L) at CSF 1.9 when maleic acid was used for pretreatment of biomass.

Marzialetti et al. (2008) studied the effect of different acids viz. TFA, HCl, H_2SO_4, HNO_3, and H_3PO_4 on loblolly pine in a batch reactor. TFA yielded highest amount of soluble monosaccharides at 150 °C and pH 1.65.

Mild-alkali

In contrary to acid treatment, alkali pretreatment methods are in general performed at ambient temperature and pressure. The most commonly used alkali reagents are the hydroxyl derivatives of sodium, potassium, calcium, and ammonium salts. Among these hydroxyl derivatives, sodium hydroxide was found to be most effective (Kumar and Wyman 2009). Alkali reagents degrade the side chains of esters and glycosides leading to structural modification of lignin, cellulose swelling, cellulose decrystallization, and hemicellulose solvation (Cheng et al. 2010; Ibrahim et al. 2011; McIntosh and Vancov 2010; Sills and Gossett 2011). Sun et al. (1995) optimized the concentration, temperature, and duration of pretreatment using sodium hydroxide. The optimized condition was 1.5% sodium hydroxide at 20 °C for 144 h released 60% lignin and 80% hemicellulose. Zhao et al. (2008) showed the effect of sodium hydroxide on different biomass viz. wheat straw, hardwoods, switchgrass, and softwoods containing less than 26% lignin. However, no effect of dilute NaOH was observed on softwoods with lignin content greater than 26% (Kumar and Wyman 2009). As compared to untreated cellulose, the sodium hydroxide treated corn stover showed increase in biogas production by 37% (Zhu et al. 2010). As compared to acid pretreatment, the solubility of cellulose and hemicellulose is very low with the alkali pretreatment. The solubility improves on increasing the internal surface area of cellulose, decreasing the degree of polymerization and crystallinity, and disrupting the lignin structure (Taherzadeh and Karimi 2008). The conditions for mild alkali pretreatment are less harsh as compared to other pretreatment methods especially acid pretreatment method. Mild alkali pretreatment can be successfully carried out at ambient conditions, however, higher temperature are required if the pretreatment is needed to be carried out for longer duration. Further, a neutralizing step is required to remove the inhibitors as well as lignin (Brodeur et al. 2011). The benefit of lime pretreatment is the low cost of lime as compared to other alkaline agents. For example, in 2005, cost of hydrated lime was $70/ton as compared to $270/ton ammonia and $320/ton for 50 wt% NaOH and 45 wt% KOH (Brodeur et al. 2011). Also, it can be easily recovered from hydrolysate by reaction with CO_2. Park et al. (2010a, b) modified the lime pretreatment method by neutralizing the lime with carbon dioxide before hydrolysis. This eliminated the solid–liquid separation step resulting in 89% glucose recovery from leafstar rice straw. They also applied this modification to examine simultaneous saccharification and fermentation (SSF) by using *Saccharomyces cerevisae* and *Pichia stipitis* which found 74% increase in ethanol yield after 79 h of fermentation at 30 °C. Being an inexpensive pretreatment

method, the only drawback of alkali treatment is its high downstream processing cost because the process utilizes a large quantity of water for removing the salts from the biomass and is a cumbersome process to remove them.

Ozonolysis

Ozone treatment is mainly used for reducing the lignin content of lignocellulosic biomass as it mainly degrades lignin but negligibly affects hemicellulose and cellulose (Kumar et al. 2009). It has been used for removal of lignin in various biomass such as wheat straw (Ben and Miron 1981), bagasse, green hay, peanut, pine (Neely 1984) and poplar sawdust (Vidal and Molinier 1988). A laboratory scale ozonolysis setup was designed and developed by Vidal and Molinier for the pretreatment of different biomass. The pretreatment of wheat straw in the mentioned reactor resulted in 60% removal of lignin followed by fivefold increase in enzymatic hydrolysis. In case of poplar sawdust, lignin percentage was reduced to 8% and sugar yield increased to 57% (Vidal and Molinier 1988). Unlike other chemical pretreatment methods, ozonolysis is performed at ambient temperature and pressure. Also, it does not produce any toxic inhibitors therefore is environment friendly and does not affect the post-pretreatment processes like enzymatic hydrolysis and yeast fermentations (Quesada et al. 1999). The important factor which affects the ozone pretreatment is the moisture content of the biomass, higher the moisture content, lower the lignin oxidization. Although ozonolysis is an effective pretreatment method, the high amount of ozone required makes it an expensive pretreatment method, making it a less suitable option for pretreatment at industrial scale. In order to make an economically viable pretreatment method, research is in progress in different areas such as generation of industrially feasible ozone concentrations, development of reactors such as packed bed, fixed-bed, and stirred tank semi-batch reactors that are capable of accommodating large quantities of low-moisture (<30%) biomass residues having particle size between 1 and 200 mm.

Organosolv

This process involves addition of aqueous organic solvents such as ethanol, methanol, ethylene glycol, acetone etc. to the biomass under specific condition of temperature and pressure (Alriols et al. 2009; Ichwan and Son 2011). Commonly, this process takes place in the presence of an acid, base or salt catalyst (Bajpai 2016). Temperature in organosolv pretreatment depends on the type of biomass and catalyst involved and may reach up to 200 °C. This process is mainly used for the extraction of lignin which is a value-added product. Apart from lignin, cellulose fraction and hemicellulose syrup of C5 and C6 sugars are also produced during the course of organosolv pretreatment. Removal of lignin from the biomass exposes the cellulose fibers for enzymatic hydrolysis leading to higher conversion of biomass (Agbor et al. 2011). The physical characteristics of pretreated biomass such as fiber length, degree of cellulose polymerization, crystallinity etc. depends upon variable factors such as temperature, reaction time, solvent concentration and catalyst used. High temperatures, high acid concentrations, and long reaction time have led to the formation of inhibitors of fermentation. Park et al. (2010a, b) studied the effect of different catalysts (H_2SO_4, NaOH, and $MgSO_4$) on pine and found H_2SO_4 as the most effective catalyst in terms of ethanol yield. However, in terms of digestibility, NaOH was found to be effective when its concentration was increased by 2%. H_2SO_4 has high reactivity therefore has proven to be a very strong catalyst but at the same time it is toxic, corrosive and is inhibitory in nature. The main drawback of this process is the high cost of the solvents, though this drawback can be minimized by recovering and recycling solvents through evaporation and condensation. Removal of solvents is very important because the solvent may cause negative effect on growth of microorganisms, enzymatic hydrolysis, and fermentation (Agbor et al. 2011). Also, organosolv is less preferred due to high risk involved in handling harsh organic solvents that are highly flammable. In absence of proper safety measures, it can cause severe damage leading to large fire explosions. The Battelle is a type of organosolv method that treats the biomass with mixture of phenol, HCl, and water at temperature 100 °C and pressure 1 atm (Villaverde et al. 2010). Acid is responsible for the depolymerization of lignin as well as it hydrolyses the hemicellulose portion of the biomass. Lignin is dissolved in the phenol while the sugars (monosaccharides) are found in the aqueous phase upon cooling. Likewise formasolv is a type of organosolv involving formic acid, water, and HCl. Lignin is soluble in formic acid and the pretreatment process can be carried out at low temperature and pressure (Zhao and Liu 2012). Unlike formasolv, ethanosolv (involving ethanol) is operated at high temperature and pressure and recovers value-added products namely cellulose, hemicellulose, and pure lignin. Further purification may be carried out through ionic liquids which are discussed later in this review (Prado et al. 2012). The less toxic nature of ethanol as compared to methanol and it being the final product makes it more popular as compared to other solvents (Kim et al. 2011a, b). However, presence of ethanol inhibits the performance of hydrolytic enzymes, therefore lower ethanol: water is used for hydrolysis of hemicellulose and enzymatic degradation of pretreated biomass (Huijgen et al. 2008). Also, nearly complete recovery of ethanol and water is a major advantage which

reduces the operation cost (Koo et al. 2012; Alriols et al. 2010). Mesa et al. (2011) applied ethanosolv on sugarcane bagasse for production of reducing sugars. 29.1% reducing sugars were produced by 30% ethanol for 60 min at 195 °C. Similarly, horticultural waste was pretreated by a modified method using ethanol under mild conditions for bioethanol production. Pretreatment resulted in hydrolysate containing 15.4% reducing sugar after 72 h, which after fermentation produced 1.169% ethanol in 8 h using *Saccharomyces cerevicae* (Geng et al. 2012). Hideno et al. (2013) reported utilization of alcohol-based organosolv treatment in combination with ball milling for pretreatment of Japanese cypress (*Chamaecyparis obtusa*). They observed that combination of alcohol-based organosolv treatment in mild conditions and short time ball milling had a synergistic effect on the enzymatic digestibility. Ichwan and Son (2011) studied the effect of various solvents such as ethanol–water, ethylene glycol–water, and acetic acid–water mixture to extract cellulose from oil palm pulp. The yield of organosolv pulping with ethylene glycol–water, ethanol–water, and acetic acid–water mixture was 50.1, 48.1, and 41.7%, respectively. Panagiotopoulos et al. (2012) treated poplar wood chips with steam followed by organosolv treatment for separating hemicellulose, lignin, and cellulose components. Lignin extraction was found to increase to 66%, while 98% of the cellulose was recovered by two stage pretreatment process and 88% of cellulose was hydrolyzed to glucose after 72 h.

Ionic liquids

Ionic liquids have received great attention in last decade for the pretreatment of lignocellulosic biomass. Ionic liquids are comparatively a new class of solvents which are entirely made of ions (cations and anions), have low melting points (<100 °C), negligible vapor pressure, high thermal stabilities, and high polarities (Zavrel et al. 2009; Behera et al. 2014). Imidazolium salts are the most commonly used ILs. ILs are assumed to compete with lignocellulosic components for hydrogen bonding there by disrupting its network (Moultrop et al. 2005). Table 2 lists various ionic liquids used for the treatment of a variety of biomasses.

According to Li et al. (2011), with suitable selection of anti-solvents up to 80% lignin and hemicellulose can be fractionated. Dadi et al. (2006) used 1-butyl-3-methylimidazolium chloride (Bmim-Cl) for pretreatment of Avicel—PH-101 reported 50- and 2-fold increase in enzymatic hydrolysis rate and yield, respectively, as compared to untreated Avicel. Liu and Chen (2006) used Bmim-Cl for pretreating wheat straw and found significant improvement in enzymatic hydrolysis yield. They found that Bmim-Cl modified structure of wheat

Table 2 Different types of ionic liquids applied for the pretreatment of different biomass (Bajpai 2016)

Biomass	Ionic liquid	Abbreviated symbol
Poplar wood	1-Ethyl-3-methylimidazolium diethyl phosphate-acetate	Emim-Ac
Pine	1-Allyl-3-methylimidazolium chloride	Amim-Cl
Eucalyptus	Ethyl-3-methylimidazolium diethyl phosphate-acetate	Emim-Ac
Spruce	1-Allyl-3-methylimidazolium chloride	Amim-Cl
Bagasse	Ethyl-3-methylimidazolium diethyl phosphate-acetate	Emim-Ac
Switch grass	Ethyl-3-methylimidazolium diethyl phosphate-acetate	Emim-Ac
Bamboo	1-Ethyl-3-methylimidazolium diethyl phosphate-glycine	Emim-Gly
Wheat straw	1-Allyl-3-methylimidazolium chloride and chloride	Amim-Ac
	1-Butyl-3-methylimidazolium-acetate	Bmim-Ac
Water hyacinth	1-Butyl-3-methylimidazolium-acetate	Bmim-Ac
Rice husk	1-Butyl-3-methylimidazolium-chloride	Bmim-Cl
	1-Ethyl-3-methylimidazolium diethyl phosphate-acetate	Emim-Ac
Rice straw	Cholinium amino acids	Ch-Aa
Kenaf powder	Cholinium acetate	Ch-Ac

straw by reducing the polymerization and crystallinity and solubilizing cellulose and hemicellulose. Sugarcane bagasse pretreated with 3-N-methylmorpholine-N-oxide (NMMO) showed twofold increase in enzymatic hydrolysis yield as compared to untreated bagasse (Kuo and Lee 2009). Ionic liquids has been able to effectively pretreat lignocellulosic biomass, however, there are certain challenges that need to be addressed such as high cost of ILs, difficulty in recycling and reuse, inhibitor generation etc.

Deep eutectic solvents

These are relatively a new class of solvents having many characteristics similar to ionic liquids. A deep eutectic solvent (DES) is a fluid generally composed of two or three cheap and safe components that are capable of self-association, often through hydrogen bond interactions, to form a eutectic mixture with a melting point lower than that of each individual component (Zhang et al. 2012a, b). These DES were able to solve some of the key concerns associated with ILs. Deep eutectic solvents can be described by the general formula

$$Cat^+X^-zY$$

where Cat^+ is in principle any ammonium, phosphonium, or sulfonium cation, and X is a Lewis base, generally a halide anion. The complex anionic species are formed between X^- and either a Lewis or Brønsted acid Y (z refers to the number of Y molecules that interact with the anion) (Smith et al. 2014). Most of the DESs have used choline chloride (ChCl) as hydrogen bond acceptor. ChCl is low-cost, biodegradable, and non-toxic ammonium salt which can be extracted from biomass. ChCl is able to synthesize DESs with hydrogen donors such as

urea, carboxylic acids, and polyols. Although DESs are similar to ILs in terms of physical behavior and physical properties, DESs cannot be considered as ionic liquids due to the fact that DESs are not entirely composed of ionic species and can be obtained from non-ionic species (Zhang et al. 2012a, b).

Natural deep eutectic solvents

In the recent past, a large number of natural products have been brought into the range of ILs and DES. These products include choline, urea, sugars, amino acids, and several other organic acids (Dai et al. 2013). Such solvents obtained from natural sources are termed as Natural Deep Eutectic Solvents (NADES). Unlike ILs, NADES are cost effective, easier to synthesize, non-toxic, biocompatible, and highly biodegradable. Moreover, many studies recovered and reused these novel solvents with high efficiency. NADES are prepared by the complex formation between a hydrogen acceptor and a hydrogen bond donor. The decrease in melting point of the prepared solvent mixtures is due to the charge delocalization of the raw individual components. Foreseeing the potentiality of NADES in diverse applications, these solvents are regarded as the solvents for the twenty first century (Paiva et al. 2014). Moreover, recent research on lignocellulosic feedstock pretreatment with NADES reagents showed high specificity towards lignin solubilisation and extraction of high purity lignin from agricultural residue such as rice straw (Kumar et al. 2016). Despite having a lot of potential for the extraction of natural products, the high viscosity of NADES is an obvious disadvantage. Dai et al. (2015) studied the dilution effect on the physicochemical properties of NADES. FT-IR and ^{1}H NMR studies showed intense H-bonds between the two components of NADES system. However, the dilution with water weakened the interactions. At around 50% (v/v) dilution with water, the intense hydrogen interactions disappeared completely. The viscosity of NADES reduced to the order of water and conductivity increased up to 100 times for some NADES reagents. Along with pretreatment, NADES can prove to be a game changer concept in pharma, food processing, and enzyme industries.

Physico-chemical pretreatment
Steam explosion

Steam pretreatment is one of the most commonly used physicochemical methods for pretreatment of lignocellulosic biomass. Earlier, this method was known as steam explosion because of the belief that an explosive action on the biomass was required to prepare them for hydrolysis (Agbor et al. 2011). Due to the changes that occur during this process, this method is also called 'auto hydrolysis.' Steam pretreatment is typically a combination of mechanical forces (pressure drop) and chemical effects (autohydrolysis of acetyl groups of hemicellulose). In this process, biomass is subjected to high pressure (0.7–4.8 MPa) saturated steam at elevated temperatures (between 160 and 260 °C) for few seconds to minutes which causes hydrolysis and release of hemicellulose. The steam enters the biomass expanding the walls of fibers leading to partial hydrolysis and increasing the accessibility of enzymes for cellulose. After this the pressure is reduced to atmospheric condition (Rabemanolontsoa and Saka 2016). During this pretreatment, the hydrolysis of hemicellulose into glucose and xylose monomers is carried out by the acetic acid produced from the acetyl groups of hemicellulose; hence this process is also termed as autohydrolysis (Mosier et al. 2005). The factors that affect steam pretreatment are temperature, residence time, biomass size, and moisture content (Rabemanolontsoa and Saka 2016). Wright (1988) found low temperature and longer residence time (190 °C for 10 min) better as compared to high temperature and lower residence time (270 °C for 1 min) due to less fermentation inhibitory product formation in the earlier process. Several biomasses have shown positive effects on pretreatment with steam such as poplar wood (*Populus tremuloides*) (Grous et al. 1986), pine chips, French maritime pine (*Pinus pinaster*), rice straw, bagasse, olive stones, giant miscanthus (*Miscanthus giganteus*), and spent Shiitake mushroom media (Jacquet et al. 2012). The efficiency of steam pretreatment can be effectively enhanced in the presence of catalysts such as H_2SO_4, CO_2 or SO_2. Out of these catalysts, acid catalyst has been found to most successful in terms of hemicellulose sugar recovery, decreased production of inhibitory compounds and improved enzymatic

hydrolysis. Steam pretreatment is found to be effective for the pretreatment of hardwoods and agricultural residues, though acid catalyst is added in case of soft woods for effective pretreatment. Limited use of chemicals, low energy requirement, no recycling cost and environment friendly are some of the advantages of steam pretreatment method. On the other hand, the possibility of formation of fermentation inhibitors at high temperature, incomplete digestion of lignin-carbohydrate matrix and the need to wash the hydrolysate which decreases the sugar yield by 20% are few disadvantages associated with steam pretreatment (Agbor et al. 2011).

Liquid hot water

This method, also called as hot compressed water is similar to steam pretreatment method but as the name suggests, it uses water at high temperature (170–230 °C) and pressure (up to 5 MPa) instead of steam. This leads to hydrolysis of hemicellulose and removes lignin making cellulose more accessible. This also avoids the formation of fermentation inhibitors at high temperatures (Yang and Wyman 2004). Different researchers have described liquid hot water (LHW) with different terms such as solvolysis, hydrothermolysis, aqueous fractionation and aquasolv (Agbor et al. 2011). LHW can be performed in three different ways based on the direction of flow of water and biomass into reactor. (1) Co-current pretreatment, in which both the slurry of biomass and the water is heated to the required temperature and held at the pretreatment conditions for controlled residence time before being cooled. (2) Counter current pretreatment, in which the hot water is pumped against the biomass in controlled conditions. (3) Flow through pretreatment where the biomass acts like a stationary bed and hot water flows through the biomass and the hydrolyzed fractions are carried out of the reactor. Abdullah and coworkers (2014) conducted studies on LHW to investigate the hydrolysis performance. The optimization could not be carried out at same severity due to the difference in rate of hydrolysis of cellulose and hemicellulose. Therefore, a two-step hot compressed water treatment was proposed. First stage is carried out at low severity for hydrolyzing the hemicellulose while second stage is carried out at high severity for depolymerization of cellulose and increase sugar yield. Ogura et al. (2013) and Phaiboonsilpa (2010) applied two-step hydrolysis (I step: 230 °C-10 MPa-15 min; II step: 275 °C-10 MPa-15 min) to Japanese beech, Japanese cedar, Nipa frond and rice straw and found to solubilize 92.2, 82.3, 92.4, and 97.9% of the starting biomass, respectively. This has proved that LHW is capable of acting on a large variety of biomass including softwoods (Rabemanolontsoa and Saka 2016). Low-temperature requirement, minimum formation of inhibitory compounds and low cost of the solvent are some of the advantages associated with LHW. However, it requires large amount of energy in downstream processing due to large amount of water involved (Agbor et al. 2011).

Wet oxidation

Wet oxidation is one of the simple methods of lignocellulosic pretreatment where the air/oxygen along with water or hydrogen peroxide is treated with the biomass at high temperatures (above 120 °C for 30 min) (Varga et al. 2003). Earlier this method as also used for waste water treatment and soil remediation (Chaturvedi and Verma 2013). This method is most suitable for lignin enriched biomass residues. The efficiency of wet oxidation is dependent on three factors: oxygen pressure, temperature, and reaction time. In this process, when the temperature is raised above 170 °C, water behaves like an acid and catalyzes hydrolytic reactions. The hemicelluloses are broken down into smaller pentose monomers and the lignin undergoes oxidation, while the cellulose is least affected by wet oxidation pretreatment. Besides these, reports on addition of chemical agents like sodium carbonate and alkaline peroxide in wet oxidation reduced the reaction temperature, improved hemicellulose degradation and decreased the formation of inhibitory components such as furfurals and furfuraldehydes (Banerjee et al. 2011). This pretreatment method is unlikely to reach industrial scale of biomass pretreatment because of the high cost of the hydrogen peroxide and the combustible nature of the pure oxygen (Bajpai 2016). Szijártó et al. (2009) applied wet oxidation for the pretreatment of common reed (*Phragmites australis*). The treatment resulted in three fold increase in digestibility of reed cellulose by cellulase as compared to control. 51.7% of hemicellulose and 58.3% of lignin was solubilised and 82.4% of cellulose got converted into cellulose on enzymatic hydrolysis of the pretreated fibers. The pretreated fibers produced 0.87% ethanol when underwent simultaneous saccharification and fermentation. Banerjee et al. (2009) optimized the wet oxidation conditions for rice husk for the production of ethanol. The optimized conditions of 0.5 MPa pressure, 185 °C temperature for 15 min yielded 67% of cellulose, removed 89% lignin, and solubilised 70% hemicellulose. Reducing sugar yields up to 70% have been obtained by utilizing this pretreatment process. Alkaline Peroxide-Assisted Wet Air Oxidation (APAWAO) treatment on rice husk resulted in solubilization of 67 and 88 wt% of hemicellulose and lignin, respectively. The glucose amount increased 13-fold as compared from untreated rice husk (Banerjee et al. 2011). This pretreatment method is unlikely to reach industrial scale of biomass pretreatment because of the high cost of the hydrogen

peroxide and the combustible nature of the pure oxygen (Bajpai 2016).

SPORL treatment

Sulfite pretreatment to overcome recalcitrance of lignocellulose (SPORL) is a popular and efficient pretreatment method for lignocellulosic biomass (Xu et al. 2016). It is carried out in a combination of two steps: First, the biomass is treated with calcium or magnesium sulfite to remove hemicellulose and lignin fractions. In the second step, the size of the pretreated biomass is reduced significantly using mechanical disk miller. Zhu et al. (2009) studied the effect of SPORL pretreatment on spruce chips using 8–10% bisulfite and 1.8–3.7% sulfuric acid at 180 °C for 30 min. After 48 h of hydrolysis with 14.6 FPU cellulase +22.5 CBU β-glucosidase per gram of substrate, more than 90% substrate was converted to cellulose. Also, only 0.5% hydroxymethyl furfural (HMF) and 0.1% furfural (fermentation inhibitors), respectively, were formed as compared to 5% HMF and 2.5% furfural formation during the acid catalyzed steam pretreatment of spruce. The amount of HMF and furfural was also reported to decrease with increasing bisulfite. The possible reason is that at same acid charge, higher amount of bisulfite leads to higher pH which reduces the decomposition of sugars to HMF and furfural.

SPORL pretreatment on switchgrass was carried out by Zhang et al. (2013) with temperature ranging between 163 and 197 °C for a period ranging from 3–37 min with sulfuric acid dosage (0.8–4.2%) and sodium sulfite dosage (0.6–7.4%). The results found improved digestibility of switchgrass by removing hemicellulose, dissolving lignin partially and decreasing hydrophobicity of lignin by sulfonation. SPORL pretreated switchgrass was hydrolysed by 83% in 48 h with 15 FPU cellulase and 30 CBU β-glucosidase/g cellulose. SPORL pretreatment method when compared with dilute acid and alkali pretreatments, was found to give the highest substrate yield of 77.2% as compared to 68.1 and 66.6% by dilute acid and alkali pretreatment, respectively. Sodium sulfide and sodium sulfite along with sodium hydroxide were applied for pretreatment of corncob, bagasse, water hyacinth and rice husk. Pretreatment under optimized conditions yielded 97% lignin and 93% hemicellulose from water hyacinth and rice husk, and 75% lignin and 90% hemicellulose were removed from bagasse and rice husk (Idrees et al. 2013).

SPORL pretreatment has been popular in the recent times because of its versatility, efficiency, and simplicity. It reduces the energy consumption to 1/10 required for the reduction of size of biomass. It has very high conversion rate of cellulose to glucose and maximizes hemicellulose and lignin removal and recovery. It has the capacity to process a variety of biomass and has excellent scalability for commercial production by retrofitting into existing mills for production of biofuels. However, certain issues such as sugar degradation, requirement of large volumes of water for post-pretreatment washing and high cost of recovering pretreatment chemicals need to be addressed for making SPORL a cost effective pretreatment technology (Bajpai 2016).

Ammonia-based pretreatment
Methods that use liquid ammonia for the pretreatment of lignocellulosic biomass are Ammonia fiber explosion (AFEX), Ammonia recycle percolation (ARP) and soaking aqueous ammonia (SAA). AFEX is conducted at ambient temperature while ARP is conducted at high temperatures (Agbor et al. 2011) SAA is a form of AFEX which treats biomass through aqueous ammonia in a batch reactor at 30–60 °C which decreases the liquid through-put during process of pretreatment (Kim and Lee 2005a). In AFEX, lignocellulosic biomass is heated with liquid ammonia (in 1:1 ratio) in a closed vessel at temperature 60–90 °C and pressure above 3 MPa for 30–60 min. After holding the desired temperature in vessel for 5 min, valve is opened which explosively releases the pressure leading to evaporation of ammonia and drop in temperature of the system (Alizadeh et al. 2005). It is similar to steam explosion but ammonia is used instead of water (Rabemanolontsoa and Saka 2016). Lignocellulosic biomass when treated with ammonia at high pressure and given temperature causes swelling and phase change in cellulose crystallinity of biomass leading to increase in the reactivity of leftover carbohydrates after pretreatment. The lignin structure gets modified which increases the water holding capacity and digestibility. Unlike other pretreatment methods, AFEX treatment does not produce inhibitors, which is highly desirable for downstream processing. Besides, the overall cost of the pretreatment process is significantly low due to the absence of additional steps like water washing, detoxification, recovery, and reuse of large quantities of water. More than 90% of celluloses and hemicelluloses could be converted to fermentable sugars if pretreated with AFEX under optimized conditions of ammonia loading, temperature, pressure, moisture content and pretreatment time (Uppugundla et al. 2014). Moreover, the ammonia could be recovered and recycled to decrease the overall cost of the pretreatment process.

Another process that utilizes ammonia is ammonia recycle percolation (ARP). In this process, aqueous ammonia (5–15 wt%) is passed through a reactor containing biomass. The temperature range is between 140 and 210 °C with a reaction time of 90 min and percolation

rate is 5 mL/min after which the ammonia is recycled (Sun and Cheng 2002; Kim et al. 2008). ARP is capable of solubilizing hemicellulose but cellulose remains unaffected (Alvira et al. 2010). The disadvantage with ARP is high requirement of energy to maintain process temperature. Both AFEX and ARP have been found to effective for herbaceous plants, agricultural residues and MSW. ARP pretreatment is found effective for hardwoods also (Kim and Lee 2005b). Another technology soaking aqueous ammonia (SAA) requires less energy as it is performed at low temperature (30–75 °C).

CO$_2$ explosion
This process carries out the pretreatment of biomass through supercritical CO$_2$ which means the gas behaves like a solvent. The supercritical CO$_2$ is passed through a high pressure vessel containing the biomass (Kim and Hong 2001). The vessel is heated to the required temperature and kept for several minutes at high temperatures (Hendricks and Zeeman 2009). CO$_2$ enters the biomass at high pressure and forms carbonic acid which hydrolyses the hemicellulose. The pressurized gas when released disrupts the biomass structure which increases the accessible surface area (Zheng et al. 1995). This pretreatment method is not suitable for biomass having no moisture content. Higher the moisture content in the biomass, higher the hydrolytic yield (Kim and Hong 2001). Low cost of carbon dioxide, low temperature requirement, high solid capacity, and no toxin formation makes it an attractive process. However, high cost of reactor which can tolerate high pressure conditions is a big obstacle in its application on large scale (Agbor et al. 2011).

Oxidative pretreatment
It involves treatment of lignocellulosic biomass by oxidizing agents such as hydrogen peroxide, ozone, oxygen or air (Nakamura et al. 2004). A number of chemical reactions such as electrophilic substitution, side chain displacements, and oxidative cleavage of aromatic ring ether linkages may take place during oxidative pretreatment. This process causes delignification by converting lignin to acids, which may act as inhibitors. Therefore, these acids need to be removed (Alvira et al. 2010). A major drawback of oxidative pretreatment is that it damages a significant amount of hemicellulose making it unavailable for fermentation (Lucas et al. 2012). The most commonly employed oxidizing agent is hydrogen peroxide. It has been found that hydrolysis of hydrogen peroxide leads to formation of hydroxyl radicals which are responsible for degradation of lignin and production of low molecular weight products. Removal of lignin from lignocellulose exposes cellulose and hemicellulose leading to increased enzymatic hydrolysis (Hammel et al. 2002). The following

enzymatic hydrolysis yield could reach up to 95%. Yu et al. (2009) combined oxidative pretreatment followed by biological treatment with *Pleurotus ostreatus*. At optimum conditions of hydrogen peroxide pretreatment i.e., 2% H_2O_2 for 48 h followed by biological pretreatment for 18 days yielded 39.8% total sugar and 49.6% glucose. This was about 5.8 and 6.5 times more as compared to fungal pretreatment alone for 18 days. Saha and Cotta (2007) has reported that peroxide pretreatment under alkaline conditions (addition of NaOH) increased the production of reducing sugars with more than 96% cellulosic conversion as compared to absence of alkali. Cao et al. (2012) performed pretreatment of sweet sorghum bagasse through different pretreatment processes and found the highest yield with dilute NaOH followed by H_2O_2 pretreatment. The highest cellulose hydrolysis yield was 74.3%, total sugar yield was 90.9% and ethanol concentration was 0.61% which was 5.9, 9.5, and 19.1 times higher as compared to control.

Biological pretreatment

In comparison to conventional chemical and physical pretreatment methods, biological pretreatment is considered as an efficient, environmentally safe and low-energy process. Nature has abundant cellulolytic and hemicellulolytic microbes which can be specifically targeted for effective biomass pretreatment (Vats et al. 2013). Biological pretreatments are carried out by microorganisms such as brown, white, and soft-rot fungi which mainly degrade lignin and hemicellulose and little amount of cellulose (Sánchez 2009). Degradation of lignin by white-rot fungi occurs due to the presence of peroxidases and laccases (lignin degrading enzymes) (Kumar et al. 2009). The white-rot fungi species commonly employed for pretreatment are *Phanerochaete chrysosporium, Ceriporia lacerata, Cyathus stercolerus, Ceriporiopsis subvermispora, Pycnoporus cinnarbarinus* and *Pleurotus ostreaus*. Besides these other basidiomycetes species were also studied for breakdown of several lignocellulosic feedstocks. Among these *Bjerkandera adusta, Fomes fomentarius, Ganoderma resinaceum, Irpex lacteus, Phanerochaete chrysosporium, Trametes versicolor,* and *Lepista nuda* are well studied. These species have been reported to show high delignification efficiency (Kumar et al. 2009; Shi et al. 2008). Table 3 summarizes different microorganisms involved in pretreatment strategies and their effects on various biomasses. Pretreatment of wheat straw by fungi (fungal isolate RCK-1) for 10 days resulted in increase of fermentable sugars and decrease in fermentation inhibitors. Although the biological pretreatment is highly intriguing, the rate of hydrolysis of lignocellulosic fractions is too slow which severely hampers to be foreseen as a potential pretreatment method at an industrial

scale (Sun and Cheng 2002). In order to make biological pretreatment at par with other pretreatment methods, more basidiomycetes fungi should be tested for its ability to delignify the biomass effectively at a faster rate.

Combined biological pretreatment

Studies have found that a combination of another pretreatment process with biological pretreatment process is more effective as compared to a single pretreatment process. Wang et al. (2012) combined biological pretreatment with liquid hot water pretreatment method for better enzymatic saccharification of *Populus tormentosa*. This combination reported highest hemicellulose removal (92.33%) resulting in 2.66-fold increase in glucose yield as compared to pretreatment carried out with liquid hot water alone. Yu et al. (2009) studied the novel combination of either physical or chemical pretreatment with biological pretreatment on rice husk. Physical pretreatment was carried out using ultrasound while chemical pretreatment was carried out using H_2O_2. Biological pretreatment was carried out using *P. ostreatus*. The combined pretreatment of rice husk carried out using 2% H_2O_2 for 48 h along with *P. ostreatus* was found more effective as compared to single step pretreatment using *P. ostreatus* for 60 days. Lignin removal was also found significantly higher as compared to one step treatment. Balan et al. (2008) studied and found that pretreatment of rice husk with *P. ostreatus* followed by AFEX pretreatment produced high glucan and xylan conversion as compared to a single pretreatment with AFEX. The combination of mild acid pretreatment (0.25% H_2SO_4) and biological pretreatment using *Echinodontium.taxodii* on water hyacinth was found more effective than one step pretreatment. The reducing sugars yield doubled as compared to single step acid pretreatment method (Ma et al. 2010). Sawada et al. (1995) combined steam explosion and pretreatment by *P. chrysosporium* for the enzymatic saccharification of plant wood. The saccharification of wood increased when treated with *P. chrysosporium* prior to steam explosion. Maximum production of reducing sugar was observed when wood was treated with *P. chrysosporium* for 28 days followed by steam explosion at 215 °C for 60–65 min.

Applications of biomass pretreatment

Biomass pretreatment results in production of several value-added products. Although, here we have described in brief but this topic is beyond the scope of this review and readers are suggested to refer recent review on various products obtained from pretreated lignocellulosic biomass (Putro et al. 2016). Several valuable products can be obtained through lignocellulosic biomass. Among

Table 3 Different biological pretreatment strategies involved for pretreatment of lignocellulosic biomass and its advantages (adapted from Sindhu et al. 2016)

Microorganism	Biomass	Major effects	References
Punctualaria sp. TUFC 20056	Bamboo culms	50% lignin removal	Suhara et al. (2012)
Irpex lacteus	Corn stalks	82% of hydrolysis yield	Du et al. (2011)
Fungal consortium	Straw	20-fold increase in hydrolysis	Taha et al. (2015)
P. ostreatus/P. pulmonarius	Eucalyptus grandis saw dust	20-fold increase in hydrolysis	Castoldi et al. (2014)
P. chrysosporium	Rice husk	–	Potumarthi et al. (2013)
Fungal consortium	Corn stover	43.8% lignin removal/sevenfold increase in hydrolysis	Song et al. (2013)
Ceriporiopsis subvermispora	Wheat straw	Minimal cellulose loss	Cianchetta et al. (2014)
Ceriporiopsis subvermispora	Corn stover	2- to 3-fold increase in reducing sugar yield	Wan and Li (2011)
Fungal consortium	Plant biomass	Complete elimination of use of hazardous chemicals	Dhiman et al. (2015)

which biofuel and chemicals are well known and widely studied.

Biofuels

Several biofuels are obtained through lignocellulosic biomass such as bio-oil, bioethanol, biohydrogen, biogas, syngas etc. Bio-oil is produced through pyrolysis along with biochar, tar and gases. Bio-oil is produced by fast depolymerisation of lignocellulose components viz. hydroxyaldehydes, sugars, hydroxyketones, carboxylic acids, and phenols. Bioethanol can be produced through 5 different methods: separate hydrolysis and fermentation (SHF), simultaneous saccharification and fermentation (SSF), simultaneous and saccharification co-fermentation (SSCF), consolidated bioprocessing (CBP), and integrated bioprocessing (IBP) (Sarkar et al. 2012; Jagmann and Philipp 2014). SSF is the most promising among these processes because of its low-cost and high product yield. IBP is another promising process which involves treatment with microorganisms at every step in a single step. However, there is no reported work on pretreatment through IBP (Chandel et al. 2015). Biohydrogen can be produced from lignocellulosic biomass through thermochemical (gasification and pyrolysis) or biological routes (Ni et al. 2006). Through pyrolysis, hydrogen can be produced through fast or flash pyrolysis (Putro et al. 2016). Hydrogen can produced through gasification by partial oxidation and steam reformation followed by waster-gas shift reaction. Two processes to produce biohydrogen through biological route are: photo fermentation which is light dependant and dark fermentation which is light independent (Sivagurunathan et al. 2016). Although biogas and syngas have similar composition (CO_2, CH_4, H_2, and N_2), they are produced through two different processes. Biogas is produced through anaerobic digestion which comprises of four steps: hydrolysis, acidogenesis, acetogenesis, and methanogenesis (Taherzadeh

and Karimi 2008) while syngas is produced by gasification carried out at lower temperature due to high reactivity of biomass. Biomass gasification has three types of processes namely: (1) pyrolysis which involves anaerobic decomposition of biomass at high temperature, (2) partial oxidation which requires less amount of oxygen as compared to oxidation, and (3) steam gasification which involves the reaction of water with biomass.

Bioproducts

The chemicals from lignocellulosic biomass can be derived either through carbohydrate source or through lignin. (1) The simplest chemical derived from carbohydrate is furfural and 5-hydroxymethylfurfural (HMF), produced through acid catalyzed dehydration of C_5 and C_6 sugars (Delidovich et al. 2016). Sugar alcohols such as sorbitol and xylitol are obtained by the hydrogenation of hexose and pentose (Romero et al. 2016). Also, glycerol, widely used for making bio-solvents, polymers, surfactants etc. can be produced by hydrogenolysis of sorbitol and xylitol (Choi et al. 2015). Also lactic acid and succinic acid can be obtained by the biological conversion caused by bacteria and mold. (2) Lignin has been used to generate heat in the earlier days. In the recent times lignin has been a rich source of valuable products like phenolic compounds. The basic principle behind the conversion of lignin to phenolic compounds is depolymerization. Different ways to convert lignin to phenolics compounds are liquefaction (Kang et al. 2013), oxidation (Ma et al. 2015), solvolysis (Kleinert and Barth 2008), hydrocracking (Yoshikawa et al. 2013) and hydrolysis (Roberts et al. 2011). Lignocellulose biomass has also been used for development of advanced technology products for energy storage, transportation, medical applications, biosensing, environmental remediation etc. (Wang et al. 2013; Brinchi et al. 2013; Yang et al. 2013).

Conclusion

The presence of lignin in the biomass inhibits the hydrolysis of cellulose and hemicellulose. Therefore, extensive research has been carried out for developing various pretreatment techniques for delignification of biomass. However, critical analysis of pretreatment methods bring us to a conclusion that pretreatment method is a 'tailor-made' process for every individual biomass which should be meticulously selected and planned based on the characteristic properties of biomass. Also, it can be concluded that till date a single pretreatment method has not been established which can carry out complete delignification of biomass in an economic and environment friendly manner. Though, combined pretreatment methods have been successful to an extent, still a lot of research needs to be done in developing combined pretreatment methods to their full potential. This critical review comprising of physical, chemical, physicochemical and biological pretreatment processes along with their advantages and disadvantages will help the researcher in planning, selection, and development of pretreatment process for various lignocellulosic biomass.

Authors' contributions

AKK and SS designed and wrote the review. Both authors read and approved the final manuscript.

Acknowledgements

The authors are thankful to the Director, Sardar Patel Renewable Energy Research Institute, Gujarat, India, for support of this research.

Competing interests

The authors declare that they have no competing interests.

Funding

The research work is financially supported by Indian Council of Agricultural Research (ICAR), under All India Co-ordinated Research Project (AICRP) –EAAI program, Govt. of India with Grant Number VVN/RES/DRET-LBT/2014/3, Department of Biotechnology with Grant Number BT/PR12368/PDB/26/431/2014.

References

Abdullah R, Ueda K, Saka S (2014) Hydrothermal decomposition of various crystalline celluloses as treated by semi-flow hot-compressed water. J Wood Sci 60:278–286

Agbor VB, Cicek N, Sparling R, Berlin A, Levin DB (2011) Biomass pretreatment: fundamentals toward application. Biotechnol Adv 29:675–685

Aimin T, Hongwei Z, Gang C, Guohui X, Wenzhi L (2005) Influence of ultrasound treatment on accessibility and regioselective oxidation reactivity of cellulose. Ultrason Sonochem 12:467–472

Alizadeh H, Teymouri F, Gilbert TI, Dale BE (2005) Pretreatment of switchgrass by ammonia fibre explosion (AFEX). Appl Biochem Biotechnol 121:1133–1141

Alriols MG, Garcia A, Llano Ponte R, Labidi J (2010) Combined organosolv and ultrafiltration lignocellulosic biorefinery process. Chem Eng J 157:113–120

Alriols MG, Tejado A, Blanco M, Mondragon I, Labidi J (2009) Agricultural palm oil tree residues as raw material for cellulose, lignin and hemicelluloses production by ethylene glycol pulping process. Chem Eng J 148:106–114

Alvira P, Tomas-Pejo E, Ballesteros M, Negro MJ (2010) Pretreatment technologies for an efficient bioethanol production process based on enzymatic hydrolysis: a review. Biores Technol 10:4851–4861

Bajpai P (2016) In Pretreatment of lignocellulosic biomass for biofuel production. Springer Briefs in Molecular Science, pp 17–70

Balan V, Souca LDC, Chundawat SPS, Vismeh R, Jones AD, Dale BEJ (2008) Mushroom-spent straw: a potential substrate for an ethanol based biorefinery. Ind Microbiol Biotechnol 35:293–301

Banerjee S, Sen R, Mudliar S, Pandey RA, Chakrabarti T, Satpute D (2011) Alkaline peroxide assisted wet air oxidation pretreatment approach to enhance enzymatic convertibility of rice husk. Biotechnol Prog 27:691–697

Banerjee S, Sen R, Pandey RA, Chakrabarti T, Satpute D, Giri BS, Mudliar S (2009) Evaluation of wet air oxidation as a pretreatment strategy for bioethanol production from rice husk and process optimization. Biomass Bioenerg 33:1680–1686

Behera S, Arora R, Nandhagopal N, Kumar S (2014) Importance of chemical pretreatment for bioconversion of lignocellulosic biomass. Renew Sustain Energy Rev 36:91–106

Ben GD, Miron J (1981) The effect of combined chemical and enzyme treatment on the saccharification and in vitro digestion rate of wheat straw. Biotechnol Bioeng 23:823–831

Boonmanumsin P, Treeboobpha S, Jeamjumnunja K, Luengnaremitchai A, Chaisuwan T, Wongkasemjit S (2012) Release of monomeric sugars from Miscanthus sinensis by microwave assisted ammonia and phosphoric acid treatments. Bioresour Technol 103:425–431

Brinchi L, Contana F, Fortunati E, Kenny JM (2013) Production of nanocrystalline cellulose from lignocellulosic biomass: technology and applications. Carbohydr Polym 94:154–169

Brodeur G, Yau E, Badal K, Collier J, Ramachandran KB, Ramakrishnan S (2011) Chemical and physicochemical pretreatment of lignocellulosic biomass: A review. Enzyme Res 2011:787532. doi:10.4061/2011/787532

Bussemaker MJ, Zhang D (2013) Effect of ultrasound on lignocellulosic biomass as a pretreatment for biorefinery and biofuel applications. Ind Eng Chem Res 52:3563–3580

Cao W, Sun C, Liu R, Yin R, Wu X (2012) Comparison of the effects of five pretreatment methods on enhancing the enzymatic digestibility and ethanol production from sweet sorghum bagasse. Bioresour Technol 111:215–221

Castoldi R, Bracht A, de Morais GR, Baesso ML, Correa RCG, Peralta RA, Moreira RFPM, Polizeli MT, de Souz CGM, Peralta RM (2014) Biological pretreatment of Eucalyptus grandis sawdust with white-rot fungi: study of degradation patterns and saccharification kinetics. Chem Eng J 258:240–246

Chandel AK, Goncalves BC, Strap JL, de Silva SS (2015) Biodelignification of lignocellulose substrates: an intrinsic and sustainable pretreatment strategy for clean energy production. Crit Rev Biotechn 35:281–293

Chang VS, Burr B, Holtzapple MT (1997) Lime pretreatment of switchgrass. Appl Biochem Biotechnol 63–65:3–19

Chaturvedi V, Verma P (2013) An overview of key pretreatment processes employed for bioconversion of lignocellulosic biomass into biofuels and value added products. 3. Biotech 3:415–431

Chen W, Yu H, Liu Y, Chen P, Zhang M, Hai Y (2011a) Individualization of cellulose nanofibers from wood using high-intensity ultrasonication combined with chemical pretreatments. Carbohy Polym 83:1804–1811

Chen WH, Tu YJ, Sheen HK (2011b) Disruption of sugarcane bagasse lignocellulosic structure by means of dilute sulfuric acid pretreatment with microwave-assisted heating. Appl Energy 88:2726–2734

Cheng YS, Zheng Y, Yu CW, Dooley TM, Jenkins BM, Gheynst JSV (2010) Evaluation of high solids alkaline pretreatment of rice straw. Appl Biochem Biotech 162:1768–1784

Choi S, Song CW, Shin JH, Lee SY (2015) Biorefineries for the production of top building block chemicals and their derivatives. Metab Eng 28:223–239

Cianchetta S, Maggio BD, Burzi PL, Galletti S (2014) Evaluation of selected white-rot fungal isolates for improving the sugar yield from wheat straw. Appl Biochem Biotechnol 173:609–623

Dadi AP, Varanasi S, Schall CA (2006) Enhancement of cellulose saccharification kinetics using an ionic liquid pretreatment step. Biotechnol Bioeng 95:904–910

Dai Y, van Spronsen J, Witkamp G-J, Verpoorte R, Choi YH (2013) Natural deep eutectic solvents as new potential media for green technology. Anal Chim Acta 766:61–68

Dai Y, Witkamp GJ, Verpoorte R, Choi YH (2015) Tailoring properties of natural deep eutectic solvents with water to facilitate their applications. Food Chem 187:14–19

Delidovich I, Hausoul PJC, Deng L, Pfutzenreuter R, Rose M, Palkovits R (2016) Alternative monomers based on lignocellulose and their use for polymer production. Chem Rev 116:1540–1599

Dhiman SS, Haw J, Kalyani D, Kalia VC, Kang YC, Lee J (2015) Simultaneous pretreatment and saccharification: green technology for enhanced sugar yields from biomass using a fungal consortium. Bioresour Technol 179:50–57

Digman MF, Shinners KJ, Casler MD (2010) Optimizing on-farm pretreatment of perennial grasses for fuel ethanol production. Bioresour Technol 101:5305–5314

Du W, Yu H, Song L, Zhang J, Weng C, Ma F, Zhang X (2011) The promising effects of by-products from Irpex lacteus on subsequent enzymatic hydrolysis of bio-pretreated corn stalks. Biotechnol Biofuels 4:37

Fan LT, Gharpuray MM, Lee YH (1987) Cellulose hydrolysis biotechnology monographs. Springer, Berlin, p 57

Geng A, Xin F, Ip JY (2012) Ethanol production from horticultural waste treated by a modified organosolv method. Bioresour Technol 104:715–721

Gogate PR, Sutkar VS, Pandit AB (2011) Sonochemical reactors: important design and scale up considerations with a special emphasis on heterogeneous systems. Chem Eng J 166:1066–1082

Grous WR, Converse AO, Grethlein HE (1986) Effect of steam explosion pretreatment on pore size and enzymatic hydrolysis of poplar. Enzyme Microb Technol 8:274–280

Hammel KE, Kapich AN, Jensen KA, Ryan ZC (2002) Reactive oxygen species as agents of wood decay by fungi. Enz Microb Technol 30:445–453

Hendricks AT, Zeeman G (2009) Pretreatments to enhance the digestibility of lignocellulosic biomass. Bioresour Technol 100:10–18

Hideno A, Inoue H, Tsukahara K, Fujimoto S, Minowa T, Inoue S, Endo T, Sawayama S (2009) Wet disk milling pretreatment without sulfuric acid for enzymatic hydrolysis of rice straw. Bioresour Technol 100:2706–2711

Hideno A, Kawashima A, Endo T, Honda K, Morita M (2013) Ethanol-based organosolv treatment with trace hydrochloric acid improves the enzymatic digestibility of Japanese cypress (Chamaecyparis obtusa) by exposing nanofibers on the surface. Bioresour Technol 18:64–70

Hu ZH, Wen ZY (2008) Enhancing enzymatic digestibility of switchgrass by microwave-assisted alkali pretreatment. Biochem Eng J 38:369–378

Huijgen WJJ, Van der Laan RR, Reith JH (2008) Modified organosolv as a fractionation process of lignocellulosic biomass for coproduction of fuels and chemicals. In: Proceedings of the 16th European biomass conference and exhibition, Valencia

Ibrahim MM, El-Zawawy WK, Abdel-Fattah YR, Soliman NA, Agblevor FA (2011) Comparison of alkaline pulping with steam explosion for glucose production from rice straw. Carbohydr Polym 83:720–726

Ichwan M, Son TW (2011) Study on organosolv pulping methods of oil palm biomass. In: International seminar on chemistry. pp 364–370

Idrees M, Adnan A, Qureshi FA (2013) Optimization of sulfide/sulfite pretreatment of lignocellulosic biomass for lactic acid production. BioMed Research International 2013:1–11

Jacquet N, Vanderghem C, Danthine S, Quiévy N, Blecker C, Devaux J, Paquot M (2012) Influence of steam explosion on physicochemical properties and hydrolysis rate of pure cellulose fibers. Bioresour Technol 121:221–227

Jagmann N, Philipp B (2014) Design of synthetic microbial communities for biotechnological production processes. J Biotechnol 184:209–218

Kang S, Li X, Fan J, Chang J (2013) Hydrothermal conversion of lignin: a review. Renew Sust Ener Rev 27:546–558

Karunanithy C, Muthukumarappan K (2010) Influence of extruder temperature and screw speed on pretreatment of corn stover while varying enzymes and their ratios. Appl Biochem Biotechnol 162:264–279

Karunanithy C, Muthukumarappan K, Gibbons WR (2013) Effect of extruder screw speed, temperature, and enzyme levels on sugar recovery from different biomasses. ISRN Biotechnol 942810:1–13

Karunanithy C, Muthukumarappan K, Julson JL (2008) Influence of high shear bioreactor parameters on carbohydrate release from different biomasses. American Society of Agricultural and Biological Engineers, Annual International Meeting 2008. ASABE 084114. ASABE, St. Joseph

Karunanithy V, Muthukumarappan K (2011) Optimizing extrusion pretreatment and big bluestem parameters for enzymatic hydrolysis to produce biofuel using response surface methodology. Int J Agric Biol Eng 4:61–74

Keshwani DR, Cheng JJ (2010) Microwave-based alkali pretreatment of switchgrass and coastal bermudagrass for bioethanol production. Biotechnol Prog 3:644–652

Kilzer FJ, Broido A (1965) Speculations on the nature of cellulose pyrolysis. Pyrodynamics 2:151–163

Kim HJ, Chang JH, Jeong BY, Lee JH (2013) Comparison of milling modes as a pretreatment method for cellulosic biofuel production. J Clean Energy Technol 1:45–48

Kim HK, Hong J (2001) Supercritical CO_2 pretreatment of lignocellulose enhances enzymatic cellulose hydrolysis. Bioresour Technol 77:139–144

Kim JS, Kim H, Lee JS, Lee JP, Park SC (2008) Pretreatment characteristics of waste oak wood by ammonia percolation. Appl Biochem Biotechnol 148:15–22

Kim JW, Kim KS, Lee JS, Park SM, Cho HY, Park JC, Kim JS (2011a) Two-stage pretreatment of rice straw using aqueous ammonia and dilute acid. Bioresour Technol 102:8992–8999

Kim TH, Lee YY (2005a) Pretreatment and fractionation of corn stover by soaking in aqueous ammonia. Appl Biochem Biotechnol 121:1119–1131

Kim TH, Lee YY (2005b) Pretreatment of corn stover by ammonia recycle percolation process. Bioresour Technol 96:2007–2013

Kim Y, Yu A, Han M, Choi GW, Chung B (2011b) Enhanced enzymatic saccharification of barley straw pretreated by ethanosolv technology. Appl Biochem Biotechnol 163:143–152

Kleinert M, Barth T (2008) Phenols from lignin. Chem Eng Technol 31:736–745

Koo BW, Min BC, Gwak KS (2012) Structural changes in lignin during organosolv pretreatment of Liriodendron tulipifera and the effect on enzymatic hydrolysis. Biomass Bioenerg 42:24–32

Kootstra AM, Beeftink HH, Scott EL, Sanders JPM (2009) Comparison of dilute mineral and organic acid pretreatment for enzymatic hydrolysis of wheat straw. Biochem Eng J 46:126–131

Kumar AK, Parikh BS, Pravakar M (2016) Natural deep eutectic solvent mediated pretreatment of rice straw: bioanalytical characterization of lignin extract and enzymatic hydrolysis of pretreated biomass residue. Environ Sci Pollut Res Int 23:9265–9275

Kumar P, Barrett DM, Delwiche MJ, Stroeve P (2009) Methods for pretreatment of lignocellulosic biomass for efficient hydrolysis and biofuel production. Ind Eng Chem Res 48:3713–3729

Kumar P, Barrett DM, Delwiche MJ, Stroeve P (2011) Pulsed electric field pretreatment of switchgrass and woodchips species for biofuels production. Ind Eng Chem Res 50:10996–11001

Kumar R, Wyman CE (2009) Effects of cellulase and xylanase enzymes on the deconstruction of solids from pretreatment of poplar by leading technologies. Biotechnol Prog 25:302–314

Kuo CH, Lee CK (2009) Enhanced enzymatic hydrolysis of sugarcane bagasse by N methylmorpholine-N-oxide pretreatment. Bioresour Technol 100:866–871

Lamsal B, Yoo J, Brijwani K, Alavi S (2010) Extrusion as a thermo-mechanical pre-treatment for lignocellulosic ethanol. Biomass Bioenerg 34:1703–1710

Laser M, Larson E, Dale B, Wang M, Greene N, Lynd LR (2009) Comparative analysis of efficiency, environmental impact, and process economics for mature biomass refining scenarios. Biofpr 3:247–270

Lee J, Houtman CJ, Kim HY, Choi IG, Jeffries TW (2011) Scale-up study of oxalic acid pretreatment of agricultural lignocellulosic biomass for the production of bioethanol. Bioresour Technol 102:7451–7456

Lee J, Jeffries TW (2011) Efficiencies of acid catalysts in the hydrolysis of lignocellulosic biomass over a range of combined severity factors. Bioresour Technol 102:5884–5890

Lee JW, Rodrigues RCLB, Jeffries TW (2009) Simultaneous saccharification and ethanol fermentation of oxalic acid pretreated corncob assessed with response surface methodology. Bioresour Technol 100:6307–6311

Li L, Yu ST, Liu FS, Xie CS, Xu CZ (2011) Efficient enzymatic in situ saccharification of cellulose in aqueous-ionic liquid media by microwave treatment. BioResources 6:4494–4504

Lin Z, Huang H, Zhang H, Zhang L, Yan L, Chen J (2010) Ball milling pretreatment of corn stover for enhancing the efficiency of enzymatic hydrolysis. Appl Biochem Biotechnol 162:1872–1880

Liu LY, Chen HZ (2006) Enzymatic hydrolysis of cellulose materials treated with ionic liquid [BMIM] Cl. Chin Sci Bull 51:2432–2436

Lu X, Xi B, Zhang Y, Angelidaki I (2011) Microwave pretreatment of rape straw for bioethanol production: focus on energy efficiency. Bioresour Technol 102:7937–7940

Lu Y, Mosier NS (2007) Biomimetic catalysis for hemicelluloses hydrolysis in corn stover. Biotechnol Progr 23:116–123

Lucas M, Hanson SK, Wagner GL, Kimball DB, Rector KD (2012) Evidence for room temperature delignification of wood using hydrogen peroxide and manganese acetate as a catalyst. Bioresour Technol 119:174–180

Luengo E, Martínez JM, Coustets M, Álvarez I, Teissié J, Rols MP, Raso J (2015) A comparative study on the effects of millisecond and microsecond-pulsed electric field treatments on the permeabilization and extraction of pigments from Chlorella vulgaris. J Membrane Biol 248:883–891

Lynd LR, Elander RT, Wyman CE (1996) Likely features and costs of mature biomass ethanol technology. App Biochem Biotechnol 57:741–761

Ma F, Yang N, Xu C, Yu H, Wu J, Zhang X (2010) Combination of biological pretreatment with mild acid pretreatment for enzymatic hydrolysis and ethanol production from water hyacinth. Bioresour Technol 101:9600–9604

Ma R, Xu Y, Zhang X (2015) Catalytic oxidation of biorefinery lignin to value-added chemicals to support sustainable biofuel production. Chem Sus Chem 8:24–51

Marzialetti T, Olarte MBV, Sievers C, Hoskins TJC, Agrawal PK, Jones CW (2008) Dilute acid hydrolysis of loblolly pine: a comprehensive approach. Ind Eng Chem Res 47:7131–7140

McIntosh S, Vancov T (2010) Enhanced enzyme saccharification of Sorghum bicolor straw using dilute alkali pretreatment. Bioresour Technol 101:6718–6727

Mesa L, González E, Cara C, González M, Castro E, Mussatto SI (2011) The effect of organosolv pretreatment variables on enzymatic hydrolysis of sugarcane bagasse. Chem Eng J 168:1157–1162

Montalbo LM, Johnson L, Khanal SK, Leeuwene JV, Grewell D (2010) Sonication of sugary-2 corn: a potential pretreatment to enhance sugar release. Bioresour Technol 101:351–358

Mosier N, Wyman CE, Dale BE, Elander R, Lee YY, Holtzapple MT, Ladisch M (2005) Features of promising technologies for pretreatment of lignocellulosic biomass. Bioresour Technol 96:673–686

Mosier NS, Ladisch CM, Ladisch MR (2002) Characterization of acid catalytic domains for cellulose hydrolysis and glucose degradation. Biotechnol Bioeng 79(6):610–618

Moultrop JS, Swatloski RP, Moyna G, Rogers RD (2005) High resolution 13-C NMR studies of cellulose and cellulose oligomers in ionic liquid solutions. Chem Commun 2005:1557–1559

Nakamura Y, Daidai M, Kobayashi F (2004) Ozonolysis mechanism of lignin model compounds and microbial treatment of organic acids produced. Water Sci Technol 50:167–172

Neely WC (1984) Factors affecting the pretreatment of biomass with gaseous ozone. Biotechnol Bioeng 26:59–65

Ni M, Leung DYC, Leung MKH, Sumathy K (2006) An overview of hydrogen production from biomass. Fuel Process Technol 87:461–472

Nomanbhay SM, Hussain R, Palanisamy K (2013) Microwave assisted enzymatic saccharification of oil palm empty fruit bunch fiber for enhanced fermentable sugar yield. J Sustain Bioenergy Syst 3:7–17

Ogura M, Phaiboonsilpa N, Yamauchi K, Saka S (2013) Two-step decomposition behavior of rice straw as treated by semi-flow hot-compressed water (in Japanese). J Jpn Inst Energy 92:456

Paiva A, Craveir R, Aroso I, Martins M, Reis RL, Duarte ARC (2014) Natural deep eutectic solvents—solvents for the 21st century. ACS Sustain Chem Eng 2:1063–1071

Panagiotopoulos IA, Chandra RP, Saddler JN (2012) A two-stage pretreatment approach to maximise sugar yield and enhance reactive lignin recovery from poplar wood chips. Bioresour Technol 130:570–577

Park JY, Shiroma R, Al-Haq MI, Zhang Y, Ike M, Arai-Sanoh Y, Ida A, Kondo M, Tokuyasu K (2010a) A novel lime pretreatment for subsequent bioethanol production from rice straw—calcium capturing by carbonation (CaCCO) process. Bioresour Technol 101:6805–6811

Park N, Kim HY, Koo BW, Yeo H, Choi IG (2010b) Organosolv pretreatment with various catalysts for enhancing enzymatic hydrolysis of pitch pine (Pinus rigida). Bioresour Technol 101:7046–7053

Phaiboonsilpa N (2010) Chemical conversion of lignocellulosics as treated by two step semi-flow hot-compressed water. In: Graduate School of Energy Science, Doctoral dissertation, Kyoto University, Kyoto

Potumarthi R, Baadhe RR, Nayak P, Jetty A (2013) Simultaneous pretreatment and saccharification of rice husk by Phanerochete chrysosporium for improved production of reducing sugars. Bioresour Technol 128:113–117

Prado R, Erdocia X, Serrano L, Labidi J (2012) Lignin purification with green solvents. Cellul Chem Technol 46:221–225

Putro JN, Soetaredjo FE, Lin SY, Ju YH, Ismadi S (2016) Pretreatment and conversion of lignocellulose biomass into valuable chemicals. RSC Adv 6:46834–46852

Quesada J, Rubio M, Gomez D (1999) Ozonation of lignin rich solid fractions from corn stalks. J Wood Chem Technol 19:115–137

Rabemanolontsoa H, Saka S (2016) Various pretreatments of lignocellulosics. Bioresour Technol 199:83–91

Rehman MSU, Kim I, Chisti Y, Han JI (2013) Use of ultrasound in the production of bioethanol from lignocellulosic biomass. EEST Part A Energy Sci Res 30:1391–1410

Roberts VM, Stein V, Reiner T, Lemonidou A, Li X, Lercher JA (2011) Towards quantitative catalytic lignin depolymerization. Chem Eur J 17:5939–5948

Romero A, Alonso A, Sastre A, Marquez AN (2016) Conversion of biomass into sorbitol: cellulose hydrolysis on MCM-48 and D-Glucose hydrogenation on Ru/MCM-48. Micropor Mesopor Mat 224:1–8

Saha BC, Cotta MA (2007) Enzymatic saccharification and fermentation of alkaline peroxide pretreated rice hulls to ethanol. Enzyme Microb Technol 41:528–532

Saha BC, Iten BL, Cotta M, Wu YV (2005) Dilute acid pretreatment, enzymatic saccharification, and fermentation of rice hulls to ethanol. Biotechnol Prog 21:3816–3822

Salerno MB, Lee HS, Parameswaran P, Rittmann BE (2009) Using a pulsed electric field as a pretreatment for improved biosolids digestion and methanogenesis. Water Environment Federation WEFTEC. 2005–2018

Sánchez C (2009) Lignocellulosic residues: biodegradation and bioconversion by fungi. Biotechnol Adv 27:185–194

Sant' Ana daSilva A, Inoue H, Endo T, Yano S, Bon EPS (2010) Milling pretreatment of sugarcane bagasse and straw for enzymatic hydrolysis and ethanol fermentation. Biores Technol 101:7402–7409

Sarkar N, Ghosh SK, Bannerjee S, Aikat K (2012) Bioethanol production from agricultural wastes: an overview. Renew Energy 37:19–27

Sassner P, Martensson CG, Galbe M, Zacchi G (2008) Steam pretreatment of H_2SO_4 impregnated salix for the production of bioethanol. Biores Technol 99:137–145

Sawada T, Nakmura Y, Kobayashi F, Kuwahara M, Watanabe T (1995) Effects of fungal pretreatment and steam explosion pretreatment on enzymatic saccharification of plant biomass. Biotechnol Bioeng 48:719–724

Shafizadeh F, Bradbury AGW (1979) Thermal degradation of cellulose in air and nitrogen at low temperatures. J Appl Poly Sci 23:1431–1442

Shi J, Chinn MS, Sharma-Shivappa RR (2008) Microbial pretreatment of cotton stalks by solid state cultivation of Phanerochaete chrysosporium. Bioresour Technol 99:6556–6564

Shuai L, Yang Q, Zhu JY, Lu FC, Weimer PJ, Ralph J, Pan XJ (2010) Comparative study of SPORL and dilute-acid pretreatments of spruce for cellulosic ethanol production. Biores Technol 101:3106–3114

Sills DL, Gossett JM (2011) Assessment of commercial hemicellulases for saccharification of alkaline pretreated perennial biomass. Biores Technol 102:1389–1398

Sindhu R, Binod P, Pandey A (2016) Biological pretreatment of lignocellulosic biomass—an overview. Bioresour Technol 199:76–82

Sivagurunathan P, Kumar G, Bakonyi P, Kim SH, Kobayashi T, Xu KQ, Lakner G, Toth G, Nemestothy N, Bako KB (2016) A critical review on issues and overcoming strategies for the enhancement of dark fermentative hydrogen production in continuous systems. Int J Hydrogen Energy 41:3820–3836

Smith EL, Abbott AP, Ryder KS (2014) Deep eutectic solvents (DESs) and Their applications. Chem Rev 114:11060–11082

Song L, Yu H, Ma F, Zhang X (2013) Biological pretreatment under non-sterile conditions for enzymatic hydrolysis of corn stover. BioResources 8:3802–3816

Suhara H, Kodama S, Kamei I, Maekawa N, Meguro S (2012) Screening of selective lignin-degrading basidiomycetes and biological pretreatment for enzymatic hydrolysis of bamboo culms. Int Biodeter Biodegr 75:176–180

Sun R, Lawther JM, Banks WB (1995) Influence of alkaline pre-treatments on the cell wall components of wheat straw. Industrial Crop Prod 2:127–145

Sun RC, Tomkinson J (2002) Comparative study of lignins isolated by alkali and ultrasound-assisted alkali extractions from wheat straw. Ultrason Sonochem 9:85–93

Sun Y, Cheng J (2002) Hydrolysis of lignocellulosic materials for ethanol production: a review. Bioresour Technol 83:1–11

Sun YE, Cheng JJ (2005) Dilute acid pretreatment of rye straw and Bermuda grass for ethanol production. Bioresour Technol 96:1599–1606

Szijártó N, Kádár Z, Varga E, Thomsen AB, Costa-Ferreira M, Réczey K (2009) Pretreatment of reed by wet oxidation and subsequent utilization of the pretreated fibers for ethanol production. Appl Biochem Biotechnol 155:386–396

Taha M, Shahsavari E, Al-Hothaly K, Mouradov A, Smith AT, Ball AS, Adetutu EM (2015) Enhanced biological straw saccharification through co-culturing of lignocellulose degrading microorganisms. Appl Biochem Biotechnol 175:3709–3728

Taherzadeh MJ, Karimi K (2008) Pretreatment of lignocellulosic wastes to improve ethanol and biogas production: a review. Int J Mol Sci 9:1621–1651

Tang J, Chen K, Huang F, Xu J, Li J (2013) Characterization of the pretreatment liquor of biomass from the perennial grass, Eulaliopsis binata, for the production of dissolving pulp. Bioresour Technol 129:548–552

Uppugundla N, Da Costa Sousa L, Chundawat SPS, Yu X, Simmons B, Singh S, Gao X, Kumar R, Wyman CE, Dale BE, Balan V (2014) A comparative study of ethanol production using dilute acid, ionic liquid and AFEXTM pretreated corn stover. Biotechnol Biofuel 7:72–85

Varga E, Schmidt AS, Réczey K, Thomsen AB (2003) Pretreatment of corn stover using wet oxidation to enhance enzymatic digestibility. Appl Biochem Biotechnol 104:37–50

Vats S, Maurya DP, Shaimoon M, Negi S (2013) Development of a microbial consortium for the production of blend enzymes for the hydrolysis of agricultural waste into sugars. J Sci Ind Res 72:585–590

Vidal PF, Molinier J (1988) Ozonolysis of lignin—improvement of in vitro digestibility of poplar sawdust. Biomass 16:1–17

Villaverde JJ, Ligero P, De Vega A (2010) Miscanthus x giganteus as a source of biobased products through organosolv fractionation: a mini review. Open Agric J 4:102–110

Wan C, Li Y (2011) Effectiveness of microbial pretreatment by Ceriporiopsis subvermispora on different biomass feed stocks. Bioresour Technol 102:7507–7512

Wang L, Mu G, Tan C, Sun L, Zhou W, Yu P, Yin J, Fu H (2013) Porous graphitic carbon nanosheets derived from cornstalk biomass for advanced supercapacitors. Chem Sus Chem 6:880–889

Wang W, Yuan T, Wang K, Cui B, Dai Y (2012) Combination of biological pretreatment with liquid hot water pretreatment to enhance enzymatic hydrolysis of Populus tomentosa. Bioresour Technol 107:282–286

Wright JD (1988) Ethanol from biomass by enzymatic hydrolysis. Chem Eng Prog 84:62–74

Wyman CE (1999) Biomass ethanol: technical progress, opportunities, and commercial challenges. Ann Rev Energy Environ 24:189–226

Xu H, Li B, Mu X (2016) Review of alkali-based pretreatment to enhance enzymatic saccharification for lignocellulosic biomass conversion. Ind Eng Chem Res 55:8691–8705

Xu J, Chen H, Kadar Z, Thomsen AB, Schmidt JE, Peng H (2011) Optimization of microwave pretreatment on wheat straw for ethanol production. Biomass Bioenerg 35:385–386

Xu J, Thomsen MH, Thomsen AB (2009) Pretreatment on corn stover with low concentration of formic acid. J Microbiol Biotechnol 19:845–850

Yachmenev V, Condon B, Klasson T, Lambert A (2009) Acceleration of the enzymatic hydrolysis of corn stover and sugar cane bagasse celluloses by low intensity uniform ultrasound. J Biobased Mater Bioenergy 3:25–31

Yang B, Wyman CE (2004) Effect of xylan and lignin removal by batch and flow through pretreatment on enzymatic digestibility of corn stover cellulose. Biotechnol Bioeng 86:88–95

Yang J, Christiansen K, Luchner S (2013) Renewable, low-cost carbon fiber for light weight vehicles. Detroit, U.S. Department of Energy

Yoo JY (2011) Technical and economical assessment of thermo-mechanical extrusion pretreatment for cellulosic ethanol production. Ph.D. Thesis, Kansas State University, Manhattan

Yoshikawa T, Yagi T, Shinohara S, Fukunaga T, Nakasaka Y, Tago T, Masuda T (2013) Production of phenols from lignin via depolymerization and catalytic cracking. Fuel Process Technol 108:69–75

Yu J, Zhang J, He J, Liu Z, Yu Z (2009) Combinations of mild physical or chemical pretreatment with biological pretreatment for enzymatic hydrolysis of rice hull. Bioresour Technol 100:903–908

Yu X, Gouyo T, Grimi N, Bals O, Vorobiev E (2016) Pulsed electric field pretreatment of rapeseed green biomass (stems) to enhance pressing and extractives recovery. Bioresour Technol 199:194–201

Zakaria MR, Fujimoto S, Hirata S, Hassan MA (2014) Ball milling pretreatment of oil palm biomass for enhancing enzymatic hydrolysis. Appl Biochem Biotechnol 173:1778–1789

Zavrel M, Bross D, Funke M, Buchs J, Spiess AC (2009) High-throughput screening for ionic liquids dissolving (ligno-)cellulose. Bioresour Technol 100:2580–2587

Zhang DS, Yang Q, Zhu JY, Pan XJ (2013) Sulfite (SPORL) pretreatment of switchgrass for enzymatic saccharification. Bioresour Technol 129:127–134

Zhang SH, Xu YX, Hanna MA (2012a) Pretreatment of corn stover with twin-screw extrusion followed by enzymatic saccharification. Appl Biochem Biotechnol 166:458–469

Zhang Q, De Vigier KO, Royer S, Jérôme F (2012b) Deep eutectic solvents: syntheses, properties and applications. Chem Soc Rev 41:7108–7146

Zhao X, Liu D (2012) Fractionating pretreatment of sugarcane bagasse by aqueous formic acid with direct recycle of spent liquor to increase cellulose digestibility-the Formiline process. Bioresour Technol 117:25–32

Zhao Y, Wang Y, Zhu JY, Ragauskas A, Deng Y (2008) Enhanced enzymatic hydrolysis of spruce by alkaline pretreatment at low temperature. Biotechnol Bioeng 99:1320–1328

Zheng J, Rehmann L (2014) Extrusion pretreatment of lignocellulosic biomass: a review. Int J Mol Sci 15:18967–18984

Zheng YZ, Lin HM, Tsao GT (1995) Supercritical carbon-dioxide explosion as a pretreatment for cellulose saccharification. Biotechnol Lett 17:845–850

Zhu J, Rezende CA, Simister R, McQueen-Mason SJ, Macquarrie DJ, Polikarpov I, Gomez LD (2016) Efficient sugar production from sugarcane bagasse by microwave assisted acid and alkali pretreatment. Biomass Bioenerg 93:269–278

Zhu J, Wan C, Li Y (2010) Enhanced solid-state anaerobic digestion of corn stover by alkaline pretreatment. Bioresour Technol 101:7523–7528

Zhu JY, Pan XJ (2010) Woody biomass pretreatment for cellulosic ethanol production technology and energy consumption evaluation. Bioresour Technol 101:4992–5002

Zhu JY, Pan XJ, Wang GS, Gleisner R (2009) Sulfite pretreatment (SPORL) for robust enzymatic saccharification of spruce and red pine. Bioresour Technol 100:2411–2418

Zhu Z, Macquarrie DJ, Simister R, Gomez LD, McQueen-Mason SJ (2015a) Microwave assisted chemical pretreatment of Miscanthus under different temperature regimes. Sustain Chem Process 3:15–27

Zhu Z, Simister R, Bird S, McQueen-Mason SJ, Gomez LD, Macquarrie DJ (2015b) Microwave assisted acid and alkali pretreatment of Miscanthus biomass for biorefineries. AIMS Bioeng 2:449–468

Zhua JY, Wang GS, Pan XJ, Gleisner R (2009) Specific surface to evaluate the efficiencies of milling and pretreatment of wood for enzymatic saccharification. Chem Eng Sci 64:474–485

Zwart RWR, Boerrigter H, Van der Drift A (2006) The impact of biomass pretreatment on the feasibility of overseas biomass conversion to fischer-tropsch products. Energy Fuels 20:2192–2197

2

Advances in industrial microbiome based on microbial consortium for biorefinery

Li-Li Jiang[1], Jin-Jie Zhou[1], Chun-Shan Quan[2] and Zhi-Long Xiu[1*]

Abstract

One of the important targets of industrial biotechnology is using cheap biomass resources. The traditional strategy is microbial fermentations with single strain. However, cheap biomass normally contains so complex compositions and impurities that it is very difficult for single microorganism to utilize availably. In order to completely utilize the substrates and produce multiple products in one process, industrial microbiome based on microbial consortium draws more and more attention. In this review, we first briefly described some examples of existing industrial bioprocesses involving microbial consortia. Comparison of 1,3-propanediol production by mixed and pure cultures were then introduced, and interaction relationships between cells in microbial consortium were summarized. Finally, the outlook on how to design and apply microbial consortium in the future was also proposed.

Keywords: Industrial microbiome, Microbial consortia, Biorefinery, Biomass, Bio-based chemicals, Biofuels

Background

Human beings have always lived with microbial communities on the earth, but know little about their compositions and functions. Therefore, a group of leading US scientists proposed an Unified Microbiome Initiative (UMI) to research almost all the microbiomes in human, plants, animals, soil, and sea (Alivisatos et al. 2015). They hoped this plan would be paid the same attention with the Precision Medicine Initiative and Brain Initiative in the United States. At the same time, three scientists from Germany, China, and America called for an International Microbiome Initiative (IMI) supported by funding agencies and foundations around the world. They suggested that interdisciplinary experts should cooperate, share standards across borders and disciplines, and realize the integration of resources (Dubilier et al. 2015). Microbiome is a new developing discipline that studies the relationship between microbial consortia in the environment and the growth of animals and plants, as well as human diseases and health. Microbial consortium is referred to microbial community with diverse species on the basis of ecological selection principles. Microbiome can be applied in the fields of industry, agriculture, fishery, medicine, and so on (Fig. 1). The research object of industrial microbiome is microbial consortia applied in food, environment, energy, chemical, and other industrial areas.

The utilization of microbial resources by human has experienced two stages, from naturally mixed culture to pure culture. Human beings have used microbial metabolites for centuries, such as bread, wine, cheese, pickles, and other fermented materials, being provided by fermentation using bacteria and fungi. The bioprocesses were carried out with naturally mixed culture (Sabra and Zeng 2014), which is microbial fermentation by different specified/unspecified microorganisms. In order to avoid contamination of the fermentation process and the product with pathogenic microbes, mixed culture was gradually replaced by pure culture. Without the complicated situation of coexistence of multiple microbes, microbial pure culture allows researchers to be undisturbed for a single strain, and to have a deeper understanding about morphological, physiological, biochemical, and genetic characteristics of microorganisms. Pure culture has built up a milestone for biochemical engineering and modern biotechnology. To date, many bulk biotechnological products such as amino acids, organic acids, antibiotics, and enzymes are almost produced by pure cultures

*Correspondence: zhlxiu@dlut.edu.cn
[1] School of Life Science and Biotechnology, Dalian University of Technology, Linggong Road 2, Dalian 116024, Liaoning Province, China
Full list of author information is available at the end of the article

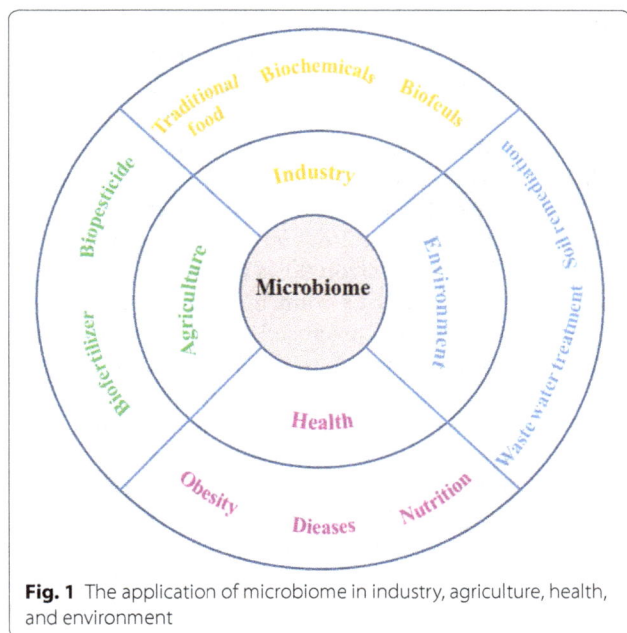

Fig. 1 The application of microbiome in industry, agriculture, health, and environment

of microorganisms (Sabra et al. 2010). However, about 90–99.8% of the microbes in natural environment cannot be cultured with currently available technologies, and hence cannot be exploited further for biotechnology with pure culture (Streit et al. 2004). The typical problems for biofuels and bio-based chemicals production with pure cultures are the high costs of substrates and product purification, high energy demand for fermentation operation, and high concentrations of by-products in the form of organic acids or alcohols which are toxic to cell growth (Xiu and Zeng 2008; Zeng and Sabra 2011).

In the face of the defects with pure culture, people rethink about the strategies of microbial fermentation. Co-culture is developed based on pure culture, which normally refers to cultures with multiple (mostly two) defined species of microorganisms under aseptic conditions (Sabra et al. 2013). It is a microbial fermentation technology utilizing the different characteristics of microbial growth and metabolism for fermentation (Bader et al. 2010). A typical application of co-culture is the production of 2-keto-L-gulonic acid (2-KLG), the precursor of vitamin C. In the co-culture system, *Ketogulonicigenium vulgare* (small strain) synthesizes 2-KLG from L-sorbose; *Bacillus megaterium* (big strain) as an associated bacterium secretes some metabolites to stimulate the growth of *K. vulgare,* and thus enhances 2-KLG production (Zhang et al. 2010). The researches on fermentation with microbial consortium have been intensive in recent years for overcoming the limitations of pure culture and adapting to the complex substrates and

environment. This biotechnology is the industrial application of naturally mixed cultures. On the basis of ecological selection principles, it is able to utilize microbial consortia which can generate a special product spectrum from mixed substrates and reduce the cost of substrates and product purification. Moreover, the processes with microbial consortia have no aseptic requirements (Dietz and Zeng 2014). Microbial consortia usually contain some unknown or non-cultured microorganisms whose effects are unclear. And microbial consortia exhibit strong superiority in the environmental remediation and energy production, such as wastewater treatment with activated sludge and biogas production.

In order to meet the needs of the sustained social and economic development, the industrial biotechnology for a conversion of renewable materials into chemicals and fuels economically has been developed to be an alternative to the traditional chemical industry with high energy consumption and high pollution. Biorefinery has been proposed as one of the key concepts for conversion of renewable materials. Biorefinery is a complex system of sustainable, environment- and resource-friendly technology for material and energy comprehensive use or recovery of renewable raw materials from green and waste biomasses (Kamm et al. 2016). The development of biorefinery is necessary to make various biological products competitive to their equivalent products based on fossil raw materials. The consolidated bioprocessing (CBP) represents an effective and feasible way to implement biorefinery. CBP is referred to integrating all bioconversion reactions in one-step biological process (Minty et al. 2013; Olson et al. 2012). The traditional strategy of CBP is the use of genetically engineered microorganisms focusing on all the required functional genes on one strain. However, many experimental results proved that it was a huge challenge to design and optimize a variety of functions in one strain (Olson et al. 2012). The synthetic biology is also facing the similar challenges in recent years. Compared with CBP based on genetically engineered strains, there are many attractive characteristics of microbial consortia in natural environment, such as composition stability, functional robustness, broad spectrum of substrates, and qualified complex tasks and so on. Therefore, industrial microbiome based on microbial consortium can play an essential role in biorefinery.

Applications of microbial consortia in industrial fermentations

The application of microbial consortia in traditional foods, such as vinegar, soy sauce, cheese, wine, bread, and pickles, has been recorded for millennia. In the fields of biofuels (biogas, biohydrogen, ethanol, butanol, etc.), bio-based chemicals (1,3-propanediol), biomaterials (polyhydroxyalkanoates), and microbial consortia were also used and studied.

Biogas

Biogas is a mixed gas containing methane, H_2, CO_2, etc., which is converted from organic waste via anaerobic digestion with anaerobic microbial consortia (Bizukojc et al. 2010). Generally, the transformation of organic wastes into biogas is considered to occur in four stages (Sabra et al. 2010). During the hydrolysis phase (Stage I), bio-polymers are degraded into monomers or oligomers which are fermented into volatile organic acids, alcohols, CO_2, and H_2 in the acidogenesis phase (Stage II). In the acetogenesis phase (Stage III), acetic acid as well as some CO_2 and H_2 is produced from the molecules formed in Stage II. In the methanogenesis phase (Stage IV), CH_4 is formed through acetate or CO_2 and H_2 by methanogens.

Because of the special growth requirement for some bacteria within microbial consortia, such as a low hydrogen partial pressure, some bacteria are difficult to cultivate using traditional culturing method, such as pure culture. The upflow anaerobic sludge bed (UASB) is the most common type of bioreactors used. In this reactor, methanogenic microbial consortia are present as granules (Diaz et al. 2006). It has been investigated that two-stage process is useful for the treatment of sugar-rich wastewater and bread wastes (Nishio and Nakashimada 2007). In the first stage, bread waste fermented by thermophilic anaerobic sludge at 55 °C was converted to hydrogen and volatile fatty acids (mainly acetate and butyrate), which were then converted to methane in the second stage. Despite of the unsterile process, the thermophilic species from the inoculated microflora were dominating in the hydrogenotrophic stage and the thermophilic process reduced the risk for contamination effectively.

Hydrogen

As a clean fuel in the future, hydrogen production by fermentation of organic waste has received significant attention in recent years. The main driving force for investigating the production of hydrogen is the economic value of hydrogen, owning to its wide range of applications in the chemical industry, such as synthesis of amines, alcohols, and aldehydes (Li and Fang 2007). And hydrogen is also an ideal fuel, which only produces water after burning. At present, the main difficulty in hydrogen production via microbial anaerobic fermentation is the low yield of hydrogen. The theoretical maximum yield of hydrogen is 4 mol/mol glucose, but in fact the yield of hydrogen from glucose is usually not more than 2 mol/mol due to the consumption of hydrogen by some microorganisms such as methanogens and homoacetogens during the mixed culture (Selembo et al. 2009). It has

been proved the pre-treatment by alkali, acid, or heat to make the above hydrogen-consuming microorganisms inactive, and the effect of heat treatment to be best except to homoacetogens (Oh et al. 2003). Due to the complexity of microbial consortium, the intracellular metabolic pathway of hydrogen is also more complex. Lee et al. (2009) developed the first model for predicting community structure in mixed-culture fermentative biohydrogen production through electron flows and $NADH_2$ balances. The clone-library analyses confirmed the model prediction, and hydrogen was produced at pH 3.5 only via the pyruvate decarboxylation-ferredoxin-hydrogenase pathway in microbial consortium. This model could easily assess the main mechanism for hydrogen formation and the dominant hydrogen-producing bacteria in mixed culture. Rahul et al. (2012) evaluated the potential of bioconversion of crude glycerol to hydrogen by an enriched microbial community from activated sludge. Hydrogen yield from raw glycerol was almost 1.1 mol-H_2/mol glycerol consumed under optimal conditions (pH 6.5, 40 °C and 1 g/l raw glycerol).

Ethanol

Ethanol is an important alternative of gasoline fuel, with the advantages of cheap, clean, environment-friendly, safe, and renewable fuel. At present, the research focused on the conversion of non-food materials, such as lignocellulose to ethanol. The main constituents of lignocellulosic hydrolysates are hexoses (glucose, mannose, galactose, etc.), pentoses (xylose, arabinose, etc.), and several toxic by-products such as phenol, acid, and aldehyde (Eiteman et al. 2008). The traditional pure culture by *Saccharomyces cerevisiae* could not convert mixture of hexoses and pentoses effectively. Du et al. (2015) selected a consortium (named HP) from 16 different natural bacterial consortia, and HP consortium exhibited relatively high ethanol production (2.06 g/l ethanol titer from 7 g/l α-cellulose at 55 °C in 6 days). They found that the community composition affected the performance of producing ethanol from cellulose. Recent studies have proved that natural microbial consortia can produce a variety of cellulases, in order to adapt the degradation requirements of different lignocelluloses. Three new anaerobic gut fungi (*Anaeromyces robustus*, *Neocallimastix californiae*, and *Piromyces finnis*) isolated from herbivores produced the biomass-degrading enzymes which exhibited strong ability to degrade lignocellulose. The relative activity for hydrolysis of xylan with these enzymes especially secreted by *Piromyces finnis* was threefold more than those optimized commercial preparation from *Aspergillus* (Solomon et al. 2016). Thus, cellulosic ethanol production by microbial consortia is a promising method.

Butanol

Butanol, a four-carbon primary alcohol, is not only an important bulk chemical feedstock, but also a promising next-generation liquid fuel because of its superior characteristics over ethanol, such as higher energy content, less hygroscopicity, better blending ability, and an energy density closer to that of gasoline (Dürre 2007). However, to date, bio-production of butanol is still not economically competitive with petrochemical production because of its major drawbacks, such as high cost of the feedstocks, low butanol concentration in the fermentation broth, and low-value by-products, i.e., acetone and ethanol (Gu et al. 2011). In order to reduce the cost of the feedstocks, biosynthesis of butanol from lignocelluloses gained popularity in recent years. Microbial conversion of lignocellulosic biomass requires multiple biological functionalities, including production of saccharifying enzymes (cellulases and hemicellulases), enzymatic hydrolysis of lignocellulose to soluble saccharides, and metabolism of soluble saccharides to desired products (Zuroff and Curtis 2012).

Consolidated bioprocessing has been suggested as an efficient and economical method of producing butanol from lignocellulose through simultaneous hydrolysis and fermentation with cellulolytic microorganisms and solventogenic bacteria in one bioreactor (Olson et al. 2012). In the consortium, microorganisms may develop the potential for synergistic utilization of the metabolic pathways from interspecies. It was very difficult to produce butanol efficiently from lignocellulose directly by pure culture. Wen et al. (2014) constructed a stable artificial symbiotic consortium by co-culturing a cellulolytic, anaerobic, butyrate-producing mesophile (*Clostridium cellulovorans* 743B) and a non-cellulolytic, solventogenic bacterium (*Clostridium beijerinckii* NCIMB 8052) to produce solvents by consolidated bioprocessing with alkali extracted deshelled corn cobs (AECC) as the sole carbon source. Under optimized conditions, the co-culture degraded 68.6 g/l AECC and produced 11.8 g/l solvents (2.64 g/l acetone, 8.30 g/l butanol, and 0.87 g/l ethanol) in less than 80 h.

Polyhydroxyalkanoates

Polyhydroxyalkanoates (PHAs) are polyesters, a kind of natural macromolecule biomaterial, which are synthesized and stored within the cell by various microorganisms. PHAs have been recognized as good candidates for biodegradable plastics because of their similar properties to conventional plastics and their complete biodegradability (Lemos et al. 2006). Industrial production processes are based on the use of pure cultures of microorganisms in their wild form or recombinant strains (Vandamme and Coenye 2004). However, due to the pure substrates

utilized and the sterile operation of the production process, the cost of PHA production with pure culture is still too high to become a competitive commodity plastic material. Therefore, in order to reduce the cost of PHA production, the interest in the use of mixed cultures for PHA production has increased in recent years (Dias et al. 2006). The production of PHA by mixed cultures could use renewable carbon sources based on agricultural or industrial wastes, and operate under non-sterile condition, which reduce the cost of substrate and equipment investment significantly. Moita et al. (2014) investigated the feasibility of PHA production by a mixed microbial community using crude glycerol as feedstock. The results showed that crude glycerol could be used to produce PHA without any pre-treatment step, leading to the overall production process more economically competitive, reducing polymer final cost.

Comparison between pure culture of single strain and mixed culture of microbial consortia

Industrial 1,3-propanediol (1,3-PD) production has attracted attention as an important monomer to synthesize a new type of polyester, polytrimethylene terephthalate (PTT), and the market demand is increasing year by year (Zeng and Sabra 2011). The traditional microbial fermentation to produce 1,3-PD is pure culture. This biotechnological method includes wild-type bacteria conversion of glycerol to 1,3-PD and gene-modified bacteria conversion of glucose to 1,3-PD directly (Chatzifragkou et al. 2011; Jolly et al. 2014; Metsoviti et al. 2013; Nakamura and Whited 2003). A surplus of crude glycerol has occurred due to large production of biodiesel; therefore, the conversion of crude glycerol into 1,3-PD was paid more and more attention. Crude glycerol usually contains many impurities such as alcohol, salts, esters or lipids, and pigments, so that it needs to be purified before used for pure culture, no doubt increased the cost of production (Johnson and Taconi 2007).

Up to date, most researches have focused on strain screening (Metsoviti et al. 2012a, b; Raghunandan et al. 2014; Rodriguez et al. 2015), genetically engineered strains (Nakamura and Whited 2003), fermentation optimization of 1,3-PD (Jun et al. 2010; Sun et al. 2010), etc., which were all based on pure cultures. The fermentation based on pure culture usually requires strict aseptic operation and purified substrates, resulting in the high cost of biological production of 1,3-PD. At the same time, in order to balance the intracellular redox state and to supply ATP during microbial production of 1,3-PD, various by-products were produced, such as acetic acid, lactic acid, succinic acid, and other organic acids as well as alcohols. The accumulation of these by-products often inhibits the growth of cells, competes for NADH against the 1,3-PD pathway to reduce the yield of 1,3-PD from glycerol, and brings difficulties for the separation and purification of target product (Xiu and Zeng 2008).

Compared with pure culture, specific advantages of fermentation with microbial consortia include the following: ① the possibility of utilizing cheaper or mixed substrates (e.g., whey, molasses, lignocellulose, and raw glycerol); ② the synergies of different enzymatic systems and combination of metabolic pathways of different microorganisms that can result in more efficient utilization of substrates and a narrow production spectrum contributing to product purification and reducing the cost; ③ due to the high microbial diversity, the operation with microbial consortia has no sterile requirement which will lower the production cost (Sabra and Zeng 2014). Thus, biotechnology based on microbial consortia could become an attractive addition or alternative to traditional biotechnology based on pure culture for the production of chemicals in industrial biotechnology (Sabra et al. 2010).

In order to overcome the shortcoming of pure culture, and reduce the cost of biological production of 1,3-PD furthermore, the fermentation with microbial consortia has been intensively studied in recent years (Dietz and Zeng 2014; Gallardo et al. 2014; Kanjilal et al. 2015; Liu et al. 2013; Temudo et al. 2008). The biological production of 1,3-PD based on pure culture of single strain was compared with that based on mixed culture of microbial consortia (Table 1). Dietz and Zeng (2014) selected microbial consortia from sludge of wastewater treatment plant. 1,3-PD can be produced as the main product in this mixed culture with typical organic acids such as acetic and butyric acids as by-products. The yield was in the range of 0.56–0.76 mol 1,3-PD/mol glycerol consumed

depending on the glycerol concentration. A final product concentration as high as 70 g/l was obtained in fed-batch cultivation with a productivity of 2.6 g/l h. This study showed that 1,3-PD production in mixed culture achieved the same levels of product titer, yield, and productivity as in typical pure cultures, especially without sterile requirement. Szymanowska-Powalowska et al. (2013) isolated bacterial strains with capability of the utilization of by-products such as butyric acid and lactic acid. The co-culture of *Clostridium butyricum* DSP1 producing 1,3-PD and *Alcaligenes faecalis* JP1 utilizing organic acids increased the volumetric productivity (1.07 g/l h) and yield of 1,3-PD (0.53 g/g). Moreover, the only by-product present was butyric acid at a concentration below 1 g/l, which significantly reduced the cost of extraction and purification for the target product. This new type of mixed culture provides a new solution to separate and purify target products in the process of bio-based chemicals production.

In the past few years, our lab selected facultative anaerobic microbial consortia from sludge in Dalian seashore. 16S rRNA gene amplicon high-throughput sequencing was performed to investigate the bacterial composition of microbial consortium DL38, and it was found that the most abundant organisms belonged to *Enterobacteriaceae* (95.57%), followed by *Enterococcaceae* (2.10%), *Moraxellaceae* (1.21%), and *Streptococcaceae* (0.64%). The results showed that mixed culture with microbial consortium DL38 (Genbank accession number: SRP066989) possessed excellent substrate tolerance and narrow product spectrum, leading to the biological production of 1,3-PD more attractive and competitive. The yield was in the range of 0.57–0.70 mol 1,3-PD/mol glycerol consumed, which depended on the glycerol

Table 1 Comparison of 1,3-propanediol production by microbial consortia and single strain

Inoculum	Fermentation type	Glycerol type	1,3-PD (g/l)	Yield (mol/mol)	References
Pure culture of single strain					
Klebsiella pneumoniae DSM 4799	Fed-batch	Raw	80.20	0.54	Jun et al. (2010)
Klebsiella oxytoca M5al	Fed-batch	Pure	83.56	0.62	Yang et al. (2007)
Citrobacter freundii FMCC-B 294	Fed-batch	Raw	68.10	0.48	Metsoviti et al. (2013)
Clostridium butyricum AKR102a	Fed-batch	Raw	93.70	0.63	Wilkens et al. (2012)
Lactobacillus reuteri ATCC 55730	Fed-batch	Pure	65.30	0.81	Jolly et al. (2014)
Mixed culture of microbial consortia					
Organic soil	Batch	Raw	3.76	0.65	Liu et al. (2013)
Wheat soil	Batch	Pure	1.71	0.69	Selembo et al. (2009)
Sludge	Batch	Raw	15.21	0.51–0.76	Dietz and Zeng (2014)
	Fed-batch	Raw	70.00	0.52–0.56	
Granular sludge	Continuous	Pure	10.74	0.52	Gallardo et al. (2014)
Marine sludge	Batch	Pure	81.40	0.63	Xiu et al. (2015)
	Fed-batch	Pure	72.15	0.70	

concentration. The initial glycerol concentration of batch fermentations with microbial consortium DL38 was up to 200 and 81.40 g/l of 1,3-PD was obtained with yield 0.63 mol/mol. In batch fermentation, a small amount of by-products were produced, especially no 2,3-butanediol was detected in favor of 1,3-PD purification (Jiang et al. 2016).

Compared with pure culture of single strain, mixed culture of microbial consortium normally showed higher efficiency or productivity and substrate tolerance. This is undoubtedly attributed to the interactions among cells in microbial consortium as discussed in the next section, although they are seldom known clearly. On the other hand, the metabolites or intermetabolites (even amino acids and nucleotides), or coenzymes (e.g., NADH/NADPH) or cofactors (e.g., ATP) produced from one strain might regulate the growth and metabolism of another strain. Besides the mechanism of mixed culture, the stability of microbial consortium structure during fermentation is also an important problem in industrial process. Some researchers aimed to bring ecological and evolutionary concepts to discussion on this question (Escalante et al. 2015). They pointed out that the system composed of cooperative consortia may be collapsed by cheaters arising during evolution (Diggle et al. 2007). We need to determine the primary strains in microbial consortia by incorporating evolutionary and ecological principles, and to design evolutionarily stable and sustainable systems by artificial structure of microbial consortia on the basis of biotechnological demand.

The interactions among cells in microbial consortia

In microbial consortium, there exist not only intraspecies interactions among the same species of microbial cells, which usually accomplish through quorum sensing (QS), but also interspecies interactions between different species cells, such as mutualism, competition for nutrition in the same ecological environment. These mutual effects based on metabolites will affect metabolisms and the yield of target product in the fermentation process.

Quorum sensing

Quorum sensing (QS) is characterized by communication information relying on bacterial density, leading to the realization of coordinated behaviors through responsive gene expression. The microbial cells can release some specific signal molecules and detect the change of their concentrations spontaneously, thus coordinating behaviors upon the establishment of a sufficient quorum (Schertzer et al. 2009). N-acyl-homoserine lactones (AHLs) are often used by Gram-negative bacteria as the QS signals (Williams 2007). In stark contrast to Gram-negative bacteria, Gram-positive bacteria make and

transport autoinducing peptides (AIPs) as communication signals (Parsek and Greenberg 2000). Each species of Gram-negative or Gram-positive bacteria produces a unique AHL (or a unique combination of AHLs) or AIPs. As a result, only the members of the same species recognize and respond to it (Federle and Bassler 2003).

The species-specific QS described above promotes intraspecies communication and apparently allows self-recognition in a mixed population. In such situations, bacteria also develop mechanisms to detect the presence of other species, and the signals of AI-2 (autoinducer-2) family are used for interspecies communication (Pereira et al. 2013). The evidence for the existence of AI-2 came from studies of the Gram-negative bioluminescent shrimp pathogen Vibrio harveyi (Bassler et al. 1997). AI-2 is synthesized by an enzyme called LuxS. However, the gene luxS is present in the genomes of a wide variety of Gram-negative and Gram-positive bacteria. Therefore, every bacterium containing a functional luxS gene is capable of producing an activity detected by an AI-2-specific V. harveyi reporter strain (Federle and Bassler 2003). AI-2 is a more universal signal that could promote interspecies bacterial communication. Quorum sensing is a key process in natural microbial interactions (Miller and Bassler 2001), and plays an important role in controlling virulence factor production, biofilm formation, improving microbial stress resistance, etc. (Park et al. 2014; Lin et al. 2016; Gambino and Cappitelli 2016). A biofilm is an group of microorganisms in which cells stick to each other and/or adhere to a surface. These adherent cells are frequently embedded within a self-produced matrix of extracellular polymeric substance (EPS). Biofilm formation can significantly improve microbial tolerance for oxygen or substrate or toxic/inhibitory substances. For example, the dissolved oxygen is consumed by one community member in biofilm, and an oxygen gradient can be established to create suitable microenvironments for anaerobic microbes (Gambino and Cappitelli 2016).

Mutualism and synergism

Mutualism refers to benefit of two or more species to one another when living together, but both of their lives will be affected badly and even die when separated. There are numerous examples of mutualisms in the fermentation processes with microbial consortia. For instance, the relationship between archaea and bacteria is mutualism during the process of anaerobic fermentation to produce methane. Stolyar et al. (2007) first used stoichiometric models through flux balance analysis to analyze mutualistic metabolite exchange between a sulfate reducer Desulfovibrio vulgaris and methanogen Methanococcus maripaludis. This study can accurately predict the relative abundances of D. vulgaris and M. maripaludis in an experimental co-culture. Shou et al. (2007) constructed a synthetic obligatory cooperative system, termed CoSMO (cooperation that is synthetic and mutually obligatory), which consists of a pair of auxotrophic yeast strains, each supplying an essential metabolite to the other strain. However, this reciprocal interaction can readily collapse, due to the evolution of "cheater" individuals that receive the benefit of the facilitation without contribution (Nowak 2006). This potential meltdown caused by cheater can be overcome or delayed depending on environmental spatial structure. The physical structure of the environment can limit the spread of cheating genotypes (Hammerschmidt et al. 2014). Synergy is one form of microbial mutualism, in which metabolites produced by one species or genotype affect the growth of other species (Escalante et al. 2015). Synergy interactions are commonly demonstrated in numerous biotechnology studies including consolidated bioprocessing of cellulose coupled with biofuel production (Du et al. 2015) and an organic acid-consuming community member scavenges inhibitory by-products from a producer population (Bizukojc et al. 2010). Kato et al. (2004) isolated two strains from the compost: one was *Clostridium straminisolvens* CSK1 which was able to degrade cellulose efficiently under anaerobic conditions; the other one was an aerobic non-cellulolytic bacterium. They successfully constructed a bacterial community with effective cellulose degradation by mixing the above two strains. The mixed culture indicated that the non-cellulolytic bacteria essentially contribute to cellulose degradation by creating an anaerobic environment, consuming metabolites, and neutralizing pH.

Competition and antagonism

Competition for limited natural resources within a microbial community is known as the selective force that promotes biosynthesis of antimicrobial compounds. Recently, it was shown that these antimicrobial molecules produced in nature are not primarily used as weapons for competition but as tools of communication that may regulate the homeostasis of microbial communities (Hibbing et al. 2010; Yim et al. 2006, 2007). For example, lactacin B produced by *Lactobacillus acidophilus* would be increased when this strain was co-cultured with the yogurt starter species *Streptococcus thermophilus* and *Lactobacillus delbrueckii* subsp. *Bulgaricus* (Tabasco et al. 2009). Antagonism is an interspecies interaction in which one species adversely affects the other one without being affected itself. It frequently occurs in food fermentations and inhibits the growth of spoilage organisms (Bas et al. 2006).

The interaction among cells in microbial consortium plays an important role to the stability of bacterial community. Recently, some researchers used mathematical models to prove that synergy between different types of microbial cells would disrupt the ecosystem stability of microbial consortium. Moreover, the competitive relationship between probiotics would offset the instability caused by the microbial diversity through negative feedback, and keep the intestinal ecosystem stable (Coyte et al. 2015). Many evidences from ecological perspectives also showed that the evolution of cheaters made the mutualism interaction more fragile than competition (Nowak 2006; Hammerschmidt et al. 2014; Escalante et al. 2015). Thus, the competitive relationship seems to be more conducive for maintaining the stability of microbial consortium.

Perspectives

Natural microbial consortia hold many appealing properties in one bioprocess, such as stability, functional robustness, and the ability to perform complex tasks (Sabra and Zeng 2014). Inspired by the powerful features of natural consortia, there are rapidly growing interests in engineered synthetic consortia for biotechnology applications (Zuroff and Curtis 2012; Bernstein and Carlson 2012). Brenner et al. (2008) reviewed researches on engineered microbial consortia by designing the communication between different microorganisms. These engineered microbial consortia can be used to study the interspecific interaction relationship (such as symbiosis, competition, and parasitism) in the smallest consortium. In addition, mathematical models can also be used to describe the defined microbial consortium, and used for development and validation of the more complex systems (Bizukojc et al. 2010). In the application of industrial biotechnology, it is more attractive and more promising to screen desired microbial strains from nature and put them together to execute new function. As people actively explore and understand the relationship of the microecology, microbial consortia will be developed and applied in many fields such as industry, agriculture, and food. In order to design and develop a successful process, it is necessary to understand the precise role and the overall contribution of each microorganism to the fermentation process. This knowledge is crucial to an inoculum with a defined co-culture or a mixture of undefined microbial consortium. There are many challenges needed to be faced in fermentation with microbial consortium, such as population dynamics and flux analysis of different species in the same reactor, the interrelationships between species, and the consistency and stability of inocula of microbial consortium during bioreactor scale-up. The most promising method for the determination of population dynamics is the molecular biological one based on the analysis and differentiation of microbial DNA, such as sequencing and metagenomics

(Röske et al. 2014). A great deal of information can be gleaned from even very complex microbial communities (Spiegelman et al. 2005). Metabolic networks and stoichiometric models can serve not only to predict metabolic fluxes and growth phenotypes of single organism, but also to capture growth parameters and composition of simple bacterial community (Stolyar et al. 2007; Sabra et al. 2015). The small microbial consortium with several and definite strains has good application prospect, which can be used as a model system in the development of methods and techniques, and is beneficial to use synthetic biology to design microbial consortia. These defined co-culture system would facilitate our understanding of the simultaneous involvement of several different microbial interactions in one and the same industrial process and controlling them (Goers et al. 2014). At the same time, the consistency and stability of inocula of microbial consortium would be maintained if the microbial behavior is understood. Therefore, the thorough research about industrial microbiome based on microbial consortium has not only profound theoretical significance, but also more extensive application potential, and can be of more benefit for humanity.

Abbreviations
1,3-PD: 1,3-propanediol; 2,3-BD: 2,3-butanediol; PHAs: polyhydroxyalkanoates; ATP: adenosine triphosphate; *K. pneumoniae*: *Klebsiella pneumoniae*; *K. vulgare*: *Ketogulonicigenium vulgare*; *V. harveyi*: *Vibrio harveyi*; *D.vulgaris*: *Desulfovibrio vulgaris*; *M. maripaludis*: *Methanococcus maripaludis*; CBP: consolidated bioprocessing; AECC: alkali extracted deshelled corn cobs; QS: quorum sensing; AHLs: *N*-acyl-homoserine lactones; AIPs: autoinducing peptides; AI-2: autoinducer-2 family.

Authors' contributions
All of them have been involved in the drafting and revision of the manuscript. All authors read and approved the final manuscript.

Author details
[1] School of Life Science and Biotechnology, Dalian University of Technology, Linggong Road 2, Dalian 116024, Liaoning Province, China. [2] Key Laboratory of Biotechnology and Bioresources Utilization, College of Life Science, Dalian Minzu University, Liaohe West Road 18, Jinzhou New District, Dalian 116600, Liaoning Province, China.

Acknowledgements
The authors acknowledge the China National Natural Science Foundation (Grant No. 21476042), Open Fund of Key Laboratory of Biotechnology and Bioresources Utilization (Dalian Minzu University), and State Ethnic Affairs Commission & Ministry of Education, China.

Competing interests
The authors declare that they have no competing interests.

References
Alivisatos AP, Blaser MJ, Brodie EL, Chun M, Dangl JL, Donohue TJ, Dorrestein PC, Gilbert JA, Green JL, Jansson JK, Knight R, Maxon ME, McFall-Ngai MJ, Miller JF, Pollard KS, Ruby EG, Taha SA (2015) A unified initiative to harness Earth's microbiomes. Science 350(6260):507–508
Bader J, Mast Gerlach E, Popovic MK, Bajpai R, Stahl U (2010) Relevance of microbial coculture fermentations in biotechnology. J Appl Microbiol 109(2):371–387
Bassler BL, Greenberg EP, Stevens AM (1997) Cross-species induction of luminescence in the quorum sensing bacterium *Vibrio harveyi*. J Bacteriol 179(12):4043–4045
Bernstein HC, Carlson RP (2012) Microbial consortia engineering for cellular factories: in vitro to in silico systems. Comput Struct Biotechnol J 3(4):1–8
Bizukojc M, Dietz D, Sun JB, Zeng AP (2010) Metabolic modelling of syntrophic-like growth of a 1,3-propanediol producer *Clostridium butyricum* and a methanogenic archeon *Methanosarcina mazei* under anaerobic conditions. Bioprocess Biosyst Eng 33:507–523
Brenner K, You L, Arnold FH (2008) Engineering microbial consortia: a new frontier in synthetic biology. Trends Biotechnol 26(9):483–489
Chatzifragkou A, Aggelis G, Komaitis M, Zeng AP, Papanikolaou S (2011) Impact of anaerobiosis strategy and bioreactor geometry on the biochemical response of *Clostridium butyricum* VPI 1718 during 1,3-propanediol fermentation. Bioresour Technol 102(22):10625–10632
Coyte KZ, Schluter J, Foster KR (2015) The ecology of the microbiome: networks, competition, and stability. Science 350(6261):663–666
Dias JM, Lemos PC, Serafim LS, Oliveira C, Eiroa M, Albuquerque MG, Ramos AM, Oliveira R, Reis MA (2006) Recent advances in polyhydroxyalkanoate production by mixed aerobic cultures: from the substrate to the final product. Macromol Biosci 6(11):885–906
Diaz EE, Stams AJM, Amils R, Sanz JL (2006) Phenotypic properties and microbial diversity of methanogenic granules from a full-scale upflow anaerobic sludge bed reactor treating brewery wastewater. Appl Environ Microbiol 72(7):4942–4949
Dietz D, Zeng AP (2014) Efficient production of 1,3-propanediol from fermentation of crude glycerol with mixed cultures in a simple medium. Bioprocess Biosyst Eng 37(2):225–233
Diggle SP, Griffin AS, Campbell GS, West SA (2007) Cooperation and conflict in quorum-sensing bacterial populations. Nature 450(7168):411–414
Du R, Yan JB, Li SZ, Zhang L, Zhang S, Li J, Zhao G, Qi P (2015) Cellulosic ethanol production by natural bacterial consortia is enhanced by *Pseudoxanthomonas taiwanensis*. Biotechnol Biofuels 8(1):1–10
Dubilier N, McFall-Ngai M, Zhao LP (2015) Great a global microbiome effort. Nature 526(7575):631–634
Dürre P (2007) Biobutanol: an attractive biofuel. Biotechnol J 2(12):1525–1534
Eiteman MA, Lee SA, Altman E (2008) A co-fermentation strategy to consume sugar mixtures effectively. J Biol Eng 2(1):3–11
Escalante AE, Rebolleda-Gómez M, Benítez M, Travisano M (2015) Ecological perspectives on synthetic biology: insights from microbial population biology. Front Microbiol 6(143):1–10
Federle MJ, Bassler BL (2003) Interspecies communication in bacteria. J Clin Invest 112(9):1291–1299
Gallardo R, Faria C, Rodrigues LR, Pereira MA, Alves MM (2014) Anaerobic granular sludge as a biocatalyst for 1,3-propanediol production from glycerol in continuous bioreactors. Bioresour Technol 155(4):28–33
Gambino M, Cappitelli F (2016) Mini-review: biofilm responses to oxidative stress. Biofouling 32(2):167–178
Goers L, Freemont P, Polizzi KM (2014) Co-culture systems and technologies: taking synthetic biology to the next level. J R Soc Interface 11(96):1058–1069
Gu Y, Jiang Y, Wu H, Liu X, Li Z, Li J, Xiao H, Shen Z, Dong H, Yang Y, Li Y, Jiang W, Yang S (2011) Economical challenges to microbial producers of butanol: feedstock, butanol ratio and titer. Biotechnol J 6(11):1348–1357
Hammerschmidt K, Rose CJ, Kerr B, Rainey PB (2014) Life cycles, fitness decoupling and the evolution of multicellularity. Nature 515(7525):75–79
Hibbing M, Fuqua C, Parsek M, Peterson SB (2010) Bacterial competition: surviving and thriving in the microbial jungle. Nat Rev Microbiol 8(1):15–25
Jiang L, Liu H, Mu Y, Sun Y, Xiu Z (2016) High tolerance to glycerol and high production of 1,3-propanediol in batch fermentations by microbial consortium from marine sludge. Eng Life Sci 1–10. doi:10.1002/elsc.201600215
Johnson DT, Taconi KA (2007) The glycerin glut: options for the value-added conversion of crude glycerol resulting from biodiesel production. Environ Prog 26(4):338–348
Jolly J, Hitzmann B, Ramalingam S, Ramachandran KB (2014) Biosynthesis of 1,3-propanediol from glycerol with *Lactobacillus reuteri*: effect of operat-

ing variables. J Biosci Bioeng 118(2):188–194

Jun SA, Moon C, Kang CH, Kong SW, Sang BI, Um Y (2010) Microbial fed-batch production of 1,3-propanediol using raw glycerol with suspended and immobilized *Klebsiella pneumoniae*. Appl Biochem Biotechnol 161(161):491–501

Kamm B, Schonicke P, Hille C (2016) Green biorefinery—industrial implementation. Food Chem 197:1341–1345

Kanjilal B, Noshadi I, Bautista EJ, Srivastava R, Parnas RS (2015) Batch, design optimization, and DNA sequencing study for continuous 1,3-propanediol production from waste glycerol by a soil-based inoculum. Appl Microbiol Biotechnol 99(5):2105–2117

Kato S, Haruta S, Cui ZJ, Ishii M, Igarashi Y (2004) Effective cellulose degradation by a mixed-culture system composed of a cellulolytic *Clostridium* and aerobic non-cellulolytic bacteria. FEMS Microbiol Ecol 51(1):133–142

Lee HS, Krajmalinik-Brown R, Zhang H, Rittmann BE (2009) An electron-flow model can predict complex redox reactions in mixed-culture fermentative BioH2: microbial ecology evidence. Biotechnol Bioeng 104(4):687–697

Lemos PC, Serafim LS, Reis MA (2006) Synthesis of polyhydroxyalkanoates from different short-chain fatty acids by mixed cultures submitted to aerobic dynamic feeding. J Biotechnol 122(2):226–238

Li CL, Fang HHP (2007) Fermentation hydrogen production from wastewater and solid wastes by mixed cultures. Crit Rev Environ Sci Technol 37(1):1–39

Lin L, Li T, Dai S, Yu JL, Chen XQ, Wang LY, Wang YG, Hua YJ, Tian B (2016) Autoinducer-2 signaling is involved in regulation of stress-related genes of *Deinococcus radiodurans*. Arch Microbiol 198(1):43–51

Liu B, Christiansen K, Parnas R, Xu Z, Li B (2013) Optimizing the production of hydrogen and 1,3-propanediol in anaerobic fermentation of biodiesel glycerol. Int J Hydrogen Energy 38(8):3196–3205

Metsoviti M, Paramithiotis S, Drosinos EH, Galiotou-Panayotou M, Nychas G-JE, Zeng A-P, Papanikolaou S (2012a) Screening of bacterial strains capable of converting biodiesel-derived raw glycerol into 1,3-propanediol, 2,3-butanediol and ethanol. Eng Life Sci 12(1):57–68

Metsoviti M, Paraskevaidi K, Koutinas A, Zeng A-P, Papanikolaou S (2012b) Production of 1,3-propanediol, 2,3-butanediol and ethanol by a newly isolated *Klebsiella oxytoca* strain growing on biodiesel-derived glycerol based media. Process Biochem 47(12):1872–1882

Metsoviti M, Zeng AP, Koutinas AA, Papanikolaou S (2013) Enhanced 1,3-propanediol production by a newly isolated *Citrobacter freundii* strain cultivated on biodiesel-derived waste glycerol through sterile and non-sterile bioprocesses. J Biotechnol 163(4):408–418

Miller MB, Bassler BL (2001) Quorum sensing in bacteria. Annu Rev Microbiol 55:165–199

Minty JJ, Singer ME, Scholz SA, Bae CH, Ahn JH, Foster CE, Liao JC, Lin XN (2013) Design and characterization of synthetic fungal-bacterial consortia for direct production of isobutanol from cellulosic biomass. Proc Natl Acad Sci 110(36):14592–14597

Moita R, Freches A, Lemos PC (2014) Crude glycerol as feedstock for polyhydroxyalkanoates production by mixed microbial cultures. Water Res 58(3):9–20

Nakamura CE, Whited GM (2003) Metabolic engineering for the microbial production of 1,3-propanediol. Curr Opin Biotechnol 14(5):454–459

Nishio N, Nakashimada Y (2007) Recent development of anaerobic digestion processes for energy recovery from wastes. J Biosci Bioeng 103(2):105–112

Nowak MA (2006) Five rules for the evolution of cooperation. Science 314(5805):1560–1563

Oh SE, Ginkel SV, Logan BE (2003) The relative effectiveness of pH control and heat treatment for enhancing biohydrogen gas production. Environ Sci Technol 37(22):5186–5190

Olson DG, Mcbride JE, Shaw AJ, Lynd LR (2012) Recent progress in consolidated bioprocessing. Curr Opin Biotechnol 23(3):396–405

Park H, Yeo S, Jia Y, Lee J, Yang J, Park S, Shin H, Holzapfel W (2014) Auto-inducer-2 associated inhibition by *Lactobacillus sakei* NR28 reduces virulence of enterohaemorrhagic *Escherichia coli* O157:H7. Food Control 45:62–69

Parsek MR, Greenberg EP (2000) Acyl-homoserine lactone quorum sensing in Gram-negative bacteria: a signaling mechanism involved in associations with higher organisms. Proc Natl Acad Sci 97(16):8789–8793

Pereira MC, Thompson JA, Xavier KB (2013) AI   mediated signalling in bacte

ria. FEMS Microbiol Rev 37(2):156–181

Raghunandan K, McHunu S, Kumar A, Kumar KS, Govender A, Permaul K, Singh S (2014) Biodegradation of glycerol using bacterial isolates from soil under aerobic conditions. J Environ Sci Health A Tox Hazard Subst Environ Eng 49(1):85–92

Rahul M, Matti K, Ville S (2012) Bioconversion of crude glycerol from biodiesel production to hydrogen. Int J Hydrogen Energy 37(17):12198–12204

Rodriguez A, Wojtusik M, Ripoll V, Santos VE, Garcia-Ochoa F (2015) 1,3-Propanediol production from glycerol with a novel biocatalyst *Shimwellia blattae* ATCC 33430: operational conditions and kinetics in batch cultivations. Bioresour Technol 200:830–837

Röske I, Sabra W, Nacke H, Daniel R, Zeng AP, Antranikian G, Sahm K (2014) Microbial community composition and dynamics in high-temperature biogas reactors using industrial bioethanol waste as substrate. Appl Microbiol Biotechnol 98(21):9095–9106

Sabra W, Zeng AP (2014) Mixed microbial cultures for industrial biotechnology: success, chance and challenges, 7th edn. Industrial Biocatalysis, Grunwald

Sabra W, Dietz D, Tjahjasari D, Zeng AP (2010) Biosystems analysis and engineering of microbial consortia for industrial biotechnology. Eng Life Sci 10(5):407–421

Sabra W, Dietz D, Zeng AP (2013) Substrate limited co-culture for efficient reduction of propionic acid from flour hydrolysate. Appl Microbiol Biotechnol 97:5771–5777

Sabra W, Röske I, Sahm K, Antranikian G, Zeng AP (2015) Metabolic and microbial characterization of high-temperature biogas reactors treating stillage from an industrial bioethanol process. Eng Life Sci 15(7):743–750

Schertzer JW, Boulette ML, Whiteley M (2009) More than a signal: non-signaling properties of quorum sensing molecules. Trends Microbiol 17(5):189–195

Selembo PA, Perez JM, Lloyd WA, Logan BE (2009) High hydrogen production from glycerol or glucose by electrohydrogenesis using microbial electrolysis cells. Int J Hydrogen Energy 34(13):5373–5381

Shou W, Ram S, Vilar JMG (2007) Synthetic cooperation in engineered yeast populations. Proc Natl Acad Sci USA 104(6):1877–1882

Solomon KV, Haitjema CH, Henske JK, Gilmore SP, Borges-Rivera D, Lipzen A, Brewer HM, Purvine SO, Wright AT, Theodorou MK, Grigoriev IV, Regev A, Thompson DA, O'Malley MA (2016) Early-branching gut fungi possess a large, comprehensive array of biomass-degrading enzymes. Science 351(6278):1192–1195

Spiegelman D, Whissell G, Greer CW (2005) A survey of the methods for the characterization of microbial consortia and communities. Can J Microbiol 51(5):355–386

Stolyar S, Dien SV, Hillesland KL, Pinel N, Lie TJ, Leigh JA, Stahl DA (2007) Metabolic modeling of a mutualistic microbial community. Mol Syst Biol 3(1):92–106

Streit WR, Daniel R, Jaeger KE (2004) Prospecting for biocatalysts and drugs in the genomes of non-cultured microorganisms. Curr Opin Biotechnol 15(4):285–290

Sun LH, Song ZY, Sun YQ, Xiu ZL (2010) Dynamic behavior of glycerol–glucose co-fermentation for 1,3-propanediol production by *Klebsiella pneumoniae* DSM 2026 under micro-aerobic conditions. World J Microbiol Biotechnol 26(8):1401–1407

Szymanowska-Powalowska D, Piatkowska J, Leja K (2013) Microbial purification of postfermentation medium after 1,3-PD production from raw glycerol. Biomed Res Int 1:949107–949114

Tabasco R, García-Cayuela T, Peláez C, Requena T (2009) *Lactobacillus acidophilus* La-5 increases lactacin B production when it senses live target bacteria. Int J Food Microbiol 132(2–3):109–116

Temudo MF, Muyzer G, Kleerebezem R, van Loosdrecht MC (2008) Diversity of microbial communities in open mixed culture fermentations: impact of the pH and carbon source. Appl Microbiol Biotechnol 80(6):1121–1130

Teusink B, Wiersma A, Molenaar D, Francke C, de Vos WM, Siezen RJ, Smid EJ (2006) Analysis of growth of *Lactobacillus plantarum* WCFS1 on a complex medium using a genome-scale metabolic model. J Biol Chem 281(52):40041–40048

Vandamme P, Coenye T (2004) Taxonomy of the genus Cupriavidus: a tale of lost and found. Int J Syst Evol Microbiol 54(6):2285–2289

Wen ZQ, Wu MB, Lin YJ, Yang LR, Lin JP, Cen PL (2014) Artificial symbiosis for acetone-butanol-ethanol (ABE) fermentation from alkali extracted deshelled corn cobs by co-culture of *Clostridium beijerinckii* and

Wilkens E, Ringel AK, Hortig D, Willke T, Vorlop KD (2012) High-level production of 1, 3-propanediol from crude glycerol by *Clostridium butyricum* AKR102a. Appl Microbiol Biotechnol 93(3):1057–1063

Williams P (2007) Quorum sensing, communication and cross-kingdom signalling in the bacterial world. Microbiol 153(12):3923–3938

Xiu ZL, Zeng AP (2008) Present state and perspective of downstream processing of biologically produced 1,3-propanediol and 2,3-butanediol. Appl Microbiol Biotechnol 78(6):917–926

Xiu ZL, Liu HF, Chen Y, Jiang LL, Sun YQ (2015) A method of fermentation of 1,3-propanediol from glycerol by mixed culture: CN, 104774879A

Yang G, Tian JS, Li JL (2007) Fermentation of 1,3-propanediol by a lactate deficient mutant of *Klebsiella oxytoca* under microaerobic conditions. Appl Microbiol Biotechnol 73(5):1017–1024

Yim G, Wang HHM, Davies J (2006) The truth about antibiotics. Int J Med Microbiol 296:163–170

Yim G, Wang HMH, Davies J (2007) Antibiotics as signalling molecules. Philos Trans R Soc London B Biol Sci 362:1195–1200

Zeng AP, Sabra W (2011) Microbial production of diols as platform chemicals: recent progresses. Curr Opin Biotechnol 22(6):749–757

Zhang J, Liu J, Shi Z, Liu L, Chen J (2010) Manipulation of *B. megaterium* growth for efficient 2-KLG production by *K. vulgare*. Process Biochem 45(4):602–606

Zuroff TR, Curtis WR (2012) Developing symbiotic consortia for lignocellulosic biofuel production. Appl Microbiol Biotechnol 93:1423–1435

3

Techno-economic analysis of extraction-based separation systems for acetone, butanol, and ethanol recovery and purification

Víctor Hugo Grisales Díaz[1]*(ID) and Gerard Olivar Tost[2]

Abstract

Background: Dual extraction, high-temperature extraction, mixture extraction, and oleyl alcohol extraction have been proposed in the literature for acetone, butanol, and ethanol (ABE) production. However, energy and economic evaluation under similar assumptions of extraction-based separation systems are necessary. Hence, the new process proposed in this work, direct steam distillation (DSD), for regeneration of high-boiling extractants was compared with several extraction-based separation systems.

Methods: The evaluation was performed under similar assumptions through simulation in Aspen Plus V7.3® software. Two end distillation systems (number of non-ideal stages between 70 and 80) were studied. Heat integration and vacuum operation of some units were proposed reducing the energy requirements.

Results: Energy requirement of hybrid processes, substrate concentration of 200 g/l, was between 6.4 and 8.3 MJ-fuel/kg-ABE. The minimum energy requirements of extraction-based separation systems, feeding a water concentration in the substrate equivalent to extractant selectivity, and ideal assumptions were between 2.6 and 3.5 MJ-fuel/kg-ABE, respectively. The efficiencies of recovery systems for baseline case and ideal evaluation were 0.53–0.57 and 0.81–0.84, respectively.

Conclusions: The main advantages of DSD were the operation of the regeneration column at atmospheric pressure, the utilization of low-pressure steam, and the low energy requirements of preheating. The in situ recovery processes, DSD, and mixture extraction with conventional regeneration were the approaches with the lowest energy requirements and total annualized costs.

Keywords: Extractive fermentation, Dual extraction, High-temperature extraction, Energy evaluation, Biobutanol

Background

The interest in biobutanol production by acetone, butanol, and ethanol (ABE) fermentation is increasing because butanol and ABE mixture are considered as an alternative biofuel (Veloo et al. 2010; Kumar et al. 2012). Butanol is the primary inhibitor in ABE fermentation and causes total inhibition at concentrations between 13 and 19 g/l (Xue et al. 2013). In order to reduce butanol inhibition, integrated fermenters have been proposed.

In these processes, butanol is selectively separated from the fermenter (Qureshi and Maddox 2005; Qureshi et al. 2005; Lu et al. 2012; González-Peñas et al. 2014b; Liu et al. 2014; Cabezas et al. 2015). An integrated fermenter allows using a higher substrate concentration. Therefore, the performance of fermenter can be increased and wastewater and energy requirement of downstream and treatment are reduced. Integrated fermenters with liquid–liquid extraction or extractive fermentations are one of the recovery options with lower energy requirements

*Correspondence: Victor.Grisales-Diaz@newcastle.ac.uk;
victor.grisales.d@gmail.com
[1] School of Chemical Engineering and Advanced Materials, Newcastle University, Newcastle upon Tyne NE1 7RU, UK
Full list of author information is available at the end of the article

reported in the literature (Groot et al. 1992; Qureshi et al. 2005; Oudshoorn et al. 2009). Solvent selection is involved because several conditions are necessary for the extractant (Kraemer et al. 2011), such as biocompatibility, non-emulsion forming, easy regeneration, high selectivity, low viscosity, high butanol distribution, availability, and low cost. Therefore, several extractive systems have been proposed (Xu and Parten 2011; Kraemer et al. 2011; Grady et al. 2013; Kurkijärvi et al. 2014).

In extractive salting-out, salt solutions or pure neutral, acid, or basic salts (Xie et al. 2013) are proposed as extractants. This is an external process; hence, the productivity of reactor does not improve. The regeneration of the salt is the main disadvantage of the salting-out process. Due to the low concentrations of butanol, in salting-out processes, high energy requirements [21.9 MJ/kg-butanol (Xie et al. 2015) or 28.5 MJ/kg-butanol (Xie et al. 2013)] are required to evaporate the water and unrecovered organic solvents from the salt solution.

Dual extraction (DEx) is proposed using toxic solvents with high butanol distribution coefficient (Kurkijärvi et al. 2014). The toxic extractant is removed with a biocompatible solvent before recirculating the aqueous phase in the fermenter. DEx has been used with a high butanol distribution extractant [Decanol (DAL) (7.1) or octanol (10)] and mesitylene as the biocompatible extractant. DAL was the most promising extractant (Kurkijärvi et al. 2014).

Low-energy fermentation with high-temperature extraction (HTE) (Kraemer et al. 2011) has been proposed for ABE production. An example of high-temperature extraction, using mesitylene as extractant, is the configuration suggested by Kraemer et al. (2011). Mesitylene has a mass partition coefficient of butanol of 0.86 at 30 °C and 3.0 at 80 °C (Kraemer et al. 2011). Therefore, in HTE, when extraction is performed at higher temperatures than fermentation (usually 30–37 °C), less extractant is needed. High selectivity (1970) and medium boiling temperature (180 °C) are the main advantages of mesitylene.

In HTE or DEx systems, the fermenter productivity probably does not increase with respect to continuous process because a high-temperature (80 °C) or toxic extractant would kill the fermenting bacteria. This effect can be avoided with the recirculation or immobilization of biomass. However, the increase in productivity will be achieved through the biomass concentration system. In fact, fermenters with biomass concentration by recirculation or immobilization achieved the highest productivity reported in the literature (Köhler et al. 2015).

Oley alcohol (OAL) is the most studied extractant to carry out in situ extractive fermentation; it has an acceptable butanol distribution coefficient [3.8 (Matsumura

et al. 1988)], high selectivity (>300), and it is biocompatible (Evans and Wang 1988). Biocompatibility is the most advantageous characteristic of the extractant because it is used in situ and butanol productivity of fermentation can be increased. However, the high boiling temperature (360 °C) of OAL hinders the extractant regeneration. Therefore, it requires high amounts of preheating, low-pressure distillation, and high-pressure steam.

The combination of toxic solvents and non-toxic OAL has been proposed to decrease the boiling temperature and viscosity of biocompatible extractants, increasing the butanol distribution coefficient (Evans and Wang 1988; Bankar et al. 2012). The mixing ratio is limited by the biocompatibility of toxic extractant. DAL has frequently been proposed in an OAL–DAL mixture of 80–20 wt%. However, non-toxic ratios as large as 60/40 wt% have been reported (Evans and Wang 1988).

Butamax (TM) Advanced Biofuels® developed processes to reduce the boiling point of high-temperature extractants in a regeneration extractant column for isobutanol production (Xu and Parten 2011; Grady et al. 2013). In these patent processes, the aqueous phase from a decanter is recycled to the top of the regeneration column. The supplementation of this aqueous phase allows the recovery of butanol and water from the top and the bottoms, respectively. The high composition of water in the bottoms of distillation column reduces the boiler temperature of the regeneration column. The energy requirement of this regeneration system was between 4.9 and 5.9 MJ/kg-isobutanol. This regeneration system has not been studied for ABE recovery.

An alternative method for regeneration of high-boiling extractants was proposed in this work. The proposed method was called direct steam distillation (DSD). Steam was fed in the bottom of the regeneration column, and water from decanter was not recirculated. DSD can operate at atmospheric pressure using low-pressure steam. Atmospheric pressure operation favors the energy integration because condensation heat can be employed in reboilers of low-pressure columns. Simultaneously, the size of the preheating unit was reduced.

The extractive systems studied in this work were HTE, DEx, conventional extraction, mixture extraction (MEx), and DSD (new process). To our knowledge, MEx using OAL–DAL (80–20 wt%) has been not evaluated economic and energetically in the literature. Due to the different assumptions of the energy requirement reported in the literature (Ezeji et al. 2005; Kraemer et al. 2011; Kurkijärvi et al. 2014; Outram et al. 2016), the selection of the lowest energy system for extractive fermentation is difficult. The main objective of this paper was to select the lowest energy and expensive extractive process for ABE recovery from fermentation. Therefore, in this work, the energy and economic evaluations were performed at similar assumptions.

Methods

ABE process was simulated in Aspen Plus V7.3®. The flash units, distillation, stripping columns, and compressor units were simulated with UNIQUAK-RK. Binary parameters of butanol–water from Aspen Plus V7.3® are not adequate for simulation of vacuum units (Mariano et al. 2011). For this reason, their binary parameters were taken from (Fischer and Gmehling 1994). Decanters were simulated with NRTL. The thermodynamic model of decanters was different because in these units the binary parameters for liquid–liquid equilibrium were used. The missing parameters of NRTL and UNIQUAC (for instance, acetone and OAL binary parameters) were estimated from UNIFAC.

In liquid–liquid extraction column, UNIFAC-LL (Table 1) was used because in Aspen Plus® the binary parameters of NRTL or UNIQUAC for the extractants studied in this paper are not based on experimental data, and UNIFAC-LL has a high accuracy in the prediction of butanol extraction by decanol and oleyl alcohol [Table 1; (Kurkijärvi et al. 2014)]. As UNIFAC is not accurate in the simulation of mesitylene extraction (Kraemer et al. 2011) (Table 1), it was simulated with a constant distribution coefficient of butanol, acetone, and ethanol of 2.2, 0.83, and 0.1 (Kraemer et al. 2011), respectively. CO_2 and H_2 were simulated as Henry's components.

The stage number and extraction efficiency (butanol) of all liquid–liquid extraction columns, based on heuristic, were five and 0.8, respectively. Similar stages were proposed evaluating DEx by Kurkijärvi et al. (2014) (four stages of extraction and 100% of efficiency). A specific substrate was not studied because ABE productivity was not calculated. A stoichiometric ABE ratio of industrial production in China was used in this paper (ABE molar basis 2/3/1) (Ni and Sun 2009). Therefore, the stoichiometric reaction is

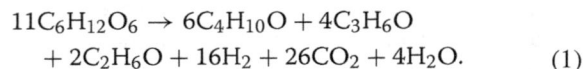

$$11C_6H_{12}O_6 \rightarrow 6C_4H_{10}O + 4C_3H_6O \\ + 2C_2H_6O + 16H_2 + 26CO_2 + 4H_2O. \quad (1)$$

The fermentation temperature in all cases was 30 °C. The operation of the process is continuous. The concentration in the feed must be limited to possible the presence of solids (e.g., lignin, cellulose, hemicellulose, or ash) and toxic compounds (e.g., furans, organic acids, or phenolic compounds), and the availability of substrate concentration of the real substrate selected (Ezeji et al. 2005; Grisales Díaz and Olivar Tost 2016a). Therefore, a substrate concentration of 200 g/l was selected for the baseline scenario. The conversion and butanol concentration in the fermenter in the baseline scenario were 80% and 10 g/l, respectively.

The comparison of energy requirements of integrated fermenters reported in the literature is difficult because the energy requirements change in reference at assumptions (Outram et al. 2016). For this reason, several concentrations and conversions and one ideal simulation were studied. The ideal evaluation was simulated as proposed by Kraemer et al. 2011: ABE (not glucose) and water were fed to the fermenter without bleed stream (therefore, the water concentration of the substrate is equivalent to water selectivity of the extractant); efficiencies for extraction and distillation columns were of 100%; and nil pressure drop.

The recycle of vinasses was obtained with a ratio of 80 kg-total/kg-ABE. Recovery of solvent from extraction column will be more feasible at lower ratios of broth/OAL. However, the fuel requirement and extractant cost increase. An adequate solvent flow of OAL must be selected through optimization of a pilot-scale system in future work. The Murphree efficiency in the distillation columns was 0.7. Distillation columns were simulated

Table 1 Solvent properties of extractants studied in this paper

Extractant	ABE	Distribution coefficient		Biocompatible
		Experimental	UNIFAC-LL	
Decanol (Kurkijärvi and Lehtonen 2014)	Ethanol	0.86–0.54 (Offeman et al. 2008)	0.56	No
	Acetone	0.6	1.2	
	Butanol	7.2	7.1	
Mesitylene (Kraemer et al. 2011)	Ethanol	0.1	0.43	Yes
	Acetone	0.83	0.43	
	Butanol	2.2	0.76	
Oleyl alcohol (Matsumura et al. 1988)	Ethanol	0.34	0.40	Yes
	Acetone	0.28	0.74	
	Butanol	3.8	3.9	

with sieve trays, and pressure drop was calculated with tray rating.

The total annualized cost (TAC) and fuel requirement were calculated with the methodology reported by Grisales Díaz and Olivar Tost (2016a). Extractant loss was included in TAC calculation. Efficiency in steam production with respect to fuel was fixed to 0.9 (Grisales Díaz and Olivar Tost 2016a). CO_2 production is directly proportional to fuel burn (Jonker et al. 2015). Therefore, fuel savings is proportional to CO_2 savings. Energy ideal efficiency of separation (IES) of the system was calculated using the following equation Grisales Díaz and Olivar Tost (2016b):

$$IES = \frac{R_S \cdot (LHV - H_S)}{LHV_{Glucose}}, \quad (2)$$

where LHV is the lower heating value of solvents and hydrogen (MJ-fuel/kg-solvent), H_S is the energy consumption of the separation (MJ-fuel/kg-solvent), R_s is the solvent yield, and $LHV_{GLUCOSE}$ is the lower heating value of glucose, 16.45 MJ/kg (Ruggeri et al. 2015). The energy efficiency was considered ideal because only the energy requirement of recovery and purification systems was calculated. The yield, R_s, was the ABE product (g) per mass (g) of substrate fed. ABE yield is calculated from stoichiometric (Eq. 1) of biocatalyst and conversion.

Installed equipment costs were calculated based on the equations reported by (Douglas 1988). Marshall and Swift equipment cost index (M&S) was 1536.5 (Kim 2015). Equipment was simulated using stainless steel materials. Installation cost of each extraction stage was performed in a pressure vessel with height/diameter ratio of three. Total residence time (aqueous and organic phase) was 0.5 h because experimentally it was found that this is the necessary contact time for an efficient extraction (Bankar et al. 2012). A minimum approach temperature of 10 °C of heat exchangers was performed. Parameters cost used in the economic evaluation are shown in Table 2 (Mussatto et al. 2013; Zauba 2015).

Stage extraction cost was not calculated for biocompatible extractants because the fermenter productivity with biocompatible extractants increases with respect to conventional fermentation. For instance, in the extractive fermentation of cane bagasse, with OAL–DAL mixture and cell immobilization, the productivity is increased to 2.5 g/l/h, fivefold higher than that for batch process (Bankar et al. 2012). In other studies, the productivity in extractive fermentation using fed-batch operation, without immobilization, a glucose concentration of 300 g/l and oleyl alcohol as extractant, increased 70% with respect to conventional batch process (Roffler et al. 1988). The fermenter productivity with 100 g/l of glucose concentration and oleyl alcohol or the mixture of oleyl

Table 2 Parameters used in economic evaluation

Unity	Valor	Unity
Low-pressure steam (3 bar)	2.2	$/tonne (Mussatto et al. 2013)
Mid-pressure steam (30 bar)	7.9	$/tonne (Mussatto et al. 2013)
High-pressure steam (105 bar)	11.8	$/tonne (Mussatto et al. 2013)
Oleyl alcohol	4.3	$/kg (Zauba 2015)
DAL	2.1	$/kg (Zauba 2015)
Mesitylene	2.9	$/kg (Zauba 2015)
Cool water	0.06	$/tonne
Electricity	0.095	$/kWh
Operation time (t_o)	8150	h
Production flow	5000	kg-ABE/h
Time of return investment (t_{ri})	5	Year

alcohol and ethyl benzoate as extractant was increased 60% with respect to batch process (Roffler et al. 1987).

Extractant selection

2-Ethyl-1-hexanol (2E1H) is proposed as an extractant for ABE production (Liu et al. 2004; van der Merwe et al. 2013). However, 2E1H toxicity is elevated (González-Peñas et al. 2014a). Additionally, the simulations reported in the literature for ABE production with 2E1H assume infinite selectivity in the extraction. This reduced the required distillation units because there are not azeotropes. However, the selectivity of 2E1H [295 (González-Peñas et al. 2014a) and 330 (Kraemer et al. 2011)] is similar to OAL (>300). For these reasons, this extractant was not studied in this paper.

Hexyl acetate is an extractant evaluated for butanol production (Sánchez-Ramírez et al. 2015; Errico et al. 2016). However, experimental data of biocompatibility or distribution coefficients of butanol extraction by hexyl acetate are not reported in the literature. Additionally, this recovery has been reported with a very high energy requirement (45 MJ/kg-ABE, calculated in this work from reboiler requirement of route D (315 kcal/s) and ABE production of 47.9 lb/h) reported by Sánchez-Ramírez et al. (2015). For these reasons, this extractant was not studied in this paper.

Biodiesel or additives of gasoline (biocompatible extractants) have been used to recover butanol from fermentation (Li et al. 2010; Kurkijärvi and Lehtonen 2014). Therefore, if butanol is used as biofuel, a final recovery system is not needed. Extraction system using gasoline additives has been proposed with DEx (Kurkijärvi and Lehtonen 2014). Methyl tert-butyl ether (MTBE) and ethyl tert-butyl ether (ETBE) were the best extraction solvents. Additional purification units were not required (Kurkijärvi and Lehtonen 2014). However, ABE obtained with gasoline additives was lower than 2.6%.

Consequently, ABE will be a minority additive in gasoline and ABE chemical market is not covered. For this reason, these fuels were not used as solvents in this paper. Alternatively, the solvents for extractive fermentation can be produced from ABE fermentation products in reactive distillation (Kurkijärvi et al. 2016). However, reactive distillation has a high energy requirement (Kurkijärvi et al. 2016), 2.2- to 2.6-fold higher than that for dual extraction (DEx). For this reason, reactive distillation was not studied in this work.

In this paper, HTE, DEx, conventional extraction, MEx, and DSD (new process) were studied. In DEx, Kurkijärvi et al. (2014) proposed mesitylene and DAL as the biocompatible solvent and toxic extractant, respectively. In this work, OAL was selected instead of mesitylene because OAL has a higher boiling point than DAL. Then, DAL can be recovered at the top of its regeneration column (EC2), and only two columns (instead of three) were required for this section. Mesitylene was used in the high-temperature extractive process (Kraemer et al. 2011). Experimentally, mesitylene toxicity is unknown. However, in this work, it will be considered biocompatible due to its low solubility, as proposed by Kraemer et al. (2011). Conventional extraction was simulated with OAL. However, in the evaluation of DSD, OAL and OAL/DAL (80–20%) were the extractants used. The main differences of extraction-based separation systems are shown in the supplemental material (Additional file 1: Table S1).

Distillation system

ABE was recovered from vinasses by distillation; it was not by extraction column, due to the low ethanol (Matsumura and Märkl 1984; Offeman et al. 2008) and acetone distribution coefficient of extractants (<1) (Table 1). Two different distillation systems (Fig. 1) were proposed to reduce the fuel requirements of purification of extractive processes. The distillation system used in each fermentation process depended on the condensation temperature of regeneration column, and the condenser or boiler temperature depended on column operation pressure.

The distillation process for HTE, DEx, and DSD had three distillation columns (3DC-1, Fig. 1a). In this system, the heat of condensation of regeneration column of the main extractant was used to apply heat to the reboiler of WC or AC column. For this reason, the columns AC and WC were operated to 0.45 and 0.27 bar, respectively. Vinasses and acetone were recovered in the AC and WC columns, respectively. In the EBC column, butanol and ethanol were recovered at 1.7 bar. In this way, the condensation heat of EBC can be used to provide heat to AC or WC boiler. The stage numbers were selected to avoid

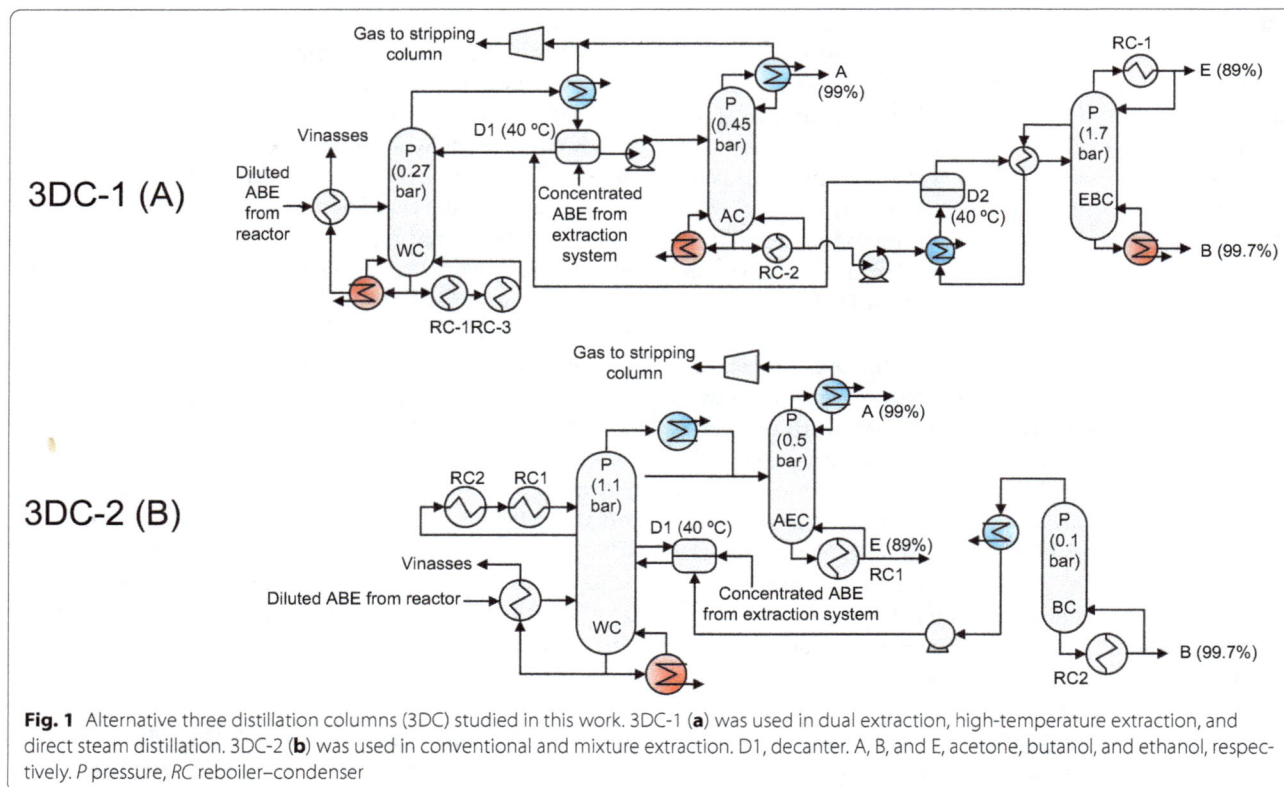

Fig. 1 Alternative three distillation columns (3DC) studied in this work. 3DC-1 (**a**) was used in dual extraction, high-temperature extraction, and direct steam distillation. 3DC-2 (**b**) was used in conventional and mixture extraction. D1, decanter. A, B, and E, acetone, butanol, and ethanol, respectively. *P* pressure, *RC* reboiler–condenser

an excess of trays. The stage numbers of columns WC, AC, and EBC were 20, 30, and 30, respectively. Organic and ABE dilute phases from the extractive process were fed to the decanter and AC column, respectively.

The design specifications to calculate the boiler ratio of WC and AC columns were the recovery of butanol and (0.999) and acetone (0.99), respectively. The design specifications to calculate the reflux ratio of AC and EBC were the purity of acetone (99 wt%) and ethanol (89 wt%), respectively. While the design specification to calculate the boiler ratio of EBC was the purity of butanol (99.9 wt%), ABE end recovery was 0.97 because ethanol has low relative volatility and acetone recovery from CO_2 was difficult.

In MEx and conventional, the condensation heat of the regeneration column (EC1) cannot be employed due to its vacuum operation (0.1 bar). For this reason, the final distillation system (3DC-2 system, Fig. 1b) was inverse to 3DC-1. Condensation heat of WC was applied in boilers of columns at vacuum, AEC, and BC. The stage numbers were selected to avoid an excess of trays. The columns trays of WC, AEC, and BC were 40, 20, and 10, respectively (Fig. 1b). The distillation systems studied in this paper have stage numbers lower than that reported in the literature. For example, a conventional five-column distillation system has been evaluated with at total ideal stages of 135 (Mariano et al. 2011; Mariano and Filho 2012) (in this work, the total non-ideal stages of 3DC-2 were 70). The pressure of WC, AEC, and BC columns were 1.3, 0.5, and 0.1 bar, respectively. In the ideal evaluation of MEx (Table 3), the pressures of columns WC and EC1 were 0.3 and 1 atm., respectively. Therefore, in the ideal evaluation of MEx, the condensation energy of EC1 was used in the boilers of AC, WC, and EBC. In a similar way to 3DC-1, the reflux and boiler ratios were fixed to design specifications.

Results
High-temperature extraction (HTE)
In external extractive systems, there are two options of the bleed stream (Additional file 1: Figure S1). In the first option (a), the bleed stream is direct from the fermenter, while in the option (b) is after the extraction. For this reason, the first part of HTE evaluation was chosen the best purge option. HTE system is shown in Fig. 2a. The extractant in HTE with the option (a) was 33.3% lower than that for the option (b). However, in the option (b), the bleed stream had a less butanol concentration, 2.7 g/l instead of 10 g/l. Therefore, the feasibility of these options depends on the amount of the extractant used and the energy requirement reduction in WC column.

Fuel requirement of WC boiler depended mainly on ethanol concentration of vinasses, not in butanol, due to the low relative volatility of ethanol (~twofold lower than butanol). Ethanol mass fraction using the option (a) and (b) was analogous (5.93 and 5.89 g/l, respectively), due to the low distribution coefficient of ethanol (0.1). Therefore, the energy consumption of WC column using the option (a) and (b) was comparable (4.3 and 4 MJ/kg-ABE, respectively).

Due to the poor butanol distribution of mesitylene (2.2) and the low butanol concentration in the fermenter, the preheating of aqueous and organic streams before extraction was 65% of the total energy. The energy requirements without integration of the options (a) and (b) were 31.3 and 33.5 MJ-fuel/kg-ABE, respectively. The integration heat was favored using an operation pressure for extractant regeneration column EC1 of 1.3 bar because the condensation heat of EC1 was applied in the boilers of vacuum columns AC and WC.

The energy requirement was reduced with energy integration to 8.3 and 8.1 MJ-fuel/kg-ABE for options (a) and (b), respectively. Due to extractant reduction of option (a) [33% lower than option (b)] and similar energy requirement, option (a) was used in all subsequent extractive processes evaluated in this work. The energy requirements of HTE in the baseline scenario were higher than that for ABE recovery from dilute solutions (12.4 g-butanol/l) by heat-integrated distillation (8 MJ-fuel/kg-ABE) (Grisales Díaz and Olivar Tost 2016a).

For comparative purposes, the effects of conversion and substrate concentration in the energy requirements and fermenter design were studied. An increase of conversion in fermenter from 80 to 100% reduced the energy requirement in 11.8%. A less feed and higher recycle, with the increase in conversion, to achieve the fixed broth/ABE ratio used in this work (80 g/g) were required. EBC column was operated at vacuum pressure at substrate concentrations higher than 500 g/l (Table 3). The EBC column was operated at vacuum pressure because the total energy requirement of reboilers of columns WC and AC was lower than that for condenser heat of the extraction column EC1. Consequently, the condensation heat of extraction column was used in the WC, AC, and EBC boilers. In HTE, high substrate concentration required higher solvent ratio and lower fuel consumption to achieve the same butanol concentration in the fermenter (10 g/l) (Table 3).

The ideal assumptions were studied to achieve the minimum energy requirements of HTE. ABE concentration in fermenter under ideal assumptions increased from 23.7 to 62.5 g/l (Table 3). The ABE concentration increased because, under ideal assumptions, bleed stream is not used and the substrate concentration is maximum. The acetone distribution coefficient of mesitylene is 8.3 times higher than that of ethanol (Table 1).

Table 3 Energy evaluation at several conditions of extractive processes for ABE recovery from fermentation

Process	Substrate (g/l)	X[a]	ABE yield[b]	Butanol titer (g/l)	ABE titer (g/l)	Pressure EBC (atm.)	H_s^c	IES	Extractant ratio[d]
DEx	200	0.8	0.311	10.2	27.4	1.7	6.9	0.56	10.1
	200	1	0.388	10.3	30.1	1.7	5.8	0.76	9.8
	200	1	0.388	8.3	23.7	1.7	6.8	0.74	18.7
	300	0.8	0.311	10.2	31.3	1.7	5.3	0.59	10.9
	300	1	0.388	10.1	32.8	1.7	5.5	0.77	10.5
	500	0.8	0.311	10.2	35.3	0.1	4.9	0.60	11.3
	500	1	0.388	10.3	37.0	0.1	4.7	0.79	10.2
	Ideal			10.1	50.0	0.1	2.6 (2.5, [Kurkijärvi et al. 2014)]	0.84	7.8
HTEx	200	0.8	0.311	10.2	23.7	1.7	8.1	0.53	24
	200	1	0.388	10.2	25.3	1.7	7.1	0.73	26
	300	1	0.388	10.2	28	1.7	6.2	0.75	29
	500	1	0.388	10.4	32.7	1.7	6.0	0.76	30
	Ideal			10.3	64.2	0.1	3.6 [3.6, (Kraemer et al. 2011)]	0.81	26.8
MEx OAL/DAL (80–20)	200	0.8	0.311	10	27	0.1	6.6	0.56	12.3
	300	1	0.388	10.2	33.5	0.1	6.2	0.75	14
	500	1	0.388	10.4	39.2	0.1	6.1	0.75	14
	Ideal			10.2	50.2	0.1	2.9	0.83	13.6
DSD OAL	200	0.8	0.311	10.4	26.2	1.7	6.4	0.57	14.4
	200	1	0.388	8.3	23.1	1.7	5.7	0.76	25.0
	300	1	0.388	10.3	29.9	1.7	4.7	0.79	16.8
	500	1	0.388	10.3	34.4	1.7	5.0	0.78	17.8
	Ideal			10.3	49.8	0.1	2.6	0.84	16.0
DSD OAL/DAL (80–20)	200	0.8	0.311	10	27	0.1	6.8	0.56	12.3
	Ideal			10.1	50.2	0.1	3.1	0.82	13.6

IES ideal efficiency of separation, *Ideal* substrate concentration as high as extractant selectivity, glucose conversion of 100%, gases in the downstream were not considered, non-pressure drop, trays, efficiency of 100% was assumed

[a] X conversion

[b] ABE yield (g-ABE/g-total-glucose)

[c] H_s (MJ-ABE/kg-fuel) energy requirement of the separation system

[d] Solvent ratio, extractant flow/solvent flow

For this reason, ethanol concentration in fermenter under ideal assumptions was 4.7 times higher than acetone (9.2 g/l). Acetone and ethanol are much less toxic than butanol (Jones and Woods 1986). In perspective, the addition of acetone at 40 g/l reduces the growth ~50%, and total growth inhibition occurs at a concentration between 50 and 60 g-ethanol/l (Jones and Woods 1986), while the fermentation is inhibited completely at butanol concentrations of approximately 15 g/l. However, this high ABE concentration in fermenter must be toxic for the biocatalyst. Therefore, biocatalysts with a low yield of ethanol and acetone are desired for the operation of HTE at high substrate concentrations.

The minimum energy requirements, under ideal evaluation, of HTE were 5.8 MJ/kg-butanol or 3.6 MJ-fuel/-kg-ABE (Table 3). Heat requirement by mesitylene regeneration was 4.9 MJ/kg-butanol. The energy requirement of the final distillation columns was 1.6 MJ/kg-butanol. However, the condensation heat of extractant regeneration column was used to supply totally the energy requirement of boilers of final ABE purification that operates under vacuum. 2.5 MJ/kg-butanol of condensation was used in the preheating of the HTE system. The reboiler temperature of mesitylene column was 171 °C. Therefore, medium-pressure steam was needed.

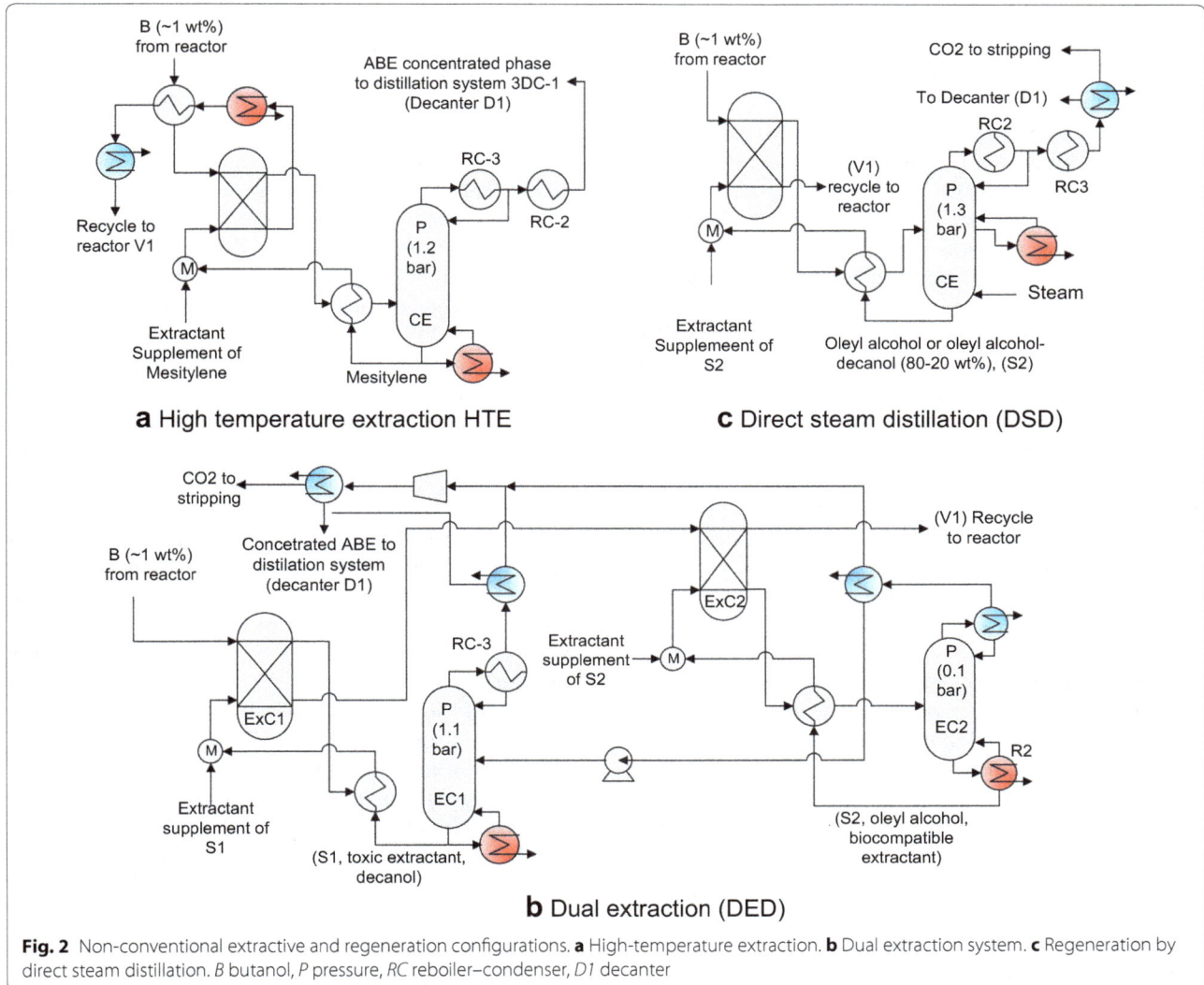

Fig. 2 Non-conventional extractive and regeneration configurations. **a** High-temperature extraction. **b** Dual extraction system. **c** Regeneration by direct steam distillation. *B* butanol, *P* pressure, *RC* reboiler–condenser, *D1* decanter

a High temperature extraction HTE

c Direct steam distillation (DSD)

b Dual extraction (DED)

Dual extraction (DEx)

In DEx, two counter-current extraction columns are used. Butanol and the toxic extractant (DAL) were recovered in the columns ExC1 and ExC2, respectively (Fig. 2b). OAL and DAL were fed into the extraction column at ratios of 1.4 and 8.3 kg-extractant/kg-ABE, respectively. Total solvent required for DEx, to achieve a butanol concentration in the fermenter of 10 g/l, was 2.4-fold lower than that for HTE process.

The OAL flow needed for 99% recovery of DAL from the broth was only 70 kg/h, a broth/OAL ratio of 5715 g/g. From a practical viewpoint, the extractants at this ratio are difficult of recovery by decantation due to the little organic fraction inside of extraction column (0.019 wt%). In this work, a flow of 7000 kg-OAL/h was chosen arbitrarily. This flow corresponds to 1.7% of total flow fed to extractant column ExC2. The energy

requirement of OAL regeneration boiler of base case was only 0.26 MJ-fuel/kg-ABE, 3.8% of total fuel requirement with integration. However, an adequate solvent flow of OAL must be selected through optimization of a pilot-scale system. The volume of each extraction stage of column ExC1 using a residence time of 0.5 h per stage for DEx was 230 m^3.

Boiler temperatures in OAL and DAL regeneration columns were 272 and 239 °C, respectively. The condensation energy of column EC1, DAL regeneration column, was used to apply heat to WC boiler. Total energy requirements of case base using DEx without integration were 21.4 MJ-fuel/kg-ABE, an energy requirement analogous to process DEx. Total energy requirement with integration was 6.9 MJ/kg-ABE (Table 3). Fuel consumption with heat integration, under similar assumptions, was between 15 and 30% lower than that for HTE.

For comparative purposes, the effect of a reduction in butanol concentration in the fermenter was studied. The extractant flow increased 1.9-fold when butanol concentration in the fermenter was reduced to 8.3 g/l (Table 3). Additionally, energy requirement was increased 14%. The reduction of butanol concentration in fermenter can prevent the strain degeneration and higher operation times can be achieved. For this reason, an optimization of fermenter cost must be performed in future works.

The ideal assumptions were studied to achieve the minimum energy requirements of DED. Energy consumption of ideal evaluation was 2.6 MJ-fuel/kg-ABE using an OAL flow of 70 kg/h. It was an energy consumption similar to that reported by Kurkijärvi et al. (2014) under ideal assumptions (3.8 MJ/kg-butanol or 2.5 MJ-fuel/kg-ABE, calculated in this work assuming 90% efficiency for steam production and A/B/E ratio of 3/6/1). Assumptions proposed by Kurkijärvi et al. (2014) are four ideal stages of extraction, mesitylene and DAL as extractants, A:B:E ratio of 3:6:1, minimum approach temperature of 3 °C, and the same energy requirement of final purification reported by Kraemer et al. (2011), 0.57 MJ/kg-butanol.

Conventional extraction with extraction mixture (MEx)
MEx achieved an energy requirement without and with heat integration of 21.4 and 6.6 MJ-fuel/kg-ABE, respectively. Energy requirements without and with heat integration were 12.6 and 1.2% lower than pure OAL, respectively. A low energy requirement without heat integration is important to reduce the exchanger area of the process. Energy requirements of MEx were between 6.6 and 6.1 MJ-fuel/kg-ABE at glucose concentrations between 200 and 500 g/l, respectively (Table 3).

Fuel consumption was reduced by 8.1% with an increase of substrate concentration from 200 to 500 g/l. MEx at a substrate concentration higher than 300 g/l was an option with higher energy requirement than that for DEx (Table 3). Given that to condensation heat of EC1 was not used in MEx, this reduction was 3.5- and 3.1-fold lower than that for DEx and HTE, respectively. Energy requirement of extraction mixture was between 1.7 and 23% lower than that for HTE (Table 3). The minimum energy requirements (2.9 MJ-fuel/kg-ABE) of mixture extraction were achieved under ideal evaluation (Table 3).

Alternative regeneration method with DSD
In the system proposed in this work, steam was fed to bottoms of extractant regeneration column and the boiler was not used (Fig. 2c). In this way, the temperature in the regeneration column decreased. Then, the exchanger area of preheating was reduced. Additionally, the low operation temperature in regeneration column can prevent

extractant degradation. Preheating is used to decrease the direct steam flow. However, the maximum temperature in column increased proportionally with respect to preheating (Fig. 3). ABE concentrated from the top of regeneration column was fed to a decanter.

An inflection point takes place at approximately 4.4 MJ-ABE/kg-fuel of preheating, using OAL without energy integration. At this preheated energy, the maximum temperature in the column was around 147 °C, and low-pressure steam (6 atm.) can be used. Without heat integration and direct bleed stream, the total energy requirement for DSD was 18.1 MJ-fuel/-kg-ABE. The energy requirement was reduced to 6.4 MJ-fuel/-kg-ABE through heat integration (Table 3). In mesitylene and DEx process, without heat integration, energy requirements were 1.7- and 1.2-fold higher than DSD with OAL extraction. The energy requirement of DSD was between 3 and 24.2% lower than that for MEx (Table 3).

In contrast to DSD, DEx required high-pressure steam to use the heat of condensation of the regeneration column EC1. In perspective, high-pressure steam is 1.5- and 5.4-fold more expensive than medium and low-pressure steam (Mussatto et al. 2013), respectively. The energy requirements for DSD and OAL of the ideal evaluation were 1.4-fold lower (2.6 MJ-fuel/kg-ABE) than that for HTE (Table 3) and analogous to DEx.

DSD can be applied with a mixture of OAL–DAL (80–20 wt%) (Fig. 4). Due to low-temperature evaporation of DAL (233 °C) with respect to OAL (357 °C), DAL was partially evaporated. For this reason, the regeneration column was proposed without condenser. Therefore, an additional column was necessary for butanol purification from the extractant–butanol mixture obtained in EBC column (Fig. 4). The minimum energy requirements of DSD without integration were 17.3 MJ-fuel/kg-ABE, 6.3% lower than that for DSD using pure OAL. Extractant was reduced by 14.8% using an OAL–DAL mixture

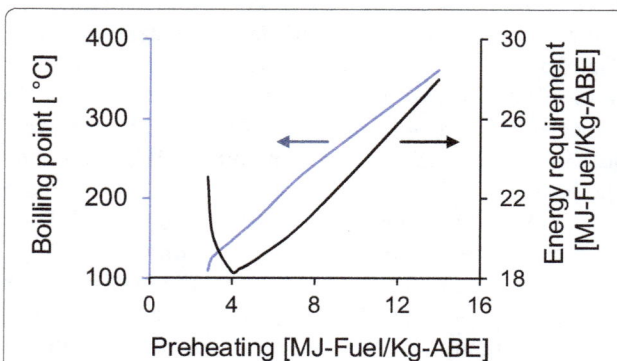

Fig. 3 Effect of preheating in energy requirement without integration of regeneration column by direct steam distillation. Top pressure is 1.3 atm.

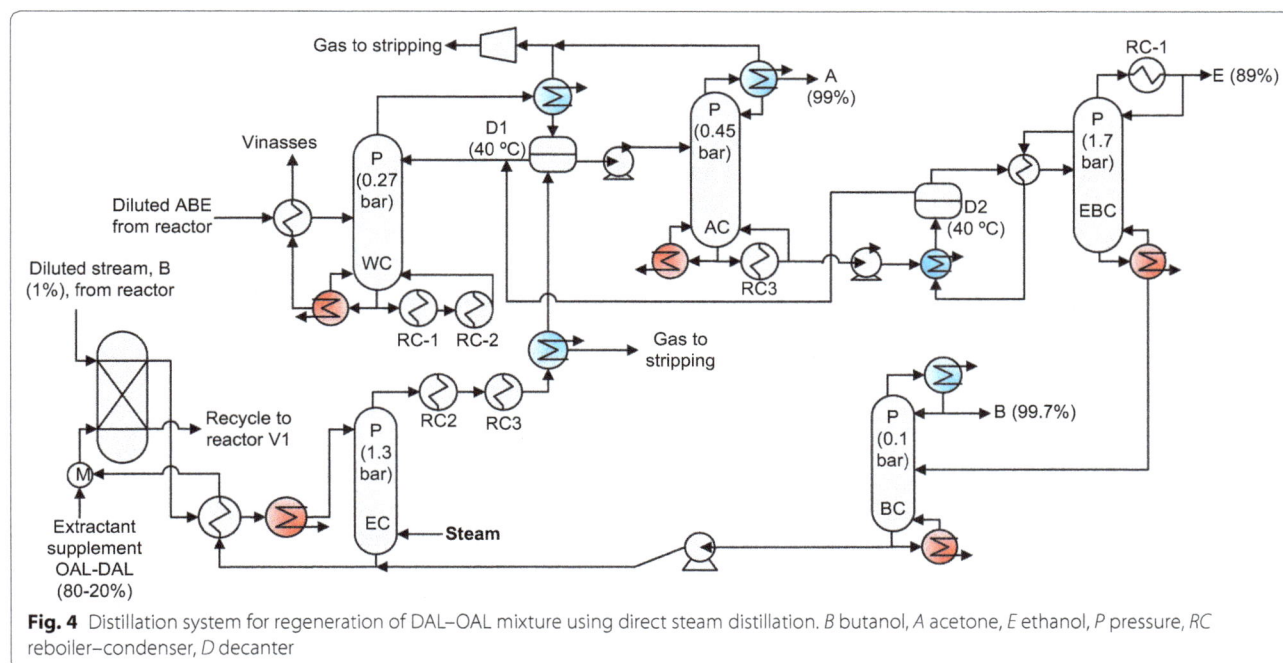

Fig. 4 Distillation system for regeneration of DAL–OAL mixture using direct steam distillation. *B* butanol, *A* acetone, *E* ethanol, *P* pressure, *RC* reboiler–condenser, *D* decanter

(80–20 wt%) with respect to pure OAL. In the regeneration column, a minimum energy requirement at a pressure of 1.4 bar was obtained with a feed temperature of 168 °C. Energy requirements were 6.6 MJ-fuel/kg-ABE with heat integration, 0.2 MJ-fuel/kg-ABE higher than the energy requirement of DSD with pure OAL.

Discussion

Energy evaluation

The energy requirements change drastically with the assumptions of operational conditions and efficiencies of units. Additionally, the energy requirements depend on the selection of final distillation system and heat integration. Hence, a comparison of extraction-based systems with literature data is difficult. In the literature, low energy requirements have been reported with HTE. However, in this work, the lowest energy-efficient system with baseline conditions was HTE (IES equal to 0.53). In this evaluation, the yield of hydrogen from glucose is 0.016 g-hydrogen/g-glucose (stoichiometric ratio of Chinese industrial process, Eq. 1). The hydrogen combustion with this yield was 15.8% of total energy produced. The IES of DSD for the base case was 0.57. The most important factor in IES evaluation was the ABE yield or glucose conversion. For instance, the IES increased from 0.57 to 0.76 when the conversion in DSD and OAL increased from 80 to 100% (ABE yield of).

The IES of ideal evaluation of all extractive systems increased to 0.81–0.84. MEx achieved a similar energy performance to DSD only in the base case. In general,

DSD achieved the lowest energy requirement and energy efficiency with and without integration. The high energy integration of DSD was possible thanks to the atmospheric operation of the regeneration column and the low-pressure columns used in ABE purification. In reference to external recovery systems, DEx required less extractant than that for DSD or MEx (Table 3). Therefore, an economic evaluation was necessary.

HTE and DEx are the only extractive processes reported in the literature with lowest energy requirements than that of DSD (6.4 MJ-fuel/kg-ABE). However, these energy requirements are under ideal evaluations. In comparison, Qureshi et al. (2005) reported energy requirements of 7.7 MJ-fuel/kg-ABE [calculated in this work assuming energy efficiency of 0.9 and ABE ratio of *C. beijerinckii* BA101 (ABE of 6/24.6/1)]. Salting-out has been reported with energy requirements between 22 and 25 M/kg-butanol (Xie et al. 2013, 2015), while extraction using hexyl acetate has been reported with energy requirements of 45 MJ/kg/ABE (Sánchez-Ramírez et al. 2015; Errico et al. 2016).

In reference at alternative biofuels, the IES achieved for ABE production by extractive fermentation (100% of conversion) were 3.9 and 5.4% greater than that for ethanol and isobutanol (alternative biofuels) dehydration with double-effect distillation. Double-effect distillation is a heat-integrated distillation system with low energy requirement. In fact, ABE recovery by double distillation has been reported with an energy requirement 20% (8 MJ-fuel/kg-ABE) lower than that for integrated fermenter by pervaporation (Grisales Díaz and Olivar Tost 2016a).

Economic evaluation

The economic performance of all configurations of extractive fermentation evaluated in this work is shown in Fig. 5. TAC of HTE was 0.097 $/kg-ABE. Total installation cost was 13.4 MM US$. Investment costs of heat exchangers and steam requirement (0.047 and 0.02 $/kg-ABE, respectively) were the most important items in the economic evaluation. Total installation cost of DEx was 10.3 MM US$. TAC and total installation cost of DEx were 4.8 and 23.1% lower than HTE, respectively. The extractant loss and extractant initial investment were the items with less effect in TAC (Fig. 5).

The cost of the extraction column with biocompatible extractant was not calculated, because the productivity with biocompatible extractants increases more than the reduction caused by the volume of stages extraction (0.31 g-ABE/l/h) (Roffler et al. 1987, 1988; Bankar et al. 2012). The volume of extraction can be reduced with the reduction of vinasses recycle or extraction stages. However, it increases the energy requirement (Kurkijärvi et al. 2014). For this reason, optimization of this item must be performed in future work.

In DEx and HTE, the extraction column cost was between 11.5 and 13.4% of TAC. TAC of DSD with pure OAL decreased in 28 and 31.4% with respect to DEx and HTE. The low cost of DSD with respect to DEx was mainly due to the non-cost estimation of extraction column in DSD, the low heat exchanger area, and the low operational costs. Total installation cost of OAL extraction (8.8 MM US$) was reduced to 8.2 MM US$ using an OAL–DAL mixture. TAC of DSD using OAL–DAL mixture was 0.065 $/kg-ABE, 3.5 and 1.4% lower than conventional regeneration using the same mixture and DSD using pure OAL, respectively.

DSD and MEx reduced the TAC in 17.1 and 15.4% with respect to conventional extraction (OAL and conventional regeneration). DSD [with OAL or OAL/DAL (80–20)] or MEx [OAL/DAL (80–20)], in situ recovery units,

was more economical than external extraction (DED and HTE). External extraction does not increase the yield or productivity of reactor. Therefore, in this work, an evaluation of fermenter cost was not necessary. The low costs of DSD were due mainly to the utilization of a biocompatible extractant and the low energy requirement without integration or the low exchanger area. An appropriated selection of fermenter conditions of DSD or MEx must be performed through economic optimization. A robust kinetic model for the economic optimization is necessary because low butanol concentrations in reactor required a high extractant flow and energy requirements.

Conclusions

At a substrate concentration of 200 g/l, HTE and DED were more expensive, and with higher energy requirements than the in situ recovery processes, MEx and DSD. In all evaluated cases, DSD was the process with lower energy requirements and the less expensive. Energy integration of DSD was higher than other extractive processes due to the atmospheric operation of the regeneration column. The less expensive cost of DSD was mainly due to the utilization of a biocompatible extractant and the low energy requirement without integration or the low exchanger area. The ABE yield was the item most important in the energy efficiency calculation of biofuel recovery.

Abbreviations

AC: column for acetone recovery; DAL: decanol; DEx: dual extraction; DSD: direct steam distillation; EBC: column for ethanol and butanol recovery; EC1: regeneration column of toxic extractant; EC2: regeneration column of biocompatible extractant; ExC1: extractive unit to recover toxic extractant; ExC2: extractive unit to recover ABE; H_S: energy requirement of separation system; HTE: high-temperature extraction; IES: ideal efficiency of separation; LHV: lower heating value; MEx: mixture extraction; OAL: oleyl alcohol; R_s: solvent yield; TAC: total annualized cost; WC: column for ABE recovery from vinasses; 2E1H: 2-ethyl-1-hexanol.

Authors' contributions

VHGD performed the simulations and wrote this manuscript. GOT contributed general advice. Both authors read and approved the final manuscript.

Author details

[1] School of Chemical Engineering and Advanced Materials, Newcastle University, Newcastle upon Tyne NE1 7RU, UK. [2] Control y Percepción Inteligente, Departamento de Ingeniería Eléctrica, Electrónica y Computación, Universidad Nacional de Colombia, Cra. 27 No. 64-60, Manizales, Colombia.

Acknowledgements

Authors thank the Colombian Administrative Department of Science, Technology and Innovation (COLCIENCIAS) for the financial support.

Competing interests

The authors declare that they have no competing interests.

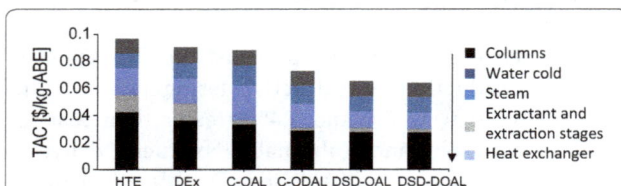

Fig. 5 TAC of extractive distillation systems evaluated in this work. C-OAL (conventional using OAL), C-DOAL (conventional using a mixture of OAL and DAL), DSD OAL (DSD using OAL), DSD DOAL (DSD using a mixture of OAL and DAL)

Funding
Colombian Administrative Department of Science, Technology, and Innovation (COLCIENCIAS) (No. 528) supported this work.

References

Bankar SB, Survase SA, Singhal RS, Granström T (2012) Continuous two stage acetone–butanol–ethanol fermentation with integrated solvent removal using *Clostridium acetobutylicum* B 5313. Bioresour Technol 106:110–116. doi:10.1016/j.biortech.2011.12.005

Cabezas R, Plaza A, Merlet G, Romero J (2015) Effect of fluid dynamic conditions on the recovery of ABE fermentation products by membrane-based dense gas extraction. Chem Eng Process Process Intensif 95:80–89. doi:10.1016/j.cep.2015.04.003

Douglas JM (1988) Conceptual design of chemical processes. McGraw-Hil, New York

Errico M, Sanchez-ramirez E, Quiroz-ramìrez JJ, Segovia-hernandez JG, Rong B (2016) Synthesis and design of new hybrid configurations for biobutanol purification. Comput Chem Eng 84:482–492. doi:10.1016/j.compchemeng.2015.10.009

Evans PJ, Wang HY (1988) Enhancement of butanol formation by *Clostridium acetobutylicum* in the presence of decanol–oleyl alcohol mixed extractants. Appl Environ Microbiol 54:1662–1667

Ezeji TC, Qureshi N, Blaschek HP (2005) Process for continuous solvent production. US Patent US2005/0089979, 28 Apr 2005

Fischer K, Gmehling J (1994) P-x and.gamma.infin. Data for the different binary butanol–water systems at 50 °C. J Chem Eng Data 39:309–315. doi:10.1021/je00014a026

González-Peñas H, Lu-Chau TA, Moreira MT, Lema JM (2014a) Solvent screening methodology for in situ ABE extractive fermentation. Appl Microbiol Biotechnol 98:5915–5924. doi:10.1007/s00253-014-5634-6

González-Peñas H, Lu-Chau TA, Moreira MT, Lema JM (2014b) Assessment of morphological changes of *Clostridium acetobutylicum* by flow cytometry during acetone/butanol/ethanol extractive fermentation. Biotechnol Lett. doi:10.1007/s10529-014-1702-3

Grady MC, Parten WD, Vrana B, Xu YT, Zaher JJ (2013) Recovery of butanol from a mixture of butanol, water, and an organic extractant. US Patent 8,373,009 B2, 12 Feb 2013

Grisales Díaz VH, Olivar Tost G (2016a) Butanol production from lignocellulose by simultaneous fermentation, saccharification, and pervaporation or vacuum evaporation. Bioresour Technol 218:174–182. doi:10.1016/j.biortech.2016.06.091

Grisales Díaz VH, Olivar Tost G (2016b) Ethanol and isobutanol dehydration by heat-integrated distillation. Chem Eng Process Process Intensif 108:117–124. doi:10.1016/j.cep.2016.07.005

Groot WJ, van der Lans RGJM, Luyben KCAM (1992) Technologies for butanol recovery integrated with fermentations. Process Biochem 27:61–75. doi:10.1016/0032-9592(92)80012-R

Jones DT, Woods DR (1986) Acetone–butanol fermentation revisited. Microbiol Rev 50:484–524

Jonker JGG, van der Hilst F, Junginger HM, Cavalett O, Chagas MF, Faaij APC (2015) Outlook for ethanol production costs in Brazil up to 2030, for different biomass crops and industrial technologies. Appl Energy 147:593–610. doi:10.1016/j.apenergy.2015.01.090

Kim YH (2015) Side-rectifier divided wall column for offshore LNG plant. Sep Purif Technol 139:25–35. doi:10.1016/j.seppur.2014.11.002

Köhler KAK, Rühl J, Blank LM, Schmid A (2015) Integration of biocatalyst and process engineering for sustainable and efficient *n*-butanol production. Eng Life Sci 15:4–19. doi:10.1002/elsc.201400041

Kraemer K, Harwardt A, Bronneberg R, Marquardt W (2011) Separation of butanol from acetone–butanol–ethanol fermentation by a hybrid extraction–distillation process. Comput Chem Eng 35:949–963. doi:10.1016/j.compchemeng.2011.01.020

Kumar M, Goyal Y, Sarkar A, Gayen K (2012) Comparative economic assessment of ABE fermentation based on cellulosic and non-cellulosic feedstocks. Appl Energy 93:193–204. doi:10.1016/j.apenergy.2011.12.079

Kurkijärvi AJ, Lehtonen J (2014) Dual extraction process for the utilization of an acetone–butanol–ethanol mixture in gasoline. Ind Eng Chem Res 53:12379–12386. doi:10.1021/ie500131x

Kurkijärvi A, Lehtonen J, Linnekoski J (2014) Novel dual extraction process for acetone–butanol–ethanol fermentation. Sep Purif Technol 124:18–25.

Kurkijärvi AJ, Melin K, Lehtonen J (2016) Comparison of reactive distillation and dual extraction processes for the separation of acetone, butanol, and ethanol from fermentation broth. Ind Eng Chem Res 55:1952–1964. doi:10.1021/acs.iecr.5b03196

Li Q, Cai H, Hao B, Zhang C, Yu Z, Zhou S, Chenjuan L (2010) Enhancing clostridial acetone–butanol–ethanol (ABE) production and improving fuel properties of ABE-enriched biodiesel by extractive fermentation with biodiesel. Appl Biochem Biotechnol 162:2381–2386. doi:10.1007/s12010-010-9010-4

Liu J, Fan LT, Seib P, Friedler F, Bertok B (2004) Downstream process synthesis for biochemical production of butanol, ethanol, and acetone from grains: generation of optimal and near-optimal flowsheets with conventional operating units. Biotechnol Prog 20:1518–1527. doi:10.1021/bp049845v

Liu G, Gan L, Liu S, Zhou H, Wei W, Jin W (2014) PDMS/ceramic composite membrane for pervaporation separation of acetone–butanol–ethanol (ABE) aqueous solutions and its application in intensification of ABE fermentation process. Chem Eng Process Process Intensif 86:162–172. doi:10.1016/j.cep.2014.06.013

Lu C, Zhao J, Yang S-T, Wei D (2012) Fed-batch fermentation for *n*-butanol production from cassava bagasse hydrolysate in a fibrous bed bioreactor with continuous gas stripping. Bioresour Technol 104:380–387. doi:10.1016/j.biortech.2011.10.089

Mariano AP, Filho RM (2012) Improvements in biobutanol fermentation and their impacts on distillation energy consumption and wastewater generation. Bioenergy Res 5:504–514. doi:10.1007/s12155-011-9172-0

Mariano AP, Keshtkar MJ, Atala DIP, Maugeri Filho F, Wolf Maciel MR, Maciel Filho R, Stuart P (2011) Energy requirements for butanol recovery using the flash fermentation technology. Energy Fuels 25:2347–2355. doi:10.1021/ef200279v

Matsumura M, Märkl H (1984) Application of solvent extraction to ethanol fermentation. Appl Microbiol Biotechnol 20:371–377. doi:10.1007/BF00261937

Matsumura M, Kataoka H, Sueki M, Araki K (1988) Energy saving effect of pervaporation using oleyl alcohol liquid membrane in butanol purification. Bioprocess Eng 3:93–100. doi:10.1007/BF00369334

Mussatto SI, Moncada J, Roberto IC, Cardona CA (2013) Techno-economic analysis for brewer's spent grains use on a biorefinery concept: the Brazilian case. Bioresour Technol 148:302–310. doi:10.1016/j.biortech.2013.08.046

Ni Y, Sun Z (2009) Recent progress on industrial fermentative production of acetone–butanol–ethanol by *Clostridium acetobutylicum* in China. Appl Microbiol Biotechnol 83:415–423. doi:10.1007/s00253-009-2003-y

Offeman RD, Stephenson SK, Franqui D, Cline JL, Robertson GH, Orts WJ (2008) Extraction of ethanol with higher alcohol solvents and their toxicity to yeast. Sep Purif Technol 63:444–451. doi:10.1016/j.seppur.2008.06.005

Oudshoorn A, van der Wielen LAM, Straathof AJJ (2009) Assessment of options for selective 1-butanol recovery from aqueous solution. Ind Eng Chem Res 48:7325–7336. doi:10.1021/ie900537w

Outram V, Lalander C-A, Lee JGM, Davis ET, Harvey AP (2016) A comparison of the energy use of in situ product recovery techniques for the acetone butanol ethanol fermentation. Bioresour Technol 220:590–600. doi:10.1016/j.biortech.2016.09.002

Qureshi N, Maddox IS (2005) Reduction in butanol inhibition by perstraction: utilization of concentrated lactose/whey permeate by *Clostridium acetobutylicum* to enhance butanol fermentation economics. Food Bioprod Process 83:43–52. doi:10.1205/fbp.04163

Qureshi N, Hughes S, Maddox IS, Cotta MA (2005) Energy-efficient recovery of butanol from model solutions and fermentation broth by adsorption. Bioprocess Biosyst Eng 27:215–222. doi:10.1007/s00449-005-0402-8

Roffler SR, Blanch HW, Wilke CR (1987) In-situ recovery of butanol during fermentation. Bioprocess Eng 2:181–190. doi:10.1007/BF00387326

Roffler SR, Blanch HW, Wilke CR (1988) In situ extractive fermentation of acetone and butanol. Biotechnol Bioeng 31:135–143. doi:10.1002/bit.260310207

Ruggeri B, Tommasi T, Sanfilippo S (2015) BioH$_2$ & BioCH$_4$ through anaerobic digestion. Springer, London

Sánchez-Ramírez E, Quiroz-Ramírez JJ, Segovia-Hernández JG, Hernández S, Bonilla-Petriciolet A (2015) Process alternatives for biobutanol purification: design and optimization. Ind Eng Chem Res 54:351–358. doi:

van der Merwe AB, Cheng H, Görgens JF, Knoetze JH (2013) Comparison of energy efficiency and economics of process designs for biobutanol production from sugarcane molasses. Fuel 105:451–458. doi:10.1016/j.fuel.2012.06.058

Veloo PS, Wang YL, Egolfopoulos FN, Westbrook CK (2010) A comparative experimental and computational study of methanol, ethanol, and n-butanol flames. Combust Flame 157:1989–2004. doi:10.1016/j.combustflame.2010.04.001

Xie S, Yi C, Qiu X (2013) Energy-saving recovery of acetone, butanol, and ethanol from a prefractionator by the salting-out method. J Chem Eng Data 58:3297–3303. doi:10.1021/je400740z

Xie S, Qiu X, Yi C (2015) Separation of a biofuel: recovery of biobutanol by salting-out and distillation. Chem Eng Technol 38:2181–2188. doi:10.1002/ceat.201500140

Xu YT, Parten WD (2011) Recovery of butanol from a mixture of butanol, water, and an organic extractant. US Patent 2011/0162954 A1, 7 July 2011

Xue C, Zhao XQ, Liu CG, Chen LJ, Bai FW (2013) Prospective and development of butanol as an advanced biofuel. Biotechnol Adv 31:1575–1584. doi:10.1016/j.biotechadv.2013.08.004

Zauba (2015) India's import and export data

4

Biofibres from biofuel industrial byproduct—*Pongamia pinnata* seed hull

Puttaswamy Manjula*, Govindan Srinikethan and K. Vidya Shetty

Abstract

Background: Biodiesel production using *Pongamia pinnata (P. pinnata)* seeds results in large amount of unused seed hull. These seed hulls serve as a potential source for cellulose fibres which can be exploited as reinforcement in composites.

Methods: These seed hulls were processed using chlorination and alkaline extraction process in order to isolate cellulose fibres. Scanning electron microscopy (SEM), dynamic light scattering (DLS), thermogravimetric analysis (TGA), X-ray diffraction (XRD), Fourier transform infrared spectroscopy (FTIR) and nuclear magnetic resonance spectroscopy (NMR) analysis demonstrated the morphological changes in the fibre structure.

Results: Cellulose microfibres of diameter 6–8 µm, hydrodynamic diameter of 58.4 nm and length of 535 nm were isolated. Thermal stability was enhanced by 70 °C and crystallinity index (CI) by 19.8% ensuring isolation of crystalline cellulose fibres.

Conclusion: The sequential chlorination and alkaline treatment stemmed to the isolation of cellulose fibres from *P. pinnata* seed hull. The isolated cellulose fibres possessed enhanced morphological, thermal, and crystalline properties in comparison with *P. pinnata* seed hull. These cellulose microfibres may potentially find application as biofillers in biodegradable composites by augmenting their properties.

Keywords: Cellulose microfibres, *Pongamia pinnata* seed hull, Hemicellulose, Lignin, Chlorination

Background

Cellulose is nature's most lavishly available polymer. Highly purified cellulose fibre is been isolated from several plant sources, such as branch barks of mulberry (Li et al. 2009), pineapple leaf fibres (Cherian et al. 2010; Mangal et al. 2003), pea hull fibre (Chen et al. 2009), coconut husk fibres (Rosa et al. 2010), banana rachis (Zuluaga et al. 2009), sugar beet (Dinand et al. 1999; Dufresne et al. 1997), wheat straw (Kaushik and Singh 2011), palm leaf sheath (Maheswari et al. 2012), *Arundo donax* L stem (Fiore et al. 2014), cotton stalk (Hou et al. 2014).

From the past two decades these biofibres are being used as filler material in the preparation of composites and have gained prodigious attention (Hubbe et al. 2008). In view of better utilization of renewable resources, there is a need to explore other renewable greener sources, which can be utilized in developing high strength light weight biocomposites for high-end applications. *Pongamia pinnata* seed hull is chosen for the present work to exploit its potential for cellulose fibres which could be utilized as reinforcement in biocomposites. In India and south East Asia, *Pongamia pinnata* (Karanja) seed is used for biodiesel production (Demirbas 2009). It is also a traditional medicinal plant with all parts having certain medicinal value (Yadav et al. 2004). Biofuel production using *P. pinnata* seeds has resulted in large-scale cultivation of these trees (Shwetha et al. 2014). The biofuel processing fallouts in significant amount of residual *P. pinnata* seed hull, in which cellulose percentage approximates to 40% and is similar as in shelly wood (Nadeem et al. 2009). Thus these underused seed hulls can find potential application as a source for cellulose fibres.

Isolation of cellulose fibres is customarily carried out by mechanical treatments such as homogenisation (Du et al. 2016; Julie et al. 2016), sonication, (Sheltami et al. 2012;

*Correspondence: manjuchintoo2@gmail.com
Department of Chemical Engineering, National Institute of Technology Karnataka, Surathkal, India

Saurabh et al. 2016), steam explosion (Saelee et al. 2014) etc.; chemical treatments such as acid hydrolysis (Abidin et al. 2015), TEMPO oxidation (Du et al. 2016), chlorination and alkaline treatments (Sheltami et al. 2012; Johar et al. 2012; Maheswari et al. 2012) etc.; enzymatic treatments (Saelee et al. 2014) and conjointly with the combination of two or more of the aforementioned processes. Chemical treatments usually act upon the binding material of the fibril structure enabling the fibres to individualize (Johar et al. 2012). Chlorination treatment being a chemical treatment is a well-established treatment which assists isolation of high quality pure cellulose fibres by bleaching and delignifies the cellulose material, while alkali treatment dissolves the wax, pectin and hemicellulose ensuring efficient isolation of cellulose microfibres. These chemical methods are used in combination to isolate cellulose fibres from different sources (Espino et al. 2014; Johar et al. 2012; Sheltami et al. 2012; Mandal and Chakrabarty 2011; Moran et al. 2008) and are also found to be efficient and economical when compared to high energy-consuming mechanical methods (Motaung and Mtibe 2015).

In the present research work, the cellulose fibres were isolated from the *P. pinnata* seed hull using chlorination and alkaline process. The isolated cellulose microfibres were characterized using scanning electron microscopy (SEM), dynamic light scattering (DLS), thermogravimetric analysis (TGA), X-ray diffraction (XRD), Fourier transform infrared spectroscopy (FTIR) and nuclear magnetic resonance spectroscopy (NMR) analysis for their morphological, thermal and crystalline properties.

Methods
Materials
Pongamia pinnata seed hulls were collected from "SEEDS" Research Centre, University of Agricultural Sciences, Bengaluru, India. All the chemicals used were of analytical grade.

Fibre processing
Pongamia pinnata seed hulls were separated from stones and other plant materials by hand picking. The dust and mud particles sticking to the seed hulls were removed by washing them extensively in tap water and finally with distilled water. Later dried under sunlight for two days and stored in sealed polythene bags for further use. Cleaned seed hulls were ground, screened (0.25 mm sieves) and oven dried at 105 °C for 8 h.

Isolation of cellulose microfibres
Cellulose microfibres were isolated from *P. pinnata* seed hull by chlorination and alkaline extraction process (Maheswari et al. 2012). Cleaned seed hull fibres were

dewaxed using toluene–ethanol mixture (2:1) for 6 h. Excess of solvent from the fibres was removed by suction and later kept for drying in hot air oven. Fibres were bleached with 7% $NaClO_2$ taken in fibre to liquor ratio of 1:50 (pH vicinity 4–4.2 was maintained using acetic acid and sodium acetate buffer) for 2 h at 100 °C and was washed successively using 2% sodium bisulphate, distilled water and ethanol. Further the extraction of holocellulose from fibres was carried out by treating with 17.5% NaOH solution at 20 °C for 45 min and subsequently washed with 10% acetic acid. Later the fibres were treated with 0.8% acetic acid and 0.7% nitric acid in the ratio 15:1 at 120 °C for 15 min. The mixture was cooled, filtered and washed sequentially with 95% ethanol and distilled water. The resulting cellulose fibres were oven dried at 105 °C until consistent weight was achieved.

Characterization
Scanning electron microscopy (SEM)
The morphological structure of gold-sputtered *P. pinnata* seed hull fibres and isolated cellulose fibres were observed under SEM (JSM-6380LA, JEOL, EVISA). The micrographs were recorded at acceleration voltage of 5–8 kV.

Dynamic light scattering (DLS)
The fibre dimension of aqueous dispersed isolated cellulose fibre (distilled water) was measured by the dynamic light scattering instrument (DLS, nanoparticle analyser, HORIBA Scientific, nano partica SZ-100, Japan).

Fourier transform infrared spectroscopy (FTIR)
Pongamia pinnata seed hull fibres and isolated cellulose fibres mixed with KBr were pressed to form transparent thin pellets. FTIR spectra of the fibres were recorded in the extent of 400–4000/cm with 4/cm resolution using FTIR instrument (Jasco 4200, Jasco analytical instruments, USA).

X-ray analysis (XRD)
XRD measurements for *P. pinnata* seed hull fibres and isolated cellulose fibres were obtained by X-ray diffractometer (X'Pert^3 Powder, PANalytical, The Netherlands) using Cu Kα radiation (1.5406 Å) with Ni filtered at 40 kV, 15 mA. Scattered radiations were recorded in the range of $2\theta = 10° - 30°$ at a scan rate of 4°/min. The Segal method [Eq. (1)] was used to calculate crystallinity index (CI) considering the intensities of (200) peak (I_{200}, $2\theta = 22.6°$) and the intensity minimum between the (200) and (110) peaks (I_{am}, $2\theta = 18^0$), where I_{200} represents the intensities of crystalline and amorphous material and I_{am} for the amorphous material.

$$CI\% = \left(1 - \frac{I_{am}}{I_{200}}\right) \times 100 \qquad (1)$$

Thermogravimetric analysis (TGA)

Thermograms for *P. pinnata* seed hull fibres and isolated cellulose fibres were determined using a thermogravimetric analyser (TGA Q50, TA instruments, USA) at a 10 °C/min heating rate in nitrogen atmosphere.

^{13}C NMR (CP-MAS) spectroscopy

Spectra of *P. pinnata* seed hull fibres and isolated cellulose fibres were run on solid-state NMR spectrometer (Bruker DSX 300 MHz). 75.46 MHz operating frequency was fixed for ^{13}C nuclei. Fibres were spun at 7.5 kHz spinning rate with filled 5 mm rotor at room temperature.

Results and discussion

Delignification of seed hull using acidified sodium chlorite was compassed, as an initial step in the isolation of cellulose. The alkaline treatment aids in the oxidation of lignin and hemicellulose, solubilizes the residual lignin and hemicellulose resulting in the isolation of cellulose fibres. These cellulose fibres were characterized for their morphological features, thermal stability and also to ensure removal of matrix components such as lignin and hemicellulose.

SEM analysis

The scanning electron microscope images of *P. pinnata* seed hull fibre after different stages of chemical treatment are as presented in Fig. 1a–c. Dewaxed seed hull fibre presented in Fig. 1a show irregular appearance due to cellulose fibre embedded between waxes and cementing materials such as lignin and hemicellulose (Reddy and Yang 2005; Haafiza et al. 2013). The fibres after sodium chlorite bleaching show cellulose fibres emerging out of the matrix as shown in Fig. 1b. This could be accounted to oxidation and solubilisation of matrix components viz. lignin and hemicellulose. The cementing components—lignin and hemicellulose isolated from the fibres are dissolved by mild alkali treatment (Elanthikkal et al. 2010). As a result, the SEM image of the isolated cellulose fibres as presented in Fig. 1c illustrates individualized single strand of cellulose fibres of diameter 6–8 μm, which in turn is a bundle of cellulose microfibres (Chen et al. 2011) having diameter of 270–370 nm. The cellulose fibres isolated in this work were of smaller diameter compared to that of the other cellulose fibres obtained from different sources such as soybean straw (Reddy and Yang 2009) yielding fibres of diameter 15.6 μm and coconut palm sheath (Maheswari et al. 2012) yielding fibres of 10–15 μm diameter.

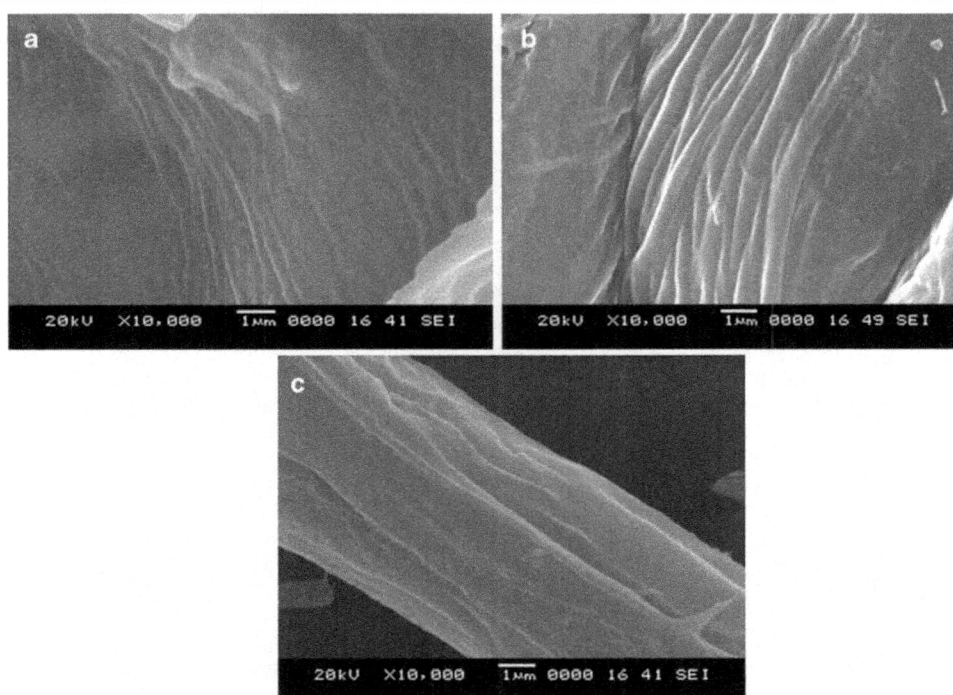

Fig. 1 Scanning electron microscope images **a** dewaxed *Pongamia pinnata* seed hull, **b** sodium chlorite-treated fibres and **c** isolated cellulose fibres

Table 1 Particle size distribution values of isolated cellulose fibre

Peak No.	S.P. area ratio[a]	Mean (nm)	SD[b] (nm)	Mode (nm)
1	0.20	58.4	3.3	58.9
2	0.80	536.3	44.1	535.0
Total	1.00	441.9	194.3	535.00

[a] Specific particle surface area ratio

[b] Standard deviation

DLS analysis

The aqueous dispersion of cellulose fibres was analysed by the dynamic light scattering technique in order to find their size distribution. DLS analysis results are summarized in Table 1. The histogram presented in Fig. 2 shows the presence of two distributions, indicating the presence of two dimensions (Kavitha et al. 2013; Srinivas et al. 2012) which is owing to the fibrous structure of cellulose representing both length and diameter. de Carvalho Mendes et al. (2015) also reported such two peaks in DLS histogram of the aqueous dispersion of cellulose fibrous structure. Dimensions determined by DLS epitomise hydrodynamic size (sphere size) having same diffusional coefficient as the fibres being measured (Horiba knowledgebase 2017). The mean hydrodynamic size of isolated cellulose fibres for shorter dimension (diameter) was observed to be 58.4 nm, whereas the longer dimension (length) of the fibres was observed to be 536.3 nm with the standard deviation of 3.3 and 44.1 nm, respectively. The diameter of the fibre obtained by SEM analysis is

lesser than that obtained by DLS technique. The sizes of the fibre obtained by SEM and DLS are not comparable, as the diameter obtained by SEM presents the dry fibre size, whereas that obtained by DLS signifies the hydrodynamic diameter in aqueous dispersion. The difference in size estimated by the two methods is generally higher for the nonspherical particles.

Fourier transform infrared spectroscopy (FTIR)

FTIR spectroscopy monitors the functional groups present in the fibres. Figure 3a and b present the spectra obtained for *P. pinnata* seed hull fibres and isolated cellulose fibres. The band around 3600–3000/cm assigned to stretching vibrations of O–H and C–H is observed in both *P. pinnata* seed hull fibres and isolated cellulose fibre, indicating the presence of cellulose-related functional groups (Qiao et al. 2016; Shin et al. 2012; Kalita et al. 2015; Sun et al. 2004a, b, c; Kaushik and Singh 2011). Peaks at 2894.63 and 2919.7/cm is generally assigned to C–H stretching vibration in lignin polysaccharide (cellulose and hemicellulose) (Shin et al. 2012; Sun et al. 2004a, b, c, d; Kaushik and Singh 2011; Zhong et al. 2013). Peak at 1735.62/cm is assigned to C=O stretching vibration of carbonyl, acetyl and uronic ester group of the ferulic and p-coumaric acids of lignin and/or xylan component of hemicellulose. The disappearance of these peaks in cellulose fibre spectra, confirms the removal of lignin and hemicellulose (Kalita et al. 2015; Kaushik and Singh 2011; Sun et al. 2004a, c, d; Elanthikkal et al. 2010; Rosa et al. 2012; Oun and Rhim 2016). Peaks at 1646.91 and

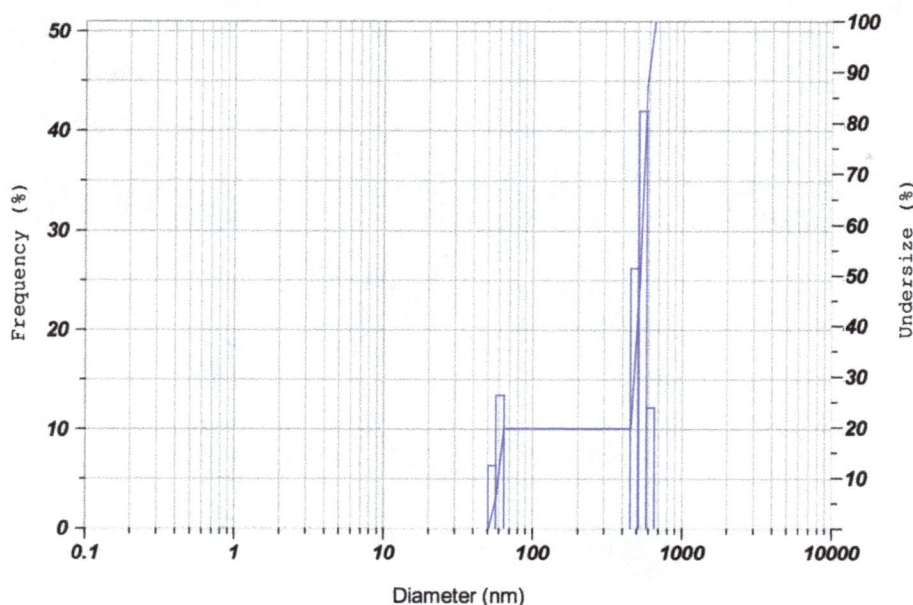

Fig. 2 DLS analysis spectra of isolated cellulose fibres

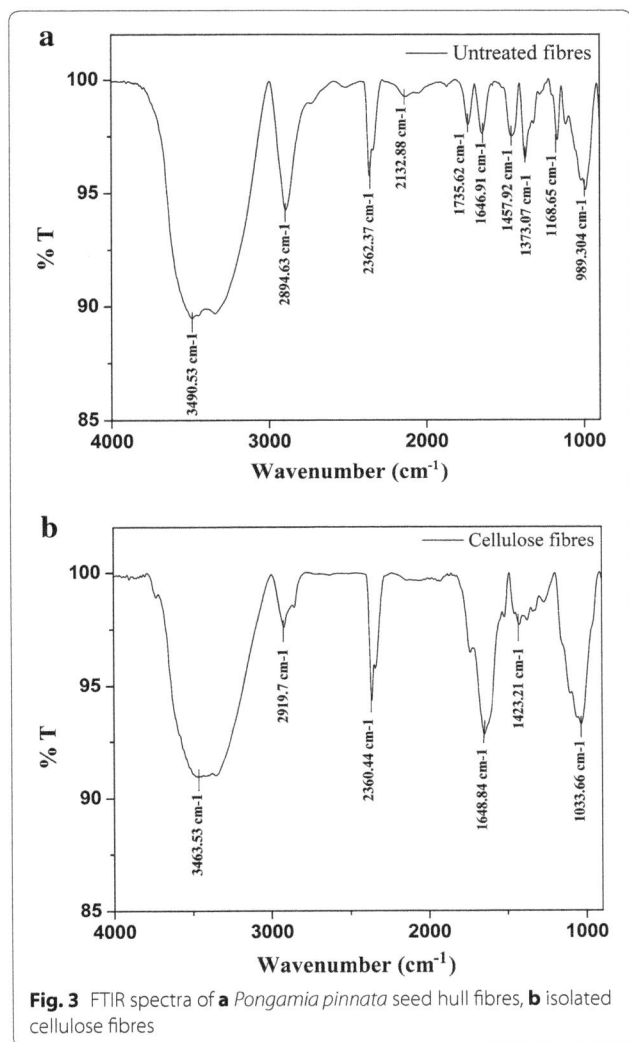

a

b

Fig. 3 FTIR spectra of **a** *Pongamia pinnata* seed hull fibres, **b** isolated cellulose fibres

2012). Finally the increase in peak 1033.66/cm, observed in isolated cellulose fibre spectra attributed to −C−O−C− pyranose ring skeletal vibration which indicates an increase in cellulose content (Sun et al. 2004a, b; Elanthikkal et al. 2010).

^{13}C NMR (CP-MAS) spectroscopy

The ^{13}C NMR spectra of untreated *P. pinnata* seed hull fibres and isolated cellulose are as shown in Fig. 4a, b. *P. pinnata* seed hull fibres spectrum in Fig. 4a, illustrates the presence of corresponding signals for the cellulose, hemicellulose and lignin, whereas in the case of isolated cellulose fibre spectrum as shown in Fig. 4b peaks of only cellulose carbon atoms were illustrated. Peaks between 107 and 60 ppm corresponding to six carbon atoms assigned to cellulose molecules are observed in both the spectra. The cellulose carbon atom peak at 107.6 is associated with C1 (Halonen et al. 2013), peaks at 77−67 ppm are assigned to C2, C3 and C5 carbon atoms (Sun et al. 2004a, d), peaks at 91.454 − 84.447 are of C4 (Bhattacharya et al. 2008) and finally 65.305 − 58 is associated with C6 carbon atom (Sun et al. 2004b, c, d). Similar observations were reported by Halonen et al. 2013, where the peaks around 109 − 101 ppm were associated with C1 atom, 80 − 68 ppm to C2, C3 and C5, 91 − 80 ppm to C4 and 68 − 58 ppm to C6 (Bhattacharya et al. 2008). In case of cellulose spectrum, the absence of peaks at 20−33 and 110−140 ppm associated with methylenes in lignin and 58.896 ppm of −OCH$_3$ groups in lignin and hemicellulose, ensures the removal of hemicellulose and lignin, the matrix components (Sun et al. 2004d; Bhattacharya et al. 2008).

Thus the removal of hemicellulose and lignin from the *P. pinnata* seed hull fibres are supported by both NMR and FTIR spectral data.

Thermogravimetric analysis (TGA)

The thermograms of untreated *P. pinnata* seed hull fibres and isolated cellulose fibres as shown in Fig. 5 have onset degradation temperature of 200 and 270 °C, respectively. The major degradation peak at around 250−350 °C observed for isolated cellulose fibre is mainly due to pyrolysis of cellulose and thermal depolymerisation of hemicellulose (Abraham et al. 2011; Li et al. 2015; Chen et al. 2011; Luduena et al. 2011), showing 75% degradation of cellulose. The increase in the decomposition temperature of the isolated cellulose fibres is related to the crystallinity of cellulose due to the removal of lignin and amorphous hemicelluloses (Abe and Yano 2009). Residual presence in both *P. pinnata* seed hull fibres and isolated cellulose fibres at 800 °C was observed to be 25 and 7%, respectively, which indicates reduction in the presence of carbonaceous materials in the nitrogen atmosphere

1648.84/cm are attributed to O−H bending of absorbed water and are observed in both the spectra; the presence of water could be related to the hydrophilic nature of cellulose component even though the samples analysed were dry (Qiao et al. 2016; Sun et al. 2004b, c; Kaushik and Singh 2011; Zhong et al. 2013; Rosa et al. 2012; Oun and Rhim 2016; Haafiza et al. 2013). Peaks at 1457.92 and 1423.21/cm are usually attributed to aromatic C=C stretch of lignin and the reduction of peak at 1423.21/cm in cellulose fibre spectra indicates the fractional delignification after the treatments (Sun et al. 2004a, b, c, d; Kaushik and Singh 2011; Elanthikkal et al. 2010; Haafiza et al. 2013). Peaks around 1373.07 and 1168.65/cm observed in *P. pinnata* seed hull fibres are assigned to C−H asymmetric deformation and C−O antisymmetric bridge stretching, respectively (Kalita et al. 2015; Sun et al. 2004a, b, c, d; Kaushik and Singh 2011; Zhong et al. 2013; Rosa et al.

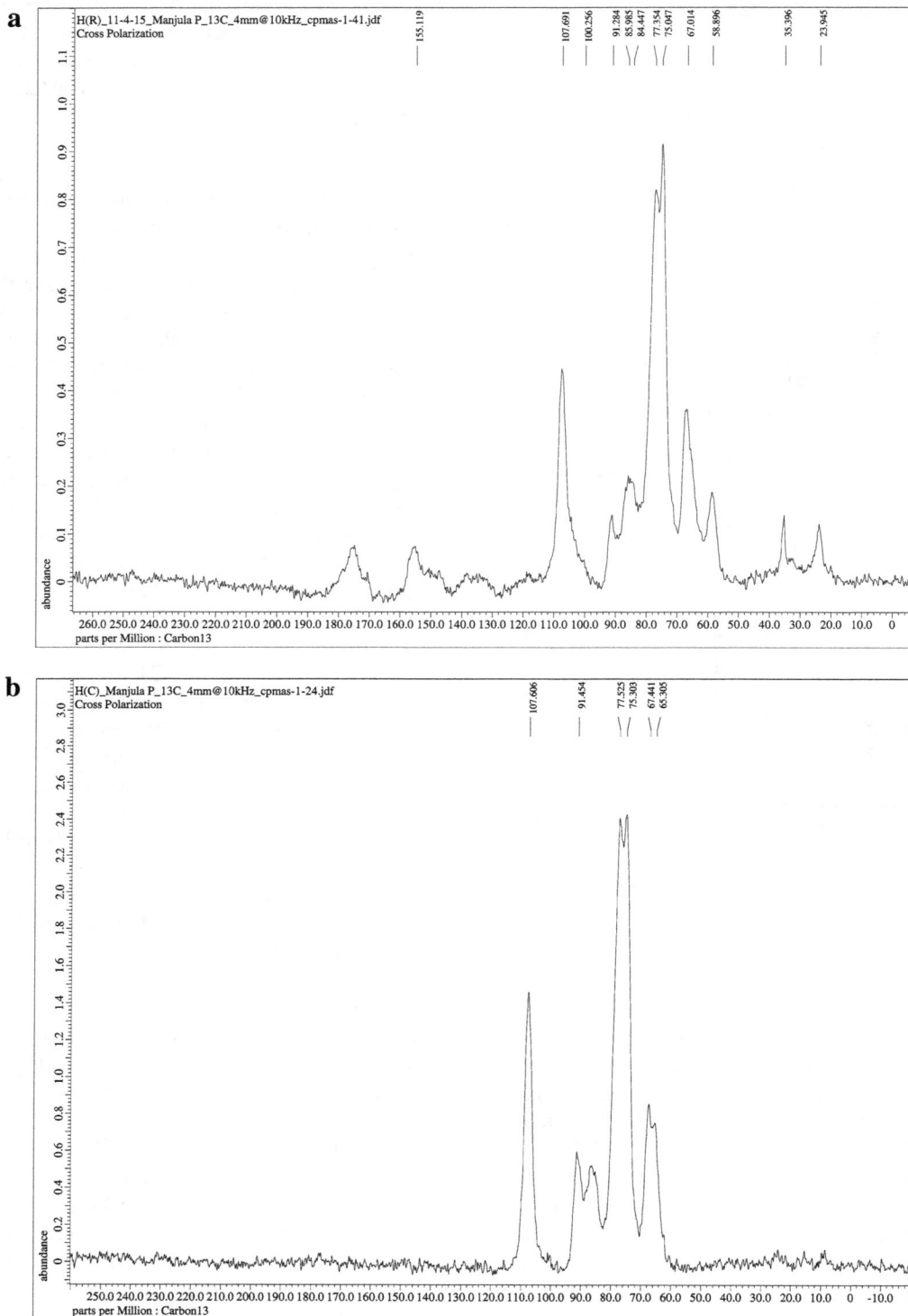

Fig. 4 The ^{13}C NMR spectra of **a** untreated *Pongamia pinnata* seed hull fibres, **b** isolated cellulose fibres

which is associated with the removal of hemicellulose (Li et al. 2015). Thus the high thermal properties perceived in case of isolated cellulose microfibres may broaden the fields of application of cellulose fibres at temperatures above 200 °C especially for biocomposite processing.

X-ray diffraction (XRD)

X-ray diffractograms of *P. pinnata* seed hull fibres and isolated cellulose fibres are presented in Fig. 6. Two peaks are observed at $2\theta = 16°$ and $22.6°$ for both the samples which is the characteristic of crystal polymorphs of cellulose I and cellulose II, respectively (Bondeson et al. 2006; Novo et al. 2015). The peak at $2\theta = 16°$ corresponds to the (110) and $2\theta = 22.6°$ corresponds to the (200). The crystallinity index (CI) obtained using Eq. (1) for *P. pinnata* seed hull fibres and isolated cellulose fibres were 27.2, and 47%, respectively. The crystallinity of the isolated cellulose microfibres was increased by 72.79%. This could be due to the presence of large amount of crystalline cellulose and removal of amorphous hemicellulose and lignin (Rosa et al. 2010) from isolated cellulose fibres by chlorination and alkaline treatment.

Thus, from the above results it can be observed that cellulose microfibres isolated from *P. pinnata* seed hull exhibited enhanced morphological, thermal and crystalline properties after chlorination and alkaline treatment. Size and increase in crystallinity of the cellulose fibres obtained from different sources and isolation methods are summarized in Table 2. The size of the fibres obtained in the present work is comparable with that obtained from other sources by different isolation methods. However, percentage increase in crystallinity for the fibres isolated from *P. pinnata* seed hull after chlorination and alkaline treatment is higher than that for the fibres isolated from other sources by chemical treatment methods obtained by other researchers. As observed in Table 2, increase in crystallinity is lower in most of the cases in spite of additional mechanical treatments. Julie et al. (2016) have obtained around 97% increase in crystallinity of the fibres isolated from Arecanut husk fibres. However, they have adopted homogenization, a mechanical process after chemical treatment. Isolation of cellulose microfibres by chlorination and alkaline treatment is economical compared to others, as enormous amount of energy is consumed in the mechanical treatments. The chlorination and alkaline treatment on *P. pinnata* seed hull resulted in the isolation of crystalline cellulose fibres of 6–8 μm diameter. It is observed that the cellulose fibres isolated from *P. pinnata* seed hull show higher percentage increase in crystallinity when compared to cellulose fibres obtained from other resources by chemical treatments. Higher crystallinity of cellulose fibres accounts to higher tensile strength of the fibres (Alemdar

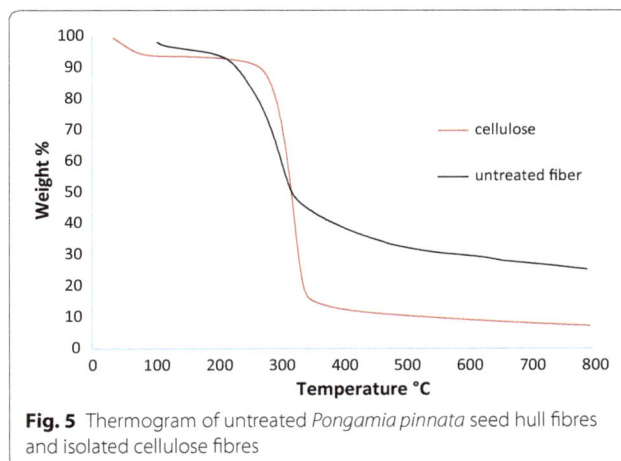

Fig. 5 Thermogram of untreated *Pongamia pinnata* seed hull fibres and isolated cellulose fibres

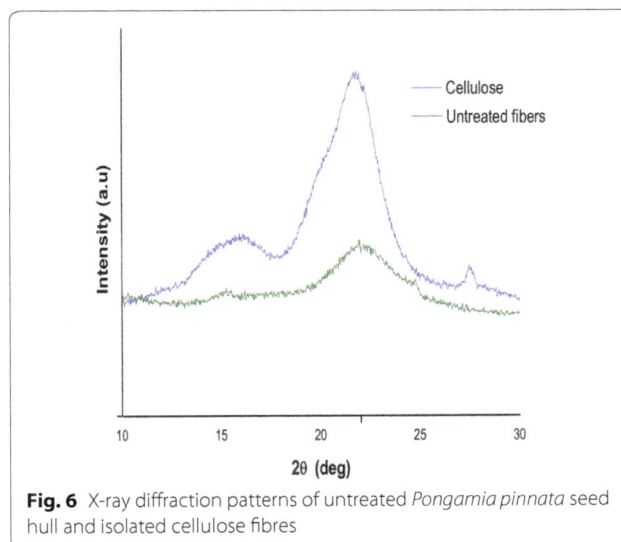

Fig. 6 X-ray diffraction patterns of untreated *Pongamia pinnata* seed hull and isolated cellulose fibres

and Sain 2008), which in turn is expected to enhance the mechanical properties of the cellulose fibre-reinforced composites.

Conclusion

Cellulose fibres were isolated from *P. pinnata* seed hull by sequential chlorination and alkaline process and the resultant microfibres were characterized by SEM, DLS, FTIR, NMR, TGA and XRD analyses. Cellulose microfibres were in diameter ranging from 6 to 8 μm and mean hydrodynamic diameter of 58.4 nm. NMR and FTIR analyses confirmed the removal of hemicellulose and lignin. Crystallinity of the fibres was increased by 72.79% after the treatment with CI of 47% for the isolated cellulose fibres. Thermal behaviour of the fibres had improved as evidenced by an increase of degradation temperature by 70 °C. Most potential observation

Table 2 Comparison of fibre size and crystallinity index (CI) of cellulose fibre isolated from different sources and isolation treatments

S. no	Source	Size (diameter) of cellulose fibres as observed under SEM	Crystallinity index (CI) isolated cellulose fibre (%)	Increase in crystallinity (%)	Treatment	Reference
1	Resak's hardwood waste	7–12 (μm)	68.1	37.33	Alkaline and acid hydrolysis	(Abidin et al. 2015)
2	Corn husk	5–8 (nm)	63.3	46.5	Alkaline, TEMPO oxidation, and homogenization	(Du et al. 2016)
3	Arecanut husk fibre	3–5 (nm)	73	97	Alkaline, acid hydrolysis, Bleaching (Chlorination), homogenization	(Julie et al. 2016)
4	Moso bamboo culms	0.5–1 (μm)	65.32	28.87	Microwave liquefaction, bleaching (Chlorination), Alkaline, homogenization and ultrasonication	(Xie et al. 2016)
5	*Gigantochloa scortechinii* bamboo culms	5.29–10.94 (nm)	65.32	36.33	Acid hydrolysis, homogenization, sonication	(Saurabh et al. 2016)
6	Sugarcane bagasse	<20 (μm)	–	–	Steam explosion, enzymatic treatment	(Saelee et al. 2014)
7	Mengkuang leaves	5–80 (μm)	69.5	26.13	Alkaline Bleaching (Chlorination), sonication	(Sheltami et al. 2012)
8	Rice husk	7 (μm)	59	26.06	Alkali, bleaching	(Johar et al. 2012)
9	Coconut palm leaf sheath	10–15 (μm)	47.7	12.7	Chlorination and alkaline	(Maheswari et al. 2012)
10	*Pongamia pinnata* seed hull	6–8 (μm)	47	72.79	Chlorination and alkaline	Present work

to be considered was the degradation temperature of the isolated cellulose fibres being higher than 200 °C, which could broaden its application potential in the fields of biocomposite processing. A notable increase in crystallinity and the dimension similar to the cellulose fibres isolated from other resources by various chemical treatments was a significant feature of the resource and the isolation method adopted in the present study. Thus the present work substantiates the success of the sequential chlorination and alkaline extraction process solely contributing to obtain smaller diameter and crystalline cellulose microfibres from *P. pinnata* seed hull. These biofibres have potential application as filler, embroiling in the process of biodegradable composites to enhance their properties.

Authors' contributions
PM, GS and KVS are the primary contributors, as this work is the result of Ph.D. work. All authors read and approved the final manuscript.

Acknowledgements
Authors gratefully acknowledge University of Agricultural Sciences, Bengaluru, India for providing *Pongamia pinnata* seed hull.

Competing interests
The authors declare that they have no competing interests.

References
Abe K, Yano H (2009) Comparison of the characteristics of cellulose microfibril aggregates of wood, rice straw and potato tuber. Cellulose 16:1017–1023

Abidin NAMZ, Aziz FA, Radiman S, Ismail A, Yunus WMZW, Nor NM, Sohaimi RM, Sulaiman AZ, Halim NA, Darius DDI (2015) Isolation of microfibrillated cellulose (MFC) from local hardwood waste, Resak (*Vatica* spp.). Mater Sci Forum 846:679–682. doi:10.4028/www.scientific.net/MSF.846.679

Abraham E, Deepa B, Pothan LA, Jacob M, Thomas S, Cvelbar U, Anandjiwala R (2011) Extraction of nanocellulose fibrils from lignocellulosic fibers: a novel approach. Carbohydr Polym 86(4):1468–1475

Alemdar A, Sain M (2008) Isolation and characterization of nanofibers from agricultural residues—wheat straw and soyhulls. Bioresour Technol 99(6):1664–1671

Bhattacharya D, Germinario LT, Winter WT (2008) Isolation, preparation and characterization of cellulose microfibers obtained from bagasse. Carbohydr Polym 73:371–377

Bondeson D, Mathew A, Oksman K (2006) Optimization of the isolation of nanocrystals from microcrystalline cellulose by acid hydrolysis. Cellulose 13:171–180

Chen Y, Liu C, Chang PR, Anderson DP, Huneault MA (2009) Pea starch-based composite films with pea hull fibers and pea hull fiber-derived nanowhiskers. Polym Eng Sci 49(2):369–378. doi:10.1002/pen.21290

Chen W, Yu H, Liu Y, Chen P, Zhang M, Hai Y (2011) Individualization of cellulose nanofibers from wood using high-intensity ultrasonication combined with chemical pretreatments. Carbohydr Polym 83:1804–1811

Cherian BM, Leão AL, de Souza SF, Thomas S, Pothan LA, Kottaisamy M (2010) Isolation of nanocellulose from pineapple leaf fibers by steam explosion. Carbohydr Polym 81:720–725

de Carvalho Mendes CA, Ferreira MS, Furtado CRG, de Sousa AMF (2015) Isolation and characterization of nanocrystalline cellulose from corn husk. Mater Lett 148:26–29

Demirbas A (2009) Heavy metal adsorption onto agro based waste materials: a review. J Hazard Mater 157(2–3):220–229

Dinand E, Chanzy H, Vignon RM (1999) Suspensions of cellulose microfibrils from sugar beet pulp. Food Hydrocolloid 13(3):275–283

Du C, Li H, Li B, Liu M, Zhan H (2016) Characteristics and properties of cellulose nanofibers prepared by TEMPO oxidation of corn husk. BioResources 11(2):5276–5284

Dufresne A, Cavaille JY, Vignon MR (1997) Mechanical behavior of sheets prepared from sugar beet cellulose microfibrils. J Appl Polym Sci 64:1185–1194

Elanthikkal S, Gopalakrishnapanicker U, Varghese S, Guthrie JT (2010) Cellulose microfibers produced from banana plant wastes: isolation and characterization. Carbohydr Polym 80:852–859

Espino E, Cakir M, Domenek S, Román-Gutiérrez AD, Belgacem N, Bras J (2014) Isolation and characterization of cellulose nanocrystals from industrial by-products of Agave tequilana and barley. Ind Crops Prod 62:552–559. doi:10.1016/j.indcrop.2014.09.017

Fiore V, Scalici T, Valenza A (2014) Characterization of a new natural fiber from *Arundo donax* L. as potential reinforcement of polymer composites. Carbohydr Polym 106:77–83

Haafiza MKM, Eichhornc SJ, Hassana A, Jawaid M (2013) Isolation and characterization of microcrystalline cellulose from oil palm biomass residue. Carbohydr Polym 93(2):628–634

Halonen H, Larsson PT, Iversen T (2013) Mercerized cellulose biocomposites: a study of influence of mercerization on cellulose supramolecular structure, water retention value and tensile properties. Cellulose 20:57–65

Horiba knowledgebase (2017) Horiba Instruments, Inc. https://www.horiba.com/fileadmin/uploads/Scientific/eMag/PSA/Guidebook/pdf/PSA_Guidebook.pdf. Accessed 05 Nov 2016

Hou X, Sun F, Zhang L, Luo J, Lu D, Yang Y (2014) Chemical-free extraction of cotton stalk fibers by steam flash explosion. BioResources 9(4):6950–6967

Hubbe MA, Rojas OJ, Lucia LA, Sain M (2008) Cellulosic nanocomposites: a review. BioResources 3:929–980

Johar N, Ahmad I, Dufresnec A (2012) Extraction, preparation and characterization of cellulose fibres and nanocrystals from rice husk. Ind Crops Prod 37:93–99

Julie CCS, George N, Narayanankutty SK (2016) Isolation and characterization of cellulose nanofibrils from arecanut husk fibre. Carbohydr Polym. doi:10.1016/j.carbpol.2016.01.015

Kalita E, Nath BK, Deb P, Agan F, Islam MR, Saikia K (2015) High quality fluorescent cellulose nanofibers from endemic rice husk: isolation and characterization. Carbohydr Polym 122:308–313

Kaushik A, Singh M (2011) Isolation and characterization of cellulose nanofibrils from wheat straw using steam explosion coupled with high shear homogenization. Carbohydr Res 346(1):76–85.

Kavitha B, Kumar KS, Narsimlu N (2013) Synthesis and characterization of polyaniline nano-fibers. Indian J Pure Appl Phys 51:207–209

Li R, Fei J, Cai Y, Li Y, Fengand J, Yao J (2009) Cellulose whiskers extracted from mulberry: a novel biomass production. Carbohydr Polym 76(1):94–99

Li W, Zhang Y, Li J, Zhou Y, Li R, Zhou W (2015) Characterization of cellulose from banana pseudo-stem by heterogeneous liquefaction. Carbohydr Polym 132:513–519

Luduena L, Fasce D, Alverez VA, Stefani PM (2011) Nanocellulose from rice husk following alkaline treatment to remove silica. BioResources 6(2):1440–1453

Maheswari UC, Obi Reddy K, Muzenda E, Guduri BR, Varada Rajulu A (2012) Extraction and characterization of cellulose microfibrils from agricultural residue—*Cocos nucifera* L. Biomass Bioenerg 46: 555–563. doi 10.1016/j.biombioe.2012.06.039. http://www.sciencedirect.com/science/journal/0961953446:555–563

Mandal A, Chakrabarty D (2011) Isolation of nanocellulose from waste sugarcane bagasse (SCB) and its characterization. Carbohydr Polym 86:1291–1299

Mangal R, Saxena NS, Sreekala MS, Thomas S, Singh K (2003) Thermal properties of pineapple leaf fiber reinforced composites. Mater Sci Eng 339:281–285

Moran JI, Alvarez VA, Cyras VP (2008) Extraction of cellulose and preparation of nanocellulose from sisal fibers. Cellulose 15:149–159. doi:10.1007/s10570-007-9145-9

Motaung TE, Mtibe A (2015) Alkali treatment and cellulose nanowhiskers extracted from maize stalk residues. Mater Sci Appl 6:1022–1032. doi:10.4236/msa.2015.611102

Nadeem R, Ansari TM, Akhtar K, Khalid AM (2009) Pb (II) sorption by pyro-
    lysed Pongamia pinnata pods carbon (PPPC). Chem Eng J 152:54–63.
    doi:10.1016/j.cej.2009.03.030
Novo LP, Bras J, García A, Belgacem N, Curvelo AA (2015) Subcritical water:
    a method for green production of cellulose nanocrystals. ACS Sustain
    Chem Eng 3:2839–2846
Oun AA, Rhim JW (2016) Characterization of nanocelluloses isolated from
    Ushar (Calotropis procera) seed fiber: effect of isolation method. Mater
    Lett 168:146–150
Qiao C, Chen G, Zhang J, Yao J (2016) Structure and rheological properties of
    cellulose nanocrystals suspension. Food Hydrocolloid 55:19–25
Reddy N, Yang Y (2005) Biofibers from agricultural byproducts for industrial
    applications. Trends Biotechnol 23(1):22–27
Reddy N, Yang Y (2009) Natural cellulose fibers from soybean straw. Bioresour
    Technol 100:3593–3598
Rosa MF, Medeiros ES, Malmonge JA, Gregorski KS, Wood DF, Mattoso LHC
    (2010) Cellulose nanowhiskers from coconut husk fibers: effect of
    preparation conditions on their thermal and morphological behavior.
    Carbohydr Polym 81(1):83–92
Rosa SM, Rehman N, de Miranda MIG, Nachtigall SM, Bica CI (2012) Chlorine-
    free extraction of cellulose from rice husk and whisker isolation. Carbo-
    hydr Polym 87:1131–1138
Saelee K, Yingkamhaeng N, Nimchua T, Sukyai P (2014) Extraction and charac-
    terization of cellulose from sugarcane bagasse by using environmental
    friendly method. In: Proceedings of The 26th Annual Meeting of the Thai
    Society for Biotechnology and International Conference, Mae Fah Lunag
    University (School of Science), Thailand, 26–29 November 2014
Saurabh CK, Mustapha A, Masri MM, Owolabi AF, Syakir MI, Dungani R, Paridah
    MT, Jawaid M, Khalil HPSA (2016) Isolation and characterization of
    cellulose nanofibers from Gigantochloa scortechinii as a reinforcement
    material. J Nanomater. doi:10.1155/2016/4024527
Sheltami RM, Abdullaha I, Ahmada I, Dufresnec A, Kargarzadeha H (2012)
    Extraction of cellulose nanocrystals from mengkuang leaves (Pandanus
    tectorius). Carbohydr Polym 88:772–779
Shin HK, Jeun JP, Kim HB, Kang PH (2012) Isolation of cellulose fibers from
    kenaf using electron beam. Radiat Phys Chem 81:936–940
Shwetha KC, Nagarajappa DP, Mamatha M (2014) Removal of copper from
    simulated wastewater using Pongamia pinnata seed shell as adsorbent.
    Int J Eng Res Appl 4(6):271–282
Srinivas CH, Srinivasu D, Kavitha B, Narsimlu N, Siva Kumar K (2012) Synthesis
    and characterization of nano size conducting polyaniline. IOSR J Appl
    Phys 1(5):12–15
Sun JX, Sun XF, Sun RC, Su YQ (2004a) Fractional extraction and structural
    characterization of sugarcane bagasse hemicelluloses. Carbohydr Polym
    56:195–204
Sun JX, Sun XF, Zhao H, Sun RC (2004b) Isolation and characterization of cel-
    lulose from sugarcane bagasse. Polym Degrad Stabil 84:331–339
Sun XF, Sun RC, Su Y, Sun JX (2004c) Comparative study of crude and purified
    cellulose from wheat straw. J Agric Food Chem 52(4):839–847
Sun XF, Sun RC, Fowler P, Baird MS (2004d) Isolation and characterisation
    of cellulose obtained by a two-stage treatment with organosolv and
    cyanamide activated hydrogen peroxide from wheat straw. Carbohydr
    Polym 55:379–391
Xie J, Hse CY, De Hoop CF, Hu T, Qi J, Shupe TF (2016) Isolation and characteri-
    zation of cellulose nanofibers from bamboo using microwave liquefac-
    tion combined with chemical treatment and ultrasonication. Carbohydr
    Polym 151:725–734
Yadav PP, Ahmed G, Maurya R (2004) Furanoflavonoids from Pongamia pinnata
    fruit. Phytochemistry 65(4):439–443
Zhong C, Wang C, Huang F, Jia H, Wei P (2013) Wheat straw cellulose dissolu-
    tion and isolation by tetra-n-butylammonium hydroxide. Carbohydr
    Polym 94:38–45
Zuluaga R, Putaux JL, Cruz J, Velez J, Mondragon I, Ganan P (2009) Cellulose
    microfibrils from banana rachis: effect of alkaline treatments on structural
    and morphological features. Carbohydr Polym 76:51–59

5

Review on the current status of polymer degradation: a microbial approach

Vinay Mohan Pathak* and Navneet

Abstract

Inertness and the indiscriminate use of synthetic polymers leading to increased land and water pollution are of great concern. Plastic is the most useful synthetic polymer, employed in wide range of applications viz. the packaging industries, agriculture, household practices, etc. Unpredicted use of synthetic polymers is leading towards the accumulation of increased solid waste in the natural environment. This affects the natural system and creates various environmental hazards. Plastics are seen as an environmental threat because they are difficult to degrade. This review describes the occurrence and distribution of microbes that are involved in the degradation of both natural and synthetic polymers. Much interest is generated by the degradation of existing plastics using microorganisms. It seems that biological agents and their metabolic enzymes can be exploited as a potent tool for polymer degradation. Bacterial and fungal species are the most abundant biological agents found in nature and have distinct degradation abilities for natural and synthetic polymers. Among the huge microbial population associated with polymer degradation, *Pseudomonas aeruginosa*, *Pseudomonas stutzeri*, *Streptomyces badius*, Streptomyces *setonii*, *Rhodococcus ruber*, *Comamonas acidovorans*, *Clostridium thermocellum* and *Butyrivibrio fibrisolvens* are the dominant bacterial species. Similarly, *Aspergillus niger*, Aspergillus *flavus*, *Fusarium lini*, *Pycnoporus cinnabarinus* and *Mucor rouxii* are prevalent fungal species.

Keywords: Polymer, Microbial degradation, Bacteria, Fungi, Natural polymers, Synthetic polymers, Polysaccharide, Hydrolytic enzyme, Pollution, Organic pollutants, Waste management, Biofilm, Surfactants, LDPE, Aerobic degradation, Anaerobic degradation, UV irradiation, Manmade compound, Plastic waste, SEM, Sturm test, FT-IR

Background

Developments in science and technology, especially over the last 2 decades, have led to the production of a number of synthetic polymers worldwide. The polymers are chains of monomers linked together by chemical bonds. Polymers such as lignin, starch, chitin, etc., are present in the environment naturally. Nowadays, synthetic polymers are used in several industries, of which packaging application covers 30% of plastic use throughout the world (Shah et al. 2008b; Dey et al. 2012; Kumar et al. 2011). In the nineteenth and twentieth centuries plastic played a revolutionary role in the packaging industries. Thereafter, approaches to transportation were changed with the introduction of carrying bags made of polyethylene (Nerland et al. 2014). Synthetic polymers are widely used

because of their durability and low cost, but disposal of packaging material has emerged as a challenge for solid waste management, and it is a major source of pollution (Song et al. 2009; Dey et al. 2012). Now such types of synthetic compounds have become a nuisance affecting natural resources like water quality and soil fertility by contaminating them (Bhatnagar and Kumari 2013; Ojo 2007; Arutchelvi et al. 2008). In the 1990s, plastic waste was found to have tripled and is continuously increasing in the marine environment (Moore 2008). The level of debris materials increased markedly from 1990 to 1995 on Bird Island of South Georgia; similarly, the garbage amount doubled in the coastline area of the UK during 1994–1998 (Walker et al. 1997; Barnes 2002). It was estimated that neuston plastic increased ten-fold between 1970 and 1980 in Japan (Moore 2008). The total demand for plastic was 107 million tons in 1993, which increased to 146 million tons in 2000. The growth rate of the plastic industry in Pakistan is 15% per annum (Shah et al.

*Correspondence: vinaymohanpathak@gmail.com
Department of Botany and Microbiology, Gurukul Kangri University, Haridwar, Uttarakhand 249-404, India

2008b). Plastic waste is being generated rapidly worldwide. The UK, China and India contribute 1 million tons, 4.5 million tons and 16 million tons, respectively (Kumar et al. 2011). India generates around 10 thousand tons of plastic waste (Puri et al. 2013). The annual production of plastic was estimated as 57 million tons in Europe in 2012. Polyethylene is one of the common forms of plastic compared to others (polyvinyl chloride, polypropylene, etc.) (Nerland et al. 2014). Plastic materials have become versatile, competitive and reliable substitutes for traditionally used metal, leather and wood materials in the past 5 decades because of their toughness, flexibility and physical properties (Sivan 2011; Singh and Sharma 2008). Durability and undesirable accumulation of synthetic polymers are major threats to the environment. Plastic waste recycling has largely unsuccessful outcomes; of the over 1 trillion plastic bags dumped per annum in the US, only 5% are recycled. Apparently, waste management (bioremediation) is one of the ways to reduce the adverse effects and can serve as a potential tool (Shah et al. 2008b; Ojo 2007; Ali et al. 2014).

In-vitro degradation of synthetic polymers is a time-consuming process (Schink et al. 1992; Bhatnagar and Kumari 2013). Production of synthetic polymers, especially polyethylene (140 million tons per annum), is causing problems with the waste management, and their consumption is increasing day by day at a rate of 12% per annum (Kumar et al. 2011; Sivan 2011; Shah et al. 2008b; Koutny et al. 2006).

Plastic waste in the form of litter enters running water in different ways according to nature and ultimately contaminates the marine environment (Obradors and Aguilar 1991). The proliferation rate of plastic materials is very fast, and the marine environment is affected by such wastes throughout the world. Plastic waste causes eight intricate problems in the marine environment: (1) plastic trash pollutes, (2) plastic entangles marine life, (3) ingestion of plastic items, (4) biodegradation of petroleum-based plastic polymers is time-consuming, (5) broken plastic and its pellets disturb the food web, (6) interference with sediment inhabitants, (7) marine litter destroying the primary habitat of new emerging life and (8) marine plastic litter causes major damage to vessels. In a 1970s study on 247 plankton samples in the Atlantic Ocean, 62% of the samples found plastic matter. Similarly, in the North Atlantic during the 1960s–1990s sampling of plankton showed a considerable increase of microscopic plastics in the marine environment (Moore 2008).

Distribution of different types of polymers

Polymers are made up from non-renewable as well as renewable feedstock. These polymers are well known for their diverse applications in industries, domestic

appliances, transportation, construction, shelters, storage and packaging practices. Such polymers are differentiated according to their chemical nature, structural arrangement, physical properties and applications as shown in Table 1 (Shah et al. 2008b; Dey et al. 2012; Kumar et al. 2011; Smith 2005).

Natural polymers

Natural polymers are found abundantly in nature in the forms of biopolymers and dry material of plants as shown in Table 2 (Leschine 1995). The constitution of the plant cell wall differs with the composition of the lignocellulosic biomass (cellulose, hemicellulose and lignin), which provides strength (Premraj and Doble 2005). Lignocelluloses play a critical role in developing plant biomass, in which cellulose, hemicellulose and lignin are the major building blocks of the natural polymer (Perez et al. 2002).

Synthetic polymers

Plastics are manmade compounds that consist of a long chain of polymeric molecules and unusual bonds, with excessive molecular mass and halogen substitutions. Nowadays plastic manufacturing involves different inorganic and organic materials, including carbon, hydrogen, chloride, oxygen, nitrogen, coal and natural gases (Shah et al. 2008b). The most widely used polymers contributing to plastic waste are low-density polyethylene (LDPE), high-density polyethylene (HDPE), polyvinyl chloride, polystyrene and polypropylene with 23, 17.3, 10.7, 12.3 and 18.5%, respectively, and the remaining 9.7% of other types of polymer (Puri et al. 2013). The polymer production in 2012 was estimated as polyethylene 30% (LLDPE and LDPE 18%, HDPE 12%), polypropylene 19%, polyvinyl chloride 11%, polystyrene 7%, polyethylene terephthalate 7% and polyurethanes 7% worldwide (Nerland et al. 2014). The sales distribution and amount in percentage of synthetic polymer consumed in North America during 1995 and 2004 are shown in Table 3 and Fig. 1, respectively (Summers 1996; Zheng and Yanful 2005).

Standards for polymer degradation

Literature and information on biodegradable products are organized by the US government, with the help of the Biodegradable Products Institute (BPI). BPI is an organization that deals with academia, industry and government bodies that encourage recycling of polymeric materials (biodegradable). Production of the biodegradable polymer involves the addition of starch and plant fiber extract. BPI provides matter to the ASTM (American Society for Testing and Materials) for assembling ASTM standards (ASTM D6400, D6866). These are the principle databases of degradation used to supervise industry. The logo for the compostable product was introduced by the USCC (US

Table 1 Types of polymer (Averous and Pollet 2012; Babul et al. 2013)

Type of polymer	Structure	R group	Structure	T_m (°C)	Application	References
Bio-based polymers						
Poly (3-hydroxyvalerate) (PHV)	Homo-polymer	Ethyl		118	Industrial, drug delivery	(Averous and Pollet 2012; Liu et al. 2014; Ojumu et al. 2004; Turesin et al. 2000; Bonartsev et al. 2007)
Poly (3-hydroxybutyrate) (PHB)	Homo-polymer	Methyl		168–182	Pharmaceutical and drug delivery	(Averous and Pollet 2012; Liu et al. 2014; Velde and Kiekens 2002; Nurbas and Kutsal 2004; Ojumu et al. 2004; Jirage et al. 2011; Turesin et al. 2000)
Poly (3-hydroxyoctadecanoate) (PHOd)	Homo-polymer	Penta decanoyl		54–55	Medicine area	(Averous and Pollet 2012; Guo et al. 2013; Dhar et al. 2008; Giudicianni et al. 2013; Bonilla and Perilla 2011)
Poly (3-hydroxyoctanoate) (PHO)	Homo-polymer	Pentyl		40–60	Medical applications	(Averous and Pollet 2012; Souza 2013; Basnett et al. 2012; Liu et al. 2011; Basnett et al. 2013)
Poly (3-hydroxydecanoate) (PHD)	Homo-polymer	Heptyl		54	Fiber industry	(Averous and Pollet 2012; Werner et al. 2014; Song et al. 1998; Rameshwari and Meenakshisundaram 2014)
Poly (3-hydroxybutyrate-co-3-hydroxyhexanoate) (PHBH$_x$)	Co-polymer	Methyl, with propyl		10%HH×120	Medical applications	(Babul et al. 2013; Averous and Pollet 2012; Chang et al. 2014; Coen and Dehority 1970; Xie et al. 2009)
Poly (hydroxybutyrate-co-hydroxyvalerate) (PHBV)	Co-polymer	Methyl with ethyl		165–175	Pharmaceutical (drug delivery)	(Nerland et al. 2014; Gerard et al. 2014; Nwachkwu et al. 2010; Hatakka 2005; Danis et al. 2015; Zembouai et al. 2014)
Synthetic polymers						
Polyethylene	Homo-polymer	Hydrogen		140–143	In wires (as insulating matter), bags	(Tobin 2010; Nakayama et al. 1991; Menon et al. 2010; Petre et al. 1999)
Polyvinyl chloride	Homo-polymer	Chorine		115–245	Leather, pipe, bottles	(Wilkes et al. 2005; Summers 2008; Summers 1996; Menon et al. 2010)
Polypropylene	Homo-polymer	Methyl		165	Fabric material, carpets	(Ruiyun et al. 1994; Tripathi 2002; Mccallum et al. 2007; Yam 2009; Perez et al. 2014; Menon et al. 2010; Petre et al. 1999)
Polyethylene terephthalate	Homo-polymer	Carboxyl and hydroxyl		280	Packaging applications, bottles, food wrappers, pipes	(Zheng and Yanful 2005; Perez et al. 2002; Liu et al. 2011; Jeffrie 1994; Kwon et al. 2009)
Polyurethane	Hetero-polymer	Isocyanate and polyol		400	Fibers, foams, paints, coating, packaging	(Zheng and Yanful 2005; Slade et al. 1964; Zafar 2013; Zembouai et al. 2014)

Table 1 continued

Type of polymer	Structure	R group	Structure	T_m (°C)	Application	References
Polystyrene	Homo-polymer	Phenyl	$\left(\!-CH_2-CH-\!\right)_n$ with phenyl group	240	Cups, containers, pharmaceutical, plates, cosmetics	(Zheng and Yanful, 2005; Sharma et al. 2000; Flavel et al. 2006; Nakayama et al. 1991; Carvalheiro et al. 2008; Mcalpine et al. 2001)
Polycarbonate	Homo-polymer	Carbonate	$\left[-O-\underset{\underset{O}{\parallel}}{C}-O-R-\right]_n$	52–150	Heat-resistant coating, optical instruments and automotives	(Sweileh et al. 2010; Koutsos 2009; Jeon and Baek 2010; Cheah and Cook 2003; Akola and Jones 2003; Scheller and Ulvskov 2010; Takanashi et al. 1982)
Nylon	Homo-polymer	Amide	$R-CONH-CH_2-CH_2-R$	190, 276	Fiber manufacturing	(Leja and Lewandowicz 2010a, b; Chao and Hovatter 1987; Wan et al. 1995; Kubokawa and Hatakeyama 2002; Hasegawa and Mikuni 2014)

Table 2 Types of bio-based polymers (Babul et al. 2013; Averous and Pollet 2012)

Microorganism based	Biotechnology based	Agro-based	
Polyhydroxyalkanoates (mcl-PHA, PHB, PHB-co-V)	Polylactides, PBS, PE, PTT, PPP	Polysaccharides and lipids (starch, cellulose, alginates)	Proteins–animal proteins (casein, whey, colagen/gelatin), plant protein (zein, soya, gluten)

Table 3 Plastic sales in North America, 1995 (Summers 1996)

Type of polymer	Billions of pounds
LDPE/LLDPE	14–16
PVC	12–14
HDPE	12–14
PP	10–12
PS	6–7
Polyester	4
PC	<2

Fig. 2 Symbol representing biodegradable grade compostable polymers (http://www2.congreso.gob.pe/sicr/cendocbib/con2_uibd.nsf/4EF8A31F2BF5D3480525772A0053CD80/$FILE/Ensayo_biodegradables_pl%C3%A1sticos_by.pdf)

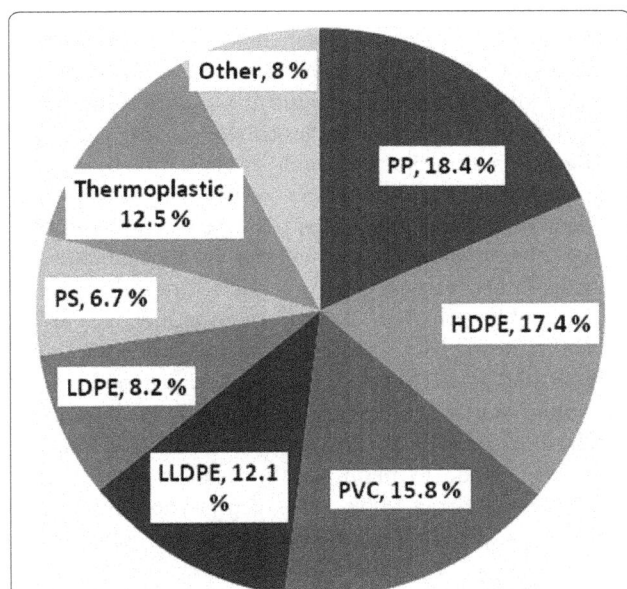

Fig. 1 Percentage distribution of synthetic polymer (*PP* polypropylene, *HDPE* high-density polyethylene, *PVC* polyvinyl chloride, *LLDPE* linear low-density polyethylene, *LDPE* low-density polyethylene, *PS* polystyrene, thermoplastics and others) sales in North America in 2004 (Zheng and Yanful 2005)

Composting Council) and BPI, shown in Fig. 2 (Kolybaba et al. 2003; http://www2.congreso.gob.pe/sicr/cendocbib/con2_uibd.nsf/4EF8A31F2BF5D3480525772A0053CD80/$FILE/Ensayo_biodegradables_pl%C3%A1sticos_by.pdf).

Diversity of polymer degradation

Living organisms are involved in the breakdown of plastic material, and consequently the recycled form reverses back to the environment. Anaerobic microbial degradation releases greenhouse gas (methane) in landfills, which increases global warming. Aerobic conditions are essential for fungal degradation while bacterial degradation proceeds in aerobic as well as anaerobic conditions (Kumar et al. 2011; Chandra and Rustgi 1998). Plastic can be reduced in an eco-friendly manner with the help of soil bacteria and proper water availability. Decomposition of the polymer depends on its chemical composition, which supports the growth of microorganisms in the form of nutrient sources. The starch-based polymer is favorable for microbial attack, and hydrolytic enzymes act on the polymer matrix to reduce their weight. Polymer made from starch or flax fiber shows greater biodegradability as compared to other synthetic polymers. Microorganisms also play an important role in the degradation of petroleum-based polymers. Petroleum-based polymers such as polyolefins are degraded through photo-degradation (Kumar et al. 2011; Sen and Raut 2015).

Emerging technology is continuously involved in improving the processing of biopolymers by using an additive (benzophenone) during their construction. Additives play a significant role in the chemical process during photo-degradation. Such amendments affect their thermal sensitivity and UV-absorbing capacities. Chemically sensitive polymers have a better biodegradability rate compared to other polymers. Similarly, thermal exposure is also involved in the breakdown of a polymer into simpler forms that increase the availability of microorganisms. Nodax is alkaline in nature and generally involves in the structural change of the polymer (Kumar et al. 2011; Augusta et al. 1993).

Renewable resources are also used in the formation of biodegradable materials. Animal and plant originated compounds are susceptible to microbial degradation (Schink et al. 1992). Development of bio-based materials is beneficial for our environment's sustainability, maintenance of greenhouse gas emissions, etc. (Song et al. 2009). This type of material's manufacturing practice plays a significant role in the environment by reducing the amount of dumped polymer waste (Leja and Lewandowicz 2010a, b). Many synthetic polymers are degraded under exposure to solar ultraviolet (UV) radiation, photo-oxidative, thermo-oxidative and photolytic reactions (Singh and Sharma 2008). In the natural environment, hydrolytic properties of seawater, oxidative properties of the atmosphere and sunlight radiation (UVB) make the polymers fragile and eventually break them into smaller pieces (Moore 2008). The American Society for Testing and Materials (ASTM) and International Standards Organization (ISO) provided the analytical protocol for plastic degradation on the basis of alteration in chemical structure and loss of physical properties of plastic (Kumar et al. 2011; http://www2.congreso.gob.pe/sicr/cendocbib/con2_uibd.nsf/4EF8A31F2BF5D3480525772A0053CD80/$FILE/Ensayo_biodegradables_pl%C3%A1sticos_by.pdf).

Biodegradation of polymers

Degradation of polymers is a process that alters the strength and color of polymeric material under controlled conditions. Disruption of the chain length initiates the primary breakdown (aging), and several external factors such as temperature and chemicals also enhance the rate of degradation. The term "aging" is used for the change in properties. It is utilized as polymer recycling, which reduces the outcome of pollution load (Kumar et al. 2011; Bhardwaj et al. 2012a). Anaerobic degradation is another way to dispose of plastic materials through landfilling (Schink et al. 1992; Shah et al. 2008b). Currently, the recycling process is increasing but the recycling rate is very low for most plastic materials because of the use of more additives in their manufacturing (Song et al. 1998). The recycling rate of thermosets is very low but thermoplastics can be easily recycled (Moore 2008). Plastic comprises 60–80% of litter, and its persistence and discharge to the surroundings create harmful effects on wildlife as well as agriculture and forest land. Furans and dioxins are persistent organic pollutants (POPs) and are formed through the burning of polyvinylchloride (PVC).

Plastic waste may come from post consumption and different stages of production (Nerland et al. 2014; Shah et al. 2008b).

Worldwide research for the last 3 decades has focused on the biodegradation of plastic (Shimao 2001). Biodegradation is compatible (microbial mineralization) compared to other waste management techniques (Schink et al. 1992). Bioremediation serves as the best way to manage waste material in an eco-friendly manner. Polluted sites are increasing constantly because of improper waste management strategies; such waste comes from industrial areas and community activities. Biological agents, both prokaryotic (bacteria) and eukaryotic (fungi, algae and plant), are involved in the bioremediation process. *Pseudomonas*, *Streptomyces*, *Corynebacterium*, *Arthrobacter*, *Micrococcus* and *Rhodococcus* are the prominent microbial agents being used for bioremediation as illustrated in Table 4 (Bhatnagar and Kumari 2013; Kathiresan 2003; Dussud and Ghiglione 2014; Shah et al. 2008a; Kale et al. 2015a, b; Grover et al. 2015; Restrepo-Flórez et al. 2014; Bhardwaj et al. 2012a).

Different steps in the polymer biodegradation mechanism

Microorganisms break down the compounds into a simpler form through biochemical transformation. Biodegradation of polymer is described as any alteration of the polymer properties such as digestion by microbial enzymes, reduction in molecular weight, and loss of mechanical strength and surface properties, in other words, the breakdown of material into fragments via microbial digestion. Degraded particles are redistributed and probably non-toxic to the environment. In nature, microorganisms form catalytic enzymes for biodegradation (Hadad et al. 2005). This approach is proficient for environmental waste management, and microorganisms involved in this process for oxidation serve as a tangible alternative mode to maintain the healthy environment (Singh and Sharma 2008). The degradation process is accomplished by microorganisms via different enzymatic activities and bond cleavage. This degradation occurs in sequential steps, bio-deterioration (altering the chemical and physical properties of the polymer), bio-fragmentation (polymer breakdown in a simpler form via enzymatic cleavage) and assimilation (uptake of molecules by microorganisms) and mineralization (production of oxidized metabolites (CO_2, CH_4, H_2O) after degradation), which are shown in Fig. 3. Mineralization of polymers takes place in both aerobic and anaerobic conditions. In the aerobic condition, CO_2 and H_2O are formed, while under anaerobic conditions, CH_4, CO_2 and H_2O are produced (Singh and Sharma 2008). The biodegradation procedure of a few polymers is known (Shimao 2001). Most of the microbial communities are

Table 4 List of microorganisms associated with polymer degradation

Type of polymer	Microorganisms	References
Polyethylene	*Brevibacillus borstelensis, Comamonas acidovorans* TB-35, *Pseudomonas chlororaphis, P. aeruginosa, P. fluorescens, Rhodococcus erythropolis, R. rubber, R. rhodochrous, Staphylococcus cohnii, S. epidermidis, S. xylosus, Streptomyces badius, S. setonii, S. viridosporus, Bacillus amyloliquefaciens, B. brevis, B. cereus, B. circulans, B. circulans, B. halodenitrificans, B. mycoides, B. pumilus, B. sphaericus, B. thuringiensis, Arthrobacter paraffineus, A. viscosus, Acinetobacter baumannii, Microbacterium paraoxydans, Nocardia asteroides, Micrococcus luteus, M. lylae, Lysinibacillus xylanilyticus, Aspergillus niger, A. versicolor, A. flavus, Cladosporium cladosporioides, Fusarium redolens, Fusarium* spp. AF4, *Penicillium simplicissimum* YK, *P. simplicissimum, P. pinophilum, P. frequentans, Phanerochaete chrysosporium, Verticillium lecanii, Glioclodium virens, Mucor circinelloides, Acremonium Kiliense, Phanerochaete chrysosporium*	(Dussud and Ghiglione 2014; Shah et al. 2008a; Kale et al. 2015a, b; Grover et al. 2015; Restrepo-Flórez et al. 2014; Bhardwaj et al. 2012a)
Polyvinyl chloride	*Pseudomonas fluorescens* B-22, *P. putida* AJ, *P. chlororaphis, Ochrobactrum* TD, *Aspergillus niger*	(Dussud and Ghiglione 2014; Shah et al. 2008a; Shah et al. 2008a; Kale et al. 2015a, b; Bhardwaj et al. 2012a)
Polyurethane	*Comamonas acidovorans* TB-35, *Curvularia senegalensis, Fusarium solani, Aureobasidium pullulans, Cladosporium* sp., *Trichoderma* DIA-T spp., *Trichoderma* sp., *Pestalotiopsis microspora*	
Poly(3-hydroxybutyrate)	*Pseudomonas lemoignei, Alcaligenes faecalis, Schlegelella thermodepolymerans, Aspergillus fumigatus, Penicillium* spp., *Penicillium funiculosum*	(Dussud and Ghiglione 2014; Shah et al. 2008a; Kale et al. 2015a, b; Bhardwaj et al. 2012a)
Poly(3-hydroxybutyrate-co-3-hydroxyvalerate)	*Clostridium botulinum, C. acetobutylicum, Streptomyces* sp. SNG9	(Dussud and Ghiglione 2014; Dussud and Ghiglione 2014; Shah et al. 2008a; Bhardwaj et al. 2012a)
Polycaprolactone	*Bacillus brevis, Clostridium botulinum, C. acetobutylicum, Amycolatopsis* sp., *Fusarium solani, Aspergillus flavus*	
Polylactic acid	*Penicillium roquefort, Amycolatopsis* sp., *Bacillus brevis, Rhizopus delemar*	(Shah et al. 2008a)

able to utilize polyester and polyurethane at a slower rate (Dey et al. 2012; Schink et al. 1992). Starch- or cellulose-based plastics are biodegradable; they degrade easily through composting, which can reduce landfilling and solve the waste management problem. Biodegradation with the help of microorganisms is an approachable way to clean up such plastic waste. Microorganisms are able to utilize synthetic polymers, but the composition of the polymer and manufacturing process need to be defined for the biological activity on the polymer material (Sivan 2011; Song et al. 2009; Leja and Lewandowicz 2010a, b; Kumar et al. 2011). Biodegradability of synthetic polymers with chemical groups that are susceptible to microbial attack can be carried out with polycaprolactone, poly-β-hydroxyalkanoates and oil-based polymers (Leja and Lewandowicz 2010a, b; Song et al. 2009). Enzymes of microbial origin are employed to control pollution and contribute to developing an eco-friendly environment. Diverse forms of microflora are known to utilize them through the mineralization process.

Microbial metabolism and physiological processing of polymer degradation

Bacteria and fungi are a widely distributed group of microorganisms that play a significant role in the processing of polymer compounds in the natural environment (Upreti and Srivastava 2003). These microorganisms are used to convert the insoluble biopolymer into a soluble biopolymer. Naturally occurring polymers consist of lipids, carbohydrates and proteins. Microbial enzymes are the ultimate source to hydrolyze only low molecular weight and soluble macromolecules. These soluble compounds are exploited by microorganisms for energy production (Gallert and Winter 2005). Microbial degradation of polymers leads to alteration of the physicochemical properties of materials. The bioconversion or degradation of biomaterials is well understood by studying the mechanical properties, degradation kinetics and recognition of the degraded products. The bioconversion process also alters the efficiency of the host response, cellular growth, material function, etc. (Azevedo and Reis 2005).

The tricarboxylic acid (TCA) cycle serves as one of the main metabolic pathways for energy generation from most of the organic compounds. In the TCA cycle, acetyl-CoA acts as the key intermediate and is exploited in cellular activities like CO_2 formation by oxidation, acetate formation, biosynthesis, etc. The major contribution of the TCA cycle is to generate ATP and provide energy to the cell. Two molecules of ATP are synthesized by oxidation of 2 mol of acetate, while 34 mol of ATP is synthesized by the electron transport chain (ETC) through phosphorylation. The metabolism and efficiency of energy production vary according to the microbial

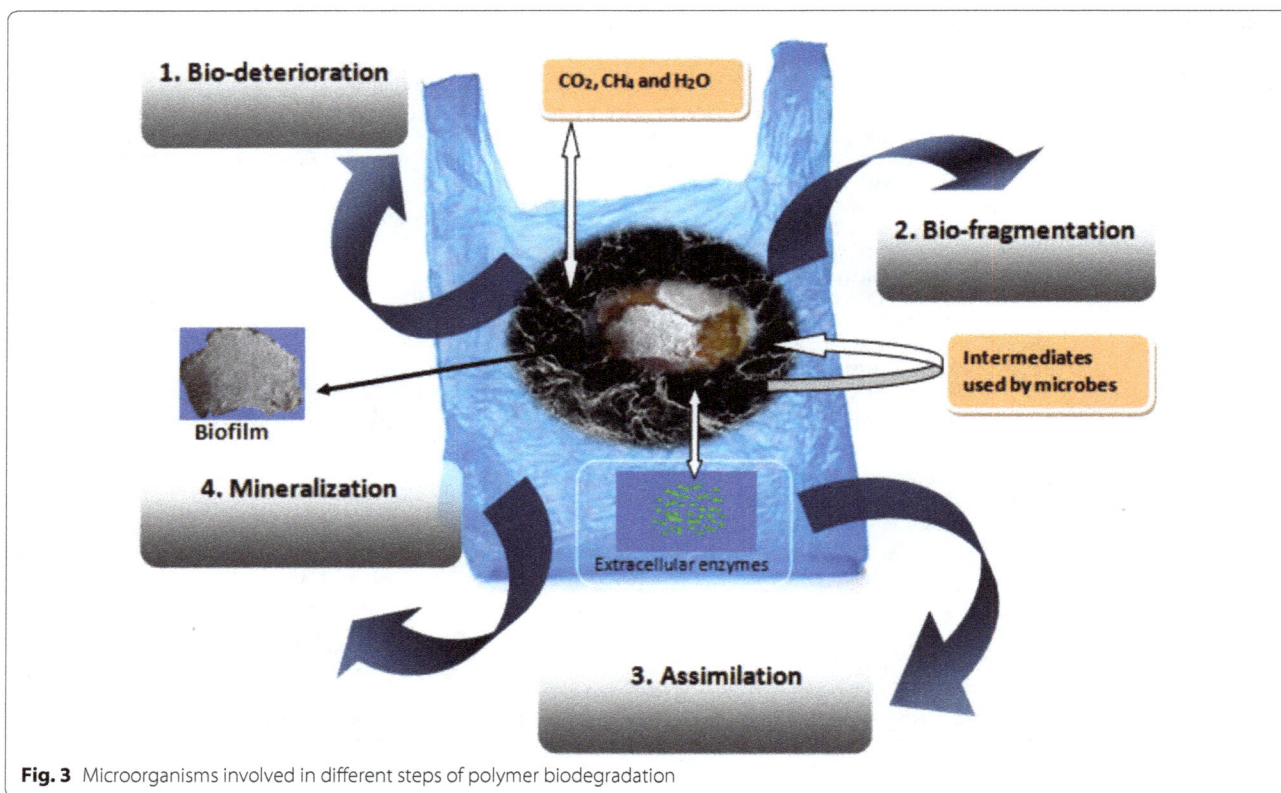

Fig. 3 Microorganisms involved in different steps of polymer biodegradation

growth conditions. Aerobic bacteria are able to respire carbohydrates, but one-third of the starting energy is not utilized by the cell and is lost in the form of heat; the remaining energy is conserved biochemically. During the processing of wastewater, activated sludge reactors lose much of their energy as heat. Under growth-limiting conditions, the ATP consumption rate is increased and less energy is available for cellular growth and metabolism (Gallert and Winter 2005).

Bode et al. (2000) investigated the physiological and chemical process of biodegradation of synthetic poly (cis-1,4-isoprene) polymer and found that two bacterial strains, i.e., *Streptomyces coelicolor* 1A and *Pseudomonas citronellolis*, were able to utilize degraded vulcanized natural rubber and synthetic poly(cis-1,4-isoprene). They observed the growth of these bacteria on polymer was better as compared to *Streptomyces lividans* 1326, and they were exploited under controlled conditions. Three degraded products have been identified from the culture suspension of *S. coelicolor* 1A, and vulcanized rubber was used as substrate. These degraded products were determined as (5Z,9Z)-6,10-dimethyl-penta-dec-5,9-diene-2,13-dione, (5Z)-6-methyl-undec-5-ene-2,9-dione and (6Z)-2,6-dimethyl-10-oxo-undec-6-enoic acid. They also

proposed the oxidative pathway for conversion of poly(cis-1,4-isoprene) into methyl-branched diketones by following different steps, i.e., aldehyde intermediate to carboxylic acid oxidation, β-oxidation, oxidation of the conjugated bond (double bond) to β-keto acid and the decarboxylation process. The authors proposed a hypothetical model for poly (cis-1, 4-isoprene) degradation as shown in Fig. 4.

Similarly, Mooney et al. (2006) studied the microbial degradation of styrene. Styrene is one of the identical xenobiotic compounds and serves as a potent toxic pollutant to the environment as well as human health. It comes from industrial practices that involve polymer and petrochemical processing. Mooney et al. (2006) reported bacterial enzymes involved in styrene biodegradation. The detailed process of styrene bioconversion to its metabolites and degraded products is shown in Fig. 5. Pyruvate, acetaldehyde 2-phenylethanol and 2-vinylmuconate are some of the degradative metabolites obtained during styrene biodegradation. Similarly, phenyl acetyl-CoA obtained via styrene degradation enters into the tricarboxylic acid (TCA) cycle. The TCA cycle thus plays a vital role in energy production, essential for cellular and metabolic events, and also produces CO_2 during the oxidation process.

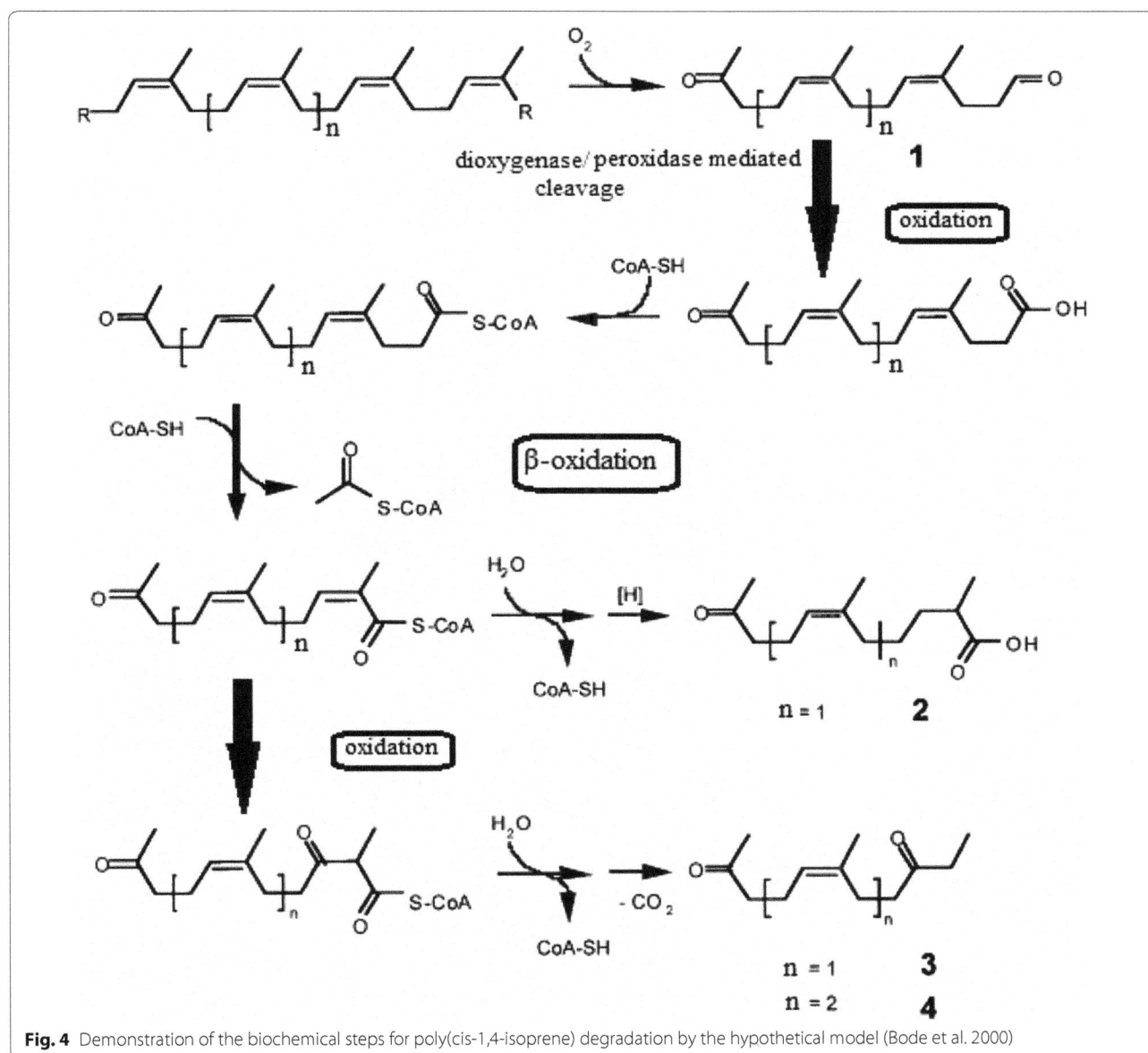

Fig. 4 Demonstration of the biochemical steps for poly(cis-1,4-isoprene) degradation by the hypothetical model (Bode et al. 2000)

Aerobic and anaerobic biodegradation of polymers

Polymer biodegradation depends on the physical and chemical properties of the polymer. Molecular weight and crystallinity are key properties of polymers that affect the biodegradation efficiency of microorganisms. The enzymes responsible for polymer degradation are categorized into two groups, i.e., extracellular depolymerase and intracellular depolymerase (Gu 2003). Exoenzymes are generally involved in the degradation of complex polymers to simple units like monomers and dimers. These are further exploited by microorganisms as energy and carbon sources as shown in Fig. 6.

Polymer degradation (mineralization) forms new products during or at the end of processes, e.g., CO_2, H_2O or CH_4. Natural polymers like cellulose, PHB and chitin are susceptible to microbial degradation and serve as biodegradable polymers (Gu 2003; Chahal et al. 1992; Brune et al. 2000). The degradation process depends on the availability of O_2. Polymer degradation accomplished under anaerobic conditions produces organic acids, H_2O and gases (CO_2 and CH_4) (Gu 2003). Under aerobic conditions, the biodegradation of the polymer forms CO_2 and H_2O in addition to the cellular biomass of microorganisms. Similarly, under sulfidogenic conditions

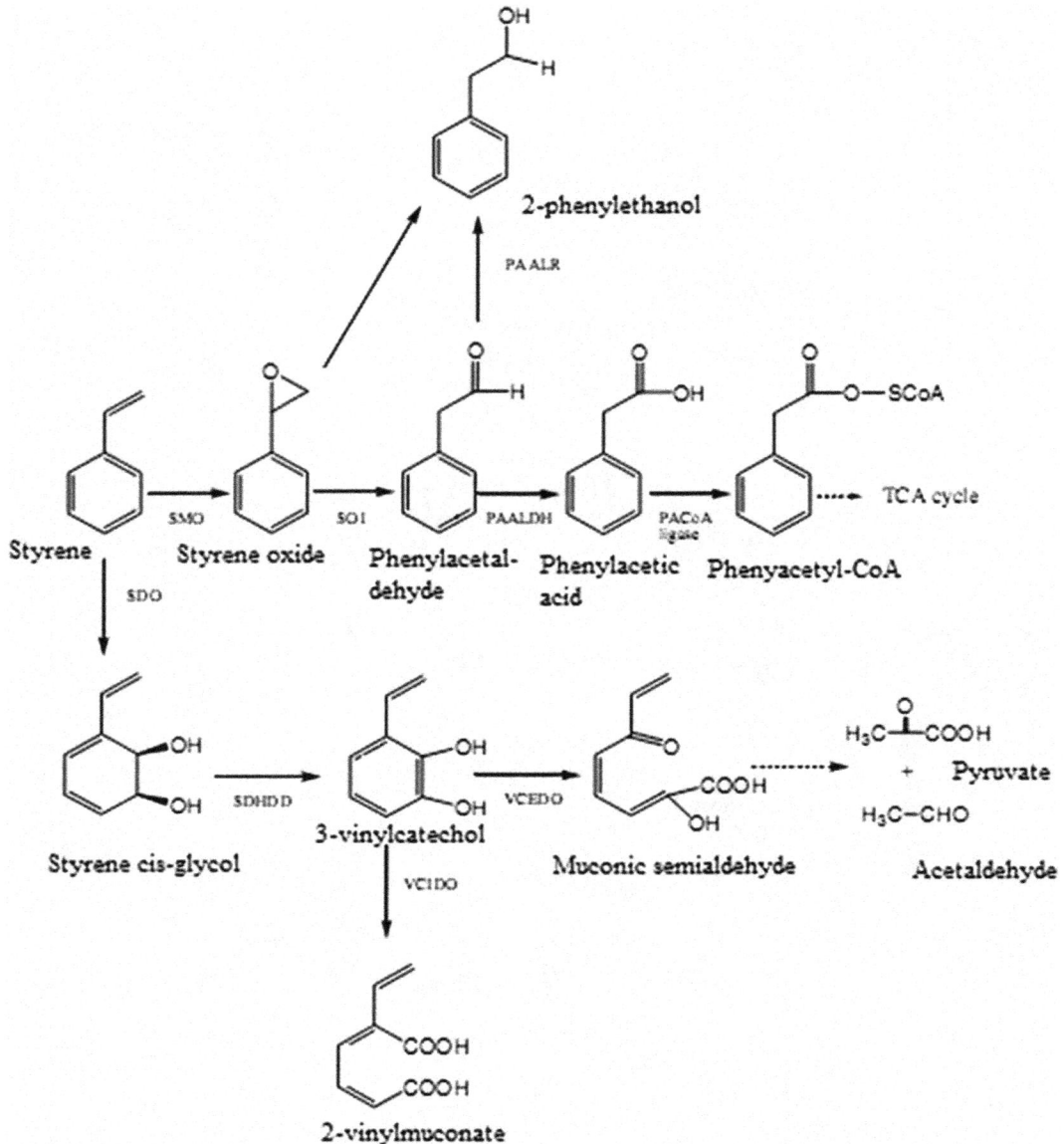

Fig. 5 Systematic diagram of the bacterial conversion of styrene and degraded products. The enzymes associated with styrene degradation are styrene monooxygenase, styrene oxide isomerase, phenylacetaldehyde dehydrogenase, phenylacetyl-CoA ligase, styrene 2,3-dioxygenase, styrene 2,3-dihydrodiol dehydrogenase, 2,3-vinylcatechol extradiol dioxygenase and 2,3-vinylcatechol intra-diol dioxygenase and designated as SMO, SOI, PAALDH, PA-CoA ligase, SDO, SDHDD, VCEDO and VCIDO, respectively (Mooney et al. 2006)

polymer degradation forms H_2S, CO_2 and H_2O (Gu and Mitchell 2006; Barlaz et al. 1989; Gu 2003; Merrettig-Bruns and Jelen 2009). The aerobic process is more efficient than the anaerobic process by means of energy production as less energy is produced in anaerobic processes because of the lack of O_2, which serves as an electron acceptor, and this is more efficient in comparison to CO_2 and SO_4^{2-} (Gu 2003).

In solid waste treatments, denitrifiers are categorized as aerobic organisms and used for nitrate or nitrite exploitation. Under anoxic conditions, nitrate or nitrite serves as a terminal electron acceptor in the metabolic respiration process. Extracellular solubilized biopolymers (natural polymers like carbohydrates, lipids and proteins) from CO_2 and H_2O form during the respiration process, while in anaerobic conditions CO_2, CH_4, NH_3,

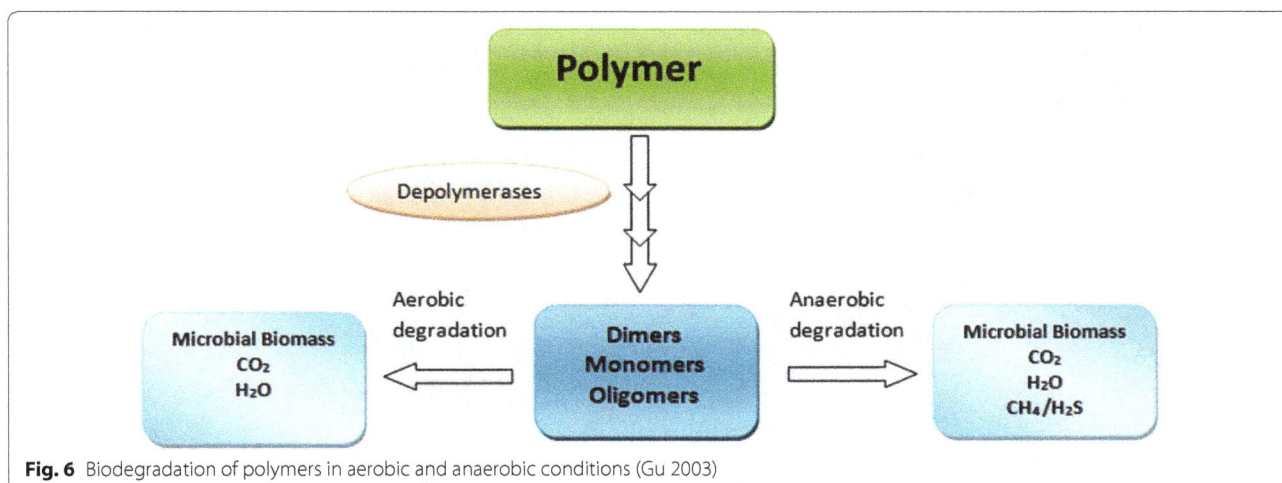

Fig. 6 Biodegradation of polymers in aerobic and anaerobic conditions (Gu 2003)

and H_2S are produced from the degradation of soluble carbon compounds via methanogens or sulfate reducers. In anaerobic conditions, if SO_4^{2-} is present, then SO_3^{2-} and CO_2 will form. On the other hand, if SO_4^{2-} is absent, then CH_4 and CO_2 will form under anaerobic conditions (Gallert and Winter 2005; Mohan and Srivastava 2011; Leja and Lewandowicz 2010a, b). A larger amount of CO_2 is produced at the lowest pH with lignolytic fungi (Kale et al. 2015a, b).

Huang et al. (2005) investigated sludge microorganisms for degradation of polyethylene glycols (PEGs) under aerobic and anaerobic conditions. They studied the effect of molecular weight (MW) on the efficiency of aerobic as well as anaerobic biodegradation processes of polyethylene glycols (MW 600, 6000 and 20,000) and found that aerobic degradation is more efficient than anaerobic degradation by means of PEG degradation abilities; 80% of biodegradability was reported in 5 days under aerobic conditions with diverse molecular weight PEGs. While the PEG biodegradation under the anaerobic condition showed 50% biodegradability of PEG 600 in 9 days, 40% biodegradability of PEG 6000 in 10 days and 80% biodegradability of PEG 20,000 obtained after 6 days of incubation. Huang et al. (2005) further investigated the nutrition effect on the degradation process under anaerobic conditions. The enriched organic media showed its positive effect on PEG 6000 biodegradation by increasing the biodegradability. Similarly, Sridevi et al. (2012) studied the metabolic pathways for phenol biodegradation and found that under aerobic conditions phenol biodegradation can form catechols and serves as intermediate products by ring cleavages at the ortho or meta position. In aerobic conditions, initially phenol is converted into catechol and then intermediates of the central metabolism by ortho or meta fission and catalyzed by catechol 1,

2 dioxygenases or catechol 2,3 dioxygenase, respectively. They found that anaerobic biodegradation of phenol is slower as compared to aerobic biodegradation. In anaerobic biodegradation the end product is CH_4 and CO_2, and under aerobic biodegradation the end product is CO_2.

Microbial development as biofilms on polymer

The growth of microorganisms is observed under different environmental conditions. These microorganisms are able to adapt to conditions as experience reflects changes, and they undergo a transition. In normal conditions, plankton is found in free swimming form, and under transition state these assemblages of microorganisms attach irreversibly to the surface and form biofilms. *Escherichia coli* O517:H7 and *Myxococcus xanthus* attach to the surface under nutrient-rich environments (O'Toole et al. 2000; Sharma et al. 2015; Simoes et al. 2010). Microorganisms in biofilms show novel phenotypic characteristics, specific mechanisms to attach to surfaces and respond to the external signals. Biofilm formation is a highly synchronized and intricate process, as shown in Fig. 7. Some microorganisms are reported to change their morphology (e.g., *Bacillus subtilis, Caulobacter crescentus*) under harsh conditions, which allows them to exist in a hostile environment (O'Toole et al. 2000). In the initial stage, microorganisms produce some proteins such as collagen, fibrin, fibronectin and laminin as a coating material, which help in cell-to-cell adhesion in biofilm (Bryers et al. 2006; Simoes et al. 2010). Biofilms exist as a matrix of extracellular polymeric substance (EPS) that depends on the surrounding conditions; it consists of a series of biological and physical changes during matrix formation and from the three-dimensional structure (Prakash et al. 2003; Bryers et al. 2006; Simoes et al. 2010).

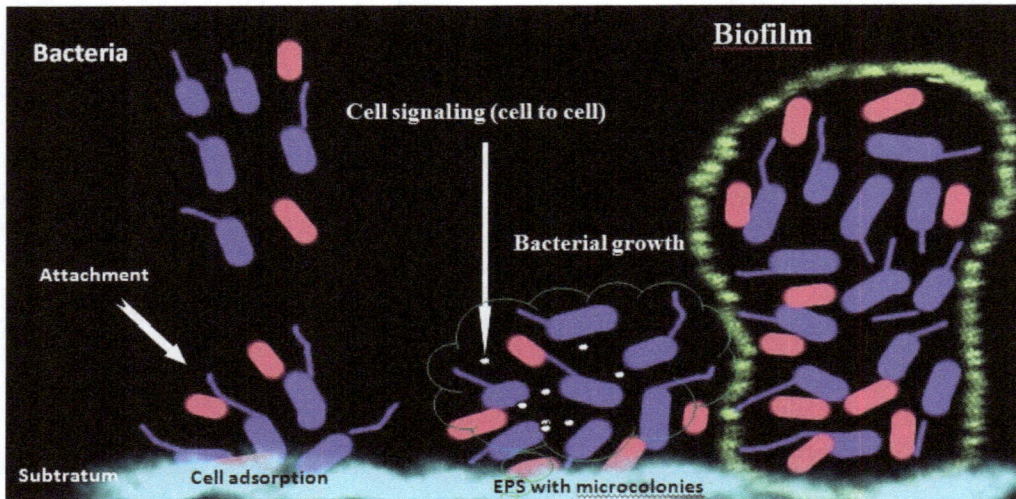

Fig. 7 Diagrammatic representations of the different stages of biofilm formation

Microbial species grow on a wide variety of surfaces, i.e., inert or living. Inert surfaces include piping systems (biofilms on noncellular materials), medical devices (biofilms composed of microorganisms with blood components) and living tissues, which have served as surfaces for the development of biofilms and consist of single or mixed populations of microorganisms. Single species of gram-positive (*Staphylococcus aureus*, *Staphylococcus epidermides* and enterococci) and gram-negative (*E. coli*, *Pseudomonas aeruginosa*, *Pseudomonas fluorescens* and *Vibrio cholerae*) bacteria were widely reported for biofilm formation. However, multiple species populations of microorganisms have been reported preponderantly in biofilms (O'Toole et al. 2000; Prakash et al. 2003). Microorganisms that form biofilms show elaborate growth by increasing their antibiotic resistance through gene regulation. Microorganisms reflect unique inhabitance in biofilms with enhanced defense mechanisms against therapeutic agents and drugs, and they relieve their survival under sessile conditions (osmotic stress, desiccation, UV radiation and pH changes) (Prakash et al. 2003; Sharma et al. 2015; Bogino et al. 2013; Simoes et al. 2010). Some dyes are used to visualize biofilm formation, i.e., safarinin and crystal violet. The term surface attachment defective (*sad*) is used for the group of mutant microorganisms unable to form biofilms (O'Toole et al. 2000). Heterogeneity of the biofilm matrix is one of the distinctive features that maintain the nutritional requirement of biofilms. The matrix is composed of polysaccharides, macromolecules and water, which provide heterogeneity and form enclosed exopolysaccharides in cells; these structures protect the biofilms from the external environment (O'Toole et al. 2000; Prakash et al.

2003). In biofilms, microorganisms are hugely differentiated compared to free-swimming organisms and form multicellular (microcolony) complex structures with interstitial voids (Sivan et al. 2006). Microorganisms are present on the upper layer of biofilms and actively dividing compared to microbes in proximity to the surface and also differ in metabolism. Interstitial voids (water channels) in the matrix help to separate the microbial colonies and facilitate diffusion of nutrients, water, gases, enzymes, waste, signals, etc., throughout the biofilms. The heterogeneity in a matrix is visible in both kinds of biofilms, i.e., single species populations of microbial biofilms and mixed species populations of microbial biofilms (Prakash et al. 2003; Bogino et al. 2013). The structural components, i.e., curli fimbriae, extracellular polymeric substances, outer membrane proteins, flagella and pili, hugely participate and are important for biofilm formation. The relieve form biofilms of flagellated motile bacteria compared to non-motile bacteria and other structural components such as curli and mannose sensitive type I, IV pili (TfP) encoding genes also play an important role in cell-cell or cell-surface attachments. Type IV pili play a significant role in bacterial motility on the surface and are important for quorum sensing. OmpR is reported to be a functional gene that promotes biofilm formation (Prakash et al. 2003; O'Toole et al. 2000; Bogino et al. 2013).

Biofilm formation is a highly regulated event, and *alg*C was reported as an alginate-producing gene that enhances the biofilm formation potential by fourfold and is positively regulated by the sigma factor (Kokare et al. 2009). Similarly, some autoinducers play a significant role in biofilm formation and have intra- and inter-species communication, e.g., AI 1, AI 2 and N-acetyl homoserine lactones (*Vibrio fisheri* or *Aliivibrio fisheri*). Such cellular communication accomplished in addition to quorum sensing signaling systems, i.e., the Rh1I/Rh1R system observed in *P. aeruginosa* and the AfeI/AfeR system in *Acidithiobacillus ferroxidans* (Vu et al. 2009).

In addition to extracellular appendages, cellular hydrophobicity plays a key role in surface attachment and biofilm formation (Simoes et al. 2010). Synthetic polymers are resistant to water because of their hydrophobic properties and non-polar nature that prevents water absorption (Khoramnejadian 2013). Usually non-polar polymers show limitations in biodegradation because of their hydrophobic surface properties, and the formation of biofilms overcomes this barrier. Microorganisms are easily attached to non-polar surfaces (plastics and Teflon), which support microbial attachment compared to hydrophilic materials. Non-polar polymer shows an enhanced surface hydrophobicity. The fimbrial structure contains hydrophobic amino acid and mycolic acid of gram-positive bacteria, and the fimbriae are important in establishing the cell attachment to the hydrophobic surface (Prakash et al. 2003; Sarjit et al. 2015). The hydrophobic nature of the surface is one of the major concerns during synthetic polymer degradation and microbial attachment to the surface.

In gram-negative bacteria O-antigen has a significant role in governing the hydrophilic properties that hinder the cell attachment to the hydrophobic surface. Inversely, O-antigen of lipopolysaccharides and extracellular polymeric substance (EPS) promotes the cell attachment to the hydrophilic surface (Prakash et al. 2003). Mutants for O-antigen also showed biofilm attachment on the plastic surface, e.g., *Bradyrhizobium japonicum* mutant (Bogino et al. 2013). In addition to fimbriae, other structures such as the EPS serve as the primary material in the biofilm matrix; it is a hydrated structure and chemically found as colonic acid and alginate in *E. coli* and *P. aeruginosa*, respectively (Prakash et al. 2003). Biofilm formation is important for degradation in the natural environment. Biofilm formation on the surface of synthetic polymers facilitates biodegradation (Sivan et al. 2006). Low-density polyethylene (LDPE) serves as the most widely used synthetic polymer. Still, many microorganisms (bacteria, fungi) are known for polymer degradation. Tribedi and Sil (2013a) reported that the degradation of LDPE is done by the *Pseudomonas* sp. AKS2 strain; it forms a biofilm on the surface of LDPE. Microbial adaptation is a key factor for bioremediation. Biofilm formation depends on the adaptable nature of *Pseudomonas* sp. AKS2 and shows positive results for LDPE degradation by means of increased hydrolytic activity up to 31%, and the cellular surface hydrophobicity increases by

about 26% (Tribedi et al. 2015). Tribedi and Sil (2013a) reported 5 ± 1% LDPE degradation with *Pseudomonas* sp. AKS2 and enhanced LDPE biodegradation were achieved with biofilm formation. Tribedi and Sil (2013a) conducted an experiment on hydrophobicity and found that Tween 80 had negative effects on hydrophobicity. Sivan et al. (2006) isolated the biofilm-producing bacterial strain *Rhodococcus ruber* C208 and reported a 0.86% per week biodegradation rate. Orr et al. (2004) reported earlier that mineral oil accelerated the degradation rate up to 50% with *R. ruber* C208 and enhanced the biofilm potential by increasing microbial colonization on polyethylene surface. Similarly, *Penicillium frequentans* and *Bacillus mycoides* exhibited biofilm formation and polyethylene degradation. *B. mycoides* colonized in a mycelial network formed by *P. frequentans* on a polyethylene surface. Degradation was determined by percent weight loss, CO_2 evolution and gas chromatography (Sangale et al. 2012; Arutchelvi et al. 2008). Metabolic activities of cells in biofilms were studied with the help of fluorescein diacetate (FDA) analysis and ATP assay (Arutchelvi et al. 2008). The development of biofilms was observed and its three-dimensional hydrated microstructure and metabolism examined under confocal scanning laser microscopy with fluorescent in situ hybridization and 16-23S rRNA hybridization (Kokare et al. 2009). Biofilm formation also played an important role in biocontrol. Gene transfer was seen within the biofilms of *Streptococcus* sp. as conjugative transposons and confers the antibiotic (tetracycline) resistance.

Cellulose

The terrestrial system is known for the plentiful production of cellulose biopolymers. The occurrence of cellulose in decaying microorganisms takes place in sundry environments (soil, aquatic, compost, anaerobic digestors and also the carbon cycle) (Leschine 1995). Several microorganisms (bacteria and fungi) were reported to have cellulose-degrading activity, and such cellulose degrading microbes exist in nature with non-cellulolytic microorganisms. Biodegradation of cellulose proceeds under aerobic (form CO_2 and H_2O) as well as anaerobic (form CO_2, CH_4 and H_2O) conditions (Perez et al. 2002; Nwachkwu et al. 2010). Cellulose is composed of β-1,4 glycosidic linkages between D-glucose subunits. It is found in different forms such as crystalline and amorphous (cellulose chains are non-organized and less resistant to enzymatic degradation) (Perez et al. 2002). Cellulases are the enzymes responsible for the breakdown of β-1,4 glycosidic bonding of cellulose. These cellulases are classified as endoglucanases (EGs or endo-1, 4-b- glucanases) and cellobiohydrolases (CBHs or exo-1,4-b-glucanases) that hydrolyze internal bonding and the end of the chains, respectively. Cellobiose molecules generated from cellulose by the endoglucanases and cellobiohydrolases are hydrolyzed by β-glucosidases and releases two units of glucose. CBHs act on crystalline cellulose effectively. Breakdown of cellulose makes carbon available to microorganisms. *Pseudomonas, Streptomyces* and *Cellulomonas* are reported for cellulase activity (Perez et al. 2002; Leschine 1995; Souza 2013; Kameshwar and Qin 2016). Fungal degradation of cellulose was reported with members of Basidiomycetes and Chytridiomycetes (Souza 2013). Fungi are also reported to have cellulase activity, including *Phanerochaete chrysosporium* and *Trichoderma reesei*. Anaerobic biodegradation of cellulose is very complex compared to aerobic degradation (Leschine 1995; Souza 2013; Kameshwar and Qin 2016). Larroque et al. (2012) studied CBM1 (carbohydrate-binding module) containing proteins from fungi and Oomycetes (*Phytophthora parasitica*). In the case of Oomycetes, CBM1 is involved in plant immunity by adhesion to polysaccharide (cellulose) substrates, while CBM1 of fungi is reported for cellulose degradation.

Anaerobic degradation contributes 5–10% of the total cellulose-degradation. In the case of anaerobic degradation of gram-positive bacteria, *Clostridium thermocellum* was reported to have cellulolytic activity, and cellulosome is a functional hydrolytic enzyme unit. *Clostridium thermocellum* has an excellent cellulose degradation ability in cellulosomal and free-enzyme systems. Carbohydrate-active enzymes (CAZymes) play an important role in their activity and form large protein complexes (primary and secondary scaffoldings) that bond with the cell wall of bacteria. Xu et al. (2016) studied the "cell-free" cellulosomal system and characterized two types of cellulosomal systems by using mutants with scaffolding gene deletions. CipA (primary scaffolding) has a significant role in cellulose degradation compared to other scaffoldings. Xu et al. (2016) also reported that primary scaffolding (CipA) plays a key role in *C. thermocellum* for cellulose degradation.

Pretreatment of cellulose material also makes it susceptible to hydrolysis by reducing the barrier [chemical and physical (UV)] effects (Jonsson and Martín 2016; Arutchelvi et al. 2008). Pretreatments were employed in milling, acid, alkali and other treatments (Karimi and Taherzadeh 2016). Some cellulase genes (9, 45, GH5 family gene) were reported in insects belonging to the taxa Crustacea, Gastropoda and Annelida (Cragg et al. 2015). Blouzard et al. (2007) reported a hydrolytic system for cellulose degradation in *Clostridium cellulolyticum* and found that hydrolase belongs to the GH2, GH9, GH10, GH26, GH27, and GH59 gene families and cellulases encoding the gene from the GH9 family. GH9 enzymes were purified from *E. coli* and characterized from one of the enzymes,

Cel9 V, belonging to this family, which showed similarity with cellulosomal cellulase, i.e., Cel9E (Ravachol et al. 2014). Berlemont and Martiny (2013) examined 21,985 genes in 5123 bacterial genomic sequences that encode for cellulases and found 24% potential opportunistic strains with 56% β-glucosidases activity throughout all sequences.

Maamar et al. (2004) studied the mutant strain *C. cellulolyticum cip CMut1* and concluded that the cellulose (crystalline) hydrolyzing genes were located on the 'cel' cluster that encodes for cellulases. Lopez-Mondejar et al. (2016) isolated *Paenibacillus* O199 reported for cellulose deconstruction and studied the O199 enzyme system. The researcher suggested the exploitation of *Paenibacillus* O199 as a second-generation technology for agricultural waste (lignocellulosic material) management strategies and biofuel production.

Dimarogona et al. (2012) reported polysaccharide monooxygenase (PMO) (or GH61) enzymes that enhanced the degradation capacity of cellulases; external electrons are important to accelerate the PMO activity. Some hydrolytic enzymes also play a key role in cellulose degradation, e.g., lytic polysaccharide monooxygenases (LPMOs), CBM33 (carbohydrate-binding module family 33) and cellobiose dehydrogenases (CDHs). Similarly, Zhang et al. (2016) reported carbohydrate-binding module 1 (CBM1) for cellulose degradation; it contains cellulases and is found in coprophilic *Podospora anserine* (coprophilous fungus) and *P. chrysosporium* (typical rot fungus).

Hemicellulose

Hemicellulose is a linear and branched heteropolymer polysaccharide that consists of sugars like D-xylose, L-arabinose, D-glucose, D-galactose, D-mannose, D-glucuronic acid, D-galacturonic acid and 4-O-methyl-glucuronic acid, which are linked to each other by β-1,4 glycosidic bonds or rarely by β-1,3-glycosidic linkage (Perez et al. 2002; Leschine 1995; Werner et al. 2014; Giudicianni et al. 2013). Hemicelluloses are found in primary as well as secondary walls of the plant and contribute one-third of the total dry weight of wood (Perez et al. 2002; Jeffrie 1994). In combination with cellulose, it makes up the major part of the plant; thus, it is known as cellulosics or cellulosic waste, which comes from agricultural practices (Petre et al. 1999). Xylans are the most common and abundant hemicelluloses; they are found in hardwood and softwood parts of angiosperms and gymnosperms, respectively (Horn et al. 2012; Hatakka 2005). Xylan polysaccharide consists of a β-1,4-linked xylose backbone with side groups like acetate, arabinofuranose in high amounts and 4-O-methyl-glucuronic acid in lesser amounts (Scheller and Ulvskov 2010; Jeffrie

1994; Menon et al. 2010; Werner et al. 2014; Carrier et al. 2012). Higher xylan degradation was found with carbonic acid and was easier to hydrolyze enzymatically compared to cellulose (Carvalheiro et al. 2008; Perez et al. 2002; Horn et al. 2012). Xylanases remove the side chain wall by attacking the backbone (Jeffrie 1994). Xylan fermentation is accomplished by strains of *Ruminococcus flavefaciens*, *Ruminococcus albus* and *Clostridium* sp. (Coen and Dehority 1970). Xylanases exist in two forms: endo 1,4 β-xylanase and xylan 1,4 β-xylosidase. *P. chrysosporium* is a white-rot fungus known to produce hemicelluloses that degrade endo-xylanases, acetyl xylan esterases and mannosidases. Heat-stable endo-xylanases are drived from thermophilic fungi, e.g., *Thermoascus aurantiacus* and *Talaromyces emersonii*, while *Butyrivibrio fibrisolvens* (ruminal bacterium) and *Bacillus sterothermophilus* possess β-xylosidase. *Bacillus* spp. produces two types of alkaline xylanases, i.e., xylanase N and xylanase A, with an active pH range of 5, 7 and 11, respectively (Perez et al. 2002; Jeffrie 1994; Kameshwar and Qin 2016). Some helper enzymes like xylan esterases, α-1 arabinofuranosidases and α-4 O-methyl glucuronosidases are required for complete degradation (Perez et al. 2002). Endo-xylanase works in combination with acetyl xylan esterase for efficient xylan degradation; these acetyl xylan esterases are derived from *Aspergillus niger*, *Schizophylum commune*, *Rhodotorula mucilaginosa*, *Trichoderma reesei* and *Fibrobacter succinogenes*. Similarly, *Streptomyces* spp. are also reported for α-O-methylglucuronidases production. *Phytophthora* also produces hydrolytic enzymes that target the hemicellulose (Jeffrie 1994; Souza 2013; Cragg et al. 2015). *Mastotermes darwiniensis* gut contains approximately 106–107 xylan degrading bacteria (Cazemier 1969). Glycoside hydrolases (GHs) are the key enzymes for hemicellulose degradation by the breakdown of the glycosidic linkages and remove the phenolic, methyl and acetyl esters. Lytic polysaccharide monooxygenases also reported this on hemicelluloses (Cragg et al. 2015). In *T. reesei*, 200 genes encode for glycosyl hydrolases and are involved in the degradation of hemicellulose and cellulose (Saloheimo and Pakula 2012). Sun et al. (2012) identified 353 genes and 34 proteins (secretary) from *Neurospora crassa*, which were involved in the breakdown of hemicellulose in the plant cell wall by hydrolytic enzymes. They reported TF and XLR-1 (xylan degradation regulator 1) played a central role in hemicellulose biodegradation by regulating some of the hemicellulase-encoding genes in *N. crassa* and orthologs to genes found in *Trichoderma* and *Aspergillus*, i.e., XYR1 and XlnR, respectively. In addition to the five hemicellulases reported in *C. thermocellum*, *C. fimi* and *Nasutitermes species*, they also have genes involved in the hydrolysis of hemicellulose (French 2009; Kameshwar and Qin 2016).

Gene prediction analysis showed 587 gene-encoded enzymes involved in hemicellulose and pectin hydrolysis and 259 ORFs reported for gene (endo-xylanases gene) expression for hemicellulose degradation (Kameshwar and Qin 2016).

Lignin

Lignin is a relatively water-insoluble aromatic heteropolymer complex. It is a phenolic polymer that provides strength to the plant cell wall via increased internal bonding. With its distinct characteristics, it confers resistance against biological stresses (Perez et al. 2002; Leschine 1995; Souza 2013; Petre et al. 1999; Hatakka 2005; Werner et al. 2014). It consists of three monolignol phenyl propionic alcohols, i.e., coniferyl, sinapyl and p-coumaryl. These alcohols present in lignin in the form of guaiacyl, syringyl and p-hydroxyphenyl propanol (Perez et al. 2002; Horn et al. 2012; Giudicianni et al. 2013; Carrier et al. 2012). Softwood lignin has coniferyl alcohol or guaiacyl propanol as the principal component, while lignin of hardwoods consists of guaiacyl and syringyl propanol as the dominant constituents (Perez et al. 2002; Horn et al. 2012; Hatakka 2005). Lignin present in grass contains all three types of propanol units. Polymerization of lignin consists of carbon-carbon linkages, aryl ether linkages and an aryl glycerol β aryl ether structure. At the stage of cell wall maturation, lignin polymers accumulate with a carbohydrate cross-linked complex (Perez et al. 2002; Jeffrie 1994). Degradation of lignin polymer is carried out by specific extracellular enzymatic systems, such as oxidoreductases (Perez et al. 2002; Horn et al. 2012). Alkaline-based solubilization is also effective in lignin degradation (Carvalheiro et al. 2008). *P. chrysosporium*, *Chrysonilia sitophila*, *P. radiata* and *Streptomyces* sp. are well-recognized producers of peroxidases, which are effective against lignin degradation (Hatakka 2005; Jeffrie 1994). Instead, the well-studied *P. chrysosporium*, *Sporotrichum pulverulentum* and *P. radiata* were also reported for lignin biodegradation (Hatakka 2005; Souza 2013). *P. radiata*, *Coriolis versicolor*, *P. chrysosporium*, brown-rot fungus (*Postia placenta*) and white-rot fungi (*Cyathus bulleri* and *C. cinnabarinus*) produce the extracellular protein that increases the OH groups and makes it susceptible to degradation (Jeffrie 1994; Souza 2013). These hydrolytic enzymes are able to degrade β-aryl ether and methylated lignin (Jeffrie 1994). Manganese peroxidases and laccases are also known to degrade lignin in addition to lignin peroxidases. Manganese peroxidases are synthesized by *T. versicolor*, *Ceriporiopsis subvermispora*, *Phanerochaete radiate*, *P. tremellosa* and *P. chrysosporium*. Similarly, *D. squalens* and *Rigidoporous lignosus* are the known organisms that synthesized both enzymes, i.e., manganese peroxidases and laccases (Hatakka 2005;

Jeffrie 1994; Souza 2013). *Pyrococcus* (archaeon) having GH12 endoglucanase is reported for lignocellulose degradation at elevated temperature (Cragg et al. 2015). Basidiomycetes (white-rot fungi) and actinomycetes are known for lignin biodegradation by forming peroxidases; these hydrolytic enzymes are accountable for lignin degradation (Souza 2013). Junnarkar et al. (2016) studied the lignin peroxidase (LiP) production at 39 °C by using *P. chrysosporium* MTCC 787. *Enterobacter lignolyticus* SCF1 was reported as a novel strain for anaerobic lignin degradation, and lignin was used as the carbon source (Fisher and Fong 2014). In biotechnological studies *C. subvermispora* (Polyporales) is widely used for selective lignin biodegradation. *Agaricales* species also serve as a prominent lignin biodegrader. Fernandez-Fueyo et al. (2016) reported lignin degradation with *Pleurotus ostreatus*. According to Ma et al. (2016), *P. ananatis* Sd-1 produces ligninolytic relevant enzymes in addition to cellulases and hemicellulases. Pereira et al. (2016) studied the catabolic pathway for β-aryl ether subunits of lignin. This catabolic pathway identified in *Sphingobium* sp. SYK-6 NAD-dependent dehydrogenases and glutathione-dependent lyase is important for the β-ether degradation pathway by approaching the early and late hydrolytic enzymes. Lignin peroxidases (LiPs) and versatile peroxidases (VPs) are able to hydrolyze the non-phenolic configuration of lignin and laccases and affect the linkage of β-aryl ether bonds (Álvarez et al. 2016). Kato et al. (2015) conducted the anaerobic degradation of lignin-derived aromatics (syringate and vanillate) by methanogenic microorganisms.

Polyethylene succinate

Polyethylene succinate (PESu) is an aliphatic polyester polymer and is synthesized by the polycondensation of succinic acid and ethylene glycol. In the polycondensation method, initially succinic acid and ethylene glycol undergo esterification in the presence of catalyst (tetrabutoxytitanium) and then polycondensation with a heat stabilizer (polyphosphoric acid) (Chrissafis et al. 2006; Seretoudi et al. 2002). Such polyethylene succinate polymers are hydroxyl terminated elastomers (Shanks and Kong 2012). Wu and Qiu (2012) reported that polyethylene succinate is miscible with other polymers, e.g., polyvinyl phenol, polyethylene oxide and polyhydroxybutyrate. Polyethylene succinate has a melting point of 103–106 °C, glass transition at −11.5 °C and cold crystallization at 55 °C, showing similarity with non-biodegradable polymers like polypropylene (Chrissafis et al. 2005; Tribedi and Sil 2013c). The crystallization nature of polyethylene altered the biodegradation rates of PESu (Seretoudi et al. 2002; Qiu et al. 2003).

Polyethylene succinate serves as a biodegradable polyester; it contains ester bonds, which are hydrolyzable in nature and susceptible to microbial attack. The *Pseudomonas* sp. AKS2 strain was reported for polyethylene succinate degradation due to its enhanced cell surface hydrophobicity, which is serving as a determining factor for surface attachment. *Pseudomonas* sp. AKS2 cell surface hydrophobicity is important to the interaction between the cell and hydrophobic surface of polyethylene succinate (Liu et al. 2012; Qiu et al. 2003; Tribedi and Sil 2013b). Biostimulation (in situ bioremediation) has constructive effects on PESu degradation by establishing a new community and shows a founder effect (Tribedi and Sil 2013c).

Thermophilic actinomycetes were used for polyethylene succinate degradation, and *Microbispora* were able to degrade polyethylene succinate film in 6 days at 50 °C (Duddu et al. 2015; Seretoudi et al. 2002; Hoang et al. 2007). Tseng et al. (2007) reported thermophilic actinomycetes from Taiwan with 29.0% polyethylene succinate degradation efficiency. Tseng et al. (2007) reported 31 isolates belonging to the genus *Saccharomonospora*, *Streptomyces*, *Microbispora*, *Thermoactinomyces* and *Actinomadura* for polymer degradation. Biodegradation of plastic employs different ways, and thermophilic composting serves as a promising method for removal of degradable plastic from the natural environment. Calabia and Tokiwa (2004) isolated a thermophilic microorganism (*Streptomyces* sp.) from soil and reported poly (D-3-hydroxybutyrate) degradation at 50 °C; *Streptomyces* sp. also showed a capacity for polyethylene succinate degradation. The strain *Bacillus* sp. TT96 also serves as a thermophilic polyethylene succinate degrader (Tokiwa et al. 2009).

On the other hand, Tezuka et al. (2004) reported the mesophilic microorganisms *Bacillus* and *Paenibacillus* for polyethylene succinate biodegradation at a temperature range of 40–45 °C. These bacteria were isolated from aquatic environments. Hazen et al. (2010) also reported that oil-contaminated marine areas supported the growth of hydrocarbon-degrading microorganisms. Similarly, Ishii et al. (2007) suggested some mesophilic microorganisms, *Bacillus pumilus*, *B. subtilis* and *Paenibacillus amylolyticus*, were involved in PESu degradation. Seretoudi et al. (2002) used *Rhizopus delemar* lipase for enzymatic degradation of polyethylene succinate and found that the crystalline structure altered the biodegradation rate while the molecular weight had no significant effect on degradation. Lipases are the endo enzymes that randomly break the ester bonds. Similarly, the property of microbial serine proteases was reported for polyethylene succinate biodegradation (Lim et al. 2005).

Polyhydroxyalkanoates

Polyhydroxyalkanoates (PHAs) are well-known bacterial-originated polymers and serve as storage polyesters. In abnormal growth conditions, bacterial cells are committed to PHA [3-hydroxybutyrate (PHB) and 3-hydroxyvalerate (PHBV)] production. The 3-hydroxybutyrate (PHB) used in various medical devices was found to accumulate in several bacteria, e.g., *Azotobacter vinelandii* and *Alcaligenes eutrophus*. The use of polyhydroxybutyrate in pharmaceutical practices is due to their bio-acceptance in patients (Bonartsev et al. 2007; Leja and Lewandowicz 2010a, b). Polyhydroxybutyrate (PHB) is a polyester and extremely crystalline (>50%) in nature. The melting point of polyhydroxybutyrate is high in contrast to polyesters (Averous and Pollet 2012). Polyhydroxybutyrate serves as an energy storage source for microorganisms. Polyhydroxybutyrate and its copolymer (3-hydroxyvalerate, etc.) are utilized in the production of degradable plastics (Ohura et al. 1999; Shimao 2001; Premraj and Doble 2005). Kadouri et al. (2002) reported three genes in *Azospirillum brasilense* strain Sp7 that are involved in PHB synthesis, i.e., *phb*A, *phb*B and *phb*C genes for β-ketothiolase, acetoacetyl coenzyme A reductase and PHB synthase, respectively. Biodegradation of PHA is accomplished by microbiological mineralization and generates CO_2 and H_2O (Bonartsev et al. 2007; Leja and Lewandowicz 2010a, b). Bacterial-originated PHA depolymerases are known, which may help in the breakdown of polymer (polyhydroxyalkanoate). PHA depolymerases are multi-subunit enzymes, and each subunit has a specific property including a signal sequence, substrate-binding domain, catalytic domain and interconnecting domain (fibronectin type III or threonine similar sequence). Instead, the interconnecting domain of PHB depolymerase contains a cadherin-like sequence found in *Pseudomonas stutzeri*. PHA depolymerases and PHB depolymerases also differ in substrate-binding domains; PHA depolymerases have one domain, while PHB depolymerases have two domains (for a giant substrate). Polymers that contain cyclic or chain (R) oligomers are sensitive to enzymatic (PHB depolymerases) degradation. *Alcaligenes faecalis* T1 utilizes the oligomer of 3-hydroxybutanoate, and their depolymerase activity was also studied (Ohura et al. 1999; Shimao 2001; Premraj and Doble 2005). Degradation of PHB was also done by *Ilyobacter delafieldii* in the absence of oxygen (Schink et al. 1992; Jirage et al. 2011). During the cell lysis process, PHB granules are released outside and hydrolyzed by extracellular PHB depolymerase. PHB depolymerases are enzymes that initiate the intracellular degradation of PHBs. In addition to other genes (*pha*Y, *pha*X, *pha*W), *pha*Z is known as the first structural gene for PHB degradation and releases the D-3-hydroxybutyrate

monomer, which can be further oxidized by 3-hydroxybutyrate dehydrogenase and the final product assimilated by the TCA cycle (Sharma and Dhingra 2016). Korotkova and Lidstrom (2001) reported the *dep*A and *dep*B genes, which encode PHB depolymerases for PHB degradation. Volova et al. (2010) reported an enhanced biodegradation rate of polymer films under the marine environment (seawater) instead of compact pellets, and *Gracilibacillus* sp., *Enterobacter* sp. and *Bacillus* sp. were identified. Similarly, Mabrouk and Sabryb (2001) found that the marine bacterium *Streptomyces* sp. SNG9 is involved in PHB utilization. Weng et al. (2011) investigated the biodegradability of PHAs (PHB, PHBV) under controlled conditions (composting) and found that the degradation rate decreases with decreasing percentages of HV (40, 20 and 3%). Oda et al. (1995) reported a *Paecilomyces lilacinus* fungal isolate and observed the degradation of PHB by secreting PHB depolymerases and glucose/lactose. Aburas (2016) isolated 20 fungal isolates to examine their PHB degradation potential; out of 20 isolates 11 (belonging to *Aspergillus, Penicillium, Fusarium, Trichoderma* and *Alterneria* genera) showed the PHB hydrolysis, with *Aspergillus oryzae* having an optimal PHB degradation potential at 30 °C and pH 6.5 in 1% PHB containing minimal medium. Foster and Tighe (2005) introduced a new hydrolytic system for PHB degradation that served as an accelerated model (70 °C and pH 10.6) and modified the degradation characteristics.

Shah et al. (2007) purified the 37-kDa enzyme (PHBV depolymerase) from *Bacillus* sp. AF3 and found a positive impact on PHBV degradation. Shah et al. (2010) conducted similar studies with *Actinomadura* sp. AF-555 for PHBV biodegradation and achieved reproducible results for degradation, determined by SEM analysis (roughness and pits on the surface) and FTIR analysis [(cleavage of −C–H bonds and O–R groups or ester (>C=O)].

Polyesters

Biodegradation is carried out on polyesters, formed through ester linkage, and esterases are the enzymes responsible for the breakdown of these linkages found in microorganisms. Some synthetic polyesters and polyhydroxyalkanoates (bacterial polyester polymer) are known for biodegradation (Shimao 2001). The rate of biodegradation depends on the type of polyester. Aliphatic polyesters susceptible to microbial degradation and ester bonds cleave easily under the aqueous condition that releases hydroxyl and carboxylic acid containing monomers. These aliphatic polyesters include microbial susceptible polymers, e.g., PCL, PGA, PLGA and PLA (Sathiskumar and Madras 2011). Medium length monomer units are rapidly mineralized as compared to smaller or longer monomers with the help of *A. flavus, A. niger* and

Pseudomonas sp. (Chandra and Rustgi 1998; Flavel et al. 2006). Microorganisms are capable of breaking the urethane or ester bonds because of the capability of microorganisms to attack the thermoset polymer and utilize it as a carbon source as well as nitrogen source (Zheng and Yanful 2005). The reliability of the polymer degradation rate is greatly impacted by the environmental conditions. The marine environment consists of different trophic levels; it is observed that the polyester (PCL, PHB/V) sample was ruined after 12 months incubation in the deep sea (Sekiguchi et al. 2011). Hazen et al. (2010) reported that the plume of hydrocarbon in sea depths stimulated the growth of Ƴ-proteobacteria. Ƴ-Proteobacteria was reported as an indigenous petroleum degrading bacteria from the depth of the sea, and such oil contaminated places are known sources for hydrocarbon degrading genes.

Sekiguchi et al. (2011) reported five bacterial species from the *Tenacibaculum*, *Alcanivorax* and *Pseudomonas* genus for polyester (PCL) degradation. Yagi et al. (2014) reported eubacteria and archaea are involved in anaerobic degradation of polyesters (PCL, PLA). *Mesorhizobium* sp. and *Xanthomonadaceae* bacteria were identified as PLA degrading eubacteria, and *A. thereius* was involved in PCL degradation. Archaea *Methanosaeta concilii* and *Methanobacterium petrolearium* are involved in anaerobic biodegradation of PLA and PCL, respectively. Zafar et al. (2014) studied commercial composting for polyester (polyurethane) degradation at maturation phase and found that low temperature supports the fungal growth on the surface of the polymer and also in compost. Polyester (polyurethane foam) also is a huge environmental problem and one of the major challenges to waste management. *Pseudomonas chlororaphis* ATCC 55729 effectively degraded polyester polyurethane foam in in vitro conditions (Gautam et al. 2007).

Nakajima-Kambe et al. (1999) reported using strain *Comamonas acidovorans* TB-35 for PUR esterase production, which is involved in polyester-polyurethane degradation. Degradation was accomplished via hydrophobic substrate binding and the catalytic domain of PUR esterase. Both of the domains play key roles in polyester-polyurethane hydrolysis. Gene *pud*A from *C. acidovorans* TB-35 encoded the enzyme PUR esterase.

Polylactic acid (PLA)

Polylactic acid (PLA) is a polymer used in medicine and also biodegradable plastics. *Bacillus brevis*, *Amycolatopsis* sp. *Fusarium moniliforme*, *Penicillium roquefort* and *B. brevis* participate in polylactic acid biodegrading (Ikada and Tsuji 2000; Tomita et al. 1999; Shah et al. 2008b). Production of polylactic acid through the polymerization of the monomer of lactic acid is shown in Table 5.

Table 5 **Effect of L-lactic acid or D-lactic acid copolymer composition of polylactic acid on melting temperature and glass transition (Babul et al. 2013)**

Composition of copolymer	Melting temperature of copolymer (°C)	Glass transition temperature of copolymer (°C)
L-/D-Lactic acid-PLA (80/20)	125	56
L-/D-Lactic acid-PLA (85/15)	140	56
L-/D-Lactic acid-PLA (90/10)	150	56
L-/D-Lactic acid-PLA (95/5)	164	59
L-/D-Lactic acid-PLA (100/0)	178	63

Lactic acid (hydroxyl carboxylic acid) is a basic component of aliphatic polyester polymer (polylactic acid). It is a bio-based bacterial fermented product. Fermentation dependent L or D forms of lactic acid contribute to the formation of several forms of polylactic acid and its derivative. T16-1 strain of *Actinomadura keratinilytica* NBRC 104111 reported for production of PLA-degrading enzyme *Amycolatopsis* sp. also shows ~60% degradation of PLA film degradation within 14 days (Babul et al. 2013; Garlotta 2002; Sukhumaporn et al. 2012; Pranamuda et al. 1997). Similarly, Prema and Uma (2013) reported mesophilic bacteria *Bacillus amyloliliquefaciens* for the degradation of polylactic acid. Iovino et al. (2008) investigated aerobic biodegradation of polylactic acid under a composting environment and checked the effect of the presence of thermoplastic starch (TPS) and short natural fibers (coir). In the initial stage, TPS degrades faster than PLA, and TPS is most susceptible to degradation as compared to PLA. Particle size is also an important factor for biodegradation; fine particles (nano-particles) involve a complex mechanism as compared to microparticles (Kunioka et al. 2006). Pranamuda et al. (1997) isolated *Amycolatopsis* and studied it for PLA degradation, finding 60% degradation of PLA film within 14 days. PLA depolymerase of *Amycolatopsis* sp. serves as a protease instead lipase and degrades casein silk fibroin and Suc-(Ala) 3-pNA. In addition to *Amycolatopsis*, *Saccharotrix* was also reported for PLA degradation (Tokiwa et al. 2009). According to Hidayat and Tachibana (2012) 12, 21, 30 and 48% polylactic acid degradations in 1, 2, 3 and 6 months, respectively, were achieved with *P. ostreatus*. Biodegradation of lactic acid polymers and its copolymers was widely affected by the amount of lactic acid content in polymers. Different fungal species were reported for lactic acid and its copolymer degradation, e.g., *Rhizopus* sp., *Mucor* sp., *Aspergillus* sp. and *Alternaria* sp. Eubeler et al. (2009) found the effect of UV irradiation on enzymatic hydrolysis of polylactic acid. The researchers expected that UV irradiation caused a chain alteration by C=C bonds formation. Proteinase K seemed to cause enzymatic breakdown of polylactic acid. Copinet et al. (2004) reported that higher temperature and relative humidity increase the hydrolytic rate and found UV irradiation (315 nm) also improves the biodegradation of PLA. Masaki et al. (2005) used the lipase for polylactic acid degradation; this hydrolytic enzyme was purified from strain *Cryptococcus* sp. S-2 and lipase showed similarity to cutinase family proteins. *R. delemar* lipase also boosts the PLA breakdown process (Tokiwa et al. 2009). Chaisu et al. (2015) examined dye containing polylactic acid films and found that enzymes from *Aneurinibacillus migulanus* show the same activity as proteinase K for polylactic acid degradation at 30 °C. Proteinase K was purified from *Tritirachium album* and served as a potent hydrolyzing enzyme for PLA degradation including L-PLA and DL-PLA (Tokiwa et al. 2009). Strain *Pseudomonas tamsuii* TKU015 was isolated from Taiwanese soil and reported for the production of PLA depolymerase; it is an important enzyme responsible for polylactic acid waste recycling. Optimum temperature and pH for PLA depolymerase activity are 60 °C and 10, respectively. Liang et al. (2016) purified the 58-kDa PLA depolymerase from *Pseudomonas tamsuii* TKU015. Sangwan and Wu (2008) exploited molecular techniques and reported genera *Thermopolyspora*, *Thermomonospora* and *Paecilomyces* for PLA degradation under controlled (aerobic) composting.

Polylcaprolactone

Polycaprolactone (PCL) was widely used in the field of medical science as an elastic biomaterial. Polycaprolactone is a semicrystalline polyester prepared ring-opening polymerization or via free radical ring-opening polymerization of ε-caprolactone and 2-methylene-1-3-dioxepane, respectively. Degradation of polycaprolactone homopolymer depends on its molecular weight. The glass transition temperature and melting point are −60 °C and 55–60 °C, respectively (Woodruff and Hutmacher 2010; Gajanand et al. 2014). PCL containing aliphatic ester linkages is susceptible to hydrolytic degradation but it seems that the PCL biodegradation process takes a lot of time (Gajanand et al. 2014). Mineralization and hydrolysis of polycaprolactone were done by the enzymatic activity of fungi (Chandra and Rustgi 1998). Structural arrangement of PCL affects the enzymatic degradation potential. The growth and enzymatic hydrolysis by *Penicillium funiculosum* and *A. flavus* were seen in an amorphous region of PCL (Tokiwa et al. 2009). Polylcaprolactone is sensitive to enzymes, e.g., esterases and lipases; degradation was achieved by these enzymes (Ianuzzo et al. 1977; Shimao 2001). Lipase derived from *R. delemar* showed a slow degradation rate for PCL because of having a higher molecular weight. Other lipase producing species, *R.*

delemar, R. arrizus Achromobacter sp. and *Candida cylindracea. Penicillium* sp., are also reported for degradation (Tokiwa et al. 2009). PCL depolymerases are also responsible for the degradation of polylcaprolactone (synthetic polymer) found in bacteria (Nishida and Tokiwa 1993; Suyama et al. 1998). Blended PCLs have a greater degradation rate when employed with 5% sebacic acid (Salgado et al. 2011; Tokiwa et al. 2009). Fungi are widely reported for the biodegradation of PCL and hydrolysis. PCL showed susceptibility to enzymatic degradation; however, the hydrolysis of the homopolymer is a time-consuming process (Averous and Pollet 2012). *Aspergillus* sp. strain ST-01 investigates for PCL degradation and absolute biodegradation of PCL accomplished at 50 °C within 6 days (Tokiwa et al. 2009).

Fungal (pathogenic for the plant) derived cutinases are responsible for plant cuticle (polymer) degradation, and phytopathogenic fungi are also known for their activity against polylcaprolactone, which may be due to the presence of cutinases and structural similarity between two cutin monomer units and PCL trimer. Breakdown of polymers (polylactate and polymalate) involves substrate-specific enzymatic cleavage through extracellular enzymes (Shimao 2001; Schink et al. 1992; Murphy et al. 1996). Benedic et al. (1983) reported yeast *Cryptococcus laurentii* for the biodegradation of PCL and concluded that the presence of casamino acids enhanced the degradation. Similarly, Motiwalla et al. (2013) reported *Bacillus pumilus* for the production of proteases, lipases and PCL degradation. Sekiguchi et al. (2011) identified genus *Tenacibaculum*, *Alcanivorax* and *Pseudomonas* from the deep sea environment and reported in vitro biodegradation of PCL. *Clostridium botulinum* and *Alcaligenes faecalis* were also recognized for PCL degradation (Caruso 2015; Tokiwa et al. 2009). Oda et al. (1995) observed 10% degradation of the PCL by *P. lilacinus* D218 within 10 days. *P. lilacinus* D218 releases PCL depolymerases in addition to PHB depolymerases in PCL and PHB containing media. Optimum activity of PCL depolymerases was observed at 30 °C and a pH range between 3.5 and 4.5 (Oda et al. 1995). Yagi et al. (2009) conducted anaerobic biodegradation of PCL at 55 °C with sludge (diluted 0.86% and undiluted 1.73%) and observed the 92% degradation of 10 g PCL having particle size 125–250 µm. Yagi et al. (2009) concluded the particle size was inverse to the biodegradation rate of PCL.

Polyurethane

Polyurethane (PUR) was prepared through polyaddition of diisocynate. The urethane bond involves a chain linkage of large molecular weight polymers (~200 as well as 6000). These large molecular weight polymers are polyether PUR and polyester PUR susceptible to microbial degradation (Shimao 2001). Polyethylene adipate (PEA) also serves as a polymer that comprises urethane bonding, which is originally a pre-polymer of PUR (Bhardwaj et al. 2012b). Ureases, esterases, proteases and lipases were classified for PUR hydrolysis via ester bond cleavage. *Trichoderma* sp. utilized the PUR as a substrate by producing ureases. Similarly, *Aspergillus terreus* and *Chaetomium globosum* were observed for the production of urethane and esterase that hydrolyzed the PUR (Bhardwaj et al. 2012b; Howard 2012). Loredo-Trevino et al. (2011) studied 22 fungal strains and found maximum urease activity in most of thm (95%), while the enzymatic activity of laccase, esterase and protease was observed in 36, 50 and 86% strains, respectively. Monooxygenase cleaves the ether linkages of the polymer of polyether. Acetaldehyde forms during ether linkage conversion with the help of sulfate-reducing bacteria. *Fusarium solani* and *Candida ethanolica* were identified as dominant species on the surface of polyurethanes (Schink et al. 1992; Zafar et al. 2013). *Curvularia senegalensis* showed more degrading activity as compared to others (Howard 2002). Proteolytic enzymes, i.e., urease and papain, were found to degrade polyurethane (medical polyester) (Bhardwaj et al. 2012b). In addition to papain, subtilisin was also reported for its involvement in polyurethanes degradation (Chandra and Rustgi 1998). The growth of *C. acidovorans* was observed on a PUR substrate and exploited as nitrogen and sole carbon; the extracellular membrane bound enzymes have a key role in PUR hydrolysis (Howard 2012). *C. acidovorans* TB-35 led to the biodegradation of polyester PUR and diethylene glycol and adipic acid were found as degraded products. Cell bound enzymes are responsible for PUR degradation and isolated from *C. acidovorans*. The gene for such enzymes encodes 548 amino acid protein and contains hydrophobic domains for polymer surface attachment, the catalytic domain, lipase box and signal sequence. In previous studies, bacterial species were less reported for PUR degradation in comparison to fungi, e.g., *Arthrographis kalrae*, *Aspergillus fumigatus*, *A. niger*, *Emericella*, *Lichthemia*, *Fusarium solanii*, *Thermomyces*, *Corynebacterium* sp., *Neonectria*, *Plectosphaerella*, *Phoma*, *Nectria*, *Alternaria* and *P. aeruginosa*, and bacteria from genera *Bacillus* and *Comamonas* also served as polyurethanes degrader (Shimao 2001; Chandra and Rustgi 1998; Flavel et al. 2006; Zafar 2013; Akutsu et al. 1998; Bhardwaj et al. 2012b). Shah et al. (2008a) isolated the bacteria from soil samples and examined them for polyurethane degradation. Isolates were characterized and identified *Bacillus* sp. AF8, *Pseudomonas* sp. AF9, *Micrococcus* sp. 10, *Arthrobacter* sp. AF11, and *Corynebacterium* sp. AF12. All the

isolates exploited PUR as carbon source, and degradation was determined via esterase activity and CO_2 estimation through the Sturm test, and further structural and chemical changes were examined by means of SEM and FT-IR analysis, respectively. Upreti and Srivastava (2003) observed the growth of *A. foetidus* on polyurethane and investigated the PUR biodegradation by means of reduction in its tensile strength and tensile modulus. Ma and Wong (2013) studied the esterase activity of *A. flavus* for polyester polyurethanes degradation. Ma and Wong (2013) exploited esterase genes from *A. flavus* transfer to the *P. pastoris* for mass production of esterase; it may also speed up the PUR degradation rate as compared to *A. flavus*. Genes *pue*A and *pue*B played an extreme role in PUR degradation, as determined by conducting different gene silencing and cloning analyses with *P. chlororaphis* and *E. coli* (Howard 2012). These genes are located along with the cluster of the ABC transporter gene and seven open reading frames. According to gene silencing experiments, Howard et al. (2007) suggested *pue*A played a significant role in increasing the cellular density, which enhanced the PUR degradation. Enzymatic secretion of these enzymes followed the Type I secretion system that employed C-terminal hydrophobic secretion signals and glycine-rich RTX motifs, which played the critical role in stabilizing the Ca^{2+} roll structure. The exact mechanism of the ion roll structure is contentious; it may help with proper alignment of the signal and their secretion (Howard 2012). Gene *pue*A cloned from the *P. chlororaphis* to *E. coli* is encoded for the extracellular enzyme that is secreted from the cloned *E. coli* cell. The recombinant product of *pue*A showed similarity with the Group I lipases. Group I lipases and serine hydrolases are widely distributed PUR hydrolyzing enzymes that form a serine triad structure in addition to histidine and aspartate/glutamate (Howard 2012). Stern and Howard (2000) exploited the pT7-6 vector for subcloning the gene transfer that encoded a 65 kDa protein and displayed serine hydrolase characteristics. *Pestalotiopsis microspora* is another instance of the production of serine hydrolase with PUR exploited as the carbon source (Bhardwaj et al. 2012b). Howard (2012) isolated soil microorganism P7, which was identified as *Acinetobacter gerneri* by 16S rRNA sequencing and reported for polyurethane degradation. Howard (2012) characterized and determined the 66-kDa enzymes involved in polyurethane hydrolysis and enhanced substrate specificity was increased with of p-nitrophenylpropanate, while ethylenediamine-tetra acetic acid and phenylmethylsulfonylfluoride showed a negative effect by inhibiting the enzyme activity. Polyurethanase (48 kDa) encoding genes were reported in *E. coli* for successfully cloning and expression.

Polyvinyl alcohol

Polyvinyl alcohol (PVA) is a typical polymer like polyethylene and polystyrene made up by linkage of the carbon–carbon bond. It can be used in several tasks as it is converted into different shapes due to its thermoplasticity and also can be used as a biodegradable transportation system due to its water-soluble nature (Shimao 2001). The *Pseudomonas* strain having alcohol peroxidase activity is involved in enzymatic mineralization of polyvinyl alcohol. *Acinetiobacter* and *Flavobacterium* bacterial strains show similar activity on polyvinyl alcohol (Chandra and Rustgi 1998). Fungi *Fusarium lini* is known to synthesize dehydratase, which is responsible for polyvinyl alcohol degradation and liberates carbon dioxide in addition to water. *Pseudomonas* O-3, *Pseudomonas vesicularis* PD, *Sphingopyxis* sp. PVA3, *Pseudomonas* sp., *Bacillus megaterium*, *Bacillus* sp., *Alcaligenes faecalis* and *Alcaligenes* sp. were also identified as a polyvinyl alcohol degrader (Pajak et al. 2010; Rong et al. 2009; Raghul et al. 2013). Biodegradation of polyvinyl alcohol through *Pycnoporus cinnabarinus* (white rot fungus) was higher when polyvinyl alcohol was employed with a combination of chemical pretreatments, which enhance the degradation (Larking et al. 1999). Raghul et al. (2013) isolated bacteria and investigated PVA-LLDPE degradation. These bacteria were used as a consortium (*Vibrio parahemolyticus*, *Vibrio alginolyticus*), polymer degradation by means of reduction of the tensile strength and surface erosion analysis determined by SEM micrographs. Husarova et al. (2010b) employed TGGE (temperature gradient gel electrophoresis) for the analysis of bacterial populations during the degradation of PVA. PVA was found as a potent pollutant in wastewater, and its clearance from wastewater is an important part of maintaining the water quality. Rong et al. (2009) discussed bacterial isolation from activated sludge and identified *Novosphingobium* sp. P7. *Novosphingobium* sp. P7 showed PVA degradation in the presence of methionine in medium. On the other hand, *Novo-sphingobium* sp. P7 is capable of degrading PVA without methionine when cultured with *Xanthobacter flavus* B2. *Sphingopyxis* sp. PVA3 and *Pseudomonas* sp. A-41 form activated sludge, also seen for PVA degradation (Agrawal and Shahi 2015; Fukae et al. 1994). Similarly, Marusincova et al. (2013) reported on *Steroidobacter* sp. PD for a municipal wastewater treatment plant, and their growth was also observed in denitrifying conditions. PVA degradation with *Steroidobacter* sp. PD was accomplished in both environmental conditions (aerobic and denitrifying conditions). Matsumura et al. (1994) isolated *Alcaligenes faecalis* KK314 for PVA breakdown degradation determined by means of stereoregularity.

Tsujiyama et al. (2011) cultivated *Flammulina velutipes* in liquid and quartz sand for PVA utilization. Patil and Bagde (2015) isolated and characterized the bacteria from plastic wastes as *Bacillus* sp. and *Pseudomonas* sp. degradation analysis involved the CO_2 evolution test and spectrophotometric techniques. Biodegradation of PVA with *Bacillus* sp. and *Pseudomonas* sp. were observed as 65 and 42% in 20 days, respectively. Larking et al. (1999) examined the effect of chemical treatment on PVA biodegradation and found a constructive effect of Fenton's reagent (chemical treatment) on the PVA biodegradability rate of *Pycnoporus cinnabarinus* (white rot fungus). Chen et al. (2007) investigated the mixed culture of microorganisms from sludge samples and found their great potential to degrade PVA (low polymerization and high saponification). Intracellular and extracellular PVA hydrolyzing enzymes are capable of degrading PVA and form molecules that were detected and examined in the cellular extract. They studied two strains from the mixed culture and found their capability to degrade PVA1799; instead PVA124 and PVA124 showed resistance to degradation by these two strains. Hirota-Mamoto et al. (2006) identified the gene *pva*A involved in PVA degradation by encoding the PVA dehydrogenase. They used the *Sphingomonas* sp. 113P3 strain for purification of PVA dehydrogenase (*pva*A) and cloned the *pva*A gene in *E. coli* with a hexahistidine tag that showed similar features as the *pva*A gene reported from the *Sphingomonas* sp. 113P3 strain. PVA dehydrogenase was also reported from some other strains, e.g., *Pseudomonas* sp. VM15C, *Xanthomonas* sp. and *Azoarcus* sp. EbN1. Shimao et al. (2000) investigated the genes *pva*A and *pva*B of *Pseudomonas* sp. VM15C that encoded for PVA hydrolyzing enzyme, i.e., PVA dehydrogenase and oxidized PVA hydrolase, respectively. The protein was encoded by gene *pva*B, having serine hydrolase characteristics. Oxidized PVA hydrolase degrades PVA in combination with the gene product of *pva*A. PVA degradation was reported for both extracellular PVA oxidase and periplasmic PVA dehydrogenase. Periplasmic PVA dehydrogenase is widely distributed in gram-negative bacteria (Kawai and Hu, 2009). Similarly, Hu et al. (2007) reported PVA degradation by the *Sphingopyxis* sp. 113P3 strain and observed that PVA with lower molecular size degrades faster as compared to average molecular size PVA. Periplasmic PVA dehydrogenase was found as an active enzyme of *Sphingopyxis* sp. 113P3 that hydrolyzes PVA.

Polyethylene glycols

Polyethylene glycols (PEGs) are extensively used in biomedical applications, drug delivery, biosensor materials and fabrication applications (Datta 2007). Polyethylene glycols act as plasticizers with molecular weight of 6000. *P. aeruginosa*, *P. stutzeri* and *Sphingomonas* sp. are involved in polyethylene glycol mineralization and symbiotically by *Rhizobium* sp. and *Sphingomonas terrae* (Smith 2005). Aerobic biodegradation of polyethylene glycols in fragmented form is carried out through the extracellular enzymatic activity of *P. aeruginosa* strains. *P. stutzeri* JA1001 strain is recognized as a biodegrader of polyethylene glycols (M.W. 13,000–14,000). *Desulfovibrio desulfuricans*, *Bacteroides* sp., *Flavobacterium* sp. and the consortium of *Pseudomonas* sp. with *Flavobacterium* sp. are also able to degrade polyethylene glycols (Ojo 2007; Flavel et al. 2006).

Sphingomonas sp., *P. aeruginosa* and *P. stutzeri* microorganisms are reported to have high molecular weight (4000–20,000) PEG degradation. Alcohol dehydrogenase of strain *Rhodopseudomonas acidophilia* M402 seemed to hydrolyze PEGs. Kawai et al. (2010) identified five genes in *peg* operon system genes that are responsible for PEG degradation, and they were recognized and cloned. These genes, in the presence of PEG and *araC*, act as a regulator. The PEG carboxylate dehydrogenase encoding gene also helped in PEG metabolism and was found in the downstream region of *peg* operon. Ether linkage showed susceptibility to several bacterial enzymes, e.g., ether hydrolase, peroxidase, monooxygenase, glycolate oxidase and laccase.

Charoenpanich et al. (2006) studied the dual regulation operon system *Sphingopyxis macrogoltabida* 103 for PEG degrading genes (*peg*B, C, D, A and E) and found regulator gene *peg*R encoding AraC-type regulator. PEGs serves as an inducer for *peg*A promoter and *peg*R promoter; it also induces the *peg*B promoter in additional oligomeric ethylene glycols. The *peg*B promoter having a regulator (AraC/XylS) binding site played the key role in the transcription of *peg* operon. Instead, AraC/XylS-type regulator and GalR/LacI-type regulator served to bind *peg*A promoter and *peg*R promoter, and the *peg* operon worked actively. Tani et al. (2007) reported *S. macrogoltabida* 103, *S. macrogoltabida* 203 and *S. terrae* for PEG degradation. The genes *peg* B, C, D, A, E and R observed from these strains served as PEG hydrolyzing genes that encoded the receptor protein and enzymes, e.g., dehydrogenase, permease and ligase. Gene *peg*C reported in strain *S. macrogoltabidus* 103 encoded the enzyme aldehyde dehydrogenase located upstream of *peg*A. Ohta et al. (2005) discussed the expression of aldehyde dehydrogenase in *Escherichia coli*, and a cloned enzyme showed similarity to NAD (P)-dependent aldehyde dehydrogenase. They reported this enzyme as a novel nicotinoprotein aldehyde dehydrogenase from *S. macrogoltabidus* 103, and, activated by Ca^{2+}, that enzyme was involved in PEG degradation. On the other hand, Ohtsubo et al. (2015) sequenced the complete genome of *Sphingopyxis* macrogoltabida (NBRC 15033); they did not find the gene *peg*A in *peg* operon is involved in PEG

degradation. Sugimoto et al. (2001) purified and characterized the dye-linked dehydrogenase from *S. terrae* and initiated the enzymatic degradation of PEG by oxidizing the terminal alcohol. They cloned the *peg*A gene in *E. coli* and purified the recombinant enzyme as a homodimeric protein (58.8 kDa) that bound to flavin adenine dinucleotide. PEG dye-linked dehydrogenases serve as a novel type of flavoprotein alcohol dehydrogenase. Frings et al. (1992) discussed two bacteria, i.e., *Pelobacter venetianus* and *Bacteroides* PG1, for anaerobic degradation of PEG. Diol dehydratase and PEG acetaldehyde lyase are hydrolyzing enzymes identified in *P. venetianus*; oxygen and citrate/sulfhydryl compounds showed their sensitivity and optimum effect, respectively, for both enzymes. *Bacteroides* PG1 has PEG acetaldehyde lyase similar to *P. venetianus*.

Polyethylene

Polyethylene is the most utilized form of synthetic polymer and highly hydrophobic in nature (Mahalakshmi et al. 2012). In the case of polyethylene, the biodegradability is inverse to its molecular weight. Less than 620 molecular weight hydrocarbon oligomers favor the growth of microorganisms; high molecular weight polyethylene is resistant to biodegradation. Its hydrophobic nature hinders its bioavailability. Some physical and chemical treatments used before biodegradation increase its effectiveness. Such treatments include UV irradiation, photo-oxidation, thermal treatment and oxidation with nitric acid. Oxidation of polyethylene raises the surface hydrophilicity, which ultimately increases the biodegradation (Hadad et al. 2005). Chemical treatments (0.5 M HNO_3 and 0.5 M NaOH) of polyethylene accelerate the biodegradation by *Pseudomonas* sp. (Nwachkwu et al. 2010). Polyethylenes and polystyrenes with ether linkage are susceptible to monooxygenases attack (Schink et al. 1992). Biodegradation of polyethylene enhanced by the physical treatments causes pre-ageing via light or heat exposure. A hot air oven was used for abiotic oxidation and created a molecular weight distribution at 60 °C followed by incubated with polymer degrading microorganisms. Genera *Gordonia* and *Nocardia* are associated with the biodegradation process (Bonhommea et al. 2003).

Some genera are also reported for polyethene degradation (*Bacillus*, *Lysinibacillus*, *Pseudomonas*, *Staphylococcus*, *Streptococcus*, *Micrococcus*, *Streptomyces*, *Rhodococcus*, *Proteus*, *Listeria*, *Vibrio*, *Bravibacillus*, *Serratia*, *Nocardia*, *Diplococcus*, *Moraxella*, *Penicillium*, *Arthrobacter*, *Aspergillus*, *Phanerochaete*, *Chaetomium* and *Gliocladium*) (Arutchelvi et al. 2008; Grover et al. 2015; Koutny et al. 2006; Bhardwaj et al. 2012b; Restrepo-Flórez et al. 2014). Polyethylene is utilized as a carbon source by microorganisms, and biofilm formation on it shows their effectiveness. Biofilm formation improved with the addition of mineral oil (0.05%) to the medium. Nonionic surfactants can promote the polymer biodegradation by increasing the hydrophilicity of the polymer, which helps in the adhesion of microorganisms on the polymer (Hadad et al. 2005).

Polyethylene can be degraded by hydro- or oxo-biodegradation. The biodegradation method depends upon the ingredients used in the formation. Fungi (*Mucor rouxii* NRRL 1835, *A. flavus*) and strains of *Streptomyces* are also involved in starch-based polyethylene degradation. Degraded polymers are explained by surface fractures, bond scratching and other changes like color, etc., and such alterations examined by scanning electron microscopy and FT-IR. Degraded polymer shows conformational changes in its texture that support the wideness of the microbial population (Mahalakshmi et al. 2012). *Streptomyces badius* 252 and *Streptomyces setonii* 75Vi2 have the ability to degrade lignocelluloses while *Streptomyces viridosporus* T7A acts on heat-treated degradable plastics (Pometto et al. 1992). Abiotic factors play an important role in increasing the surface availability for microbial growth on polymers; these factors include photo-oxidation, physical disintegration and hydrolysis that cause decreasing molecular weight (Singh and Sharma 2008). *R. ruber* degrades polyethylene via colonizing on them and forms a biofilm (Basnett et al. 2012). Peroxidant additives are employed in polyethylene manufacturing for agriculturally used plastic, this type of polyethylene showing susceptibility to thermal and photochemical mineralization in vitro. In addition to UV and heat treatments, it reduces the strength of hydroxyls and carbonyls by changing their structure (Feuilloley et al. 2005; Li 2000). *P. chrysosporium* and *Streptomyces* sp. are known to degrade starch-blended polyethylene (Flavel et al. 2006). Similar results were reported by Psomiadou et al. for starch-blended LDPE degradation by means of a reduction in mechanical properties (Psomiadou et al. 1997). Arvanitoyannis et al. (1998) attempted to make biodegradable blended LDPE with starch and found that starch 10% (w/w) content in blended LDPE enhanced its biodegradation rate by altering the mechanical properties.

Microorganisms take part in degradation via modification of their metabolic functional pathways according to environmental conditions to utilize xenobiotic compounds. A bioremediation process is more affordable by discovering novel catabolic mechanisms (Ojo 2007). The biodegradation process is slow but this does not indicate that ingredients in plastic material and polymers are not bioactive. Polycarbonate plastics undergo leaching of bioactive bisphenol-A monomer when undergoing salt exposure in seawater. Commercial use of several

synthetic polymers made with bioactive additives monomers, which are non-stick compounds; softeners and UV stabilizers are found in nature. Their degradation rate depends on the environmental circumstances. Symphony is a type of polymeric material that is used in polyethylene formation and degradable in nature (Moore 2008; Kumar et al. 2011).

Yoon et al. (2012) isolated low molecular weight polyethylene degrading bacteria and identified *Pseudomonas* sp. E4. They found 28.6, 14.9, 10.3 and 4.9% carbon mineralized from different molecular weight (1700, 9700, 16,900, 23,700) polyethylene samples within 80 days at 37 °C that evolved CO_2. The *alk*B gene encoded the enzyme alkane hydroxylase, and Yoon et al. (2012) cloned the *alk*B from *Pseudomonas* sp. E4 to *E. coli* strain BL21 and found their carbon mineralization potential as 19.3% in 80 days at 37 °C. Yoon et al. (2012) concluded that *alk*B gene played a key role in polyethylene degrading. Santo et al. (2013) reported bacterial originated copper-binding laccase from *R. ruber* for enzymatic degradation of polythene. Nowadays different groups of microorganisms are reported for biofilm formation. Similarly, Tribedi et al. (2015) studied *Pseudomonas* sp. AKS2 for biofilm formation and reported LDPE degradation. They found enhanced microbial growth with 26% surface hydrophobicity and 31% hydrolytic activity.

Odusanya et al. (2013) observed the surface deformities of plastic when treated with *S. marcescens*; further degradation was evaluated by the reduction in glass transition temperature (Tg) and reduction in crystallinity via DSC analysis. Similarly, Ambika et al. (2015) identified a marine bacterial strain as *Achromobacter denitrificans* S1 for LDPE degradation that was determined by NMR, XRD, TGA and GCMS analysis. Das and Kumar (2015) investigated a sample of municipal solid soil for distribution of polymer degrading microorganisms and isolated *Bacillus amyloliquefaciens* BSM-1 and *B. amyloliquefaciens* BSM-2. Das and Kumar (2015) found enhanced biodegradation with strain *B. amyloliquefaciens* BSM-2 in comparison to *B. amyloliquefaciens* BSM-1. The rate and efficiency of polymer degradation were determined by pH alteration in media, CO_2 evolution, weight loss, SEM and FT-IR analysis. Similarly, Gajendiran et al. (2016) isolated a fungus from landfill soil, identified as *A. clavatus* (strain JASK1). They exploited *A. clavatus* for LDPE degradation by incubating the samples for 90 days and evaluated the biodegradation by means of CO_2 evolution, SEM and AFM analysis.

Restrepo-Flórez et al. (2014) discussed the effect of biotic and abiotic factors on the initiation of polyethylene degradation by including enzymatic hydrolysis, photo-oxidation (UV light), etc. The number of carbons present on the polymer influenced the rate of degradation via enzyme interactions. Structural arrangements are also important in degradation; it was observed that amorphous regions are more susceptible to microbial attack. Thomas et al. (2015) reported three microbial species for polythene degradation, i.e., *P. fluorescens, S. aureus* and *A. niger*. They used *P. fluorescens* and evaluated the degradation in laboratory conditions as well as the field environment; 8.06% degradation was found within a month under laboratory conditions, and 8 and 16% degradations were observed in 9 and 12 months, respectively, in field conditions. Pramila et al. (2012) reported LDPE degrading bacterial strains isolated from the municipal landfill, i.e., *Brevibacillus parabrevis* PL-1, *Acinetobacter baumannii* PL-2, *A. baumannii* PL-3 and *P. citronellolis* PL-4. The polymer (LDPE) degradation potentials of bacterial strains were determined by the biofilm formation ability on polymers and CO_2 evolution from the carbon mineralization of polymer. Muenmee et al. (2015) used a bacterial consortium (heterotrophs, autotrophs and methanotrophs) for HDPE and LDPE degradation and found *Methylobacter* sp. or *Methylocella* sp. (methanotrophs) predominantly on the surface of deteriorated samples. The LDPE degradation rate was reported as lowest in comparison to HDPE degradation.

Muenmee et al. (2016) conducted an experiment for HDPE and LDPE degradation with type I and type II methanotrophs, heterotrophs and nitrifying bacteria. Degradation was accomplished under lysimeters in different environmental conditions. Muenmee et al. (2016) reported these microorganisms as prominent polymer (HDPE and LDPE) degraders, identified as *Nitrosomonas* sp. AL212, *Nitrobacter winogradkyi, Burkholderia* sp., *Methylobactor* sp., *Methylococcus capsulatus, Methylocystic* sp. and *Methylocella* sp. Duddu et al. (2015) studied 83 microbial isolates for biosurfactant production from oil-contaminated sites. Out of 83 isolates, the NDYS-4 isolate was identified as *Streptomyces coelicoflavas* 15399[T] and selected for LDPE degradation, showing 30% weight loss in 4 months and metabolic activity of isolates evaluated by TTC reduction test.

Soil-buried LDPE showed active microbial growth on LDPE in 7–9 months, and surface deterioration was confirmed within 17–22 months as determined by SEM analysis (Mumtaz et al. 2010). Abrusci et al. (2011) tested the biodegradability of photo-degraded polyethylene through *Bacillus cereus, B. megaterium, B. subtilis* and *Brevibacillus borstelensis* at 30 and 45 °C. These microorganisms were isolated from soil-buried polyethylene films and *B. borstelensis* obtained from a German collection of microorganisms and cell cultures. Degradation of polymers was observed by means of biofilm formation and other deformities, characterized and confirmed by ATR, FTIR, chemiluminescence and GC-product analysis. The

maximum carbon mineralization was found at 11.5% and 7–10% with *B. borstelensis* and MIX (a mixed culture of *Bacillus cereus, B. megaterium* and *B. subtilis*), respectively, at 45 °C.

Devi et al. (2015) reported two fungal strains for polyethylene (HDPE) degradation via biofilm formation. These fungal strains were identified as *Aspergillus tubingensis* VRKPT1 and *A. flavus* VRKPT2. Biofilm formation and surface deformation as a result of fungal degradation were determined by epifluorescent microscope and SEM.

Hadad et al. (2005) isolated *B. borstelensis* 707 from soil and exploited it for polyethylene degradation. They found maximum biodegradation with pretreated (photooxidation) polyethylene samples by means of reduction in molecular weight and gravimetric weight, i.e., 30 and 11%, respectively. Similarly, Fontanella et al. (2010) observed the effect of oxidants on LLDPE, LDPE and HDPE degradation. They used manganese, iron and cobalt as prooxidants in polythene; samples were placed for pretreatment by thermal method and photooxidation. They exploited *Rhodococcus rhodochrous* in the degradation of pretreated polymer samples. Bacterial activity and biodegradation were determined by adenosine triphosphate content and ^1H NMR spectroscopy, respectively; surface alteration was confirmed by SEM analysis.

It was found that accelerated degradation of polyethylene was observed for abiotic treatments that included thermo- and photo-oxidation. These treatments oxidized the prooxidants and helped polyethylene degradation. Exploitation of prooxidants as blended in polyethylene formation make it further susceptible to microbial deterioration.

Husarova et al. (2010a) investigated biodegradation of calcium carbonate (prooxidant) containing LDPE. They found that prooxidant containing LDPE degraded (carbon mineralization) 16% in 80 days, while samples that did not contain any prooxidant were mineralized 7% in 13 months under soil and 23% in compost conditions. Mehmood et al. (2016) isolated bacterial strains from a solid waste dump and identified *P. aeruginosa* CA9, *Burkholderia seminalis* CB12 and *Stenotrophomonas pavanii* CC18; these strains were tested for modified LDPE [blend of titania (TiO_2) and starch]. Out of three strains, CC18 showed enhanced viability and degradation. The reproducibility of degradation was examined by SEM, TGA, XRD and FTIR analysis, and bacterial growth was scrutinized by biofilm formation, salt aggregation test and cell surface hydrophobicity. *P. citronellolis* EMBS027 isolated from landfill soil of municipal sites showed LDPE deterioration (Bhatia et al. 2014). Sheik et al. (2015) reported *Aspergillus* sp., *P. lilacinus* and *L. theobromae* as endophytic fungi for laccase production and polymer (LDPE) degradation. They incubated fungal cultures with radiation (with different range 0–1000 kGy) irradiated LDPE samples and observed the reduction in intrinsic viscosity and average molecular weight. On the other hand, Shahnawaz et al. (2016) isolated *Lysinibacillus fusiformis* VASB14/WL and *Bacillus cereus* strain VASB1/TS from rhizospheric soil of *Avicennia marina* and reported them for polythene degradation.

Conclusion

Microorganisms are capable of degrading inorganic and organic materials, and interest has been aroused to study microbes for their ability to degrade plastic polymers. *P. aeruginosa, P. stutzeri, S. badius, S. setonii, R. ruber, C. acidovorans, C. thermocellum* and *B. fibrisolvens* are the dominant bacterial spp. associated with polymer degradation. *P. aeruginosa* is one of the widely reported microorganisms for polymer degradation via biofilm formation with the help of alginate-like chemicals and quorum sensing signaling systems, i.e., Rh1I/Rh1R. Biofilm formation improves the degradation efficiency followed by the mineralization (polyethylene glycols mineralization) process. *Pseudomonas aeruginosa* CA9 is reported to have better biodegradation with LDPE, *Pseudomonas* sp. AKS2 is reported for biofilm formation on LDPE and biodegradation of LDPE via enhancing microbial growth with 26% surface hydrophobicity and 31% hydrolytic activity. *P. stutzeri* is reported for high molecular weight (4000–20,000) PEG degradation, and *Streptomyces badius* 252 and *Streptomyces setonii* 75Vi2 were more effective against heat-treated degradable plastics. *Rhodococcus ruber* has been reported to colonize and degrade polyethylene by forming a biofilm and hydrolyting enzymes. Polyethylene biodegradation was improved by introducing peroxidant additives in manufacturing processes that make it susceptible to in vitro thermal and photochemical mineralization. *C. acidovorans* TB-35 is also useful for polyester–polyurethane degradation through PUR esterase production and enzymatic hydrolysis. Gene *pud*A is the key gene from *Comamonas acidovorans* TB-35 encoding the enzyme PUR esterase. Fungi like *A. niger, A. flavus, F. lini, P. cinnabarinus* and *M. rouxii* are prevalently found for polymer degradation. *A. niger* produces acetyl xylan esterase, which works with the combination of endo-xylanase for efficient xylan degradation. *A. niger* and *A. flavus* are also suitable for the rapid mineralization of medium-length monomer units. *A. niger* is effective in polythene degradation, while *Aspergillus flavus* is reported for PCL as well as polythene biodegradation. Similarly, *Mucor rouxii* NRRL 1835, *Aspergillus flavus* and *Streptomyces* are also involved in starch-based polyethylene degradation. *Fusarium lini* is involved in synthesizing dehydratase, which is responsible for polyvinyl alcohol degradation with CO_2 and H_2O formation. *P. cinnabarinus* is also known as white rot

fungus and is involved in PVA biodegradation in the presence of Fenton's reagent. The above discussion illustrates the occurrence of polymer-degrading microorganisms. Hence, further studies on the screening of effective microbial strains are essential to minimize polymer risks for the environment.

Abbreviations
LDPE: low-density polyethylene; HDPE: high-density polyethylene; PVA: polyvinyl alcohol; PEA: polyethylene adipate; PUR: polyurethane; PCL: polycaprolactone; TPS: thermoplastic starch; PLA: polylactic acid; PHB: polyhydroxybutyrate; PHAs: polyhydroxyalkanoates; PESu: polyethylene succinate; EPS: extracellular polymeric substance; ISO: International standards Organization; ASTM: American Society for Testing and Materials; SEM: scanning electron microscopy; FTIR: Fourier transform infrared.

Authors' contributions
Both authors have contributed equally to the manuscript. Author VMP has carried out the data study. Author N guided him during the study. Both authors read and approved the final manuscript.

Competing interests
The authors declare that they have no competing interests.

Funding
This work was supported by research Grant No.F. 25-1/2013-14(BSR)/11-13/2008 (BSR), University Grants Commission (UGC), India.

References
Abrusci C, Pablos JL, Corrales T, López-Marín J, Marín I, Catalina F (2011) Biodegradation of photo-degraded mulching films based on polyethylenes and stearates of calcium and iron as pro-oxidant additives. Int Biodeterior Biodegrad 65(3):451–459

Aburas MMA (2016) Degradation of poly (3-hydroxybuthyrate) using *Aspergillus oryzae* obtained from uncultivated soil. Life Sci J 13(3):51–56

Agrawal N, Shahi SK (2015) An environmental cleanup strategy-Microbial transformation of xenobiotic compounds. Int J Curr Microbiol App Sci 4(4):429–461

Akola J, Jones RO (2003) Branching reactions in polycarbonate: a density functional study. Macromolecules 36:1355–1360

Akutsu Y, Kambe TN, Nomura N, Nakahara T (1998) Purification and properties of a polyester polyurethane-degrading enzyme from *Comamonas acidovorans* TB-35. Appl Environ Microbiol 64(1):62–67

Ali MI, Ahmed S, Robson G, Javed I, Ali N, Atiq N, Hameed A (2014) Isolation and molecular characterization of polyvinyl chloride (PVC) plastic degrading fungal isolates. J Basic Microbiol 54:18–27

Álvarez C, Reyes-Sosa FM, Díez B (2016) Enzymatic hydrolysis of biomass from wood. Microbiol Biotechnol 9(2):149–156

Ambika DK, Lakshmi BKM, Hemalatha KPJ (2015) Degradation of low density polythene by *Achromobacter denitrificans* strain s1, a novel marine isolate. Int J Rec Sci Res 6(7):5454–5464

Arutchelvi J, Sudhakar M, Arkatkar A, Doble M, Bhaduri S, Uppara PV (2008) Biodegradation of polyethylene and polypropylene. Ind J Biotechnol 7(1):9–22

Arvanitoyannis I, Biliaderis CG, Ogawa H, Kawasaki N (1998) Biodegradable films made from low-density polyethylene (LDPE), rice starch and potato starch for food packaging applications: part 1. Carbohydr Polym 36(2):89–104

Augusta J, Muller RJ, Widdecke H (1993) A rapid evaluation plate-test for the biodegradability of plastics. Appl Microbiol Biotechnol 3:673–678

Averous L, Pollet E (2012) Biodegradable polymers. Environ Sil Nano Biol Gre Energy Technol: 13–39

Azevedo HS, Reis RL (2005) Understanding the enzymatic degradation of biodegradable polymers and strategies to control their degradation rate Biodegradable systems in tissue engineering and regenerative medicine. CRC Press, Boca Raton, pp 177–201

Babul RP, O'Connor K, Seeram R (2013) Current progress on bio-based polymers and their future trends. Prog Biomater 2(8):1–16

Barlaz MA, Ham RK, Schaefer DM (1989) Mass-balance analysis of anaerobically decomposed refuse. J Environ Eng 115(6):1088–1102

Barnes DK (2002) Biodiversity: invasions by marine life on plastic debris. Nature 416(6883):808–809

Basnett P, Knowles JC, Pishbin F, Smith C, Keshavarz T, Boccaccini AR, Roy I (2012) Novel biodegradable and biocompatible poly (3-hydroxyoctanoate)/bacterial cellulose composites. Adv Eng Mater 14(6):330–343

Basnett P, Ching KY, Stolz M, Knowles JC, Boccaccini AR, Smith C, Locke IC, Keshavarz T, Roy I (2013) Novel poly (3-hydroxyoctanoate)/poly (3-hydroxybutyrate) blends for medical applications. React Funct Polym 73:1340–1348

Benedic CV, Cameron JA, Huang SJ (1983) Polycaprolactone degradation by mixed and pure cultures of bacteria and a yeast. J Appl Polym Sci 28(1):335–342

Berlemont R, Martiny AC (2013) Phylogenetic distribution of potential cellulases in bacteria. Appl Environ Microbiol 79(5):1545–1554

Bhardwaj H, Gupta R, Tiwari A (2012a) Microbial population associated with plastic degradation. Sci Rep 1(2):1–4

Bhardwaj H, Gupta R, Tiwari A (2012b) Communities of microbial enzymes associated with biodegradation of plastics. J Polym Environ 21(2):575–579

Bhatia M, Girdhar A, Tiwari A, Nayarisseri A (2014) Implications of a novel Pseudomonas species on low density polyethylene biodegradation: an in vitro to in silico approach. SpringerPlus 3(497):1–10

Bhatnagar S, Kumari R (2013) Bioremediation: a sustainable tool for environmental management—a review. Ann Rev Res Biol 3(4):974–993

Blouzard JC, Bourgeois C, De Philip P, Valette O, Bélaïch A, Tardif C, Belaich JP, Pagès S (2007) Enzyme diversity of the cellulolytic system produced by Clostridium cellulolyticum explored by two-dimensional analysis: identification of seven genes encoding new dockerin-containing proteins. J Bact 189(6):2300–2309

Bode HB, Zeeck A, Plückhahn K, Jendrossek D (2000) Physiological and chemical investigations into microbial degradation of synthetic poly (cis-1, 4-isoprene). Appl Environ Microbiol 66(9):3680–3685

Bogino PC, Oliva MDLM, Sorroche FG, Giordano W (2013) The role of bacterial biofilms and surface components in plant-bacterial associations. Int J Mol Sci 14(8):15838–15859

Bonartsev AP, Myshkina VL, Nikolaeva DA, Furina EK, Makhina TA, Livshits VA, Boskhomdzhiev AP, Ivanov EA, Iordanskii AL, Bonartseva GA (2007) Biosynthesis, biodegradation, and application of poly (3-hydroxybutyrate) and its copolymers-natural polyesters produced by diazotrophic bacteria. Commun Curr Res Educ Top Trends Appl Microbiol 1:295–307

Bonhommea S, Cuerb A, Delort AM, Lemairea J, Sancelmeb M, Scott G (2003) Environmental biodegradation of polyethylene. Polym Degrad Stab 81:441–452

Bonilla CEP, Perilla JE (2011) The past, present and near future of materials for use in biodegradable orthopaedic implants. Ing Investig 31(2):124–133

Brune A, Frenzel P, Cypionka H (2000) Life at the oxic–anoxic interface: microbial activities and adaptations. FEMS Microbiol Rev 24(5):691–710

Bryers JD, Jarvis RA, Lebo J, Prudencio A, Kyriakides TR, Uhrich K (2006) Biodegradation of poly (anhydride-esters) into non-steroidal anti-inflammatory drugs and their effect on Pseudomonas aeruginosa biofilms in vitro and on the foreign-body response in vivo. Biomaterials 27(29):5039–5048

Calabia BP, Tokiwa Y (2004) Microbial degradation of poly (D-3-hydroxybutyrate) by a new thermophilic Streptomyces isolate. Biotechnol Lett 26(1):15–19

Carrier M, Serani AL, Absalon C, Aymonier C, Mench M (2012) Degradation pathways of holocellulose, lignin and a-cellulose from Pteris vittata fronds in sub- and super critical conditions. Biomass Bioenergy 43:65–71

Caruso G (2015) Plastic degrading microorganisms as a tool for bioremediation of plastic contamination in aquatic environments. J Pollut Eff Cont 3(3):1–2

Carvalheiro F, Duarte LC, Girio FM (2008) Hemicellulose biorefineries: a review on biomass pretreatments. J Sci Ind Res 67:849–864

Cazemier AE (1969) (Hemi) cellulose degradation by microorganisms from the intestinal tract of arthropods. Wageningen, Ponsen & Looijen, pp 1–135

Chahal PS, Chahal DS, André G (1992) Cellulase production profile of Trichoderma reesei on different cellulosic substrates at various pH levels. J Ferment Bioeng 74(2):126–128

Chaisu K, Siripholvat V, Chiu CH (2015) New method of rapid and simple colorimetric assay for detecting the enzymatic degradation of poly lactic acid plastic films. Int J Life Sci Biotechnol Pharm 4(1):57–61

Chandra R, Rustgi R (1998) Biodegradable polymers. Perg 23:1273–1335

Chang HM, Wang ZH, Luo HN, Xu M, Ren XY, Zheng GX, Wu BJ, Zhang XH, Lu XY, Chen F, Jing XH, Wang L (2014) Poly (3-hydroxybutyrate-co-3-hydroxyhexanoate)-based scaffolds for tissue engineering. Braz J Med Bio Res 47(7):533–539

Chao HSI, Hovatter TW (1987) Preparation and characterization of polyphenylene ether and nylon-6 block co polymer. Polym Bull 17:423–430

Charoenpanich J, Tani A, Moriwaki N, Kimbara K, Kawai F (2006) Dual regulation of a polyethylene glycol degradative operon by AraC-type and GalR-type regulators in Sphingopyxis macrogoltabida strain 103. Microbiology 152(10):3025–3034

Cheah K, Cook WD (2003) Structure-property relationships of blends of polycarbonate. Polym Eng Sci 43(11):1727–1739

Chen J, Zhang Y, Du GC, Hua ZZ, Zhu Y (2007) Biodegradation of polyvinyl alcohol by a mixed microbial culture. Enzyme Microbiol Technol 40(7):1686–1691

Chrissafis K, Paraskevopoulos KM, Bikiaris DN (2005) Thermal degradation mechanism of poly (ethylene succinate) and poly (butylene succinate): comparative study. Thermochim Act 43(2):142–150

Chrissafis K, Paraskevopoulos KM, Bikiaris DN (2006) Effect of molecular weight on thermal degradation mechanism of the biodegradable polyester poly (ethylene succinate). Thermochim Acta 440(2):166–175

Coen JA, Dehority BA (1970) Degradation and utilization of hemicellulose from intact forages by pure cultures of rumen bacteria. Appl Microbiol 20(3):362–368

Copinet A, Bertrand C, Govindin S, Coma V, Couturier Y (2004) Effects of ultraviolet light (315 nm), temperature and relative humidity on the degradation of polylactic acid plastic films. Chemosphere 55(5):763–773

Cragg SM, Beckham GT, Bruce NC, Bugg TD, Distel DL, Dupree P, Etxabe AG, Goodell BG, Jellison J, McQueen-Mason SJ, Schnorr K, Walton PH, Watts JE, Zimmer M (2015) Lignocellulose degradation mechanisms across the tree of life. Curr Opin Chem Biol 29:108–119

Danis O, Ogan A, Tatlican P, Attar A, Cakmakci E, Mertoglu B, Birbir M (2015) Preparation of poly (3-hydroxybutyrate-co-hydroxyvalerate) films from halophilic archaea and their potential use in drug delivery. Extr. doi:10.1007/s00792-015-0735-4

Das MP, Kumar S (2015) An approach to low-density polyethylene biodegradation by Bacillus amyloliquefaciens. 3 Biote 5(1):81–86

Datta A (2007) M.Sc. Thesis. B.E. University of Pune, Pune

Devi RS, Kannan VR, Nivas D, Kannan K, Chandru S, Antony AR (2015) Biodegradation of HDPE by Aspergillus spp. from marine ecosystem of Gulf of Mannar. India Marine Pollut Bull 96(1):32–40

Dey U, Mondal NK, Das K, Dutta S (2012) An approach to polymer degradation through microbes. J Pharm 2(3):385–388

Dhar M, Sepkovic DW, Hirani V, Magnusson RP, Lasker JM (2008) Omega oxidation of 3-hydroxy fatty acids by the human CYP4F gene subfamily enzyme CYP4F11. J Lipid Res 49:612–624

Dimarogona M, Topakas E, Christakopoulos P (2012) Cellulose degradation by oxidative enzymes. Comput Struct Biotechnol J 2(3):1–8

Duddu MK, Tripura KL, Guntuku G, Divya DS (2015) Biodegradation of low density polyethylene (LDPE) by a new biosurfactant-producing thermophilic Streptomyces coelicoflavus NBRC 15399T. Afr J Biotechnol 14(4):327–340

Dussud C, Ghiglione JF (2014) Bacterial degradation of synthetic plastics. In CIESM Workshop Monogr (No. 46)

Eubeler JP, Zok S, Bernhard M, Knepper TP (2009) Environmental biodegradation of synthetic polymers I. Test methodologies and procedures. TrAC Trends Anal Chem 28(9):1057–1072

Fernandez-Fueyo E, Ruiz-Dueñas FJ, López-Lucendo MF, Pérez-Boada M, Rencoret J, Gutiérrez A, Pisabarro AG, Ramírez L, Martínez AT (2016) A secretomic view of woody and nonwoody lignocellulose degradation by Pleurotus ostreatus. Biotechnol Biofuels 9(49):1–18

Feuilloley P, Cesar G, Benguigui L, Grohens Y, Pillin I, Bewa H, Lefaux S, Jama M (2005) Degradation of polyethylene designed for agricultural purposes. J Polym Environ 13(4):349–355

Fisher AB, Fong SS (2014) Lignin biodegradation and industrial implications. AIMS Bioeng 1(2):92–112

Flavel BS, Shapter JG, Quinton JS (2006) Nanosphere lithography using thermal evaporation of gold. IEEE, New York, pp 578–581

Fontanella S, Bonhomme S, Koutny M, Husarova L, Brusson JM, Courdavault JP, Pitterif S, Samuelg G, Pichonh G, Lemairea J, Delort AM (2010) Comparison of the biodegradability of various polyethylene films containing pro-oxidant additives. Polym Degrad Stab 95(6):1011–1021

Foster LJR, Tighe BJ (2005) Centrifugally spun polyhydroxybutyrate fibres: accelerated hydrolytic degradation studies. Polym Degrad Stab 87(1):1–10

French CE (2009) Synthetic biology and biomass conversion: a match made in heaven? J R Soc Interface: 1–12

Frings J, Schramm E, Schink B (1992) Enzymes involved in anaerobic polyethylene glycol degradation by Pelobacter venetianus and Bacteroides strain PG1. Appl Environ Microbiol 58(7):2164–2167

Fukae R, Fujii T, Takeo M, Yamamoto T, Sato T, Maeda Y, Sangen O (1994) Biodegradation of poly (vinyl alcohol) with high isotacticity. Polym J 26(12):1381–1386

Gajanand E, Soni LK, Dixit VK (2014) Biodegradable polymers: a smart strategy for today's crucial needs. Crit Rev Pharm Sci 3(1):1–70

Gajendiran A, Krishnamoorthy S, Abraham J (2016) Microbial degradation of low-density polyethylene (LDPE) by Aspergillus clavatus strain JASK1 isolated from landfill soil. 3 Biote 6(1):1–6

Gallert C, Winter J (2005) Bacterial metabolism in wastewater treatment systems. Wiley-VCH, Weinheim, pp 1–48

Garlotta D (2002) A literature review of poly (lactic acid). J Polym Environ 9(2):63–84

Gautam R, Bassi AS, Yanful EK, Cullen E (2007) Biodegradation of automotive waste polyester polyurethane foam using Pseudomonas chlororaphis ATCC55729. Int Biodeterior Biodegrad 60(4):245–249

Gerard T, Budtova T, Podshivalov A, Bronnikov S (2014) Polylactide/poly (hydroxybutyrate-co-hydroxyvalerate) blends: morphology and mechanical properties. Expr Polym Lett 8(8):609–617

Giudicianni P, Cardone G, Ragucci R (2013) Cellulose, hemicellulose and lignin slow steam pyrolysis: thermal decomposition of biomass components mixtures. J Anal Appl Pyrolysis 100:213–222

Grover A, Gupta A, Chandra S, Kumari A, Khurana SP (2015) Polythene and environment. Int J Environ Sci 5(6):1091–1105

Gu JD (2003) Microbiological deterioration and degradation of synthetic polymeric materials: recent research advances. Int Biodeterior Biodegrad 52(2):69–91

Gu JD, Mitchell R (2006) Biodeterioration. "The Prokaryotes". Springer, New York, pp 864–903

Guo W, Duan J, Geng W, Feng J, Wang S, Song C (2013) Comparison of medium-chain-length polyhydroxyalkanoates synthases from Pseudomonas mendocina NK-01 with the same substrate specificity. Microbiol Res 168:231–237

Hadad D, Geresh S, Sivan A (2005) Biodegradation of polyethylene by the thermophilic bacterium Brevibacillus borstelensis. J Appl Microbiol 98(5):1093–1100

Hasegawa T, Mikuni T (2014) Higher-order structural analysis of nylon-66 nanofibers prepared by carbon dioxide laser supersonic drawing and exhibiting near-equilibrium melting temperature. J Appl Polym Sci 40361:1–8

Hatakka A (2005) Biodegradation of lignin. University of Helsinki, Viikki Biocenter, Helsinki, pp 129–145

Hazen TC, Dubinsky EA, DeSantis TZ, Andersen GL, Piceno YM, Singh N, Jansson JK, Probst A, Borglin SE, Fortney JL, Stringfellow WT, Bill M, Conrad ME, Tom LM, Chavarria KL, Alusi TR, Lamendella R, Joyner JK, Spier C, Baelum J, Auer M, Zemla ML, Chakraborty R, Sonnenthal EL, D'haeseleer P, Holman HN, Osman S, Lu Z, Nostrand JDV, Deng P, Zhou J, Mason OU (2010) Deep-sea oil plume enriches indigenous oil-degrading bacteria. Science 330(6001):204–208

Hidayat A, Tachibana S (2012) Characterization of polylactic acid (PLA)/kenaf composite degradation by immobilized mycelia of Pleurotus ostreatus. Int Biodeterior Biodegrad 71:50–54

Hirota-Mamoto R, Nagai R, Tachibana S, Yasuda M, Tani A, Kimbara K, Kawai F (2006) Cloning and expression of the gene for periplasmic poly (vinyl alcohol) dehydrogenase from Sphingomonas sp. strain 113P3, a novel-type quinohaemoprotein alcohol dehydrogenase. Microbiology 152(7):1941–1949

Hoang KC, Tseng M, Shu WJ (2007) Degradation of polyethylene succinate (PES) by a new thermophilic Microbispora strain. Biodegradation 18(3):333–342

Horn SJ, Kolstad GV, Westereng B, Eijsink VG (2012) Novel enzymes for the degradation of cellulose. Biotechnol Biofuel 5(45):1–12

Howard GT (2002) Biodegradation of polyurethane: a review. Int Biodeterior Biodegrad 49:245–252

Howard GT (2012) Polyurethane biodegradation. Microbiol Degrade Xenobiot. 371–394

Howard GT, Mackie RI, Cann IKO, Ohene-Adjei S, Aboudehen KS, Duos BG, Childers GW (2007) Effect of insertional mutations in the pueA and pueB genes encoding two polyurethanases in Pseudomonas chlororaphis contained within a gene cluster. J Appl Microbiol 103(6):2074–2083

Hu X, Mamoto R, Shimomura Y, Kimbara K, Kawai F (2007) Cell surface structure enhancing uptake of polyvinyl alcohol (PVA) is induced by PVA in the PVA-utilizing Sphingopyxis sp. strain 113P3. Arch Microbiol 188(3):235–241

Huang YL, Li QB, Deng X, Lu YH, Liao XK, Hong MY, Wang Y (2005) Aerobic and anaerobic biodegradation of polyethylene glycols using sludge microbes. Process Biochem 40(1):207–211

Husarova L, Machovsky M, Gerych P, Houser J, Koutny M (2010a) Aerobic biodegradation of calcium carbonate filled polyethylene film containing pro-oxidant additives. Polym Degrad Stab 95(9):1794–1799

Husarova L, Ruzicka J, Marusincova H, Koutny M. 2010b. Use of temperature gradient gel electrophoresis for the investigation of poly (vinyl alcohol) biodegradation. Develop Energy Environ Econom: 157–159

Ianuzzo D, Patel P, Chen V, Obrien P, Willams C (1977) Hydrolysis of polyesters by lipases. Nature 270:76–78

Ikada Y, Tsuji H (2000) Biodegradable polyesters for medical and ecological applications. Macromol Rapid Commun 21(3):117–132

Iovino R, Zullo R, Rao MA, Cassar L, Gianfreda L (2008) Biodegradation of poly (lactic acid)/starch/coir biocomposites under controlled composting conditions. Polym Degrad Stab 93(1):147–157

Ishii N, Inoue Y, Shimada KI, Tezuka Y, Mitomo H, Kasuya KI (2007) Fungal degradation of poly (ethylene succinate). Polym Degrad Stab 92(1):44–52

Jeffrie TW (1994) Biodegradation of lignin and hemicelluloses. Biochem Microbiol Degrad: 233–277

Jeon IY, Baek JB (2010) Nanocomposites derived from polymers and inorganic nanoparticles. Materials 3:3654–3674

Jirage AS, Baravkar VS, Kate VK, Payghan SA, Disouza JI (2011) Poly-β-hydroxybutyrate: intriguing biopolymer in biomedical applications and pharma formulation trends. Int J Pharm Biol Arch 4(6):1107–1118

Jonsson LJ, Martín C (2016) Pretreatment of lignocellulose: formation of inhibitory by-products and strategies for minimizing their effects. Bioresour Technol 199:103–112

Junnarkar N, Pandhi N, Raiyani N, Bhatt N, Raiyani R (2016) Production of LiP by Phanerochaete chrysosporium MTCC 787 through solid state fermentation of wheat straw and assessing its activity against reactive black B. Int J Adv Res 4(1):812–819

Kadouri D, Burdman S, Jurkevitch E, Okon Y (2002) Identification and isolation of genes involved in poly (β-hydroxybutyrate) biosynthesis in Azospirillum brasilense and characterization of a phbC mutant. Appl Environ Microbiol 68(6):2943–2949

Kale SK, Deshmukh AG, Dudhare MS, Patil VB (2015) Microbial degradation of plastic: a review. J Biochem Technol 6(2):952–961

Kameshwar AKS, Qin W (2016) Recent developments in using advanced sequencing technologies for the genomic studies of lignin and cellulose degrading microorganisms. Int J Biol Sci 12:156–171

Karimi K, Taherzadeh MJ (2016) A critical review of analytical methods in pretreatment of lignocelluloses: composition, imaging, and crystallinity. Bioresour Technol 200:1008–1018

Kathiresan K (2003) Polythene and plastics-degrading microbes from the mangrove soil. Rev Biol Trop 51(3):629–634

Kato S, Chino K, Kamimura N, Masai E, Yumoto I, Kamagata Y (2015) Methanogenic degradation of lignin-derived monoaromatic compounds by microbial enrichments from rice paddy field soil. Sci Rep 5:1–11

Kawai F (2010) The biochemistry and molecular biology of xenobiotic polymer degradation by microorganisms. Biosci Biotechnol Biochem 74(9):1743–1759

Kawai F, Hu X (2009) Biochemistry of microbial polyvinyl alcohol degradation. Appl Microbiol Biotechnol 84(2):227–237

Khoramnejadian S (2013) Microbial degradation of starch based polypropylene. J Pure Appl Microbiol 7(4):2857–2860

Kokare CR, Chakraborty S, Khopade AN, Mahadik KR (2009) Biofilm: importance and applications. Ind J Biotechnol 8(2):159–168

Kolybaba M, Tabil LG, Panigrahi S, Crerar WJ, Powell T, Wang B (2003) Biodegradable polymers: past, present, and future. Soc Eng Agric Food Biol Syst: 1–15

Korotkova N, Lidstrom ME (2001) Connection between Poly-β-Hydroxybutyrate Biosynthesis and Growth on C1 and C2 Compounds in the Methylotroph Methylobacterium extorquens AM1. J Bact 183(3):1038–1046

Koutny M, Lemaire J, Delort AM (2006) Biodegradation of polyethylene films with prooxidant additives. Chemosphere 64(8):1243–1252

Koutsos V (2009) Polymeric materials: an introduction. ICE Man Constr Mater: 571–597

Kubokawa H, Hatakeyama T (2002) Melting behavior of nylon 6 fiber in textiles. J Therm Anal Calorim 70:723–732

Kumar AA, Karthick K, Arumugam KP (2011) Biodegradable polymers and its applications. Int J Biosci Biochem Bioinform 1(3):173–176

Kunioka M, Ninomiya F, Funabashi M (2006) Biodegradation of poly (lactic acid) powders proposed as the reference test materials for the international standard of biodegradation evaluation methods. Polym Degrad Stab 91(9):1919–1928

Kwon HJ, Jung CH, Hwang IT, Choi JH, Nho YC (2009) Surface functionalization of poly (ethylene terephthalate) for biomolecule immobilization by ion implantation. J Korea Phys Soc 54(5):2071–2075

Larking DM, Crawford RJ, Christie GBY, Lonergan GT (1999) Enhanced degradation of polyvinyl alcohol by Pycnoporus cinnabarinus after pretreatment with fenton's reagent. Appl Environ Microbiol 65(4):1798–1800

Larroque M, Barriot R, Bottin A, Barre A, Rougé P, Dumas B, Gaulin E (2012) The unique architecture and function of cellulose-interacting proteins in oomycetes revealed by genomic and structural analyses. BMC Genom 13(605):1–15

Leja K, Lewandowicz G (2010) Polymer biodegradation and biodegradable polymers—a review. Pol J Environ Stud 19(2):255–266

Leschine SB (1995) Cellulose degradation in anaerobic environments. Annu Rev Microbiol 49:399–426

Li R (2000) Environmental degradation of wood–HDPE composite. Polym Degrad Stab 70(2):135–145

Liang TW, Jen SN, Nguyen AD, Wang SL (2016) Application of chitinous materials in production and purification of a poly (l-lactic acid) depolymerase from Pseudomonas tamsuii TKU015. Polymers 8(98):2–11

Lim HA, Raku T, Tokiwa Y (2005) Hydrolysis of polyesters by serine proteases. Biotechnol Lett 27(7):459–464

Liu Q, Luo G, Zhou XR, Chen GQ (2011) Biosynthesisofpoly (3-hydroxydecanoate) and 3-hydroxydodecanoate dominating polyhydroxyalkanoatesby b-oxidation pathway inhibited Pseudomonas putida. Metab Eng 13:11–17

Liu C, Zeng JB, Li SL, He YS, Wang YZ (2012) Improvement of biocompatibility and biodegradability of poly (ethylene succinate) by incorporation of poly (ethylene glycol) segments. Polymers 53(2):481–489

Liu Q, Zhang H, Deng B, Zhao X (2014) Poly (3-hydroxybutyrate) and poly (3 hydroxybutyrate-co-3-hydroxyvalerate): structure, property, and fiber. Int J Polym Sci: 1–11

Lopez-Mondéjar R, Zühlke D, Větrovský T, Becher D, Riedel K, Baldrian P (2016) Decoding the complete arsenal for cellulose and hemicellulose deconstruction in the highly efficient cellulose decomposer Paenibacillus O199. Biotechnol Biofuel 9(104):1–12

Loredo-Treviño A, García G, Velasco-Téllez A, Rodríguez-Herrera R, Aguilar CN (2011) Polyurethane foam as substrate for fungal strains. Adv Biosci Biotechnol 2(2):52–58

Ma A, Wong Q (2013) Identification of esterase in Aspergillus flavus during degradation of polyester polyurethane. Can Young Sci J 2(2013):24–31

Ma J, Zhang K, Liao H, Hector SB, Shi X, Li J, Liu B, Xu T, Tong C, Liu X, Zhu Y (2016) Genomic and secretomic insight into lignocellulolytic system

of an endophytic bacterium Pantoea ananatis Sd-1. Biotechnol Biofuel 9(25):1–15

Maamar H, Valette O, Fierobe HP, Bélaich A, Bélaich JP, Tardif C (2004) Cellulolysis is severely affected in Clostridium cellulolyticum strain cipCMut1. Mol Microbial 51(2):589–598

Mabrouk MM, Sabry SA (2001) Degradation of poly (3-hydroxybutyrate) and its copolymer poly (3-hydroxybutyrate-co-3-hydroxyvalerate) by a marine Streptomyces sp. SNG9. Microbiol Res 156(4):323–335

Mahalakshmi V, Siddiq A, Andrew SN (2012) Analysis of polyethylene degrading potentials of microorganisms isolated from compost soil. Int J Pharm Biol Arch 3(5):1190–1196

Marusincova H, Husárová L, Růžička J, Ingr M, Navrátil V, Buňková L, Koutny M (2013) Polyvinyl alcohol biodegradation under denitrifying conditions. Int Biodeterior Biodegrad. 84:21–28

Masaki K, Kamini NR, Ikeda H, Iefuji H (2005) Cutinase-like enzyme from the yeast Cryptococcus sp. strain S-2 hydrolyzes polylactic acid and other biodegradable plastics. Appl Envir Microbiol. 71(11):7548–7550

Matsumura S, Shimura Y, Terayama K, Kiyohara T (1994) Effects of molecular weight and stereoregularity on biodegradation of poly (vinyl alcohol) by Alcaligenes faecalis. Biotechnol Lett 16(11):1205–1210

Mcalpine SR, Lindsley CW, Hodges JC, Leonard DM, Filzen GF (2001) Determination of functional group distribution within rasta resins utilizing optical analysis. J Comb Chem 3(1):1–5

Mccallum TJ, Kontopoulou M, Park CB, Muliawan EB, Hatzikiriakos SG (2007) The rheological and physical properties of linear and branched polypropylene blends. Polym Eng Sci 47:1133–1140

Mehmood CT, Qazi IA, Hashmi I, Bhargava S, Deepa S (2016) Biodegradation of low density polyethylene (LDPE) modified with dye sensitized titania and starch blend using Stenotrophomonas pavanii. Int Biodeterior Biodegrad. doi:10.1016/j.ibiod.2016.01.025

Menon V, Prakash G, Rao M (2010) Value added products from hemicellulose: biotechnological perspective. Division of Biochemical Sciences, National Chemical Laboratory, Pune, pp 1–58

Merrettig-Bruns U, Jelen E (2009) Anaerobic biodegradation of detergent surfactants. Material 2(1):181–206

Mohan SK, Srivastava T (2011) Microbial deterioration and degradation of polymeric materials. J Biochem Technol 2(4):210–215

Mooney A, Ward PG, O'Connor KE (2006) Microbial degradation of styrene: biochemistry, molecular genetics, and perspectives for biotechnological applications. Appl Microbiol Biotechnol 72(1):1–10

Moore CJ (2008) Synthetic polymers in the marine environment: a rapidly increasing, long-term threat. Environ Res 108:131–139

Motiwalla MJ, Punyarthi PP, Mehta MK, D'Souza JS, Kelkar-Mane V (2013) Studies on degradation efficiency of polycaprolactone by a naturally-occurring bacterium. J Environ Biol 34:43–49

Muenmee S, Chiemchaisri W, Chiemchaisri C (2015) Microbial consortium involving biological methane oxidation in relation to the biodegradation of waste plastics in a solid waste disposal open dump site. Int Biodeterior Biodegrad 102:172–181

Muenmee S, Chiemchaisri W, Chiemchaisri C (2016) Enhancement of biodegradation of plastic wastes via methane oxidation in semi-aerobic landfill. Int Biodeterior Biodegrad. doi:10.1016/j.ibiod.2016.03.016

Mumtaz T, Khan MR, Hassan MA (2010) Study of environmental biodegradation of LDPE films in soil using optical and scanning electron microscopy. Micron 41(5):430–438

Murphy CA, Cameron JA, Huang SJ, Vinopal RT (1996) Fusarium polycaprolactone depolymerase is cutinase. Appl Environ Microbiol 62(2):456–460

Nakajima-Kambe T, Shigeno-Akutsu Y, Nomura N, Onuma F, Nakahara T (1999) Microbial degradation of polyurethane, polyester polyurethanes and polyether polyurethanes. Appl Microbiol Biotechnol 51(2):134–140

Nakayama K, Furumiya A, Okamot T, Yag K, Kaito A, Choe CR, Wu L, Zhang G, Xiu L, Liu D, Masuda T, Nakajima A (1991) Structure and mechanical weight polyethylene deformed near melting temperature properties of ultra-high molecular. Pur Appl Chem 63(12):1793–1804

Nerland IL, Halsband C, Allan I, Thomas KV (2014). Microplastics in marine environments: occurrence, distribution and effects (Re.no.6754-2014). Norwegian Institute for Water Research, Oslo, pp 1–71. http://www.miljodirektoratet.no/Documents/publikasjoner/M319/M319.pdf. Accessed 31 Dec 2014

Nishida H, Tokiwa Y (1993) Distribution of poly (β-hydroxybutyrate) and poly (ε-caprolactone) aerobic degrading microorganisms in different environments. J Environ Polym Degrad 1(3):227–233

Nurbas M, Kutsal T (2004) Production of PHB and p (HB-co-HV) biopolymers by using *Alcaligenes eutrophus*. Iran Polym J 13(1):45–51

Nwachkwu S, Obidi O, Odocha C (2010) Occurrence and recalcitrance of poly-ethylene bag waste in nigerian soils. Afr J Biotechnol 9(37):6096–6104

Obradors N, Aguilar J (1991) Efficient biodegradation of high-molecular-weight polyethylene glycols by pure Cultures of *Pseudomonas stutzeri*. Appl Environ Microbiol 57(8):2383–2388

Oda Y, Asari H, Urakami T, Tonomura K (1995) Microbial degradation of poly (3-hydroxcybutyrate) and polycaprolctone by filamentous fungi. J Ferment Bioengine 80(3):265–269

Odusanya SA, Nkwogu JV, Alu N, Udo GE, Ajao JA, Osinkolu GA, Uzomah AC (2013) Preliminary studies on microbial degradation of plastics used in packaging potable water in Nigeria. Niger Food J 31(2):63–72

Ohta T, Tani A, Kimbara K, Kawai F (2005) A novel nicotinoprotein aldehyde dehydrogenase involved in polyethylene glycol degradation. Appl Microbiol Biotechnol 68(5):639–646

Ohtsubo Y, Nagata Y, Numata M, Tsuchikane K, Hosoyama A, Yamazoe A, Tsuda M, Fujita N, Kawai F (2015) Complete genome sequence of *Sphingopyxis macrogoltabida* type strain NBRC 15033, originally isolated as a polyethylene glycol degrader. Genome Announc 3(6):e01401–e01415

Ohura T, Kasuya KI, Doi Y (1999) Cloning and characterization of the polyhydroxybutyrate depolymerase gene of *Pseudomonas stutzeri* and analysis of the function of substrate-binding domains. Appl Environ Microbiol 65(1):189–197

Ojo OA (2007) Molecular strategies of microbial adaptation to xenobiotics in natural environment. Biotechnol Mol Biol Rev 2(1):1–13

Ojumu TV, Yu J, Solomon BO (2004) Production of polyhydroxyalkanoates, a bacterial biodegradable polymer. Afr J Biotechnol 3(1):18–24

Orr IG, Hadar Y, Sivan A (2004) Colonization, biofilm formation and biodegradation of polyethylene by a strain of *Rhodococcus ruber*. Appl Microbiol Btechnol 65(1):97–104

O'Toole G, Kaplan HB, Kolter R (2000) Biofilm formation as microbial development. Annu Rev Microbiol 54(1):49–79

Pajak J, Ziemski M, Nowak B (2010) Poly (vinyl alcohol)—biodegradable vinyl material. CHEM Nauka-Tech-Rynek 64(7–8):523–530

Patil R, Bagde US (2015) Enrichment and isolation of microbial strains degrading bioplastic polyvinyl alcohol and time course study of their degradation potential. Afr J Biotechnol 14(27):2216–2226

Pereira JH, Heins RA, Gall DL, McAndrew RP, Deng K, Holland KC, Donohue TJ, Noguera DR, Simmons BA, Sale KL, Ralph J (2016) Structural and biochemical characterization of the early and late enzymes in the lignin β-Aryl ether cleavage pathway from *Sphingobium* sp. SYK-6. J Chem 291(19):10228–10238

Perez J, Dorada JM, Rubia TDL (2002) Biodegradation and biological treatments of cellulose, hemicelluloses and lignin: an overview. Int Microbiol 5:53–63

Perez LAB, Rodriguez DN, Rodriguez FJM, Hsiao B, Orta CAA, Sics I (2014) Molecular weight and crystallization temperature effects on poly (ethylene terephthalate) (PET) homopolymers, an isothermal crystallization analysis. Polymers 6:583–600

Petre M, Zarnea G, Adrian P, Gheorghiu E (1999) Biodegradation and bioconversion of cellulose wastes using bacterial and fungal cells immobilized in radiopolymerized hydrogels. Resour Conserv Recycl 27:309–332

Pometto AL, Lee B, Johnson KE (1992) Production of an extracellular polyethylene-degrading enzyme(s) by *Streptomyces* species. Appl Environ Microbiol 58(2):731–733

Prakash B, Veeregowda BM, Krishnappa G (2003) Biofilms: a survival strategy of bacteria. Curr Sci 85(9):1299–1307

Pramila R, Padmavathy K, Ramesh KV, Mahalakshmi K (2012) *Brevibacillus parabrevis*, *Acinetobacter baumannii* and *Pseudomonas citronellolis*-Potential candidates for biodegradation of low density polyethylene (LDPE). Afr J Bacteriol Res 4(1):9–14

Pranamuda H, Tokiwa Y, Tanaka H (1997) Polylactide degradation by an *Amycolatopsis* sp. Appl Environ Microbiol 63(4):1637–1640

Prema S, Uma MDP (2013) Degradation of poly lactide plastic by mesophilic bacteria isolated from compost. Int J Res Purif Appl Microbiol 3(4):121–126

Premraj R, Doble M (2005) Biodegradation of polymer. Ind J Biotechnol 4:186–193

Psomiadou E, Arvanitoyannis I, Biliaderis CG, Ogawa H, Kawasaki N (1997) Biodegradable films made from low density polyethylene (LDPE),

wheat starch and soluble starch for food packaging applications Part 2. Carbohydr Polym 33(4):227–242

Puri N, Kumar B, Tyagi H (2013) Utilization of recycled wastes as ingredients in concrete mix. Int J Innov Technol Explor Eng 2(2):74–78

Qiu Z, Ikehara T, Nishi T (2003) Crystallization behaviour of biodegradable poly (ethylene succinate) from the amorphous state. Polymers 44(18):5429–5437

Raghul SS, Bhat SG, Chandrasekaran M, Francis V, Thachil ET (2013) Biodegradation of polyvinyl alcohol-low linear density polyethylene-blended plastic film by consortium of marine benthic vibrios. Int J Environ Sci Technol 11(7):1827–1834

Rameshwari R, Meenakshisundaram M (2014) A review on downstream processing of bacterial thermoplastic-polyhydroxyalkanoate. Int J Purif Appl Biosci 2(2):68–80

Ravachol J, Borne R, Tardif C, De Philip P, Fierobe HP (2014) Characterization of all family-9 glycoside hydrolases synthesized by the cellulosome-producing bacterium *Clostridium cellulolyticum*. J Biol Chem 289(11):7335–7348

Restrepo-Flórez JM, Bassi A, Thompson MR (2014) Microbial degradation and deterioration of polyethylene-a review. Int Biodeterior Biodegrad 88:83–90

Rong D, Usui K, Morohoshi T, Kato N, Zhou M, Ikeda T (2009) Symbiotic degradation of polyvinyl alcohol by *Novosphingobium* sp. and *Xanthobacter flavus*. J Environ Biotechnol 9(2):131–134

Ruiyun Z, Xiaolie L, Qunhua W, Dezhu M (1994) Melting behavior of low ethylene content polypropylene copolymer with and without nucleating agents. Chem J Polym Sci 12(3):246–255

Salgado CL, Sanchez EMS, Zavaglia CAC, Granja PT (2011) Biocompatibility and biodegradation of polycaprolactone-sebacic acid blended gels. J Biomed Mater Res 100A(1):243–251

Saloheimo M, Pakula TM (2012) The cargo and the transport system: secreted proteins and protein secretion in *Trichoderma reesei* (*Hypocrea jecorina*). Microbiology 158(1):46–57

Sangale MK, Shahnawaz M, Ade AB (2012) A review on biodegradation of polythene: the microbial approach. J Bioremed Biodegrad 3(10):1–9

Sangwan P, Wu DY (2008) New insights into polylactide biodegradation from molecular ecological techniques. Macromol Biosci 8(4):304–315

Santo M, Weitsman R, Sivan A (2013) The role of the copper-binding enzyme—laccase—in the biodegradation of polyethylene by the actinomycete Rhodococcus ruber. Int Biodeterior Biodegrad 84:204–210

Sarjit A, Tan SM, Dykes GA (2015) Surface modification of materials to encourage beneficial biofilm formation. AIMS Bioeng 2(4):404–422

Sathiskumar PS, Madras G (2011) Synthesis, characterization, degradation of biodegradable castor oil based polyesters. Polym Degrad Stab 96(9):1695–1704

Scheller HV, Ulvskov P (2010) Hemicelluloses. Annu Rev Plant Biol 61:263–289

Schink B, Janssen PH, Frings J (1992) Microbial degradation of natural and of new synthetic polymers. FEMS Microbiol Rev 103(2/4):311–316

Sekiguchi T, Saika A, Nomura K, Watanabe T, Fujimoto Y, Enoki M, Sato T, Kato C, Kanehiro H (2011) Biodegradation of aliphatic polyesters soaked in deep seawaters and isolation of poly (ε-caprolactone)-degrading bacteria. Polym Degrad Stab 96(7):1397–1403

Sen SK, Raut S (2015) Microbial degradation of low density polyethylene (LDPE): a review. J Environ Chem Eng 3:462–473

Seretoudi G, Bikiaris D, Panayiotou C (2002) Synthesis, characterization and biodegradability of poly (ethylene succinate)/poly (ε-caprolactone) block copolymers. Polymers 43(20):5405–5415

Shah AA, Hasan F, Hameed A, Ahmed S (2007) Isolation and characterization of poly (3-hydroxybutyrate-co-3-hydroxyvalerate) degrading bacteria and purification of PHBV depolymerase from newly isolated *Bacillus* sp. AF3. Int Biodeterior Biodegrad 60(2):109–115

Shah AA, Hasan F, Akhter JI, Hameed A, Ahmed S (2008a) Degradation of polyurethane by novel bacterial consortium isolated from soil. Anal Microbiol 58(3):381–386

Shah AA, Hasan F, Hameed A, Ahmed S (2008b) Biological degradation of plastics: a comprehensive review. Biotechnol Adv 26(3):246–265

Shah AA, Hasan F, Hameed A (2010) Degradation of poly (3-hydroxybutyrate-co-3-hydroxyvalerate) by a newly isolated *Actinomadura* sp. AF-555, from soil. Int Biodeterior Biodegrad 64(4):281–285

Shahnawaz M, Sangale MK, Ade AB (2016) Rhizosphere of *Avicennia marina* (Forsk.) Vierh. as a landmark for polythene degrading bacteria. Environ

Shanks R, Kong I (2012) Thermoplastic elastomers, 137–154. Applied sciences. RMIT University, Melbourne

Sharma M, Dhingra HK (2016) Poly-β-hydroxybutyrate: a biodegradable poly-ester, biosynthesis and biodegradation. Br Microbiol Res J 14(3):1–11

Sharma S, Rafailovich MH, Sokolov J, Liu Y, Schwarz SA, Eisenberg A (2000) Dewetting properties of polystyrene homopolymer thin films on grafted polystyrene brush surfaces. High Perform Polym 12:581–586

Sharma BK, Saha A, Rahaman L, Bhattacharjee S, Tribedi P (2015) Silver inhibits the biofilm formation of Pseudomonas aeruginosa. Adv Microbiol 5(10):677

Sheik S, Chandrashekar KR, Swarccccoop K, Somashekarappa HM (2015) Bio-degradation of gamma irradiated low density polyethylene and poly-propylene by endophytic fungi. Int Biodeterior Biodegrad 105:21–29

Shimao M (2001) Biodegradation of plastics. Curr Opin Biotechnol 12(3):242–247

Shimao M, Tamogami T, Kishida S, Harayama S (2000) The gene pvaB encodes oxidized polyvinyl alcohol hydrolase of Pseudomonas sp. strain VM15C and forms an operon with the polyvinyl alcohol dehydrogenase gene pvaA. Microbiology. 146(3):649–657

Simoes M, Simões LC, Vieira MJ (2010) A review of current and emergent biofilm control strategies. LWT-Food Sci Technol 43(4):573–583

Singh B, Sharma N (2008) Mechanistic implications of plastic degradation. Polym Degrad Stab 93:561–584

Sivan A (2011) New perspectives in plastic biodegradation. Curr Opin Biotech-nol 22:422–426

Sivan A, Szanto M, Pavlov V (2006) Biofilm development of the polyethylene-degrading bacterium Rhodococcus ruber. Appl Microbiol Biotechnol 72(2):346–352

Slade PE et al (1964) Thermal analysis of polyurethane elastomers. J Polym Sci 6:27–32

Smith R (2005) Biodegradable polymers for industrial applications. CRC Press, Boca Raton, pp 1–516

Song JJ, Yoon SC, Yu SM, Lenz RW (1998) Differential scanning calorimetric study of poly (3-hydroxyoctanoate) inclusions in bacterial cells. Int J Biol Macromol 23:165–173

Song JH, Murphy RJ, Narayan R, Davies GBH (2009) Biodegradable and com-postable alternatives to conventional plastics. Philos Trans R Soc Biol 364:2127–2139

Souza WRD (2013) Microbial degradation of lignocellulosic biomass. InTech, West Palm Beach, pp 207–247

Sridevi V, Lakshmi MVVC, Manasa M, Sravani M (2012) Metabolic pathways for the biodegradation of phenol. Int J Eng Sci Adv Technol 2:695–705

Stern RV, Howard GT (2000) The polyester polyurethanase gene (pueA) from Pseudomonas chlororaphis encodes a lipase. FEMS Microbiol Lett 185(2):163–168

Sugimoto M, Tanabe M, Hataya M, Enokibara S, Duine JA, Kawai F (2001) The first step in polyethylene glycol degradation by sphingomonads proceeds via a flavoprotein alcohol dehydrogenase containing flavin adenine dinucleotide. J Bact 183(22):6694–6698

Sukhumaporn S, Tokuyama S, Kitpreechavanich V (2012) Poly (L-Lactide)-degrading enzyme production by Actinomadura keratinilytica T16-1 in 3 L airlift bioreactor and its degradation ability for biological. J Microbiol Biotechnol 22(1):92–99

Summers JW (1996) A review of vinyl technology for non-scientists in the vinyl industry. Regional technical meeting, fort Mitchell. Soc Plas Eng. https://www.researchgate.net/publication/237644319

Summers JW (2008) The melting temperature (or not melting) of poly (vinyl chloride). J Vinyl Addit Technol 14:105–109

Sun J, Tian C, Diamond S, Glass NL (2012) Deciphering transcriptional regula-tory mechanisms associated with hemicellulose degradation in Neuros-pora crassa. Eukaryot Cell 11(4):482–493

Suyama T, Tokiwa Y, Ouichanpagdee P, Kanagawa T, Kamagata Y (1998) Phylogenetic affiliation of soil bacteria that degrade aliphatic polyesters available commercially as biodegradable plastics. Appl Environ Micro-biol 64(12):5008–5011

Sweileh BA, Hiari YMA, Kailani MH, Mohammad HA (2010) Synthesis and char-acterization of polycarbonates by melt phase interchange reactions of alkylene and arylene diacetates with alkylene and arylene diphenyl dicarbonates. Molecules 15:3661–3682

Takanashi M, Nomura Y, Yoshida Y, Inoue S (1982) Functional polycarbonate by copolymerization of carbon dioxide and epoxide: synthesis and

Tani A, Charoenpanich J, Mori T, Takeichi M, Kimbara K, Kawai F (2007) Struc-ture and conservation of a polyethylene glycol-degradative operon in sphingomonads. Microbiology 153(2):338–346

Tezuka Y, Ishii N, Kasuya KI, Mitomo H (2004) Degradation of poly (ethylene succinate) by mesophilic bacteria. Poly Degrad Stab 84(1):115–121

Thomas BT, Olanrewaju-Kehinde DSK, Popoola OD, James ES (2015) Degrada-tion of Plastic and Polythene materials by some selected microorgan-isms isolated from soil. World Appl Sci J. 33(12):1888–1891

Tobin E (2010) Microstructuralism and macromolecules: the case of moon-lighting proteins. Found Chem. doi:10.1007/s10698-009-9078-5

Tokiwa Y, Calabia BP, Ugwu CU, Aiba S (2009) Biodegradability of plastics. Int J Mol Sci 10(9):3722–3742

Tomita K, Kuroki Y, Nagai K (1999) Isolation of thermophiles degrading poly (l-lactic acid). J Biosci Bioeng 87(6):752–755

Tribedi P, Sil AK (2013a) Low-density polyethylene degradation by Pseu-domonas sp. AKS2 biofilm. Environ Sci Poll Res. 20(6):4146–4153

Tribedi P, Sil AK (2013b) Cell surface hydrophobicity: a key component in the degradation of polyethylene succinate by Pseudomonas sp. AKS2. J Appl Microbiol 116(2):295–303

Tribedi P, Sil AK (2013c) Founder effect uncovers a new axis in polyethylene succinate bioremediation during biostimulation. FEMS Microbial Lett 346(2):113–120

Tribedi P, Gupta AD, Sil AK (2015) Adaptation of Pseudomonas sp. AKS2 in biofilm on low-density polyethylene surface: an effective strategy for efficient survival and polymer degradation. Bioresour Bioproc 2(14):1–10

Tripathi D (2002) Practical guide to polypropylene. Rapid Technol Lim: 1–104

Tseng M, Hoang KC, Yang MK, Yang SF, Chu WS (2007) Polyester-degrading thermophilic actinomycetes isolated from different environment in Taiwan. Biodegradation 18(5):579–583

Tsujiyama SI, Nitta T, Maoka T (2011) Biodegradation of polyvinyl alcohol by Flammulina velutipes in an unsubmerged culture. J Biosci Bioeng 112(1):58–62

Turesin F, Gumusyazici Z, Kok FN, Gursel I, Alaaddinolu NG, Hasirci V (2000) Biosynthesis of polyhydroxybutyrate and its copolymers and their use in controlled drug release. Turk J Med Sci 30:535–541

Upreti MC, Srivastava RB (2003) A potential Aspergillus species for biodegrada-tion of polymeric materials. Curr Sci 84(11):1399–1402

Velde KV, Kiekens P (2002) Biopolymers: overview of several properties and consequences on their applications. Polym Test 121:433–442

Volova TG, Boyandin AN, Vasiliev AD, Karpov VA, Prudnikova SV, Mishukova OV, Boyarskikh UA, Filipenko ML, Rudnev VP, Xuân BB, Dũng VV, Gitelson II (2010) Biodegradation of polyhydroxyalkanoates (PHAs) in tropical coastal waters and identification of PHA-degrading bacteria. Polym Degrad Stab 95(12):2350–2359

Vu B, Chen M, Crawford RJ, Ivanova EP (2009) Bacterial extracellular polysac-charides involved in biofilm formation. Molecules 14(7):2535–2554

Walker TR, Reid K, Arnould JP, Croxall JP (1997) Marine debris surveys at Bird Island, South Georgia 1990–1995. Mar Poll Bull 34(1):61–65

Wan IY, Mcgrathz JE, Kashiwagi T (1995) Triaryphosphine oxide containing nylon 6, 6 copolymer. Am Chem Soc: 31–40

Weng YX, Wang XL, Wang YZ (2011) Biodegradation behavior of PHAs with different chemical structures under controlled composting conditions. Polym Test 30(4):372–380

Werner M, Pommer L, Brostrom M (2014) Thermal decomposition of hemicel-luloses. J Anal Appl Pyro 110:130–137

Wilkes CE, Daniels CA, Summers JW (2005) PVC hand book. Hansar, Bangkok, pp 315–335

Woodruff MA, Hutmacher DW (2010) The return of a forgotten polymer–polycaprolactone in the 21st century. Prog Polym Sci. doi:10.1016/j.progpolymsci.2010.04.002

Wu H, Qiu Z (2012) Synthesis, crystallization kinetics and morphology of novel poly (ethylene succinate-co-ethylene adipate) copolymers. Cryst Eng Comm 14(10):3586–3595

Xie Y, Kohls D, Noda I, Schaefer DW, Yvonne A, Akpalu YA (2009) Poly (3-hydroxybutyrate-co-3-hydroxyhexanoate) nanocomposites with optimal mechanical properties. Polymers 50:4656–4670

Xu Q, Resch MG, Podkaminer K, Yang S, Baker JO, Donohoe BS, Wilson C, Klingeman DM, Olson DG, Giannone RJ, Hettich RL, Brown SD, Lynd LR, Bayer EA, Himmel ME, Bomble YJ (2016) Dramatic performance of *Clostridium thermocellum* explained by its wide range of cellulase modalities. Sci Adv 2(e1501254):1–12

Yagi H, Ninomiya F, Funabashi M, Kunioka M (2009) Anaerobic biodegradation tests of poly (lactic acid) and polycaprolactone using new evaluation system for methane fermentation in anaerobic sludge. Poly Degrad Stab 94(9):1397–1404

Yagi H, Ninomiya F, Funabashi M, Kunioka M (2014) Mesophilic anaerobic biodegradation test and analysis of eubacteria and archaea involved in anaerobic biodegradation of four specified biodegradable polyesters. Poly Degrad Stab 110:278–283

Yam KL (2009) The Wiley encyclopedia of packaging technology (third edition). Wiley, New York, pp 1–1353

Yoon MG, Jeon HJ, Kim MN (2012) Biodegradation of polyethylene by a soil bacterium and AlkB cloned recombinant cell. J Bioremed Biodegrad 3(4):1–8

Zafar U (2013) Ph.D. Thesis. The University of Manchester, Manchester

Zafar U, Houlden A, Robson GD (2013) Fungal communities associated with the biodegradation of polyester polyurethane buried under compost at different temperatures. Appl Environ Microbiol 79(23):7313–7324

Zafar U, Nzeram P, Langarica-Fuentes A, Houlden A, Heyworth A, Saiani A, Robson GD (2014) Biodegradation of polyester polyurethane during commercial composting and analysis of associated fungal communities. Bioresour Technol 158:374–377

Zembouai I, Bruzaud S, Kaci M, Benhamida A, Corre YM, Grohens Y, Taguet A, Cuesta JML (2014) Poly (3-hydroxybutyrate-co-3-Hydroxyvalerate)/polylactide blends: thermal stability, flammability and thermo-mechanical behavior. J Polym Environ 22:131–139

Zhang W, Cheng X, Liu X, Xiang M (2016) Genome studies on nematophagous and entomogenous fungi in China. J Fungi 2(9):1–14

Zheng Y, Yanful EK (2005) A review of plastic waste biodegradation. Crit Rev Biotechnol 25:243–250

6

Lignocellulases: a review of emerging and developing enzymes, systems, and practices

Eugene M. Obeng[1], Siti Nurul Nadzirah Adam[1], Cahyo Budiman[1], Clarence M. Ongkudon[1,2]*, Ruth Maas[3] and Joachim Jose[3,4]

Abstract

The highly acclaimed prospect of renewable lignocellulosic biocommodities as obvious replacement of their fossil-based counterparts is burgeoning within the last few years. However, the use of the abundant lignocellulosic biomass provided by nature to produce value-added products, especially bioethanol, still faces significant challenges. One of the crucial challenging factors is in association with the expression levels, stability, and cost-effectiveness of the cellulose-degrading enzymes (cellulases). Interestingly, several recommendable endeavors in the bid to curb these challenges are in pursuance. However, the existing body of literature has not well provided the updated roadmap of the advancement and key players spearheading the current success. Moreover, the description of enzyme systems and emerging paradigms with high prospects, for example, the cell-surface display system has been ill-captured in the literature. This review focuses on the lignocellulosic biocommodity pathway, with emphasis on cellulase and hemicellulase systems. The paradigm shift towards cell-surface display system and its emerging recommendable developments have also been discussed. The attempts in supplementing cellulase with other enzymes, accessory proteins, and chemical additives have also been discussed. Moreover, some of the prominent and influential discoveries in the cellulase fraternity have been discussed.

Keywords: Cellulases, Lytic polysaccharide mono-oxygenases, Cellulase systems, Cell-surface display systems, Autodisplay systems, Cellulosomes

Background

The demand for cellulosic biocommodities as an alternative to fossil-based chemicals has surged within the last few decades. This burgeoning exploration could partly be attributed to the prevailing economic and environmental concerns of fossil-based chemicals. Lignocellulosic biomass is one of the abundant, low-cost, and renewable/sustainable feedstock for the production of biochemicals (including biofuels) due to its rich cellulose content (Roedl 2010; Doherty et al. 2011; Gallezot 2012). Unfortunately, the production of cellulosic biocommodities has been technically challenging owing to the recalcitrance

of lignocellulose, which comprises hemicellulose (20–30%), cellulose (30–40%), and lignin (20–30%) (Chang et al. 2011; Park et al. 2011). This recalcitrance has been identified as a major hindrance toward lignocellulose depolymerization. Technically, the resistance to enzymatic hydrolysis is ascribed to morphological and physicochemical factors such as lignin content (Hendriks and Zeeman 2009), degree of crystallinity (Park et al. 2010), degree of polymerization (Kim et al. 2015b), hemicellulose sheathing (Mosier et al. 2005), accessibility of inner microfibrils and porosity (Sharrock 1988), and moisture content and particle size of substrate (Chandra et al. 2007).

Also, the enzymatic hydrolysis of the cellulose and hemicellulose content of lignocellulosic biomass to their constituent monomeric sugars capable of use in the production of biocommodities (e.g., bioethanol and other

*Correspondence: clarence@ums.edu.my
[1] Biotechnology Research Institute, Universiti Malaysia Sabah, 88400 Kota Kinabalu, Sabah, Malaysia
Full list of author information is available at the end of the article

value-added biochemicals) has been hindered in so many ways. The hydrolysis mostly requires multiple enzymes with different specificities to deconstruct the complex lignocellulosic structure (Boyce and Walsh 2015). Specifically, a synergetic action of lignocellulases—cellulases, hemicellulases, lignases (ligninolytic enzymes) and, most recently, lytic polysaccharide mono-oxygenases (LPMO)—is required for an effective deconstruction activity. Remarkably, many efforts toward finding sustainable means of producing significant quantities of cellulosic biochemicals are in pursuance.

Consequently, various reviews focusing on lignocellulose-degrading enzymes, structure, and mode of actions have been remarkably reported (Rabinovich et al. 2002; Haki 2003; Ulrich et al. 2008; Wilson 2009; Juturu and Wu 2014; Bornscheuer et al. 2014). There are also reviews on cellulase engineering and other in vitro strategies towards improving the functionality of cellulases (Bayer et al. 2008; Himmel et al. 2010; Schoffelen and van Hest 2012). However, the fraternity still faces challenges in terms of robustness, hydrolysis efficiency, and cost of these crucial enzymes. Some exemplary accounts on cellulase improvement strategies have been reported (van den Burg 2003; Percival Zhang et al. 2006; Beckham et al. 2012; Elleuche et al. 2014). Nevertheless, these pronounced reviews individually could not provide an updated framework of the advancements and key players spearheading the current success. Moreover, the paradigmatic shift from cell-free systems to robust surface display systems has been ill-captured in the literature. Thus, the recommendable achievements have been uncoupled with the roadmap of cellulose-degrading enzymes.

This review provides an overview of lignocellulases and discusses the roadmap of enzymes and enzyme systems in ensuring that high levels of reduced sugars are obtained from the lignocellulosic biomasses for industrial use. The attempts in supplementing cellulase with other enzymes, accessory proteins, and chemical additives have also been discussed. Herein, the sterling progress in the surface display of enzymes has been emphasized. Moreover, some of the prominent and influential discoveries in the cellulase fraternity have been discussed.

Cellulases and their functional properties

Cellulases are glycoside hydrolases (GHs) that decompose cellulose—a hydrophilic, water-insoluble polymer composed of repeated units of D-glucose interlinked by β-1,4-glycosidic bonds—into shorter chain polysaccharides such as cellodextrin, cellobiose, and glucose. They commonly have a catalytic domain (CD) that cleaves the glycosidic bond; carbohydrate-binding module (CBM) that targets the CD to the polysaccharide substrate; and, in many cases, additional types of ancillary modules such as FN3-like modules (Moraïs et al. 2012; Garvey et al. 2013). Cellulases are distinctly categorized into three (i.e., endoglucanases, exoglucanases or cellobiohydrolases, and β-glucosidases or cellobiases) as per their structure and function, but work collaboratively to enforce the hydrolysis of the complex cellulose microfibrils of the plant cell wall.

The endo- and exoglucanases functionally perform the same task—the hydrolysis of glycosidic bonds—but they differ structurally in terms of the site (loop) for cellulose binding (Juturu and Wu 2014). For instance, endoglucanases (E.C.3.2.1.4) are characterized by short loops, defining open active site clefts that can bind to any accessible site (especially the amorphous sites) along cellulose chains to yield long-chain oligomers (Juturu and Wu 2014; Wilson 2015). They exhibit rapid dissociation compared with other cellulases, and their action on cellulose has been identified as the enzyme activity with greatest liquefaction ability that results in a decrease in the chain length and viscosity (Boyce and Walsh 2015).

For exoglucanases, they have long loops and affinity for the crystalline sites along cellulose chains and yield primarily cellodextrin (Segato et al. 2014). Most often, the loops form a tunnel around the catalytic residues; therefore, substrates usually are directed from the end of the tunnel to encounter the active site of the enzyme (Juturu and Wu 2014). Exoglucanases are in two forms—the reducing end (E.C.3.2.1.176) and non-reducing end (E.C. 3.2.1.91) cellobiohydrolases—but act uni-directionally on the long-chain oligomers (Juturu and Wu 2014). These classifications are based on the portion of the oligosaccharide chain each enzyme favorably attacks; however, they work "processively" to ensure the breakdown of the polysaccharide. For example, *Trichoderma reesei* cellobiohydrolases (Cel7A and Cel6A) progressively hydrolyze cellodextrin from the reducing and non-reducing chain ends, respectively (Wahlström et al. 2014). On the other hand, β-glucosidases possess a rigid structure with active site residing in a large cavity, called the active site pocket, which favors the entry of disaccharides (Nam et al. 2010); even though β-glucosidases are also capable of hydrolyzing soluble cellodextrins with degree of polymerization ≤6 (González-Candelas et al. 1989; Zhang and Lynd 2004). The active site pocket is encased in four hydrophobic loops with different conformations to enhance substrate binding (Czjzek et al. 2000; Nam et al. 2010). Like exoglucanase, β-glucosidases are classified into two sub-families, namely: sub-family A and sub-family B. Sub-family A includes plant and non-rumen prokaryotic cellobiases, and sub-family B includes fungal cellobiases (e.g., *Trichoderma reesei*, *Aspergillus niger*, and *A. aculeatus*) and rumen bacteria cellobiases, for example, from the anaerobic bovine symbiotic *Butyrivibrio fibrisolvens* (Park et al. 2011).

The complementary functions of these cellulases are crucial for efficient cellulose deconstruction. The classical hydrolysis theory explains that endoglucanases catalyze random deconstruction of cellulose chains along the amorphous regions through the cycles of adsorption and desorption, producing mainly cellodextrin; cellobiohydrolases processively hydrolyze the crystalline cellulose regions either from the reducing or non-reducing end, liberating cellobiose as their main product; and β-glucosidases finally hydrolyze the released soluble cello-oligomers to glucose (Wahlström et al. 2014). The cascading depolymerization activity is governed by (1) synergism, (2) processivity, and (3) substrate-channeling

ability of the enzyme, and the catalytic mechanism (Fig. 1) follows the classical acid-catalyst hydrolysis model (Garvey et al. 2013). Two critical amino acid residues—one as a proton donor and the other as a nucleophile—facilitate the enzymatic cleavage of glycosidic bonds by the stereochemical modification (i.e., retention or inversion) of the anomeric carbon configuration (Koshland 1953; Garvey et al. 2013). It is worth noting that the products of both endo- and exoglucanases can inhibit the respective enzyme in a process, called feedback inhibition. For this reason, exoglucanases and β-glucosidases are essentially required to relieve endo- and exoglucanases, respectively, from feedback inhibition. Similarly,

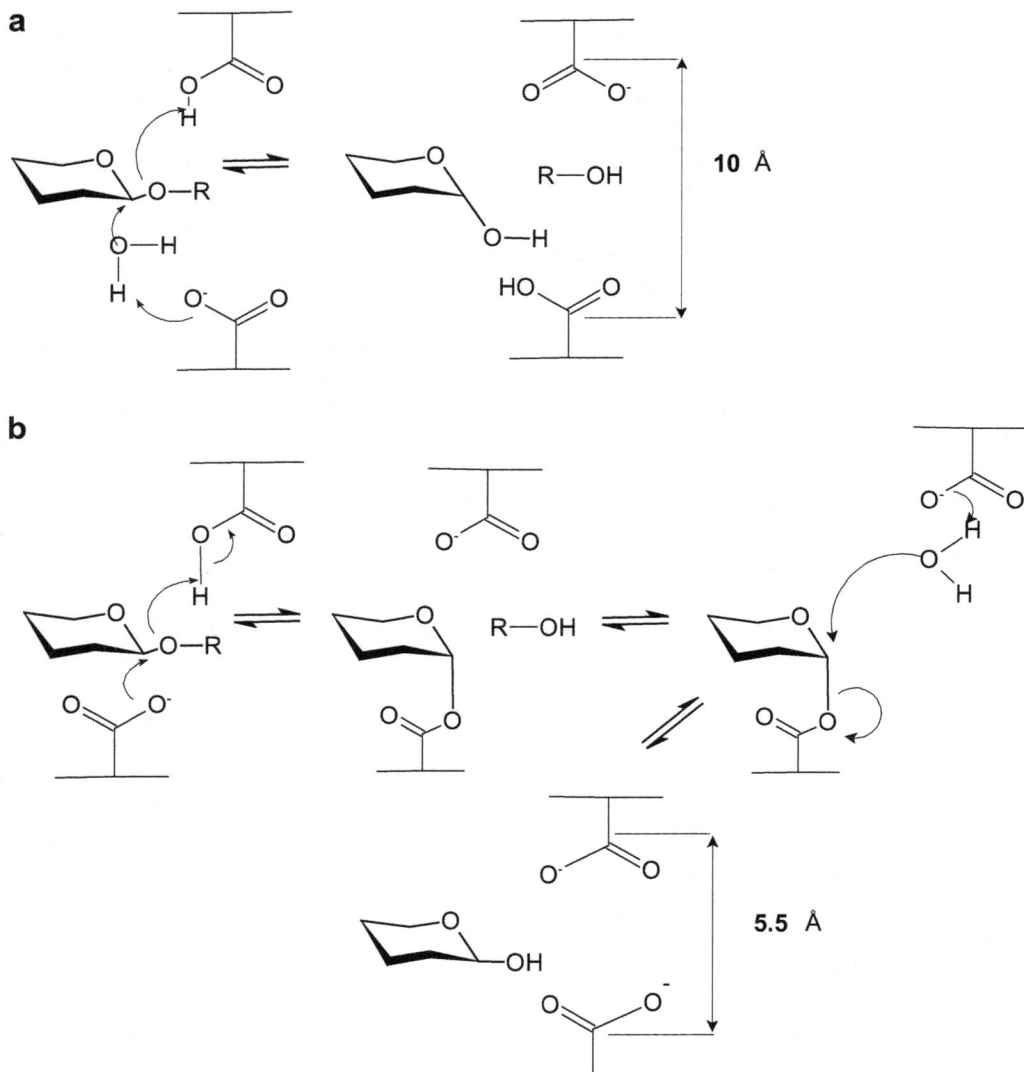

Fig. 1 The two major catalytic mechanisms of GHs, namely: the inversion (**a**) and retention (**b**) mechanisms. These two mechanisms lead to the effective hydrolysis of cellulosic substrates. The reader is referred to Zechel and Withers (2000) and Koshland (1953) for a detailed review

β-glucosidases also face glucose inhibition and, thus, the search for glucose-tolerant β-glucosidases is developing.

Recent insights have revealed oxidative enzymes, lytic polysaccharide mono-oxygenases (LPMOs), as key players in biomass decomposition. According to reports, LPMOs complement the functionality of the canonical cellulases by improving substrate accessibility and introducing chain breaks in the cellulose strand by oxidative means (Vaaje-Kolstad et al. 2010; Horn et al. 2012). The emergence of these auxiliary enzymes has critically disputed the classical concept of carbohydrate polymer saccharification and, thus, has provided additional insight into how saprophytes effectively attack cellulosic substrate (Hemsworth et al. 2013a). LPMOs have been further discussed in "Lytic polysaccharide mono-oxygenases

(LPMOs)." Figure 2 describes the contemporary understanding of cellulose degradation.

Common and developing sources of cellulases

Cellulases have been commonly sourced from different organism, mainly fungi, bacteria, and protozoans, although plant and animal cellulases are known (Kim and Kim 2012). Among the organisms, fungi and bacteria express functionally diverse multiple isoforms of cell wall degrading enzymes as a result of genetic redundancy, differential mRNA processing, or post-translational modification (Badhan et al. 2007). Therefore, fungi and bacteria have become the focus of the recent cellulase industry. Table 1 displays some cellulolytic fungi and bacteria with their sources.

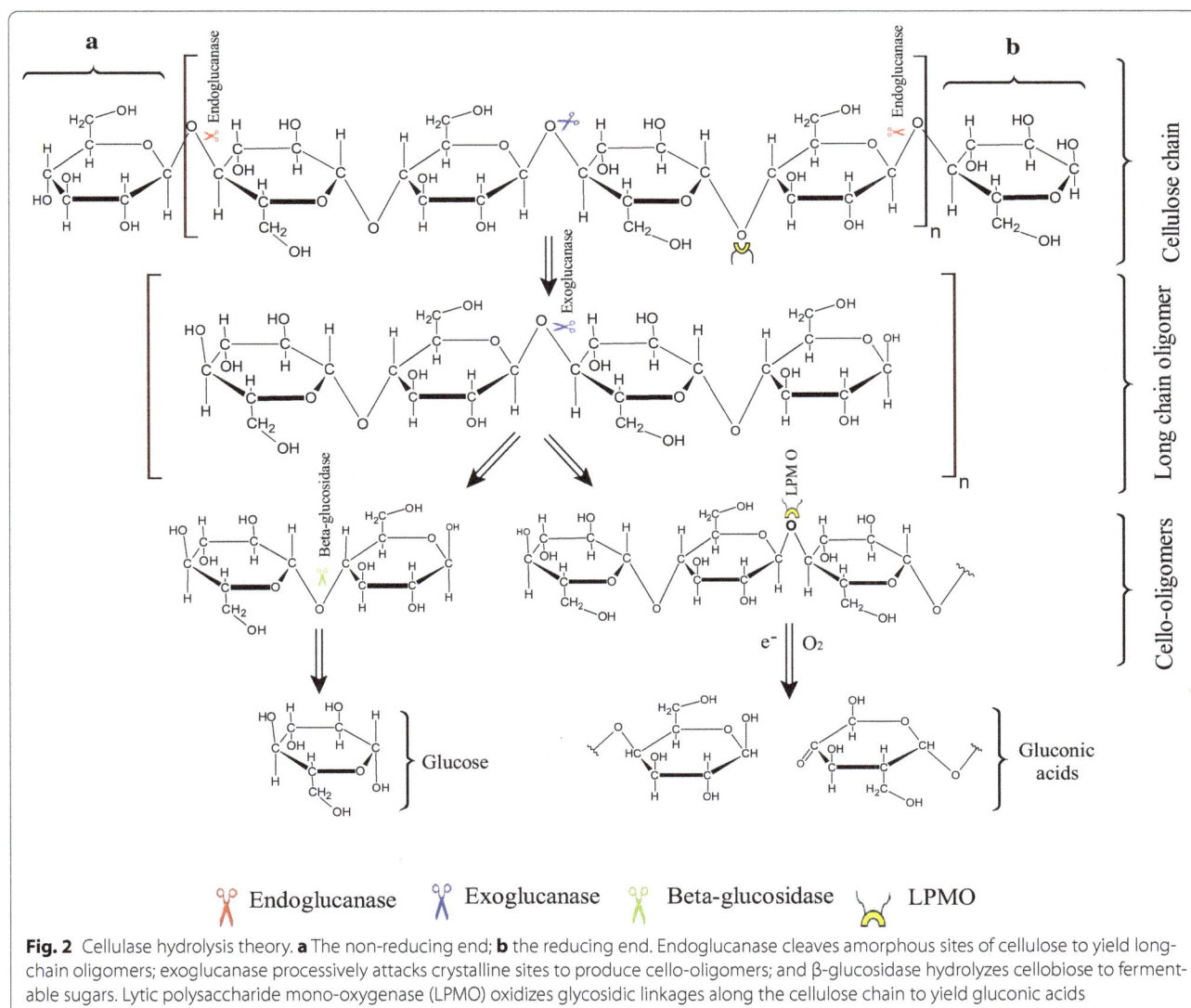

Fig. 2 Cellulase hydrolysis theory. **a** The non-reducing end; **b** the reducing end. Endoglucanase cleaves amorphous sites of cellulose to yield long-chain oligomers; exoglucanase processively attacks crystalline sites to produce cello-oligomers; and β-glucosidase hydrolyzes cellobiose to fermentable sugars. Lytic polysaccharide mono-oxygenase (LPMO) oxidizes glycosidic linkages along the cellulose chain to yield gluconic acids

Table 1 Some cellulolytic microbes and their sources. Modified from: Himmel et al. (2010)

Bacteria		Fungi	
Species	Source	Species	Source
Aerobes (free, non-complexed cellulases)			
Mesophilic bacteria		Mesophilic fungi	
Bacillus brevis[a]	Termite gut	*Aspergillus nidulans, A. niger, A. oryzae*	Soil, wood rot
B. thuringiensis[a]	Caterpillar gut		
Bacillus cereus[a], *B. subtilis*[a]	Soil, rumen	*Agaricus bisporus*	Compost
Cellulomonas fimi[a]	Soil	*Coprinus truncorum*	Soil, compost
Cellvibrio japonicas	Soil	*Geotrichum candidum*	Soil, compost
Cytophaga hutchinsonii	Soil, compost	*Penicillium chrysogenum*	Soil, wood rot
Paenibacillus polymyxa	Compost	*Phanerochaete chrysosporium*	Compost
Pseudomonas fluorescens	Soil, sludge	*Rhizopus oryzae*	Soil, dead organic matter
Pseudomonas putida	Soil, sludge	*Trichocladium canadense*	Soil
Saccharophagus degradans	Rotting marsh grass	*Trichoderma reesei*	Soil, rotting canvas
Sorangium cellulosum	Soil	*Trichoderma longibrachiatum*	Soil
Thermophilic bacteria		Thermophilic fungi	
Acidothermus cellulolyticus	Hot spring	*Chaetomium thermophilum*	Soil
Thermobifida fusca	Compost	*Corynascus thermophilus*	Mush compost
		Paecilomyces thermophile	Soil, compost
		Thielavia terrestris	Soil
Anaerobes (complexed or free, non-complexed cellulases)			
Mesophilic bacteria		Mesophilic fungi	
Acetivibrio cellulolyticus	Sewage	*Neocallimastix patriciarum*	Rumen
Bacteroides cellulosolvens	Sewage	*Orpinomyces joyonii*	Rumen
Clostridium cellulolyticum	Compost	*Orpinomyces PC-2*	Rumen
Clostridium cellulovorans	Wood fermenter	*Piromyces equi*	Rumen
Clostridium josui	Compost	*Piromyces E2*	Feces
Clostridium papyrosolvens	Mud (freshwater)		
Clostridium phytofermentans	Soil		
Fibrobacter succinogenes	Rumen		
Prevotella ruminicola	Rumen		
Ruminococcus albus	Rumen		
Ruminococcus flavefaciens	Rumen		
Thermophilic bacteria			
Anaerocellum thermophilum	Hot spring		
Caldicellulosiruptor Saccharolyticus	Hot spring		
Clostridium thermocellum	Sewage, soil, manure		
Clostridium stercorarium	Compost		
Thermotoga maritima	Mud (marine)		
Rhodothermus marinus	Hot spring		

[a] Most *Cellulomonas* and *Bacillus* strains are facultative anaerobes that can also grow anaerobically

Fungi sources

Currently, fungi are the most studied group of cellulose-degrading microorganisms, owing to their high protein secretion capabilities and multi-component, synergetic, cellulolytic, enzyme activity (Ulrich et al. 2008; Juturu and Wu 2014). The most extensively studied cellulolytic enzymes are *T. reesei* cellulases because of their application in commercial cellulase preparations (Wahlström et al. 2014). The cellulase mixtures of *T. reesei* (the 'gold standard') consist predominantly of exoglucanases, which contribute up to 80% of the total protein; endoglucanases (up to 15% of the total protein); and lesser amounts of enzymes with other hydrolytic activities (Garvey et al. 2013). According to Parisutham et al.

(2014), *T. reesei* also possesses intracellular β-glucosidase to avoid effects of cellobiose feedback inhibition during cellulose hydrolysis. However, the levels of β-glucosidases are mostly low and, thus, require supplementation from other sources such as Aspergilli.

The emergence of filamentous fungi of the genus *Aspergillus* as one of the key cellulase-producing organisms has made an outstanding impact in bioprocessing. For example, *Aspergillus oryzae* (Chandel et al. 2011; Begum and Alimon 2011), *A. unguis* (Rajasree et al. 2013), *A. tubingensis* (Decker et al. 2001), *A. fumigatus* (Watanabe et al. 1992; Anthony et al. 2003; Soni et al. 2010; Sherief et al. 2010; Liu et al. 2011; Das et al. 2013), and the most pronounced *A. niger* (Kang et al. 2004; Hanif et al. 2004; Varzakas et al. 2006; Sohail et al. 2009; Sakthi et al. 2011; Bansal et al. 2012; Oberoi et al. 2014) have been studied for their cellulolytic benefits. The *Aspergillus* species produce different isoforms of enzymes such as cellulases, xylanases, laccases, and other accessory proteins necessary for biomass depolymerization. The multiplicity is due to the presence of diverse protein encoding genes, differential glycosylation of common polypeptide chains, and post-translational modification disparities (Willick and Seligy 1985; Decker et al. 2001). Moreover, physical and nutritional factors may also account for reported differences in enzyme expression and expression levels. Enzymes from Aspergilli are mostly reported to exhibit low total cellulase activity (Falkoski et al. 2013); however, their high β-glucosidase expressing levels have made them relevant game changers for industrial applications. One remarkable property of the species in the genus is their tolerance against osmotic gradients. For example, the high glucose-tolerance of β-glucosidases from *Aspergillus* sp. has been reported (Riou et al. 1998; Günata and Vallier 1999; Rajasree et al. 2013; Das et al. 2015), and this revelation has been vital in the roadmap to the 'green' future.

Yeast has also had its use in cellulolytic investigations. Foreseeably, yeast found its application as a common expression platform for enzyme systems because of its robustness. Interestingly, a recombinant yeast has been able to express three copies each of endoglucanase and exoglucanase, and one copy of β-glucosidase for cellulose depolymerization (Matano et al. 2013; Parisutham et al. 2014). According to Juturu and Wu (2014), yeast provides numerous advantages when used as a host for recombinant protein expression. The benefits include: (1) the ability to perform eukaryotic post-translational modifications; (2) the ability to secrete recombinant proteins; (3) the ability to grow to very high cell densities; (4) the wide availability of yeast strains for recombinant protein expression; and (5) the relatively toxin-free nature of yeast cells in comparison with endotoxin-associated

bacterial strains, whose products may require purification (if ingestible or injectable). The unending stream of science has more to uncover regarding fungal cellulases, owing to their capabilities of producing copious amounts of enzymes.

Bacteria sources

Although much of the cellulases used for lignocellulosic biomass hydrolysis are derived from fungi, yet the isolation and characterization of novel carbohydrate-degrading enzymes from bacteria are now widely exploited. This is because of the efficient heterologous production, high specific activity, and less stringent pH requirement of bacterial systems. The most effective natural cellulolytic system known is produced by bacteria (Stern et al. 2015). Well-known genera for bacteria-based cellulolytic enzymes are mostly *Bacillus, Cellulomonas, Streptomyces, Cytophaga, Cellvibrio*, and *Pseudomonas*. Although many types of proteins have been produced by *Escherichia coli*, there is no report on natural cellulolytic *E. coli* in the past several years (Yamada et al. 2013). However, through metabolic engineering *E. coli* are made tractable such that they can be endowed with an efficient cellulolytic system capable of producing high-value compounds from lignocellulosic biomass.

In bacteria, cellulases are mostly present as extracellular aggregated structures attached to the cells (Juturu and Wu 2014). However, the expression of highly active cellulases of fungal origin in bacterial expression platforms has been a persisting challenge, with many resulting in diverse expression inefficiencies (Garvey et al. 2013). *E. coli* remains the most commonly used system for recombinant cellulase protein production, particularly for the expression and characterization of novel cellulolytic proteins, including those from extreme habitats or animal guts (Garvey et al. 2013). The high protein secretion capacity of *Bacillus subtilis*, with its high-activity endoglucanase, has also been used to engineer recombinant cellulase strains that thrive on cellulose as a sole carbon source without any other organic nutrient (Zhang 2011).

Remarkably, the future of cellulolytic enzyme sources is gradually shifting toward bacterial sources. The discovery of the exceptional cellulolytic properties of bacteria from the genera *Clostridium* and *Thermotoga* has contributed to the gradual shift from the dominant fungi sources to that of bacteria. The nature of cellulases from these species are thermostable and optimally active at elevated temperatures between 60 and 125 °C (Vieille and Zeikus 2001; Schiraldi and De Rosa 2002; Haki 2003); thus, making them essential candidates for improving the techno-economics of biomass saccharification (Parisutham et al. 2014). Notably, running enzymatic hydrolysis at higher temperatures has the penchant to (1) promote biomass disorganization; (2) increase substrate solubility; (3) improve rheological properties (e.g., viscosity); and (4) reduce the risk of microbial contamination (Vieille and Zeikus 2001; Boyce and Walsh 2015). Bacteria from the genera *Clostridium* and *Thermotoga* also produce self-assembled scaffolded multimodular enzyme systems, termed cellulosomes, to efficiently hydrolyze the complex and rigid structure of cellulose (Brunecky et al. 2012). The extreme stabilities (e.g., pH and thermal) and multifunctional nature of enzymes produced by these cellulosome-expressing bacteria have revolved the attention of scientist on to understanding the structure and function of their genetic makeup in order to mimic the innate abilities. On account of the obvious benefits reported in the literature, scientists have consistently investigated the gene (Yagüe et al. 1990; Zverlov et al. 2003; Koeck et al. 2013), fusion/modification of enzymes (Ciolacu et al. 2010; Lee et al. 2010; Ye et al. 2010; Lee et al. 2011; Nakashima et al. 2014), and the optimal growth (Islam et al. 2013; Reed et al. 2014) of these useful microbes to harness their inherent benefits. For example, the biochemical and biophysical characteristics of multimodular enzymes from *Clostridium thermocellum* (Zverlov et al. 2005; Tachaapaikoon et al. 2012; Brunecky et al. 2012; Hirano et al. 2013; Yuan et al. 2015) and *Thermotoga maritima* (Chhabra et al. 2002; Carvalho et al. 2004; Pereira et al. 2010; Wu et al. 2011) have been reported.

Most currently, the main interest of the biobased industries has been on the application of extremozymes (Demirjian et al. 2001; Egorova and Antranikian 2005). These enzymes derived from extremophilic microorganisms (acidophiles, alkaliphiles, halophiles, thermophiles, psychrophiles, and piezophiles) are rich sources of natural tailored enzymes, which are functionally more superior over their mesophilic counterparts for applications at extreme/harsh conditions that were long thought to be destructive to proteins (van den Burg 2003; Elleuche et al. 2014). Extremozymes are capable of catalyzing their respective reactions in non-aqueous environments, water/solvent mixtures, at extremely high pressures, acidic and alkaline pH, at temperatures up to 140 °C, or near the freezing point of water (Schiraldi and De Rosa 2002; Elleuche et al. 2014). The outstanding prospects of these enzymes have created a surge in their investigation for use in biotechnological and industrial applications. In conformity with industrial demands, the cellulolytic prospects of the anaerobic extremophile, *Caldicellulosiruptor bescii* (formerly *Anaerocellum thermophilum*, isolated from a geothermally heated pool), have been exemplified in literature (Yang et al. 2009; Kanafusa-Shinkai et al. 2013). The *C. bescii* and some of its relatives in the same genus secrete free (non-cellulosomal) biomass-degrading enzymes rich in CBMs (specifically

CBM3 family) that target the enzymes to crystalline cellulose, but show high degree of multi-modularity (Harris et al. 2014). This Gram-positive, non-spore-forming, neutrophilic, cellulolytic/hemicellulolytic bacterium grows in a temperature range of 40–90 °C, with an optimum temperature of 72–80 °C, and efficiently degrades crystalline cellulose, xylan, and non-pretreated plant biomass such as Napier grass, switch grass, and hardwood poplar (Yang et al. 2009; Kanafusa-Shinkai et al. 2013). For example, the *C. bescii* CelA (comprising a GH family 9 and a family 48 CD, as well as three type-III CBMs) and its fragments can depolymerize lignocellulosic biomass to glucose, cellobiose, and xylose via a combined surface ablation and cavity-forming mechanism without the help of accessory proteins (Brunecky et al. 2012). These abilities of *C. bescii* make it a potential candidate for thorough investigation and implementation. Table 2 shows a list of some industrially relevant thermostable cellulases that have been isolated and characterized.

Developing practices for improving the production and performance of cellulases
Some cellulase improvement techniques
Recently, there have been several attempts to acquire highly efficient cellulases with improved cellulolytic activity and stability (di Lauro et al. 2006; Mesas et al. 2012; Jagtap et al. 2013). Various improvement methods including rational design and directed evolution in complementation with techniques like DNA family shuffling and error-prone polymerase chain reaction (PCR) have been prominent. For example, Wang et al. (2014) have reported the application of random mutagenesis followed by genome shuffling to improve the cellulase production of *Trichoderma koningii* D-64. Also, structure-based protein design has been successfully used to increase thermal resistance and modify substrate specificity of glucosidases (Lee et al. 2012). The uses of random insertion domain strategies to allosterically modify enzymes have also been reported (Ribeiro et al. 2015). These allosteric enzymes present spatially distinct locations for regulation and catalysis and offer oligomeric states where tertiary and quaternary structural changes are transmitted across protein–protein interfaces to facilitate the communication between effector binding and modulation of catalytic activity (Ribeiro et al. 2015). The random insertion strategy has been relevant for curbing the hindrance of inhibition. However, the very large and costly nature of random insertion libraries, and associated bias towards certain insertion points have been challenging; therefore, the design of smaller high-quality libraries using a semi-rational approach is developing. Convincingly, the application of metagenomic techniques to exploit the functional genes in uncultured natural

Table 2 Some thermostable cellulases of industrial significance

Cellulase	Source/organism	Maximum/optimum activity	Stability ($T_{1/2}$)	References
Cellulase	*Desulfurococcus fermentans* (Hyperthermophilic archaea)	80–82 °C (pH 6)	85 °C, >3 days	(Perevalova 2005)
Endoglucanase (GH5)	Hyperthermophilic archaea	109 °C (pH 6.8)	100 °C, 5 h	(Graham et al. 2011)
Endoglucanase (GH5)	*Dictyoglomus thermophilum*	50–85 °C (pH 5)	70 °C, 336 h	(Shi et al. 2013)
β-Glucosidase (GH1)	*Thermotoga thermarum* DSM 5069T	90 °C (pH 4.8)	90 °C, 2 h	(Zhao et al. 2013)
β-Glucosidase (GH1)	Hydrothermal spring metagenome	90 °C (pH 6.5)	90 °C, >1.5 h	(Schröder et al. 2014)
β-Glucosidase (GH1)	*Alicyclobacillus acidocaldarius*	65 °C (pH 5.5)	65 °C, >3 h	(di Lauro et al. 2006)
β-Glucosidase (GH3)	*Thermofilum pendens*	90 °C (pH 3.5)	90 °C, >2 h	(Li et al. 2013)
β-Glucosidase (GH3)	*Pholiota adiposa* SKU0714	65 °C (pH 5)	65 °C, 23 h	(Jagtap et al. 2013)
β-Glucosidase	*Pyrococcus furiosus*	102–05 °C (pH 5)	100 °C, 85 h	(Kengen et al. 1993)
β-Glucosidase	*Oenococcus oeni* ST81	40 °C (pH 5)	40 °C, 50 days	(Mesas et al. 2012)
β-Glucosidase	*Sphingopyxis alaskensis*	50 °C (pH 5.5)	NA	(Shin and Oh 2014)
CMcellulase	*Bacillus pumilus* S124A	50 °C (pH 6.0)	NA	(Balasubramanian and Simões 2014)
Multi-domain (Hemi)cellulase	*Caldicellulosiruptor bescii*	72–80 °C (pH 5–6)	NA	(Kanafusa-Shinkai et al. 2013)

NA not analyzed

microorganisms could help in overcoming the limit of pure cultivation methods (Chang et al. 2011). Moreover, harnessing glycosylation—a form of post-translational modification—to improve cellulase activity looks promising (Easton 2011; Beckham et al. 2012).

Cellulase supplementations

The supplementation of cellulases with additives (biological and non-biological) for lignocellulose saccharification has been witnessed. This developing practice is based on the understanding that the effective degradation of the complex structure of lignocellulose requires not only cellulases, but also supplementary enzyme blends of ligninases (laccases), hemicellulases, and accessory proteins, depending on the morphological characteristics of the lignocellulosic biomass. Chemical additives have also been used to improve the functionality of cellulases.

Biological additives

Laccases Laccases (benzenediol: oxygen oxidoreductase; EC 1.10.3.2) are multi-copper oxidases, capable of catalyzing one-electron oxidation of various substrates such as phenolic and non-phenolic subunits of lignin (Lahtinen et al. 2009; Dwivedi et al. 2011; Chandel et al. 2013). Laccases have four copper atoms present in their active sites which are distributed at three different copper centers, namely: Type-1 (blue copper center), Type-2 (normal copper), and Type-3 (coupled binuclear copper centers) (Dwivedi et al. 2011). These copper atoms serve as a catalytic metal and reducing agent for the oxidation of various carbons (C-1, C-4, and C-6) in the polymeric

structure (Segato et al. 2014). Ascorbate oxidase, ferroxidases, ceruloplasmin, and bilirubin oxidases are examples of members in the multi-copper protein family.

The primary substrate of laccases is lignin, an amorphous, complex cross-linked polymer consisting of phenylpropane units (Claus 2004; Moilanen et al. 2011). In general, laccases break down lignin into less harmful products, using electron transfer and hydrogen atom transfer mediators. Laccases are widely distributed in plants, fungi, and bacteria and exhibit diverse functions and stability, depending on their source organism and physiology. Molecular structure elucidations and the electrochemical assessment of laccases have resulted in three classifications, namely: high, medium, and low redox potential laccases (Mot and Silaghi-Dumitrescu 2012; Mate and Alcalde 2015). Plant and bacterial laccases belong to the low redox potential category, whereas fungal laccases are categorized as either high or medium redox potential laccases. The magnitude of the redox potential correlates with the substrate range and oxidation capacity of the enzyme (Mate and Alcalde 2015). As a result, fungal laccases exhibit high wood depolymerization activity and are widely distributed in ascomycetes, deuteromycetes, and basidiomycetes; the most efficient species known is the white-rot fungus (Dwivedi et al. 2011; Pandiyan et al. 2014). Bacterial laccases are also active lignin degraders, but with high thermal and pH stability compared with fungal laccases, and hence more compatible with almost all industrial processes when immobilized (Dwivedi et al. 2011). Some examples of bacterial laccase sources are *Azospirillum lipoferum*,

Bacillus subtilis, Anabaena azollae, Streptomyces cyaneus, and *Streptomyces lavendulae.* Contrary to fungal and bacterial laccases which accelerate lignin degradation and aid in bioremediation, plant laccases typically facilitate the biosynthesis of lignin in the plant cell wall (Lahtinen et al. 2009). Some sources of plant laccases are *Rhus vernicifera, Rhus succedanea, Populus euramericana, Nicotiana tobacco,* and *Zea mays.*

Remarkably, the complex plant cell wall lignin depolymerization property of laccases (fungi and bacteria) has been vital in the deconstruction of residual lignin that may be present after pretreatment. For instance, the presence of lignin oxidases (laccases) in cellulose hydrolysis boosts cellulase activity by liberating cellulases from unproductive binding sites on lignocellulosic substrates to increase the effective concentration of free cellulases in solution (Berlin 2013). Also, laccases could possibly address issues regarding phenolic compound inhibition of cellulases. For example, Hyeon et al. (2014) achieved 2.6-fold increase in the yield of reduced sugar from pretreated barley straw using cellulase–laccase blends. Moilanen et al. (2011) employed blend of commercial cellulases and laccases on pretreated spruce and obtained 12% increase in hydrolysis yield. Furtado et al. (2013) and Ribeiro et al. (2011) have also demonstrated the improvement in synergy and catalytic performance of fused laccases–(hemi)cellulase complex for biomass hydrolysis.

Hemicellulases Hemicellulases commonly share similar activities with cellulases because of the common β-1,4-glycosidic bonds in the backbone of the hemicellulose component of plant biomass (Chang et al. 2011). The hemicellulose substrate is a complex carbohydrate structure consisting of different easy hydrolysable polymers such as pentoses (e.g., xylose and arabinose), hexoses (e.g., mannose, glucose, and galactose), and sugar acids (Hendriks and Zeeman 2009). Pretreated lignocellulosic biomass hydrolysis is strongly affected by the presence of hemicellulose—the most thermo-chemically sensitive among cellulose, hemicellulose, and lignin—which connects lignin to cellulose fibers and gives the whole cellulose–hemicellulose–lignin network more rigidity (Hendriks and Zeeman 2009). For instance, xylans—the dominant component of hemicellulose from hardwood and agricultural plant—and xylooligomers putatively have a direct inhibitory effect on cellulases (Hendriks and Zeeman 2009; Harris et al. 2014); hence, the need for its depolymerization to reduce the burden on cellulases and improve sugar yields.

Hemicellulases are mostly modular proteins possessing CDs, CMBs, and other functional modules to facilitate the cleavage of either glycosidic or esterified acid side groups (Shallom and Shoham 2003; Decker et al. 2008).

For instance, α-glucuronidases, α-arabinofuranosidases, α-D-galactosidases, and mannanases attack glycosidic bonds whereas acetyl or feruloyl esterases hydrolyze ester bonds of acetate or ferulic acid side groups in the plant cell wall structure. In most cases, hemicellulases are employed in concert with cellulases in the depolymerization of lignocellulosic biomass to fermentable sugars. Relative to the theoretical sugar content, Gao et al. (2011) reported recommendable quantities of reduced sugars from corn stover pretreated by ammonium fiber expansion (99% glucose and 55% xylose), dilute acid (97% glucose and 68% xylose), and ionic liquid (88% glucose and 53% xylose) using cellulase–hemicellulase cocktail. Since hemicellulose presents a rich source of carbon, its successful hydrolysis improves the yield of fermentable sugars.

Lytic polysaccharide mono-oxygenases (LPMOs) LPMOs are copper-dependent enzymes mostly found in saprophytic fungi (e.g., *Thermoascus aurantiacus, Gloeophyllum trabeum, Lentinus similis, Pichia pastoris, Neurospora crassa*) and bacteria (e.g., *Bacillus amyloliquefaciens, Enterococcus faecalis*) (Quinlan et al. 2011; Phillips et al. 2011; Beeson et al. 2012). They were previously grouped among GHs because of their weak endocellulase activities (Karlsson et al. 2001; Karkehabadi et al. 2008). However, modern understanding of their characteristics has resulted in their reclassification as auxiliary activity (AA) family enzymes. Based on mainly structural differences, bacterial (AA10; formerly CBM33) and fungal (AA9; formerly GH61) LPMOs have been studied and classified. Moreover, a supportive classification based on Peptide Pattern Recognition sequencing has recently been reported (Busk and Lange 2015). Nevertheless, their functional distinctions and associated mechanisms are yet to be fully elucidated to help exploit their maximum benefits. Accordingly, studies focusing on the structure (Harris et al. 2010; Aachmann et al. 2012; Hemsworth et al. 2013b; Borisova et al. 2015; Frandsen et al. 2016) and interactions (Isaksen et al. 2014; Eibinger et al. 2014; Courtade et al. 2016; Kracher et al. 2016) of LPMOs are surfacing.

According to structural discussions, the active sites of LPMOs are held in the center of an extended flat face structure—unlike the tunnel-shaped structures housing the active sites of canonical hydrolases (i.e., endo- and exoglucanase)—for an efficient interaction with substrates such as cellulose (including cello-oligosaccharides) and chitin (Hemsworth et al. 2013a; Isaksen et al. 2014). Technically, the active site is said to possess a monomeric type II copper ion (Cu^{2+}) aligned by an N-methylated N-terminal histidine in a network, termed histidine brace, to help the enzyme interact with substrates (Quinlan et al. 2011; Hemsworth et al. 2013b).

The LPMO substrate catalysis is a consequence of the binding of active oxygen molecule from the atmosphere to the monomeric Cu^{2+}, which culminates in the interaction of the active site with available chains within the polysaccharide matrix (substrate). LPMOs assist in the biomass decomposition process by oxidatively attacking the most accessible and most reactive C–H bonds (i.e., C-1 and C-4) along the cellulose strand using molecular oxygen, an external electron donor and, putatively, CBM (Hemsworth et al. 2013a; Walton and Davies 2016). In other words, the enzymes promote the abstraction of hydrogen atoms and assist in the scission of β-1,4-glycosidic linkages between C-1 and C-4 of the cellulose chain.

The role of LPMOs is dependent on substrate dynamics and process conditions. Practically, the overall saccharification yield increases when LPMOs are combined with the three common cellulases, especially in the processing of dry matter with relevant remnants of lignin (Cannella and Jørgensen 2014). Jung et al. (2015) investigated LPMO from *Gloeophyllum trabeum* in concert with cellulases and xylanase. Though no significant individual LPMO activity was observed, the work reported an accelerated synergistic degradation of pretreated kenaf and oak (Jung et al. 2015). Also, Müller et al. (2015) studied the activity of LPMOs with Celluclast® on lignocellulosic biomass of high dry matter concentration and reported an improved product generation. However, their work revealed the need to reconsider process conditions to favor the oxygen and free electron demands of LPMOs (Müller et al. 2015). Nevertheless, Westereng et al. (2015) showed that the lignin component of lignocellulosic substrates provides a reserve of electrons capable of promoting the activity of LPMOs. The effects of divalent cations on LPMO effectiveness was previously stressed by Harris et al. (2010). Also, Cannella et al. (2012) have unveiled the possible inhibition of β-glucosidase activity by the LPMO products (e.g., cellobionic and gluconic acids).

'Non-hydrolytic' accessory proteins   The common non-hydrolytic proteins known are expansins and swollenins. Expansins are phytoproteins capable of loosening the plant cell wall and disrupting the cellulose crystal structure, whereas swollenins are expansin derivatives from fungi (e.g., *T. reesei*, *Aspergillus fumigatus*, etc) and bacteria (e.g., *Bacillus subtilis*). Swollenin also exhibits crystal-disruption activity on cellulosic materials (Nakashima et al. 2014). There are proofs that these non-hydrolytic accessory proteins can enhance cellulase activity through their ability to disrupt hydrogen bonds to reduce cellulose crystallinity while increasing cellulase accessibility to enzymes (Harris et al. 2014).

In response to the known benefits, researchers are investigating the enhancing effects of these non-hydrolytic accessory proteins on cellulose degradation, especially in a reaction mixture. Nakatani et al. (2013) demonstrated, for the first time, the synergetic effect of co-displayed cellulase and expansin-like protein on a *Saccharomyces cerevisiae* cell surface, and they reported 2.9-fold higher degradation activity on phosphoric acid-swollen cellulose (PASC) compared with the activity of cellulase-expressing strain only. Nakashima et al. (2014) also studied fused *Bacillus subtilis* expansin and *Clostridium thermocellum* endoglucanase for the degradation of highly crystalline cellulose and reported about 35% digestibility by the fused proteins. The use of these accessory enzymes in cellulase blends for industrial applications is liable to improve the level of reduced sugar obtainable from lignocellulosic substrates, thus, requires more investigation.

Chemical additives

Chemical additives have been used with cellulases to provide enzymatic process enhancement in the form of metal cofactors or activators. These activators come in the form of metal ions and chelating agents, yielding significant effects on enzymatic activities by assisting in the biochemical transformations. Some of these additives (e.g., surfactants) are effective for lignocellulose depolymerization, in that they putatively prevent enzyme denaturation and inactivation by reducing the unproductive adsorption of enzymes onto the substrate via hydrophobic interactions with lignin (Eriksson et al. 2002). For a quick example, Fontes and Gilbert (2010) explained that calcium is pivotal for dockerin (a facet of most enzyme structures) stability and function, and in the presence of ethylenediaminetetraacetic acid (EDTA, a chelating agent), dockerins are unable to interact with cohesins (another facet of most enzyme structures).

Boyce and Walsh (2015) studied the effect of various additives, such as $CaCl_2$, EDTA, $MgCl_2$, Tween 20, and Triton X-100, on *Alicyclobacillus vulcanalis* endoglucanase activity by adding specified concentration of these additives to the enzyme sample and immediately measuring their influence on the enzyme activity. Relative to the control (enzyme without additives), they reported that $CaCl_2$ (10 mM) and EDTA (2 mM) yielded, respectively, 97 and 98% activities; whereas $MgCl_2$ (10 mM) yielded 86%, but exhibited a slight inhibitory effect on the activity of the endoglucanase. They further reported that the inclusion of 0.1% Tween 20 or 0.5% Triton X-100 in the enzyme solution improved the enzyme thermal stability while enhancing the enzyme activity with 124 and 126%, respectively. They attributed the significant beneficial effect of Tween 20 and Triton X-100 to (1) reduced unproductive adsorption of enzymes to lignin; (2) changes in the enzyme reaction milieu, and (3) reduced enzyme denaturation as a result of the surfactant binding on enzyme secondary and tertiary structures.

Also, Kim et al. (2015a) analyzed the effects of metal ions and a chelating agent on the activity of xylanase–cellulase fusion protein (Xyl10g GS Cel5B) and reported that the endoglucanase and xylanase activities increased by 39 and 15%, respectively, in the presence of 1 mM $CoCl_2$. They, however, reported a complete inhibition of activity of the fused protein by $HgCl_2$.

Moreover, in an experiment to characterize a β-glycosidase (Aab-gly) from the thermoacidophilic bacterium (*Alicyclobacillus acidocaldarius*), Lauro et al. (2006) reported that divalent cations, namely: Mg^{2+}, Mn^{2+}, Ca^{2+}, Zn^{2+}, Co^{2+}, Cu^{2+}, and Ni^{2+} (each at 5 mM, 65 °C, and on 2 mM 2NP-β-Glc) had significant activation effect on Aab-gly. However, Zn^{2+} and Co^{2+} inhibited the enzyme by 33 and 96%, respectively. Mesas et al. (2012) examined the effects of chloride salts ($MgCl_2$, $MnCl_2$, $FeCl_2$, $ZnCl_2$, $CoCl_2$, $CaCl_2$, and $CuCl_2$) on the activity of β-glucosidase from *Oenococcus oeni* ST81 and reported that only Mn^{2+} seemed to slightly increase the enzyme activity; whereas Cu^{2+}, Fe^{2+}, Zn^{2+}, and Co^{2+} clearly reduced the catalytic activity of the enzyme from 8 to 54%, depending on the identity and concentration of the metal ion. There are more other interesting additive-effect observations reported in literature (Kengen et al. 1993; Schülein 2000; Li et al. 2013; Zhao et al. 2013; Jagtap et al. 2013; Balasubramanian and Simões 2014).

The major concern of additive experiments has been the ill-explained discrepancies of the data obtained. The discrepancies in enzyme-additive reportage reinforce the phenomenon of enzyme selectivity in the use of cofactors. Interestingly, even cations of the same valency have yielded different results. The discrepancy could be associated with the charge density of the additive and the size of the active site pocket of the enzyme, but this point should be proved experimentally. It is rational to conclude that the effects of these metal cofactors are enzyme and/or organism depended, and hence thorough studies should be focused on this to consolidate existing understanding.

The chronology of cellulolytic GH systems

Lignocellulosic substrates require several enzymatic strategies, even after pretreatment, to ensure significant generation of fermentable sugars and subsequent production of biochemicals. These strategies may be conducted separately or in combination, and they involve the following dominant microbial paradigms: cell-free enzyme systems, multi-enzyme (cellulosome) complexes, and multifunctional enzyme systems. These underlined systems have their associated pros and cons, and hence

require continual studies and improvement. Figure 3 shows the common configurations of microbial cellulase systems.

GH cell-free systems

The concept of cell-free enzymes was presented by Buchner in 1897, where he claimed that biological processes could be carried out without living cells (Khattak et al. 2014b). Typically, cell-free systems are used for cofactor-independent reactions, and often exhibit reaction-rate-limited kinetics, resulting from the direct access to substrate in solution (Smith et al. 2015). Cell-free enzymes have been exploited both in single (Kengen et al. 1993; Kim et al. 2010; Böhmer et al. 2012) and multiple domain systems (Kanafusa-Shinkai et al. 2013). Several immobilization practices have been reported (Kazenwadel et al. 2015). The general concept of immobilization has been highlighted in a subsequent section.

Among the numerous microorganisms known for their cellulolytic potentials, few has been identified to produce significant quantities and a complete set of cell-free lignocellulases in vitro (Patagundi et al. 2014). According to Khattak et al. (2014a), GH cell-free systems are considered as possible solution for surmounting all complexities and shortcomings associated with conventional enzyme hydrolysis by providing the following advantages: (1) well-regulated, continuous, and prolonged processing of substrate conversion; (2) easy evaluation of the effect of additional cofactors; and (3) no consumption of reduced sugar for cell energy requirement. Cell-free system dramatically reduces the time and effort needed to obtain proteins since it does not require gene transfection and extensive purification procedures (Kim et al. 2010). Moreover, it provides flexible reaction conditions for the introduction of several additives (such as chaperones, detergent, and affinity tags) into the reaction mixture as compared to in vivo systems (Kim et al. 2010).

Interestingly, the cell-free system has grabbed a tremendous interest in the production of various biocommodities, not only reduced sugars but also recombinant proteins, proteinaceous antibiotics, vaccines, hormones, and dihydrofolic acid reductase, etc. (Rollin et al. 2013; Khattak et al. 2014b). However, numerous limitations have confronted cell-free systems, especially when a mixture of enzymes constituting cascade of reactions is employed to produce bioproducts. Some of these shortcomings are the subjects of instability, reusability, and inactivation during biochemical processes (Khattak et al. 2014b). Problems of overall cellulase viability in the presence of high substrate and product concentration are also possible (Khattak et al. 2012, 2014b). The development of synthetic cell-free enzyme systems, with reprogrammed

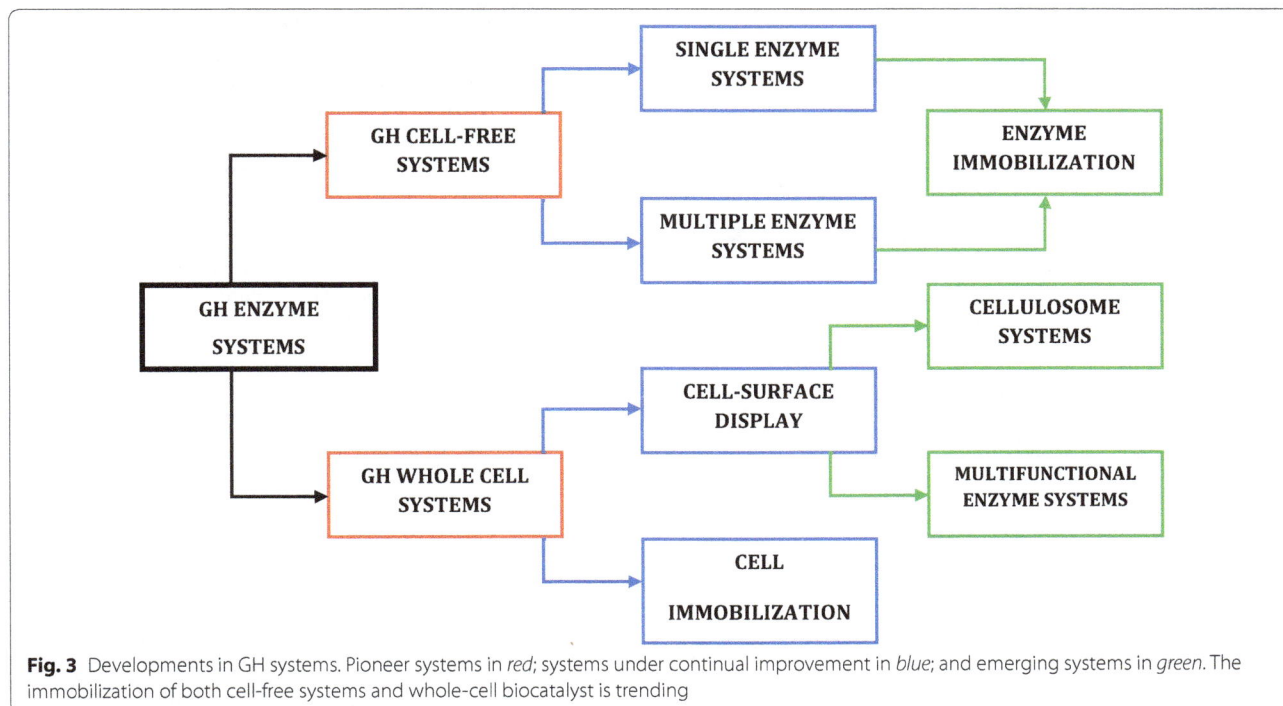

Fig. 3 Developments in GH systems. Pioneer systems in *red*; systems under continual improvement in *blue*; and emerging systems in *green*. The immobilization of both cell-free systems and whole-cell biocatalyst is trending

or newly constructed metabolic pathways to produce high-volume reduced sugars, are believed to be much more efficient due to reasons including the absence of external barriers (Percival Zhang 2010). Well-established approaches for the development of synthetic cell-free enzyme pathways include micro-compartmentalization, ionic channeling, co-polymerization, and protein fusion. Notably, the synthetic cell-free enzyme systems favor maximum enzyme–substrate interaction, product-oriented substrate utilization, and a higher concentration of biocatalyst (Khattak et al. 2014b). However, factors such as cofactor balance, thermodynamics, reaction equilibrium, and product separation and purification still need to be addressed (Zhang 2011).

GH whole-cell systems

The whole-cell biocatalyst system was developed to overcome the cost and complexities associated with enzyme purification via intracellular and extracellular localization of enzymes. In the former, the microorganism provides the most favorable working environment for the enzymes by (1) availing all necessary cofactors and regeneration networks; and (2) providing sufficient protection of enzymes from effects such as destabilization and degradation, while allowing both the substrate and product to cross the membrane barrier (Jose et al. 2012). On the other hand, the extracellular localization of enzymes involves the display

of enzymes on the surface of the microorganism (thus, the designation "cell-surface display") to avoid possible substrate–product transport complexities across the cell membrane (Schüürmann et al. 2014).

The advent of whole-cell systems has helped to overcome some of the challenges faced by cell-free systems. The whole-cell systems convey several advantages such as stability, resistance, lower cost, reusability, and reduced labor, while providing products with high purity (Brault et al. 2014; Kim et al. 2014; Khattak et al. 2014b). The reduced proneness to cell injury; improved resistance to physiological and environmental factors, such as variation in pH, elevated temperature, and system inhibition; high metabolic productivity; and reduced incubation time make the whole-cell system more promising for biotechnological implementation.

Currently, the introduction of new knowledge and techniques, including genetic engineering, peptide engineering, and metabolic engineering, with specializations such as system and synthetic biology, has successfully improved the whole-cell system in various ways (Turner 2003; de Carvalho 2011; Pearsall et al. 2015). For example, the whole-cell biocatalyst system has been enhanced to immobilize the enzymes and improve substrate–enzyme contact, while increasing the catalytic potential of the enzymes by extending their overall lifetime (Kisukuri and Andrade 2015).

Immobilized and co-immobilized systems

Conventional enzyme immobilization is the practice of restraining the movement of enzymes, for example, by direct cross-linking, covalent coupling, entrapment, micro-encapsulation, and tethering onto a solid support to improve technical performance, usability, and industrial process economy. The technical performance includes enzyme stability, substrate specificity, enantioselectivity, and reactivity (Mateo et al. 2007; Schoffelen and van Hest 2012). The target of most immobilization practices is mainly to achieve fewer side reactions, high tolerance of structural variation of the substrates, high productivity and space–time yield, and high durability of the biocatalyst (Cao et al. 2003).

Immobilization is widely practiced in both cell-free and whole-cell systems. However, concerns regarding thermal instability at elevated temperatures, ineffective substrate utilization, by-product formation, and downstream industrial processing cost of end-product make the conventional immobilized system an ineffective approach for process industrialization (Khattak et al. 2014b). These consequences possibly result from intrinsic alterations in the catalytic activity, the overall stability, and the morphological structure of the individual enzymes in the new microenvironment. However, the use of efficient catalyst base; the use of hydrophilic and inert spacer arms; and the careful selection of the enzyme residues involved in the immobilization are some of the strategies toward curbing the steric obstacles. For example, the use of affinity tags (e.g., histidine tag) to selectively immobilize enzymes onto surfaces like cells, DNA scaffolds, and chelating supports is microbiologically practicable.

Cell-surface display systems

The cell-surface display system is the practice whereby whole cells are empowered to extracellularly degrade substrates and, sometimes, internalize resulting products to produce value-added end-products. Regardless of the host organism, surface display systems often have three core features in common. These are: (1) a signal peptide to direct the protein of interest toward the secretory pathway; (2) an endogenous surface protein pliable to recombination (i.e., insertion, deletion, and fusion) to facilitate a stable surface anchorage of the target protein; and (3) an epitope tag to facilitate the detection of successful surface display (Smith et al. 2015). In the surface display system, the amount of cellulase displayed is strictly dependent on the cell surface area, unlike cell-free systems, where there are no such limits (Yamada et al. 2013).

The cell-surface display system serves as an inherent biological platform for immobilizing enzymes, and thus offers three main advantages: (1) no protein diffusion into surrounding media; (2) enhanced biomass hydrolysis stemming from the close proximity, and induced synergy of enzymes present; and (3) easy recoverability and reusability by simple sedimentation or centrifugation. According to Yamada et al. (2013), the low diffusion rate of cell-surface displayed enzymes, owing to its insolubility in the substrate, is however a disadvantage. Arguably, when these surface displayed enzymes are aligned cooperatively to work synergistically, there would be a more efficient hydrolysis via substrate channeling, resulting in high enzymatic activity with high monomer yields. Common surface display systems that have been explored are cellulosomal (multi-enzyme) systems and multifunctional enzyme systems. The most recent subset is the autotransporter display (autodisplay) system, which is described in subsequent section.

Cellulosomal (multi-enzyme) systems Cellulosomes can be described as one of nature's most elegant and elaborate nanomachines (Fontes and Gilbert 2010). They are organized multi-enzyme complexes consisting of carbohydrate-binding modules (CBMs), catalytic domains (CDs), and scaffoldin subunits, which selectively integrate different CDs of enzymes in close proximity onto their individual unified complexes through a cohesin–dockerin interaction. The embedded enzymes work cooperatively and synergistically to ensure efficient depolymerization of the cellulose material.

The cellulosome phenomenon is a mimicry of interesting in vivo activities involving co-localization of enzymes for cascading reactions. Many crucial cellular functions such as biosynthesis (e.g., Krebs TCA cycle) and cellular signaling are controlled in living organisms by multi-step simultaneous enzymatic reactions with excellent efficiency and specificity. A key characteristic of these highly efficient enzyme pathways is the cooperative and spatial organization of enzymes to ensure the sequential conversion of substrates (Fontes and Gilbert 2010; Park et al. 2014). The effect of this systematic organization of enzymes is very distinct, in that it enhances the overall efficiency of molecular activities by increasing the local enzyme–substrate concentrations and channeling intermediates between consecutive enzymes to avoid competition with other reactions present in the cell (Park et al. 2014).

The genesis of cellulosomal enzymes in microbes is linked to the discovery of *Clostridium thermocellum* and its potentials, which initiated the call to investigate the cellulosome genomics and metagenomics: cellulosomics (Bayer et al. 2008). The cellulosome architecture (Fig. 4) is dictated by a primary scaffoldin subunit, consisting of repeating units of cohesin (type I) modules that engage in high specificity and or affinity protein–protein

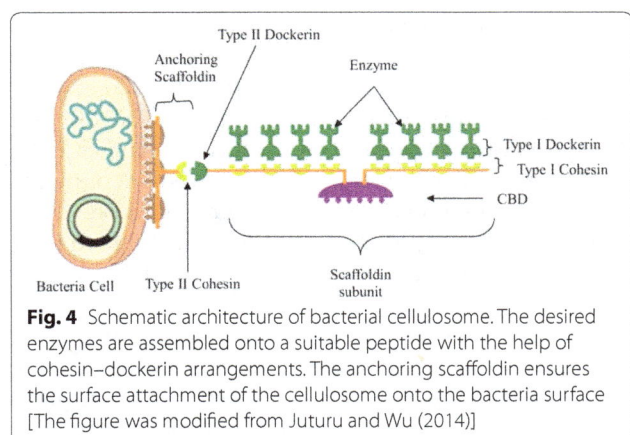

Fig. 4 Schematic architecture of bacterial cellulosome. The desired enzymes are assembled onto a suitable peptide with the help of cohesin–dockerin arrangements. The anchoring scaffoldin ensures the surface attachment of the cellulosome onto the bacteria surface [The figure was modified from Juturu and Wu (2014)]

interaction ($K_D < 10^{-9}$ M) with type I dockerin-containing enzyme, allowing the assembly of multiple enzymes in a spatially defined manner (Park et al. 2014; Stern et al. 2015). Most scaffoldins contain 6–9 different cohesins, which can bind up to 26 different cellulosomal enzymes (Juturu and Wu 2014). The primary scaffoldin interacts by means of type II cohesin–dockerin interaction with an anchoring scaffoldin to enforce cellulosome attachment to the cell surface via an S-layer homology (SLH) module (Stern et al. 2015). However, the intermodular cohesin–dockerin interaction dictates the assembly of the cellulosome complex; hence, granting the possibility of expressing different cellulosomes within a single organism, depending on the enzyme subunit compositions (Moraïs et al. 2012; Juturu and Wu 2014).

The CBM has multiple roles in the hydrolysis of cellulose: (1) to increase the concentration of cellulase close to the substrate; (2) to target the CD of the enzyme to specific sites on the substrate; and (3) to disrupt the crystalline structure of the substrate all through hydrophobic interaction between the three hydrophobic amino acid residues on the flat face of the CBM (Wahlström et al. 2014). There is also the possibility that CBMs assist in the improvement of the overall structure of the multimeric enzyme, leading to an increase in the hydrolysis yield (Fan et al. 2009). However, the number and position of CBMs in the multi-enzyme complex may cause effects (such as product inhibition) on the enzymes during cellulose degradation (Moraïs et al. 2012); thus, it requires investigation.

The CBM modules are classified into three, namely type A, type B, and type C, to define CBMs in terms of their binding specificity. Type A CBMs bind the surface of complex polysaccharides, type B CBMs (with specificity for amorphous regions) recognize internal glycan chains (endo type), and type C CBMs (with specificity for crystalline regions) bind the termini of glycans (exo type),

according to Bornscheuer et al. (2014) and Fontes and Gilbert (2010).

The practice of the use of cellulosomes is interestingly surging in cellulose degradation activities. In this case, the GH enzyme assembly is attached onto the surface of the organism (mostly fungi and bacteria) for an effective saccharification process. The repeating scaffoldin-cohesins are docked individually with different dockerin-bearing GHs to enforce efficient cascade reaction, leading to high yields of fermentable sugars.

According to Zhang (2011), the most investigated in vitro multi-enzyme complex—even in the conversion of cellulose into fermentable sugars—are cellulosomes. This stems from the highly active-synergistic hydrolytic effect of the enzymes. To effectively evaluate the proposed benefits of cellulosomal enzymes over free enzymes, it is imperative to compare the optimized-state characteristics of each system on the same substrate (Moraïs et al. 2010). Park et al. (2014) reported 23-fold glucose production enhancement over that of free enzymes after their investigation of the effects of localization, surface accessibility, and functionality of synergetic enzymes on Scaf3-decorated bacteria outer membrane vesicles (OMVs) using phosphoric acid-swollen cellulose (PASC) as substrate. Yuan et al. (2015), in an investigation to biochemically characterize and structurally analyze cellulase/xylanase from *Clostridium thermocellum*, also revealed equally insightful results. Advancement in cellulosome investigation has led to the advent of its artificial counterpart, called designer cellulosomes, described below.

Designer cellulosomes (Chimeras) Designer cellulosomes (also known as chimeras)—unlike native cellulosomes—are artificial constructs, composed of chimeric scaffoldin and enzymes with cohesins and dockerins of divergent specificities, thus providing interdomain flexibility in the enzyme complex while maintaining (to some extent) the original wild-type functionality (Fierobe et al. 2002; Stern et al. 2015). Although synthetic cellulosomes present faster hydrolysis rates than non-composite cellulase mixtures, Zhang (2011) remarked that there is a limitation in the understanding of why synthetic cellulosomes constructed to date have been much less active than their natural counterparts. This may be due to factors such as changes in the microenvironments of the active sites, possible unproductive competition between functionally similar enzymes, difficulties in component arrangement as well as the nature of the peptide linker.

Cota et al. (2013) investigated and assembled a complex xylanase–lichenase (XylLich) chimera—both enzymes from *Bacillus subtilis*—through all-atom molecular dynamics simulations. Contrary to the remark by Zhang (2011), Cota et al. (2013) reported based on comparison between the recombinant protein yield and the hydrolytic activity achieved that the production of chimeric enzymes is more efficient (in terms of cost and catalytic efficiency) than wild-type proteins and could be more profitable in streamlining biomass conversion strategies than separate production of single enzyme. Cota et al. (2013) further reported that the mode of operation of their chimera was exactly similar to that of the parental enzymes. Moreover, Moraïs et al. (2012) reported that their designer cellulosome system from *Thermobifida fusca* exhibited "equal or superior" activity to that of the free system. This presumably reflects the combined proximity effect of the enzymes and high flexibility of the designer cellulosome components to enable efficient enzymatic activity of the catalytic modules.

The higher flexibility and structural conformations of the fused CDs of designer cellulosomes explicate their more efficient enzymatic abilities (Moraïs et al. 2012). Stern et al. (2015), based on extensive combinatorial analysis, devised and developed a designer cellulosome concept consisting of chimeric scaffoldins for controlled incorporation of recombinant polysaccharide-degrading enzymes. Their results supported the argument that for a given set of cellulosomal enzymes, the relative position of enzymes within a scaffoldin can be critical for optimal degradation of microcrystalline cellulosic substrates. Liang et al. (2014) also constructed a penta-functional minicellulosome by co-expressing lytic polysaccharide mono-oxygenases (LPMOs) and cellobiose dehydrogenases (CDH) with cellobiohydrolases, endoglucanases, and β-glucosidases in *Saccharomyces cerevisiae* for simultaneous saccharification and ethanol fermentation of PASC. The synergetic activity of this penta-enzyme complex increased the ethanol titer from 1.8 to 2.7 g/l.

Engineering multi-domain enzymes that are capable of catalyzing two or more reactions is a potential strategy to reduce enzyme costs in bio-industrial processes, as multiple catalytic properties on a single polypeptide conceivably simplify production and purification operations of biochemicals (Ribeiro et al. 2011). Similarly, the cost-effective optimization of chimeras to prevent unproductive competition between functionally similar enzymes by testing the importance of both the positions of enzymes and CBMs for an efficient use in bioprocessing industries is necessary, though demanding a vigorous investigation. However, the successful expression of the essential cellulolytic enzymes (*i.e.*, endoglucanases, exoglucanases, and β-glucosidases) on a single peptide chain, in a processive order, such that their proportionate quantities favor maximum hydrolysis efficiency has been highly challenging (Tozakidis et al. 2016). Also, the determination of simple and reliable structural organization of the chimeric domains has been a significant drawback in the construction of a protein chimera, but the advent of small-angle X-ray scattering (SAXS) with flexible analytical models (e.g., molecular dynamics (MD) and Monte Carlo simulations) has provided not only successful computational data validation approaches, but also accurate fitting of the scattering profile due to their potential to explore the protein conformation in space (Cota et al. 2013).

A critical factor for success in the creation of enzyme chimeras is the compatibility and cooperativity among the involved CBMs and CDs, with respect to their physicochemical requirements such as solubility, optimum pH, and temperature (Kim et al. 2010; Ribeiro et al. 2011). Howbeit, Stern et al. (2015) suggested that the optimal order for the positioning of enzymes as per their investigation is processive endoglucanase, exoglucanase, and non-processive endoglucanase; and for overall higher enzymatic activity the CBM should not be placed in the middle of the scaffoldin.

In parallel with the designer cellulosome approach, another interesting attempt to increase enzyme synergism, in the form of multifunctional enzyme conjugates, has been reported recently, and it is believed that this strategy may provide a component cost-reducing advantage over designer cellulosomes in future industrial applications (Moraïs et al. 2010). However, the multifunctional enzyme strategy is limited to small numbers of enzymes and restricted to suboptimal equimolar ratios of enzymes. This paradigm permits the expression of single enzymes on the surface of a suitable microorganism such that blending complexities could be overcome.

Multifunctional enzyme systems Multifunctional enzymes—comprising the direct surface display of multiple enzymes in a non-complex form—are very high-molecular weight proteins of one or several CBMs and two or more CDs for effective substrate targeting and efficient degradation of plant cell walls, respectively (Moraïs et al. 2012; Smith et al. 2015). The several catalytic modules on the same polypeptide chain are assembled such that their enforced proximity account for an enhanced concerted action on substrates (Moraïs et al. 2012). The enzyme assemblies in multifunctional enzyme systems enable metabolic control and prevent metabolic crosstalk between competing pathways (Conrado et al. 2008). Multifunctional enzymes are formed by linking the CD of desired enzymes, using flexible peptide linkers or linkers containing CBMs, with that of the parent enzyme (Fan et al. 2009). The resulting enzyme may retain similar properties (example, pH and temperature profiles, kinetics, etc.) as the parent enzyme and exhibit synergetic effects in the hydrolysis of the target substrate. Though the multifunctional enzyme system is under thorough investigation, a broader understanding of (1) how the structure of an enzyme relates to its function and (2) what changes can be tolerated within a multifunctional enzyme framework are needed to promote industrial applications.

Moraïs et al. (2012) employed a synthetic biology approach to convert two different cellulases from the free enzyme system of *Thermobifida fusca* into bifunctional enzymes with different modular architectures and examined their performance compared to those of the combined parental free enzyme and equivalent designer cellulosome systems. They reported that the different architectures of the bifunctional enzymes displayed "somewhat inferior" cellulolytic activity to that of the wild-type free enzyme system. However, Ribeiro et al. (2011) created two bifunctional enzymes with xylanase–laccase activity using rational design methods and reported catalytic properties similar to the parental enzymes. Moreover, Chang et al. (2011) reported an excellent performance of a bifunctional xylanase/endoglucanase (RuCelA), which distinguishes it as an ideal candidate for industrial applications. Cho et al. (2008) reported a multifunctional enzyme Cel44C-Man26A (secreted by *Paenibacillus polymyxa* GS01) with cellulase, xylanase, lichenase, and mannanase activities. The construction of multifunctional enzymes is putatively dependent on the nature and anatomy of source organism as well as the design technique, and thus more insights are required for this justification.

The autodisplay systems The autodisplay system is an induced superior advancement of the whole-cell biocatalyst strategy. It mostly involves the recombinant surface display of proteins or peptides by means of autotransporter proteins in Gram-negative bacteria (Jose et al. 2012). The autotransporter proteins—the peptide chains that 'link' or hold the passenger protein onto the outer membrane of the organism—are common secretion proteins of most Gram-negative bacteria, and are synthesized as precursor protein containing all domains needed to transport the passenger (e.g., cellulases, proteases, lipases, esterases etc.) to the cell surface (Jose 2006; Jose et al. 2012). This provides the possibility to transport protein (recombinant or natural passenger) to the outer membrane so long as its coding region lies between a typical signal peptide and a C-terminal "β-barrel" domain (Schumacher et al. 2012). Tozakidis et al. (2016) has published a proof of concept of cellulose hydrolysis using autodisplay cellulases.

The hypothetical model of the autodisplay secretion mechanism (Fig. 5) has been described by Himmel et al. (2010). It typically involves the transport of a polyprotein precursor across the inner membrane (IM) of the cell into the periplasm, with the help of the sec signal peptide (SP). A typical precursor protein comprises a signal peptide (SP) followed by the autotransporter protein, composed of an N-terminal passenger domain (α-domain) and a C-terminal translocator domain (β-domain). The β-domain, as its name signifies, involves β-barrel and linker. Inside the periplasm, the C-terminal part of the precursor forms a porin-like structure (β-barrel) within the outer membrane (OM). Subsequently, the passenger proteins translocate to the cell surface through the pores, with anchorage from the free mobile β-barrel, unlike in other display systems where they are covalently attached to the cell envelope. Complementarily, the peptide linker ensures the full surface exposure and functionality of the passenger protein.

The recombinant expression principle of autotransporter proteins has several advantages. The flexible transfer of these proteins from one Gram-negative bacterium to the other needs little or no additional machinery for its propagation. Moreover, a large number (more than 10^5) of recombinant proteins or peptide molecules can be displayed on the surface of a single microorganism, without reducing cell viability or integrity (Jose and Meyer 2007). In addition, the relatively simple modular structure of autodisplay systems allows the easy interaction of passenger proteins on the bacterial cell surface, thus displaying desired heterologous enzymes. The autodisplayed proteins normally expressed as monomers are capable of forming multimers upon membrane interaction after expression (Schumacher et al. 2012; Smith et al. 2015). Furthermore, the autodisplay secretion method makes subsequent, often costly, purification steps to recover the enzyme of interest unnecessary (Kranen et al. 2014).

Conclusions

The bioprocess industry is constantly seeking to obtain useful products from the highly abundant lignocellulosic feedstock. Thus, lignocellulases have been vital in the production of reduced sugars for the manufacturing of biocommodities. The industrial pursuit of obtaining high level of fermentable sugars from lignocellulosic biomass depends substantially on the successful expression and blend of cellulases, hemicellulases, lignases, and other accessory proteins in a non-competing, progressive, and synergetic order, in one complex. However, the challenge has been the successful assembly of an entire suite of these enzymes that could function optimally at the same time and under different conditions to completely digest lignocellulosic biomass to simple sugars. Many cellulase improvement practices and enzyme systems (i.e., cell-free or whole-cell) have surfaced and presently the fraternity is witnessing a gradual shift towards the cell-surface display system. However, the challenge has been

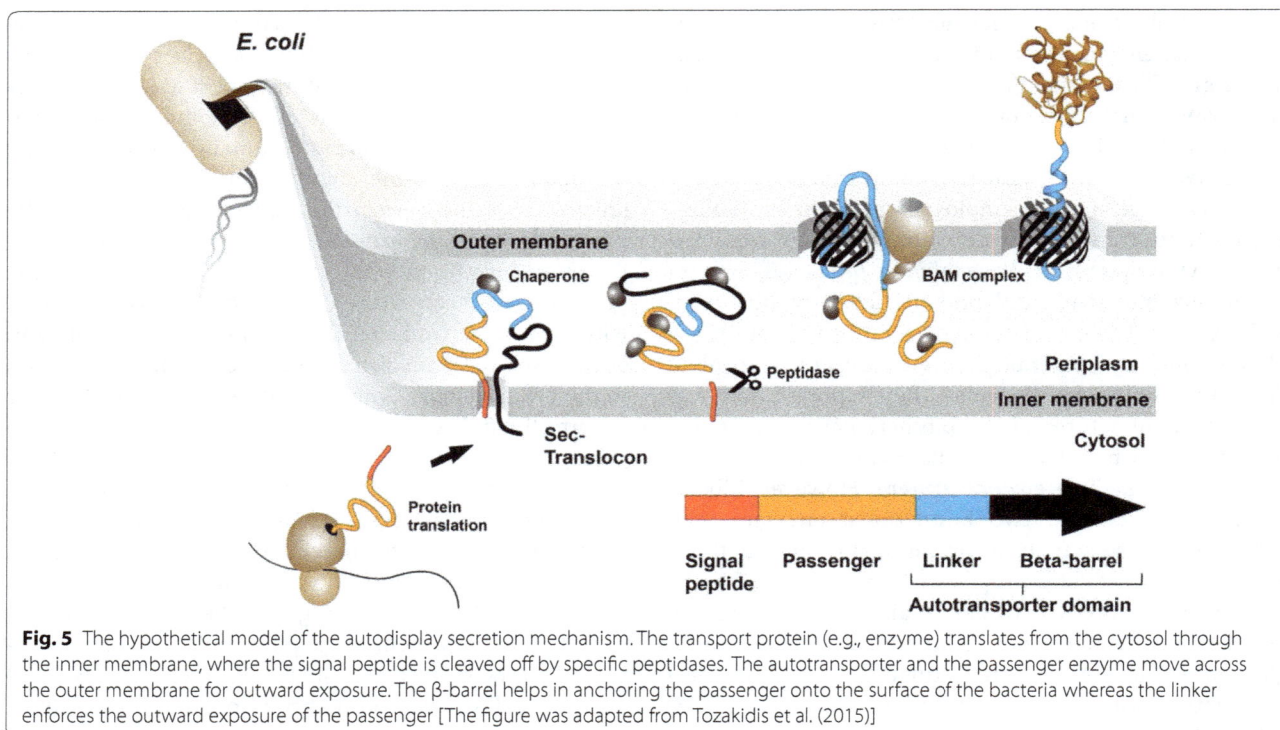

Fig. 5 The hypothetical model of the autodisplay secretion mechanism. The transport protein (e.g., enzyme) translates from the cytosol through the inner membrane, where the signal peptide is cleaved off by specific peptidases. The autotransporter and the passenger enzyme move across the outer membrane for outward exposure. The β-barrel helps in anchoring the passenger onto the surface of the bacteria whereas the linker enforces the outward exposure of the passenger [The figure was adapted from Tozakidis et al. (2015)]

the achievement of high-level expressions necessary for industrial use. Techniques such as directed evolution and rational design have been used in improving cellulases. The practice of harnessing glycosylation to improve cellulase activity looks promising. A success in these ventures would be influential to the proposed 'green' future.

Abbreviations
LPMOs: lytic polysaccharide mono-oxygenases; GHs: glycoside hydrolases; CBM: carbohydrate-binding module; CD: catalytic domain; PCR: polymerase chain reaction; EDTA: ethylenediaminetetraacetic acid; SLH: S-layer homology; IN: inner membrane; OM: outer membrane; OMV: outer membrane vesicle; PASC: phosphoric acid-swollen cellulose; CDH: cellobiose dehydrogenase; SAXS: small-angle X-ray scattering; MD: molecular dynamics; SP: signal peptide.

Authors' contributions
EMO and SNNA drafted the manuscript. CMO, CB, RM, and JJ revised and approved the content of the manuscript. All authors read and approved the final manuscript.

Author details
[1] Biotechnology Research Institute, Universiti Malaysia Sabah, 88400 Kota Kinabalu, Sabah, Malaysia. [2] Energy Research Institute, Universiti Malaysia Sabah, 88400 Kota Kinabalu, Sabah, Malaysia. [3] Autodisplay Biotech GmbH, Lifescience Center, Merowinger Platz 1a, 40225 Dusseldorf, Germany. [4] Institute of Pharmaceutical and Medicinal Chemistry, PharmaCampus, Westphalian Wilhelms-University of Münster, Corrensstraße 48, 48149 Münster, Germany.

Acknowledgements
The authors would like to thank Autodisplay Biotech GmbH (Germany) for the collaborative research Grant (GL00139) out of which emerged this review article.

Competing interests
RM and JJ are the founding members of Autodisplay Biotech GmbH (Germany); CMO is the project lead consultant; CB is the project associate consultant; EMO and SNNA are the postgraduate students at Universiti Malaysia Sabah.

References
Aachmann FL, Sorlie M, Skjak-Braek G et al (2012) NMR structure of a lytic polysaccharide monooxygenase provides insight into copper binding, protein dynamics, and substrate interactions. Proc Natl Acad Sci 109:18779–18784. doi:10.1073/pnas.1208822109

Anthony T, Chandra Raj K, Rajendran A, Gunasekaran P (2003) High molecular weight cellulase-free xylanase from alkali-tolerant *Aspergillus fumigatus* AR1. Enzyme Microb Technol 32:647–654. doi:10.1016/S0141-0229(03)00050-4

Badhan AK, Chadha BS, Kaur J et al (2007) Production of multiple xylanolytic and cellulolytic enzymes by thermophilic fungus *Myceliophthora* sp. IMI 387099. Bioresour Technol 98:504–510. doi:10.1016/j.biortech.2006.02.009

Balasubramanian N, Simões N (2014) *Bacillus pumilus* S124A carboxymethyl cellulase; a thermo stable enzyme with a wide substrate spectrum utility. Int J Biol Macromol 67:132–139. doi:10.1016/j.ijbiomac.2014.03.014

Bansal N, Tewari R, Soni R, Soni SK (2012) Production of cellulases from *Aspergillus niger* NS-2 in solid state fermentation on agricultural and kitchen waste residues. Waste Manag 32:1341–1346. doi:10.1016/j.wasman.2012.03.006

Bayer EA, Lamed R, White BA, Flints HJ (2008) From cellulosomes to cellulosomics. Chem Rec 8:364–377. doi:10.1002/tcr.20160

Beckham GT, Dai Z, Matthews JF et al (2012) Harnessing glycosylation to improve cellulase activity. Curr Opin Biotechnol 23:338–345. doi:10.1016/j.copbio.2011.11.030

Beeson WT, Phillips CM, Cate JHD, Marletta MA (2012) Oxidative cleavage of cellulose by fungal copper-dependent polysaccharide monooxygenases. J Am Chem Soc 134:890–892. doi:10.1021/ja210657t

Begum MF, Alimon AR (2011) Bioconversion and saccharification of some ligno-cellulosic wastes by *Aspergillus oryzae* ITCC-4857.01 for fermentable sugar production. Electron J Biotechnol. doi:10.2225/vol14-issue5-fulltext-3

Berlin A (2013) No barriers to cellulose breakdown. Science 342:1454–1456. doi:10.1126/science.1247697

Böhmer N, Lutz-Wahl S, Fischer L (2012) Recombinant production of hyper-thermostable CelB from *Pyrococcus furiosus* in *Lactobacillus* sp. Appl Microbiol Biotechnol 96:903–912. doi:10.1007/s00253-012-4212-z

Borisova AS, Isaksen T, Dimarogona M et al (2015) Structural and functional char-acterization of a lytic polysaccharide monooxygenase with broad substrate specificity. J Biol Chem 290:22955–22969. doi:10.1074/jbc.M115.660183

Bornscheuer U, Buchholz K, Seibel J (2014) Enzymatic degradation of (Ligno) cellulose. Angew Chemie Int Ed 53:10876–10893. doi:10.1002/anie.201309953

Boyce A, Walsh G (2015) Characterisation of a novel thermostable endo-glucanase from *Alicyclobacillus vulcanalis* of potential application in bioethanol production. Appl Microbiol Biotechnol 99:7515–7525. doi:10.1007/s00253-015-6474-8

Brault G, Shareck F, Hurtubise Y et al (2014) Short-chain flavor ester synthesis in organic media by an *E. coli* whole-cell biocatalyst expressing a newly characterized heterologous lipase. PLoS ONE 9:e91872. doi:10.1371/journal.pone.0091872

Brunecky R, Alahuhta M, Bomble YJ et al (2012) Structure and function of the *Clostridium thermocellum* cellobiohydrolase A X1-module repeat: enhancement through stabilization of the CbhA complex. Acta Crystal-logr Sect D Biol Crystallogr 68:292–299. doi:10.1107/S0907444912001680

Busk PK, Lange L (2015) Classification of fungal and bacterial lytic poly-saccharide monooxygenases. BMC Genom 16:368. doi:10.1186/s12864-015-1601-6

Cannella D, Jørgensen H (2014) Do new cellulolytic enzyme preparations affect the industrial strategies for high solids lignocellulosic ethanol production? Biotechnol Bioeng 111:59–68. doi:10.1002/bit.25098

Cannella D, Hsieh CC, Felby C, Jørgensen H (2012) Production and effect of aldonic acids during enzymatic hydrolysis of lignocellulose at high dry matter content. Biotechnol Biofuels 5:26. doi:10.1186/1754-6834-5-26

Cao L, van Langen L, Sheldon RA (2003) Immobilised enzymes: carrier-bound or carrier-free? Curr Opin Biotechnol 14:387–394. doi:10.1016/S0958-1669(03)00096-X

Carvalho AL, Goyal A, Prates JAM et al (2004) The family 11 carbohydrate-bind-ing module of *Clostridium thermocellum* Lic26A-Cel5E accommodates β-1,4- and β-1,3-1,4-mixed linked glucans at a single binding site. J Biol Chem 279:34785–34793. doi:10.1074/jbc.M405867200

Chandel AK, Singh OV, Venkateswar Rao L et al (2011) Bioconversion of novel substrate *Saccharum spontaneum*, a weedy material, into ethanol by *Pichia stipitis* NCIM3498. Bioresour Technol 102:1709–1714. doi:10.1016/j.biortech.2010.08.016

Chandel AK, Gonçalves BCM, Strap JL, da Silva SS (2013) Biodelignification of lignocellulose substrates: an intrinsic and sustainable pretreatment strategy for clean energy production. Crit Rev Biotechnol 1:1–13. doi:10.3109/07388551.2013.841638

Chandra RP, Bura R, Mabee WE et al (2007) Substrate pretreatment: the key to effective enzymatic hydrolysis of lignocellulosics? Adv Biochem Eng Biotechnol 108:67–93. doi:10.1007/10_2007_064

Chang L, Ding M, Bao L et al (2011) Characterization of a bifunctional xylanase/endoglucanase from yak rumen microorganisms. Appl Microbiol Bio-technol 90:1933–1942. doi:10.1007/s00253-011-3182-x

Chhabra SR, Shockley KR, Ward DE, Kelly RM (2002) Regulation of endo-acting glycosyl hydrolases in the hyperthermophilic bacterium *Thermotoga maritima* grown on glucan- and mannan-based polysaccharides. Appl Environ Microbiol 68:545–554. doi:10.1128/AEM.68.2.545

Cho K-M, Math RK, Hong S-Y et al (2008) Changes in the activity of the multi-functional b -glycosyl hydrolase (Cel44C-Man26A) from *Paenibacillus polymyxa* by removal of the C-terminal region to minimum size. Biotechnol Lett 30:1061–1068. doi:10.1007/s10529-008-9640-6

Ciolacu D, Kovac J, Kokol V (2010) The effect of the cellulose-binding domain from *Clostridium cellulovorans* on the supramolecular structure of cellu-lose fibers. Carbohydr Res 345:621–630. doi:10.1016/j.carres.2009.12.023

Claus H (2004) Laccases: structure, reactions, distribution. Micron 35:93–96. doi:10.1016/j.micron.2003.10.029

Conrado RJ, Varner JD, DeLisa MP (2008) Engineering the spatial organization of metabolic enzymes: mimicking nature's synergy. Curr Opin Biotech-nol 19:492–499. doi:10.1016/j.copbio.2008.07.006

Cota J, Oliveira LC, Damásio ARL et al (2013) Assembling a xylanase–lichenase chimera through all-atom molecular dynamics simulations. Bio-chim Biophys Acta Proteins Proteom 1834:1492–1500. doi:10.1016/j.bbapap.2013.02.030

Courtade G, Wimmer R, Røhr ÅK et al (2016) Interactions of a fungal lytic polysaccharide monooxygenase with β-glucan substrates and cel-lobiose dehydrogenase. Proc Natl Acad Sci 113:5922–5927. doi:10.1073/pnas.1602566113

Czjzek M, Cicek M, Zamboni V et al (2000) The mechanism of substrate (agly-cone) specificity in beta-glucosidases is revealed by crystal structures of mutant maize beta -glucosidase-DIMBOA, -DIMBOAGlc, and -dhurrin complexes. Proc Natl Acad Sci USA 97:13555–13560. doi:10.1073/pnas.97.25.13555

Das A, Paul T, Halder SK et al (2013) Production of cellulolytic enzymes by *Aspergillus fumigatus* ABK9 in wheat bran-rice straw mixed substrate and use of cocktail enzymes for deinking of waste office paper pulp. Bioresour Technol 128:290–296. doi:10.1016/j.biortech.2012.10.080

Das A, Paul T, Ghosh P et al (2015) Kinetic study of a glucose tolerant β-glucosidase from *Aspergillus fumigatus* ABK9 entrapped into alginate beads. Waste Biomass Valoriz 6:53–61. doi:10.1007/s12649-014-9329-0

de Carvalho CCCR (2011) Enzymatic and whole cell catalysis: finding new strategies for old processes. Biotechnol Adv 29:75–83. doi:10.1016/j.biotechadv.2010.09.001

Decker CH, Visser J, Schreier P (2001) β-Glucosidase multiplicity from *Aspergil-lus tubingensis* CBS 643.92: purification and characterization of four β-glucosidases and their differentiation with respect to substrate speci-ficity, glucose inhibition and acid tolerance. Appl Microbiol Biotechnol 55:157–163. doi:10.1007/s002530000462

Decker SR, Siika-Aho M, Viikari L (2008) Enzymatic depolymerization of plant cell hemicelluloses. In: Himmel ME (ed) Biomass recalcitrance: decon-structing the plant cell wall for bioenergy. Blackwell Publishing, Oxford, pp 354–378

Demirjian DC, Morís-Varas F, Cassidy CS (2001) Enzymes from extre-mophiles. Curr Opin Chem Biol 5:144–151. doi:10.1016/S1367-5931(00)00183-6

di Lauro B, Rossi M, Moracci M (2006) Characterization of a β-glycosidase from the thermoacidophilic bacterium *Alicyclobacillus acidocaldarius*. Extre-mophiles 10:301–310. doi:10.1007/s00792-005-0500-1

Doherty WOS, Mousavioun P, Fellows CM (2011) Value-adding to cellulosic ethanol: lignin polymers. Ind Crops Prod 33:259–276. doi:10.1016/j.indcrop.2010.10.022

Dwivedi UN, Singh P, Pandey VP, Kumar A (2011) Structure–function relation-ship among bacterial, fungal and plant laccases. J Mol Catal B Enzym 68:117–128. doi:10.1016/j.molcatb.2010.11.002

Easton R (2011) Glycosylation of proteins—structure, function and analysis. Life Sci Tech Bull 60:1–5

Egorova K, Antranikian G (2005) Industrial relevance of thermophilic Archaea. Curr Opin Microbiol 8:649–655. doi:10.1016/j.mib.2005.10.015

Eibinger M, Ganner T, Bubner P et al (2014) Cellulose surface degradation by a lytic polysaccharide monooxygenase and its effect on cellulase hydrolytic efficiency. J Biol Chem 289:35929–35938. doi:10.1074/jbc.M114.602227

Elleuche S, Schröder C, Sahm K, Antranikian G (2014) Extremozymes—biocata-lysts with unique properties from extremophilic microorganisms. Curr Opin Biotechnol 29:116–123. doi:10.1016/j.copbio.2014.04.003

Eriksson T, Börjesson J, Tjerneld F (2002) Mechanism of surfactant effect in enzymatic hydrolysis of lignocellulose. Enzyme Microb Technol 31:353–364. doi:10.1016/S0141-0229(02)00134-5

Falkoski DL, Guimarães VM, de Almeida MN et al (2013) *Chrysoporthe cubensis*: a new source of cellulases and hemicellulases to application in biomass saccharification processes. Bioresour Technol 130:296–305. doi:10.1016/j.biortech.2012.11.140

Fan Z, Wagschal K, Chen W et al (2009) Multimeric hemicellulases facilitate bio-mass conversion. Appl Environ Microbiol 75:1754–1757. doi:10.1128/AEM.02181-08

Fierobe Bayer EA, Tardif C et al (2002) Degradation of cellulose substrates by cellulosome chimeras. J Biol Chem 277:49621–49630. doi:10.1074/jbc.M207672200

Fontes CMGA, Gilbert HJ (2010) Cellulosomes: highly efficient nanoma-chines designed to deconstruct plant cell wall complex car-bohydrates. Annu Rev Biochem 79:655–681. doi:10.1146/annurev-biochem-091208-085603

Frandsen KEH, Simmons TJ, Dupree P et al (2016) The molecular basis of polysaccharide cleavage by lytic polysaccharide monooxygenases. Nat Chem Biol 12:298–303. doi:10.1038/nchembio.2029

Furtado GP, Ribeiro LF, Lourenzoni MR, Ward RJ (2013) A designed bifunc-tional laccase/-1,3-1,4-glucanase enzyme shows synergistic sugar release from milled sugarcane bagasse. Protein Eng Des Sel 26:15–23. doi:10.1093/protein/gzs057

Gallezot P (2012) Conversion of biomass to selected chemical products. Chem Soc Rev 41:1538–1558. doi:10.1039/c1cs15147a

Gao D, Uppugundla N, Chundawat SP et al (2011) Hemicellulases and auxiliary enzymes for improved conversion of lignocel-lulosic biomass to monosaccharides. Biotechnol Biofuels 4:5. doi:10.1186/1754-6834-4-5

Garvey M, Klose H, Fischer R et al (2013) Cellulases for biomass degradation: comparing recombinant cellulase expression platforms. Trends Biotech-nol 31:581–593. doi:10.1016/j.tibtech.2013.06.006

González-Candelas L, Aristoy MC, Polaina J, Flors A (1989) Cloning and charac-terization of two genes from *Bacillus polymyxa* expressing beta-glucosi-dase activity in *Escherichia coli*. Appl Environ Microbiol 55:3173–3177

Graham JE, Clark ME, Nadler DC et al (2011) Identification and characterization of a multidomain hyperthermophilic cellulase from an archaeal enrich-ment. Nat Commun 2:375. doi:10.1038/ncomms1373

Günata Z, Vallier MJ (1999) Production of a highly glucose-tolerant extra-cellular β-glucosidase by three *Aspergillus* strains. Biotechnol Lett 21:219–223. doi:10.1023/A:1005407710806

Haki G (2003) Developments in industrially important thermostable enzymes: a review. Bioresour Technol 89:17–34. doi:10.1016/S0960-8524(03)00033-6

Hanif A, Yasmeen A, Rajoka MI (2004) Induction, production, repression, and de-repression of exoglucanase synthesis in *Aspergillus niger*. Bioresour Technol 94:311–319. doi:10.1016/j.biortech.2003.12.013

Harris PV, Welner D, McFarland KC et al (2010) Stimulation of lignocellulosic biomass hydrolysis by proteins of glycoside hydrolase family 61: struc-ture and function of a large, enigmatic family. Biochemistry 49:3305–3316. doi:10.1021/bi100009p

Harris PV, Xu F, Kreel NE et al (2014) New enzyme insights drive advances in commercial ethanol production. Curr Opin Chem Biol 19:162–170. doi:10.1016/j.cbpa.2014.02.015

Hemsworth GR, Davies GJ, Walton PH (2013a) Recent insights into copper-containing lytic polysaccharide mono-oxygenases. Curr Opin Struct Biol 23:660–668. doi:10.1016/j.sbi.2013.05.006

Hemsworth GR, Taylor EJ, Kim RQ et al (2013b) The copper active site of CBM33 polysaccharide oxygenases. J Am Chem Soc 135:6069–6077. doi:10.1021/ja402106e

Hendriks ATWM, Zeeman G (2009) Pretreatments to enhance the digestibility of lignocellulosic biomass. Bioresour Technol 100:10–18. doi:10.1016/j.biortech.2008.05.027

Himmel ME, Xu Q, Luo Y et al (2010) Microbial enzyme systems for biomass conversion: emerging paradigms. Biofuels 1:323–341. doi:10.4155/bfs.09.25

Hirano N, Hasegawa H, Nihei S, Haruki M (2013) Cell-free protein synthesis and substrate specificity of full-length endoglucanase CelJ (Cel9D-Cel44A), the largest multi-enzyme subunit of the *Clostridium thermocellum* cel-lulosome. FEMS Microbiol Lett 344:25–30. doi:10.1111/1574-6968.12149

Horn S, Vaaje-Kolstad G, Westereng B, Eijsink VG (2012) Novel enzymes for the degradation of cellulose. Biotechnol Biofuels 5:45. doi:10.1186/1754-6834-5-45

Hyeon JE, You SK, Kang DH et al (2014) Enzymatic degradation of lignocel-lulosic biomass by continuous process using laccase and cellulases with the aid of scaffoldin for ethanol production. Process Biochem 49:1266–1273. doi:10.1016/j.procbio.2014.05.004

Isaksen T, Westereng B, Aachmann FL et al (2014) A C4-oxidizing lytic polysac-charide monooxygenase cleaving both cellulose and cello-oligosac-charides. J Biol Chem 289:2632–2642. doi:10.1074/jbc.M113.530196

Islam R, Özmihçi S, Cicek N et al (2013) Enhanced cellulose fermentation and end-product synthesis by *Clostridium thermocellum* with varied nutrient compositions under carbon-excess conditions. Biomass Bioenergy 48:213–223. doi:10.1016/j.biombioe.2012.11.010

Jagtap SS, Dhiman SS, Kim T-S et al (2013) Characterization of a β-1,4-glucosidase from a newly isolated strain of *Pholiota adiposa* and its application to the hydrolysis of biomass. Biomass Bioenergy 54:181–190. doi:10.1016/j.biombioe.2013.03.032

Jose J (2006) Autodisplay: efficient bacterial surface display of recombi-nant proteins. Appl Microbiol Biotechnol 69:607–614. doi:10.1007/s00253-005-0227-z

Jose J, Meyer TF (2007) The autodisplay story, from discovery to biotechni-cal and biomedical applications. Microbiol Mol Biol Rev 71:600–619. doi:10.1128/MMBR.00011-07

Jose J, Maas RM, Teese MG (2012) Autodisplay of enzymes—molecular basis and perspectives. J Biotechnol 161:92–103. doi:10.1016/j.jbiotec.2012.04.001

Jung S, Song Y, Kim HM, Bae H-J (2015) Enhanced lignocellulosic biomass hydrolysis by oxidative lytic polysaccharide monooxygenases (LPMOs) GH61 from *Gloeophyllum trabeum*. Enzyme Microb Technol 77:38–45. doi:10.1016/j.enzmictec.2015.05.006

Juturu V, Wu JC (2014) Microbial cellulases: engineering, production and applications. Renew Sustain Energy Rev 33:188–203. doi:10.1016/j.rser.2014.01.077

Kanafusa-Shinkai S, Wakayama J, Tsukamoto K et al (2013) Degradation of microcrystalline cellulose and non-pretreated plant biomass by a cell-free extracellular cellulase/hemicellulase system from the extreme ther-mophilic bacterium *Caldicellulosiruptor bescii*. J Biosci Bioeng 115:64–70. doi:10.1016/j.jbiosc.2012.07.019

Kang SW, Park YS, Lee JS et al (2004) Production of cellulases and hemicel-lulases by *Aspergillus niger* KK2 from lignocellulosic biomass. Bioresour Technol 91:153–156. doi:10.1016/S0960-8524(03)00172-X

Karkehabadi S, Hansson H, Kim S et al (2008) The first structure of a glycoside hydrolase family 61 member, Cel61B from *Hypocrea jecorina*, at 1.6 Å resolution. J Mol Biol 383:144–154. doi:10.1016/j.jmb.2008.08.016

Karlsson J, Saloheimo M, Siika-aho M et al (2001) Homologous expression and characterization of Cel61A (EG IV) of *Trichoderma reesei*. Eur J Biochem 268:6498–6507. doi:10.1046/j.0014-2956.2001.02605.x

Kazenwadel F, Franzreb M, Rapp BE (2015) Synthetic enzyme supercomplexes: co-immobilization of enzyme cascades. Anal Methods 7:4030–4037. doi:10.1039/C5AY00453E

Kengen SWM, Luesink EJ, Stams AJM, Zehnder AJB (1993) Purification and characterization of an extremely thermostable beta-glucosidase from the hyperthermophilic archaeon *Pyrococcus furiosus*. Eur J Biochem 213:305–312. doi:10.1111/j.1432-1033.1993.tb17763.x

Khattak WA, Ul-Islam M, Park JK (2012) Prospects of reusable endogenous hydrolyzing enzymes in bioethanol production by simultaneous saccharification and fermentation. Korean J Chem Eng 29:1467–1482. doi:10.1007/s11814-012-0174-1

Khattak WA, Ul-Islam M, Ullah MW et al (2014a) Yeast cell-free enzyme system for bio-ethanol production at elevated temperatures. Process Biochem 49:357–364. doi:10.1016/j.procbio.2013.12.019

Khattak WA, Ullah MW, Ul-Islam M et al (2014b) Developmental strategies and regulation of cell-free enzyme system for ethanol production: a molecular prospective. Appl Microbiol Biotechnol 98:9561–9578. doi:10.1007/s00253-014-6154-0

Kim S, Kim CH (2012) Production of cellulase enzymes during the solid-state fermentation of empty palm fruit bunch fiber. Bioprocess Biosyst Eng 35:61–67. doi:10.1007/s00449-011-0595-y

Kim T-W, Chokhawala HA, Nadler D et al (2010) Binding modules alter the activity of chimeric cellulases: effects of biomass pretreatment and enzyme source. Biotechnol Bioeng 107:601–611. doi:10.1002/bit.22856

Kim CS, Seo JH, Kang DG, Cha HJ (2014) Engineered whole-cell biocatalyst-based detoxification and detection of neurotoxic organophos-phate compounds. Biotechnol Adv 32:652–662. doi:10.1016/j.biotechadv.2014.04.010

Kim HM, Jung S, Lee KH et al (2015a) Improving lignocellulose deg-radation using xylanase–cellulase fusion protein with a gly-cine–serine linker. Int J Biol Macromol 73:215–221. doi:10.1016/j.ijbiomac.2014.11.025

Kim Y, Kreke T, Ko JK, Ladisch MR (2015b) Hydrolysis-determining substrate characteristics in liquid hot water pretreated hardwood. Biotechnol Bioeng 112:677–687. doi:10.1002/bit.25465

Kisukuri CM, Andrade LH (2015) Production of chiral compounds using immo bilized cells as a source of biocatalysts. Org Biomol Chem 13:10086–10107. doi:10.1039/C5OB01677K

Koeck DE, Wibberg D, Koellmeier T et al (2013) Draft genome sequence of the cellulolytic *Clostridium thermocellum* wild-type strain BC1 playing a role in cellulosic biomass degradation. J Biotechnol 168:62–63. doi:10.1016/j.jbiotec.2013.08.011

Koshland DE (1953) Stereochemistry and the mechanism of enzymatic reactions. Biol Rev 28:416–436. doi:10.1111/j.1469-185X.1953.tb01386.x

Kracher D, Scheiblbrandner S, Felice AKG et al (2016) Extracellular electron transfer systems fuel cellulose oxidative degradation. Science 352:1098–1101. doi:10.1126/science.aaf3165

Kranen E, Detzel C, Weber T, Jose J (2014) Autodisplay for the co-expression of lipase and foldase on the surface of *E. coli*: washing with designer bugs. Microb Cell Fact 13:19. doi:10.1186/1475-2859-13-19

Lahtinen M, Kruus K, Boer H et al (2009) The effect of lignin model compound structure on the rate of oxidation catalyzed by two different fungal laccases. J Mol Catal B Enzym 57:204–210. doi:10.1016/j.molcatb.2008.09.004

Lee CY, Yu KO, Kim SW, Han SO (2010) Enhancement of the thermostability and activity of mesophilic *Clostridium cellulovorans* EngD by in vitro DNA recombination with *Clostridium thermocellum* CelE. J Biosci Bioeng 109:331–336. doi:10.1016/j.jbiosc.2009.10.014

Lee H-L, Chang C-K, Teng K-H, Liang P-H (2011) Construction and characterization of different fusion proteins between cellulases and β-glucosidase to improve glucose production and thermostability. Bioresour Technol 102:3973–3976. doi:10.1016/j.biortech.2010.11.114

Lee H-L, Chang C-K, Jeng W-Y et al (2012) Mutations in the substrate entrance region of β-glucosidase from *Trichoderma reesei* improve enzyme activity and thermostability. Protein Eng Des Sel 25:733–740. doi:10.1093/protein/gzs073

Li D, Li X, Dang W et al (2013) Characterization and application of an acidophilic and thermostable β-glucosidase from *Thermofilum pendens*. J Biosci Bioeng 115:490–496. doi:10.1016/j.jbiosc.2012.11.009

Liang Y, Si T, Ang EL, Zhao H (2014) Engineered pentafunctional minicellulosome for simultaneous saccharification and ethanol fermentation in *Saccharomyces cerevisiae*. Appl Environ Microbiol 80:6677–6684. doi:10.1128/AEM.02070-14

Liu D, Zhang R, Yang X et al (2011) Thermostable cellulase production of *Aspergillus fumigatus* Z5 under solid-state fermentation and its application in degradation of agricultural wastes. Int Biodeterior Biodegrad 65:717–725. doi:10.1016/j.ibiod.2011.04.005

Matano Y, Hasunuma T, Kondo A (2013) Cell recycle batch fermentation of high-solid lignocellulose using a recombinant cellulase-displaying yeast strain for high yield ethanol production in consolidated bioprocessing. Bioresour Technol 135:403–409. doi:10.1016/j.biortech.2012.07.025

Mate DM, Alcalde M (2015) Laccase engineering: from rational design to directed evolution. Biotechnol Adv 33:25–40. doi:10.1016/j.biotechadv.2014.12.007

Mateo C, Palomo JM, Fernandez-Lorente G et al (2007) Improvement of enzyme activity, stability and selectivity via immobilization techniques. Enzyme Microb Technol 40:1451–1463. doi:10.1016/j.enzmictec.2007.01.018

Mesas JM, Rodríguez MC, Alegre MT (2012) Basic characterization and partial purification of β-glucosidase from cell-free extracts of *Oenococcus oeni* ST81. Lett Appl Microbiol 55:247–255. doi:10.1111/j.1472-765X.2012.03285.x

Moilanen U, Kellock M, Galkin S, Viikari L (2011) The laccase-catalyzed modification of lignin for enzymatic hydrolysis. Enzyme Microb Technol 49:492–498. doi:10.1016/j.enzmictec.2011.09.012

Moraïs S, Barak Y, Caspi J et al (2010) Contribution of a xylan-binding module to the degradation of a complex cellulosic substrate by designer cellulosomes. Appl Environ Microbiol 76:3787–3796. doi:10.1128/AEM.00266-10

Moraïs S, Barak Y, Lamed R et al (2012) Paradigmatic status of an endo- and exoglucanase and its effect on crystalline cellulose degradation. Biotechnol Biofuels 5:78. doi:10.1186/1754-6834-5-78

Mosier N, Wyman C, Dale B et al (2005) Features of promising technologies for pretreatment of lignocellulosic biomass. Bioresour Technol 96:673–686. doi:10.1016/j.biortech.2004.06.025

Mot AC, Silaghi-Dumitrescu R (2012) Laccases: complex architectures for one-electron oxidations. Biochem 77:1395–1407. doi:10.1134/

Müller G, Várnai A, Johansen KS et al (2015) Harnessing the potential of LPMO-containing cellulase cocktails poses new demands on processing conditions. Biotechnol Biofuels 8:187. doi:10.1186/s13068-015-0376-y

Nakashima K, Endo K, Shibasaki-kitakawa N, Yonemoto T (2014) A fusion enzyme consisting of bacterial expansin and endoglucanase for the degradation of highly crystalline cellulose. RSC Adv 4:43815–43820. doi:10.1039/c4ra05891g

Nakatani Y, Yamada R, Ogino C, Kondo A (2013) Synergetic effect of yeast cell-surface expression of cellulase and expansin-like protein on direct ethanol production from cellulose. Microb Cell Fact 12:66. doi:10.1186/1475-2859-12-66

Nam KH, Sung MW, Hwang KY (2010) Structural insights into the substrate recognition properties of β-glucosidase. Biochem Biophys Res Commun 391:1131–1135. doi:10.1016/j.bbrc.2009.12.038

Oberoi HS, Rawat R, Chadha BS (2014) Response surface optimization for enhanced production of cellulases with improved functional characteristics by newly isolated *Aspergillus niger* HN-2. Antonie Van Leeuwenhoek 105:119–134. doi:10.1007/s10482-013-0060-9

Pandiyan K, Tiwari R, Rana S et al (2014) Comparative efficiency of different pretreatment methods on enzymatic digestibility of *Parthenium* sp. World J Microbiol Biotechnol 30:55–64. doi:10.1007/s11274-013-1422-1

Parisutham V, Kim TH, Lee SK (2014) Feasibilities of consolidated bioprocessing microbes: from pretreatment to biofuel production. Bioresour Technol 161:431–440. doi:10.1016/j.biortech.2014.03.114

Park S, Baker JO, Himmel ME et al (2010) Cellulose crystallinity index: measurement techniques and their impact on interpreting cellulase performance. Biotechnol Biofuels 3:10. doi:10.1186/1754-6834-3-10

Park S, Ransom C, Mei C et al (2011) The quest for alternatives to microbial cellulase mix production: corn stover-produced heterologous multicellulases readily deconstruct lignocellulosic biomass into fermentable sugars. J Chem Technol Biotechnol 86:633–641. doi:10.1002/jctb.2584

Park M, Sun Q, Liu F et al (2014) Positional assembly of enzymes on bacterial outer membrane vesicles for cascade reactions. PLoS ONE 9:1–6. doi:10.1371/journal.pone.0097103

Patagundi BI, Shivasharan CT, Kaliwal BB (2014) Isolation and characterization of cellulase producing bacteria from soil. Int J Curr Microbiol Appl Sci 3:59–69

Pearsall SM, Rowley CN, Berry A (2015) Advances in pathway engineering for natural product biosynthesis. ChemCatChem 7:3078–3093. doi:10.1002/cctc.201500602

Percival Zhang YH (2010) Production of biocommodities and bioelectricity by cell-free synthetic enzymatic pathway biotransformations: challenges and opportunities. Biotechnol Bioeng 105:663–667. doi:10.1002/bit.22630

Percival Zhang YH, Himmel ME, Mielenz JR (2006) Outlook for cellulase improvement: screening and selection strategies. Biotechnol Adv 24:452–481. doi:10.1016/j.biotechadv.2006.03.003

Pereira JH, Chen Z, McAndrew RP et al (2010) Biochemical characterization and crystal structure of endoglucanase Cel5A from the hyperthermophilic *Thermotoga maritima*. J Struct Biol 172:372–379. doi:10.1016/j.jsb.2010.06.018

Perevalova AA (2005) *Desulfurococcus fermentans* sp. nov., a novel hyperthermophilic archaeon from a Kamchatka hot spring, and emended description of the genus *Desulfurococcus*. Int J Syst Evol Microbiol 55:995–999. doi:10.1099/ijs.0.63378-0

Phillips CM, Beeson WT, Cate JH, Marletta MA (2011) Cellobiose dehydrogenase and a copper-dependent polysaccharide monooxygenase potentiate cellulose degradation by *Neurospora crassa*. ACS Chem Biol 6:1399–1406. doi:10.1021/cb200351y

Quinlan RJ, Sweeney MD, Lo Leggio L et al (2011) Insights into the oxidative degradation of cellulose by a copper metalloenzyme that exploits biomass components. Proc Natl Acad Sci 108:15079–15084. doi:10.1073/pnas.1105776108

Rabinovich ML, Melnik MS, Bolobova AV (2002) Microbial cellulases (review). Appl Biochem Microbiol 38:305–321. doi:10.1023/A:1016264219885

Rajasree KP, Mathew GM, Pandey A, Sukumaran RK (2013) Highly glucose tolerant β-glucosidase from *Aspergillus unguis*: NII 08123 for enhanced hydrolysis of biomass. J Ind Microbiol Biotechnol 40:967–975. doi:10.1007/s10295-013-1291-5

Reed PT, Izquierdo JA, Lynd LR (2014) Cellulose fermentation by *Clostridium thermocellum* and a mixed consortium in an automated repetitive batch reactor. Bioresour Technol 155:50–56. doi:10.1016/j.biortech.2013.12.051

Ribeiro LF, Furtado GP, Lourenzoni MR et al (2011) Engineering bifunctional laccase-xylanase chimeras for improved catalytic performance. J Biol Chem 286:43026–43038. doi:10.1074/jbc.M111.253419

Ribeiro LF, Nicholes N, Tullman J et al (2015) Insertion of a xylanase in xylose binding protein results in a xylose-stimulated xylanase. Biotechnol Biofuels 8:118. doi:10.1186/s13068-015-0293-0

Riou C, Salmon JM, Vallier MJ et al (1998) Purification, characterization, and substrate specificity of a novel highly glucose-tolerant beta-glucosidase from *Aspergillus oryzae*. Appl Environ Microbiol 64:3607–3614

Roedl A (2010) Production and energetic utilization of wood from short rotation coppice—a life cycle assessment. Int J Life Cycle Assess 15:567–578. doi:10.1007/s11367-010-0195-0

Rollin JA, Tam TK, Zhang YHP (2013) New biotechnology paradigm: cell-free biosystems for biomanufacturing. Green Chem 15:1708–1719. doi:10.1039/c3gc40625c

Sakthi SS, Saranraj P, Rajasekar M (2011) Optimization for cellulase production by *Aspergillus niger* using paddy straw as substrate. Int J Adv Sci Tech Res 1:68–85

Schiraldi C, De Rosa M (2002) The production of biocatalysts and biomolecules from extremophiles. Trends Biotechnol 20:515–521. doi:10.1016/S0167-7799(02)02073-5

Schoffelen S, van Hest JCM (2012) Multi-enzyme systems: bringing enzymes together in vitro. Soft Matter 8:1736. doi:10.1039/c1sm06452e

Schröder C, Elleuche S, Blank S, Antranikian G (2014) Characterization of a heat-active archaeal β-glucosidase from a hydrothermal spring metagenome. Enzyme Microb Technol 57:48–54. doi:10.1016/j.enzmictec.2014.01.010

Schülein M (2000) Protein engineering of cellulases. Biochim Biophys Acta Protein Struct Mol Enzymol 1543:239–252. doi:10.1016/S0167-4838(00)00247-8

Schumacher SD, Hannemann F, Teese MG et al (2012) Autodisplay of functional CYP106A2 in *Escherichia coli*. J Biotechnol 161:104–112. doi:10.1016/j.jbiotec.2012.02.018

Schüürmann J, Quehl P, Festel G, Jose J (2014) Bacterial whole-cell biocatalysts by surface display of enzymes: toward industrial application. Appl Microbiol Biotechnol 98:8031–8046. doi:10.1007/s00253-014-5897-y

Segato F, Damásio ARL, de Lucas RC et al (2014) Genome analyses highlight the different biological roles of cellulases. Microbiol Mol Biol Rev 78:588–613. doi:10.1128/MMBR.00019-14

Shallom D, Shoham Y (2003) Microbial hemicellulases. Curr Opin Microbiol 6:219–228. doi:10.1016/S1369-5274(03)00056-0

Sharrock KR (1988) Cellulase assay methods: a review. J Biochem Biophys Methods 17:81–105. doi:10.1016/0165-022X(88)90040-1

Sherief AA, El-Tanash AB, Atia N (2010) Cellulase production by *Aspergillus fumigatus* grown on mixed substrate of rice straw and wheat bran. Res J Microbiol 5:199–211. doi:10.3923/jm.2010.199.211

Shi R, Li Z, Ye Q et al (2013) Heterologous expression and characterization of a novel thermo-halotolerant endoglucanase Cel5H from *Dictyoglomus thermophilum*. Bioresour Technol 142:338–344. doi:10.1016/j.biortech.2013.05.037

Shin KC, Oh DK (2014) Characterization of a novel recombinant B-glucosidase from *Sphingopyxis alaskensis* that specifically hydrolyzes the outer glucose at the C-3 position in protopanaxadiol-type ginsenosides. J Biotechnol 172:30–37. doi:10.1016/j.jbiotec.2013.11.026

Smith MR, Khera E, Wen F (2015) Engineering novel and improved biocatalysts by cell surface display. Ind Eng Chem Res 54:4021–4032. doi:10.1021/ie504071f

Sohail M, Siddiqi R, Ahmad A, Khan SA (2009) Cellulase production from *Aspergillus niger* MS82: effect of temperature and pH. N Biotechnol 25:437–441. doi:10.1016/j.nbt.2009.02.002

Soni R, Nazir A, Chadha BS (2010) Optimization of cellulase production by a versatile *Aspergillus fumigatus* fresenius strain (AMA) capable of efficient deinking and enzymatic hydrolysis of Solka floc and bagasse. Ind Crops Prod 31:277–283. doi:10.1016/j.indcrop.2009.11.007

Stern J, Kahn A, Vazana Y et al (2015) Significance of relative position of cellulases in designer cellulosomes for optimized cellulolysis. PLoS ONE

Tachaapaikoon C, Kosugi A, Pason P et al (2012) Isolation and characterization of a new cellulosome-producing *Clostridium thermocellum* strain. Biodegradation 23:57–68. doi:10.1007/s10532-011-9486-9

Tozakidis IEP, Quehl P, Schüürmann J, Jose J (2015) Let's do it outside: neue Biokatalysatoren mittels surface display. BIOspektrum 21:668–671. doi:10.1007/s12268-015-0628-1

Tozakidis IEP, Brossette T, Lenz F et al (2016) Proof of concept for the simplified breakdown of cellulose by combining *Pseudomonas putida* strains with surface displayed thermophilic endocellulase, exocellulase and β-glucosidase. Microb Cell Fact 15:103–114. doi:10.1186/s12934-016-0505-8

Turner NJ (2003) Directed evolution of enzymes for applied biocatalysis. Trends Biotechnol 21:474–478. doi:10.1016/j.tibtech.2003.09.001

Ulrich A, Klimke G, Wirth S (2008) Diversity and activity of cellulose-decomposing bacteria, isolated from a sandy and a loamy soil after long-term manure application. Microb Ecol 55:512–522. doi:10.1007/s00248-007-9296-0

Vaaje-Kolstad G, Westereng B, Horn SJ et al (2010) An oxidative enzyme boosting the enzymatic conversion of recalcitrant polysaccharides. Science 330:219–222. doi:10.1126/science.1192231

van den Burg B (2003) Extremophiles as a source for novel enzymes. Curr Opin Microbiol 6:213–218. doi:10.1016/S1369-5274(03)00060-2

Varzakas T, Arapoglou D, Israilides C (2006) Kinetics of endoglucanase and endoxylanase uptake by soybean seeds. J Biosci Bioeng 101:111–119. doi:10.1263/jbb.101.111

Vieille C, Zeikus GJ (2001) Hyperthermophilic enzymes: sources, uses, and molecular mechanisms for thermostability. Microbiol Mol Biol Rev 65:1–43. doi:10.1128/MMBR.65.1.1-43.2001

Wahlström R, Rahikainen J, Kruus K, Suurnäkki A (2014) Cellulose hydrolysis and binding with *Trichoderma reesei* Cel5A and Cel7A and their core domains in ionic liquid solutions. Biotechnol Bioeng 111:726–733. doi:10.1002/bit.25144

Walton PH, Davies GJ (2016) On the catalytic mechanisms of lytic polysaccharide monooxygenases. Curr Opin Chem Biol 31:195–207. doi:10.1016/j.cbpa.2016.04.001

Wang Z, Bay H, Chew K, Geng A (2014) High-loading oil palm empty fruit bunch saccharification using cellulases from *Trichoderma koningii* MF6. Process Biochem 49:673–680. doi:10.1016/j.procbio.2014.01.024

Watanabe T, Sato T, Yoshioka S et al (1992) Purificication and properties of *Aspergillus niger* beta-glucosidase. Eur J Biochem 209:651–659. doi:10.1111/j.1432-1033.1992.tb17332.x

Westereng B, Cannella D, Wittrup Agger J et al (2015) Enzymatic cellulose oxidation is linked to lignin by long-range electron transfer. Sci Rep 5:18561. doi:10.1038/srep18561

Willick GE, Seligy VL (1985) Multiplicity in cellulases of *Schizophyllum commune*. Derivation partly from heterogeneity in transcription and glycosylation. Eur J Biochem 151:89–96

Wilson DB (2009) Cellulases and biofuels. Curr Opin Biotechnol 20:295–299. doi:10.1016/j.copbio.2009.05.007

Wilson DB (2015) Processive cellulases. Elsevier B.V

Wu T, Huang C, Ko T et al (2011) Diverse substrate recognition mechanism revealed by *Thermotoga maritima* Cel5A structures in complex with cellotetraose, cellobiose and mannotriose. Biochim Biophys Acta Proteins Proteom 1814:1832–1840. doi:10.1016/j.bbapap.2011.07.020

Yagüe E, Béguin P, Aubert JP (1990) Nucleotide sequence and deletion analysis of the cellulase-encoding gene celH of *Clostridium thermocellum*. Gene 89:61–67

Yamada R, Hasunuma T, Kondo A (2013) Endowing non-cellulolytic microorganisms with cellulolytic activity aiming for consolidated bioprocessing. Biotechnol Adv 31:754–763. doi:10.1016/j.biotechadv.2013.02.007

Yang S-J, Kataeva I, Hamilton-Brehm SD et al (2009) Efficient degradation of lignocellulosic plant biomass, without pretreatment, by the *Thermophilic Anaerobe "Anaerocellum thermophilum"* DSM 6725. Appl Environ Microbiol 75:4762–4769. doi:10.1128/AEM.00236-09

Ye X, Rollin J, Zhang YP (2010) Thermophilic α-glucan phosphorylase from *Clostridium thermocellum*: cloning, characterization and enhanced thermostability. J Mol Catal B Enzym 65:110–116. doi:10.1016/j.molcatb.2010.01.015

Yuan S-F, Wu T-H, Lee H-L et al (2015) Biochemical characterization and structural analysis of a bifunctional cellulase/xylanase from *Clostridium ther-*

mocellum. J Biol Chem 290:5739–5748. doi:10.1074/jbc.M114.604454

Zechel DL, Withers SG (2000) Glycosidase mechanisms: anatomy of a finely tuned catalyst. Acc Chem Res 33:11–18. doi:10.1021/ar970172+

Zhang YHP (2011) Substrate channeling and enzyme complexes for biotechnological applications. Biotechnol Adv 29:715–725. doi:10.1016/j.biotechadv.2011.05.020

Zhang Y-HP, Lynd LR (2004) Toward an aggregated understanding of enzymatic hydrolysis of cellulose: noncomplexed cellulase systems. Biotechnol Bioeng 88:797–824. doi:10.1002/bit.20282

Zhao L, Xie J, Zhang X et al (2013) Overexpression and characterization of a glucose-tolerant β-glucosidase from *Thermotoga thermarum* DSM 5069T with high catalytic efficiency of ginsenoside Rb1 to Rd. J Mol Catal B Enzym 95:62–69. doi:10.1016/j.molcatb.2013.05.027

Zverlov VV, Velikodvorskaya GA, Schwarz WH (2003) Two new cellulosome components encoded downstream of cell in the genome of *Clostridium thermocellum*: the non-processive endoglucanase CelN and the possibly structural protein CseP. Microbiology 149:515–524. doi:10.1099/mic.0.25959-0

Zverlov VV, Schantz N, Schwarz WH (2005) A major new component in the cellulosome of *Clostridium thermocellum* is a processive endo-β-1,4-glucanase producing cellotetraose. FEMS Microbiol Lett 249:353–358. doi:10.1016/j.femsle.2005.06.037

7

Industrial cellulase performance in the simultaneous saccharification and co-fermentation (SSCF) of corn stover for high-titer ethanol production

Qiang Zhang and Jie Bao[*]

Abstract

Background: Cellulase enzymes contribute to the largest portion of operation cost on production of cellulosic ethanol. The industrial cellulases available on the industrial enzyme market from different makers and sources vary significantly in hydrolysis and ethanol, and finally lead to the changes of enzyme cost. Therefore, the selection of the proper industrial cellulase enzymes for commercial-scale production of cellulosic ethanol is crucially important in terms of high performance and cost reduction.

Results: In this study, three major cellulase enzyme products available on the Chinese industrial enzyme market were selected and evaluated as the biocatalysts for the biorefining process of lignocellulose biomass into high-titer ethanol. The cellulase enzymes included Cellic CTec 2.0 from Novozymes (Beijing), and LLC 4 from Vland (Qingdao), as well as # 7 from an industrial enzyme maker. The detailed assays on the filter paper activity, the cellobiase activity, and the total protein contents of the enzymes were conducted according to the standard protocols. When the cellulase enzymes were applied to the practical hydrolysis and ethanol-fermentation operation under the conditions of high solids loading and low range of cellulase dosage, the hydrolysis yield shows the significant difference, and the difference was narrowed in the final ethanol yield.

Conclusions: The commercially available cellulase enzymes showed different performances in the activities, the cellulose hydrolysis yield, and the ethanol fermentation yields based on the protein dosage per gram of cellulose of corn stover. In general, the industrial cellulase products give satisfactory performance and can be applied for the practical cellulosic ethanol production on commercial scale.

Keywords: Industrial cellulase enzyme, Activity, Ethanol, Hydrolysis, Lignocellulose

Background

Currently cellulosic ethanol is on the way to its large scale commercialization in USA, Europe, and China. Beta renewables first launched the first commercial cellulosic ethanol plant with the annual ethanol production of 40,000 metric tons from corn stover in Oct 2013, Italy (Ramesh 2013). DuPont biofuels solution started the plant for production of 89,600 metric tons of ethanol annually from corn stover and switchgrass. Abengoa Bioenergy produced 74,900 metric tons of ethanol annually from corn stover and other non-feed energy crops. Poet-DSM produced 59,700 metric tons of ethanol annually from corn stover (Chiaramonti et al. 2013). In China, Shandong Longlive Co. used corncob residue from xylitol industry and produced 50,000 metric tons of ethanol (Lei et al. 2014). However, due to the reasons of low petroleum price and relatively high cost, the factories are not operated in full scale and further modifications on the plants are still going on.

*Correspondence: jbao@ecust.edu.cn
State Key Laboratory of Bioreactor Engineering, East China University
of Science and Technology, 130 Meilong Road, Shanghai 200237, China

In the trend of commercialization of cellulosic ethanol, a proper cellulase enzyme with high hydrolysis performance and low cost is crucially important because cellulase enzyme contributes to the largest portion of the cost on lignocellulose biorefining process (Klein-Marcuschamer et al. 2012; Gang et al. 2016). In the past several decades, great efforts were made by the enzyme industry worldwide, and several high-performance cellulase enzymes had been developed and introduced into the market for practical use in cellulosic ethanol production. The latest cellulase enzyme products include the CTec series from Novozymes, such as the recent products of Cellic CTec 2, Cellic CTec 3, and HTec 3 (Chen et al. 2016); and the Accellerase series from the former Genencor (now part of DuPont), such as Accellerase 1500 (Marcos et al. 2013). On the Chinese industrial enzyme market, several homemade cellulase enzymes are available for use in cellulosic ethanol production (Zhang et al. 2015, 2016; Liu and Wang 2014).

In this study, we selected three major cellulase enzyme products available on the China industrial market including Cellic CTec 2.0 from Novozyme (Beijing), LLC 4 from Vland (Qingdao), and # 7 from an industrial enzyme maker as the saccharification biocatalyst of corn stover. The lignocellulose feedstock, corn stover, was pretreated by dry dilute acid pretreatment (DDAP) and biologically detoxified to remove the inhibitors, then hydrolyzed at the high solids loading (30%, w/w) of the pretreated and detoxified corn stover. The ethanol fermentation was performed under the simultaneous saccharification and co-fermentation (SSCF) by a xylose utilizing yeast strain to achieve the high-titer ethanol and yield. The results indicate that the industrial enzymes available as cellulase products in the market give the satisfactory performance in general and can be applied for the practical cellulosic ethanol production on commercial scale.

Methods

Corn stover feedstock

Corn stover (CS) was harvested from Bayan Nur League, Inner Mongolia Autonomous Region, China in fall 2015. The collected corn stover was milled coarsely and screened through a mesh with the circle diameter of 10 mm. Then the milled corn stover was water washed to remove the field dirts, stones and metal pieces, and air dried. The composition of corn stover was determined by the two-step acid hydrolysis method according to National Renewable Energy Laboratory (NREL) protocols (Sluiter et al. 2008, 2012). On dry weight base (w/w), corn stover contained 35.4% of cellulose, 24.6% of hemicellulose, 16.1% of lignin, and 3.5% of ash.

Cellulase enzymes

We selected three representative commercial cellulases from the Chinese industrial enzyme market, by the availability, the production capacity, and the representativeness in activity and cost, including Cellic CTec 2.0 [kindly donated by Novozymes (China), Beijing, China], Vland LLC 4 (purchased Vland Biotech, Qingdao, China), and # 7 (purchased from an industrial enzyme maker). CTec 2.0 and LLC 4 are liquid enzymes, and # 7 is solid enzyme. The liquid enzyme was taken as water and the mass of the total liquid of the saccharification process was calculated. The # 7 enzyme is the solid enzyme produced by the adsorption of the liquid enzyme onto the solid bran particles and dried for the purpose of long-term storage. There is no need for the extraction step because the bran is hydrolyzed very quickly when the enzyme is put into the hydrolysis system and the cellulase enzyme adsorbed on the bran is releases into the hydrolysate. The filter paper activity was determined according to the NREL protocol LAP-006 (Adney and Baker 1996). The cellobiase activity was determined using the method of Ghose (1987). The total protein concentration was determined by Bradford method using bovine serum albumin (BSA) as protein standard (Bradford 1976). The cellulase enzyme was used based on the total protein weight per gram of cellulose substrate in the biomass feedstock.

The reagents KH_2PO_4, $(NH_4)_2SO_4$, $MgSO_4$, and H_2SO_4 were purchased from Lingfeng chemical reagent, Shanghai, China. Yeast extract was procured from Angel Yeast Co., Yichang, Hubei, China. Agar was purchased from Aladdin BioChem, Shanghai, China.

Strains and medium

Biodetoxification fungus *Amorphotheca resinae* ZN1 was isolated in our previous works and stored in China General Microorganism Collection Center (CGMCC), Beijing, China with the registration number 7452 (Zhang et al. 2010). *A. resinae* ZN1 was maintained on a potato dextrose agar medium (PDA) slant. The PDA medium was prepared by boiling 200 g of peeled and sliced potatoes in 1 L deionized water for 30 min. 15 g/L of agar was added for preparation of PDA slant for *A. resinae* ZN1 growth.

Ethanol fermentation strain *Saccharomyces cerevisiae* XH7 was an engineered strain from the wild diploid *Saccharomyces cerevisiae* BSIF (Li et al. 2015). The strain was cultured in YPD medium containing 20 g/L of glucose, 20 g/L of peptone, and 10 g/L of yeast extract. The culture vial was stored in the YPD medium containing 30% of glycerol at −80 °C freezer.

The media used included (1) activation medium, 20 g/L of glucose, 20 g/L of peptone, 10 g/L of yeast extract; (2) seed culture medium, 5% (w/w) of the pretreated and biodetoxified corn stover, the cellulase dosage of 10 mg protein per gram of cellulose, 2 g/L of KH_2PO_4, 2 g/L of $(NH_4)_2SO_4$, 1 g/L of $MgSO_4$, 10 g/L of yeast extract; (3) adaptation seed medium, 10% (w/w) of the pretreated and biodetoxified corn stover, the cellulase dosage of 10 mg protein per gram of cellulose, 2 g/L of KH_2PO_4, 2 g/L of $(NH_4)_2SO_4$, 1 g/L of $MgSO_4$, 10 g/L of yeast extract. The seed culture medium contained only low concentration of xylose from the pretreated corn stover feedstock and no glucose or pre-hydrolysate in it at the beginning of the culture. As the seed culture proceeds, the added pretreated and detoxified corn stover is hydrolyzed into soluble glucose by the cellulase, and the glucose is acting as the carbon source for the seed cell growth. In this way, the pure glucose sugar is saved by corn stover feedstock; and (4) SSCF medium, 2 g/L of KH_2PO_4, 2 g/L of $(NH_4)_2SO_4$, 1 g/L of $MgSO_4$, 10 g/L of yeast extract.

Pretreatment, biodetoxification, and SSCF

Corn stover was pretreated using the dry dilute acid pretreatment (DDAP) method (Zhang et al. 2011; He et al. 2014). The major components' contents and the inhibitors' contents in the pretreated biomass feedstocks were determined according to NREL protocols (Sluiter et al. 2008, 2012). The pretreated biomass was briefly disk milled to remove the long cellulose fibers to avoid the blockage of the slurry flow of the downstream hydrolysate and broth.

The pretreated biomass solids were biodetoxified in a 15-L bioreactor at 28 °C and aeration for 36 h to remove the inhibitors generated during the dry dilute acid pretreatment operation (Zhang et al. 2010; He et al. 2016).

The pretreated solids were firstly converted into liquid hydrolysate slurry containing both monosaccharides (glucose and xylose) and oligomer sugars (oligo-glucan and oligo-xylan) in the specially designed 5 L bioreactors equipped with helical ribbon impeller (Zhang et al. 2011) at 50 °C, pH 4.8 for 12 h. Then glucose and xylose were co-fermented into ethanol simultaneously with the further hydrolysis of cellulose and oligomer sugars (simultaneous saccharification and co-fermentation, SSCF) by the engineered *Saccharomyces cerevisiae* XH7 at the high solids loading (30%, w/w) in the same bioreactor at 30 °C for 96 h by inoculating the shortly adapted yeast seed cells into the hydrolysate at 10% (v/v). The nutrients added included 2 g/L of KH_2PO_4, 2 g/L of $(NH_4)_2SO_4$, 1 g/L of $MgSO_4$, 10 g/L of yeast extract. Samples were periodically withdrawn and for analysis of glucose, xylose, ethanol, glycerol, acetic acid, furfural, and 5-hydroxymethylfurfural (HMF).

The *S. cerevisiae* XH7 seed broth was prepared in a two-step adaption procedure using the pretreated biomass feedstock as the carbon source instead of glucose sugar (Qureshi et al. 2015).

Calculation of ethanol yield

The ethanol yield (%) in SSCF and the xylose utilization (%) were calculated based on the method proposed by Zhang and Bao (2012):

$$\text{Ethanol yield (\%)} = \frac{[\text{Ethanol}] \times W}{976.9 - 0.804 \times [\text{Ethanol}]}$$
$$\cdot \frac{1}{0.511 \times ([\text{Cellulose}] \times 1.111 + [\text{Xylose}]) \times [\text{Solids}] \times M}$$
$$\times 100\%$$

where [Ethanol] is the concentration of ethanol in the fermentation broth at the end of the SSCF (g/L); W is the total water input into the hydrolysis or the SSCF system (g); M is the total weight of the hydrolysis or the SSCF system at the beginning of the operation (g); [Cellulose] is the cellulose content in the dry pretreated solids (g/g); [Xylose] is the xylose content in the dry pretreated solids (g/g); [Solids] is the pretreated solids loading of the hydrolysis and SSCF system on the dry-weight base (g/g); 976.9 is the ethanol correction factor (g/L) between the mass concentration (g/g) and the volumetric concentration (g/L); 0.804 is the dimensionless factor in calculating water loss in SSCF; 1.111 is the dimensionless conversion factor for cellulose to equivalent glucose; 0.511 is the dimensionless conversion factor for glucose to ethanol based on the stoichiometric biochemistry of yeast.

Xylose conversion is calculated by measuring the percentage ratio of the decreased xylose concentrations in the hydrolysate at the beginning and the end of the SSCF operation over the total xylose concentration.

Analysis

Sugars, ethanol, acetic acid, 5-HMF, Furfural, and Glycerol were analyzed on HPLC (LC-20AD, Shimazu, Kyoto, Japan) equipped with a Bio-rad Aminex HPX-87H column (Bio-rad, Hercules, CA, USA) and RID-10A detector (Shimadzu, Kyoto, Japan). 5 mM H_2SO_4 solution was used as flow phase at the flow rate of 0.6 mL/min. Furans were analyzed on HPLC (LC-20AT, Shimazu, Kyoto, Japan) equipped with a YMC-Pack ODS-A column (YMC, Tokyo, Japan) and an SPD-20A UV detector (Shimadzu, Kyoto, Japan).

The yeast cell viability in the simultaneous saccharification and co-fermentation (SSCF) was assayed by counting the colony-forming units (CFU) on the YPD (Gu et al. 2015) petri dish when the 100 μL of the 10^{-5} or 10^{-6} diluted fermentation broth withdrawn at different time points were stretched and cultured for 48 h at 30 °C.

Results and discussion

Enzyme assays of filter paper unit, cellobiase activity, and total protein content

The activities and protein contents of the three industrial cellulase enzymes from different makers were assayed. The filter paper unit (FPU/mL), the cellobiase activity or β-glucosidase activity (CBU/mL), and the total protein content (mg/mL) of Cellic CTec 2.0, # 7, and LLC 4 were determined as shown in Table 1. The results show that the filter paper activity and the cellobiase activity of CTec 2.0 and Vland LLC 4 were similar based on the volumetric basis, while # 7 is relatively low (only 30% of the filter paper activity and 2% in the cellobiase activity of CTec 2.0). The total protein concentrations of CTec 2.0 and LLC 4 were also similar at 75–90 mg/mL, while that of # 7 was less than half of the two. The specific filter paper activities of CTec 2.0 and LLC 4 were also close, and that of # 7 was about half of the first two enzymes. # 7 was absorbed on wheat bran solid particles; thus, its specific activity was relatively lower than that of the other two enzymes.

SSCF assay under the same cellulase protein additions

The hydrolysis performances of the three industrial cellulase enzymes were evaluated under the high solids loading of pretreated corn stover and the simultaneous saccharification and co-fermentation (SSCF). The corn stover feedstock was pretreated by dry dilute acid pretreatment (DDAP) and then biologically detoxified to remove the inhibitors. The moderate cellulase dosages of the pretreated corn stover and a xylose utilizing *S. cerevisiae* XH7 strain were used. The pre-hydrolysis lasted for 12 h at 50 °C, and then the SSCF was started and performed for 96 h (Fig. 1; Table 2).

The 12-h pre-hydrolysis of 30% (w/w) solid content system generated 82.4, 74.3, and 63.6 g/L of glucose by CTec 2.0, LLC 4, and # 7, respectively. The hydrolysis result indicates that the hydrolysis capacity of CTec 2.0 is advantage to LLC 4 and # 7. The 96-h SSCF generated 76.3, 73.0, 76.0 g/L of ethanol and 73.8, 73.7, 76.4% of conversion yields by CTec 2.0, LLC 4, and # 7,

respectively. Similar ethanol titer and yield were obtained in the given cellulase dosage (10 mg protein per gram of cellulose), although the considerable difference in hydrolysis capacity existed. The SSCF results indicate that the all the three cellulase worked well in the SSCF to achieve the high ethanol titer and yield, but certain differences existed. The xylose conversions achieved were 83.4, 85.6, and 89.5%, respectively. when CTec 2.0, LLC 4, and # 7 were used. The cell viability result shows that the lower cellulose conversion and glucose accumulation by # 7 led to the better cell growth of the fermenting yeast cells, while CTec 2.0 and LLC 4 with the high hydrolysis yields

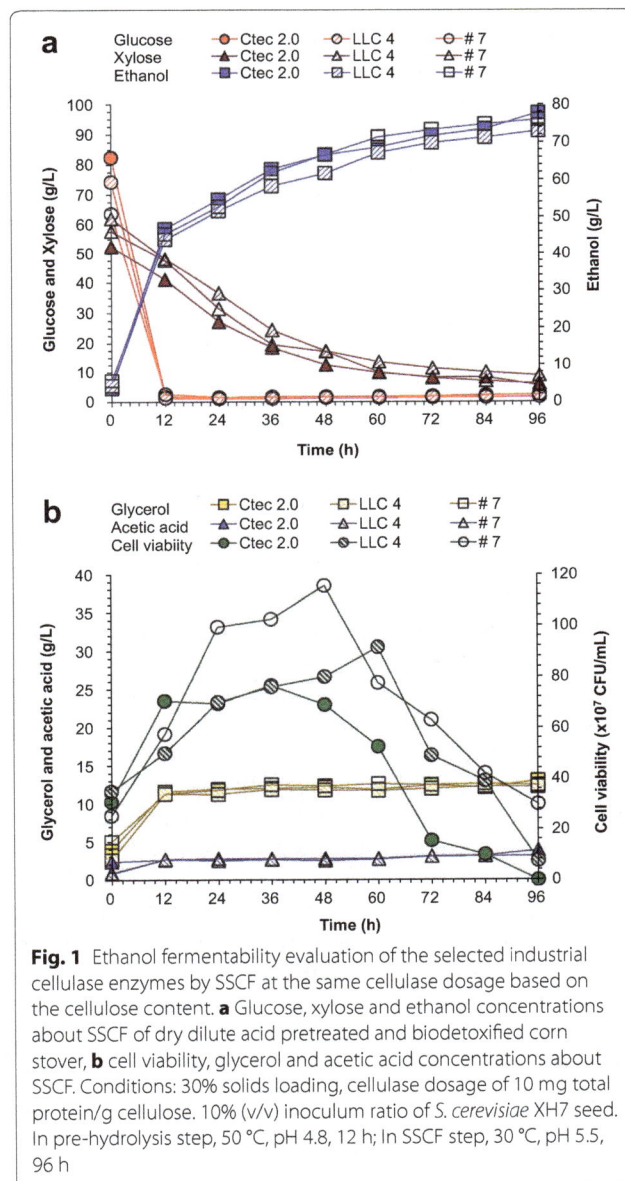

Fig. 1 Ethanol fermentability evaluation of the selected industrial cellulase enzymes by SSCF at the same cellulase dosage based on the cellulose content. **a** Glucose, xylose and ethanol concentrations about SSCF of dry dilute acid pretreated and biodetoxified corn stover, **b** cell viability, glycerol and acetic acid concentrations about SSCF. Conditions: 30% solids loading, cellulase dosage of 10 mg total protein/g cellulose. 10% (v/v) inoculum ratio of *S. cerevisiae* XH7 seed. In pre-hydrolysis step, 50 °C, pH 4.8, 12 h; In SSCF step, 30 °C, pH 5.5, 96 h

Table 1 Activity assays of the industrial cellulase enzymes

Enzyme type	Cellic CTec 2.0	Vland LLC 4	#7
Total protein content (mg/mL or mg/g)	87.3	75.8	46.7
Filter paper unit (FPU/mL or FPU/g)	203.2	199.4	63.0
β-glucosidase unit (CBU/mL or CBU/g)	4900	5500	99.9
FPU and CBU ratio (FPU/CBU)	1:24.1	1:27.6	1:1.6
Specific filter paper unit (FPU/mg protein)	2.33	2.62	1.35

Table 2 SSCF of corn stover using different industrial cellulase enzymes

	Pre-hydrolysis		SSCF			
	Glucose (g/L)	Xylose (g/L)	Ethanol titer (g/L)	Ethanol yield (%)[a]	Xylose conversion (%)	Glycerol (g/L)
Cellic CTec 2.0	82.39	52.38	78.50	77.72	90.83	12.79
Valnd LLC 4	74.26	62.00	77.25	78.32	86.29	12.84
# 7	63.58	57.70	77.09	77.62	90.08	12.87

[a] The sugars in the enzyme solution or solids were not taken into account

did not help in making the xylose conversion. Glycerol formation accumulated to about 12 g/L and considerably reduced the ethanol yield, revealing that the fermenting yeast was under the stress of high glucose concentration then easily led to the glycerol formation.

Cellulase enzymes may also contain some soluble carbohydrates (Zhang et al. 2007), which can be directly used for ethanol production in SSCF. Three industrial cellulases, CTec 2.0, LLC 4, and # 7, contained 277.1, 13.4, and 12.9 mg/g of glucose, respectively, as well as 36.2, 4.4, and 0.9 mg/g of xylose. At 30% solids loading, and with the cellulase dosage of 10 mg protein per gram of cellulose, the three enzymes would provide 4.88, 0.32, and 0.40 g/L of carbohydrates (including glucose and xylose) which theoretically generated 2.5, 0.16, and 0.20 g/L of ethanol, individually. Compared with the other two enzymes, CTec 2.0 supplied more fermentable sugars for SSCF. However, the ethanol obtained from the sugars was quite lower, and it was difficult to calculate it accurately in practice. As a result, the effect of the carbohydrates from the cellulase on the ethanol production was not considered in this study.

SSCF assays under varying cellulase enzyme dosage

Following the same enzyme dosages for use in SSCF, different cellulase dosages of each industrial enzyme on the SSCF performance were tested in the ranges of 5, 10, 15, and 25 mg total protein per gram cellulose of the pretreated corn stover as shown in Fig. 2.

The 12-h pre-hydrolysis assay shows the similar tendency at varying cellulase dosages. The hydrolysis capacity of CTec 2.0 showed the maximum hydrolysis yield at each cellulase dosage from 5 to 25 mg/g, followed by the LLC 4 and then # 7. The hydrolysis yield increased with the increasing cellulase dosage but the increase almost ceased or slowed down due to the increasing glucose inhibition on the cellulase enzyme activity. When the cellulase dosage was in the range of 5–10 mg protein per gram of cellulose, the glucose concentrations increased by 66, 21, and 17% with the use of # 7, CTec 2.0, and LLC 4, respectively. When the cellulase dosage was in the range of 10–15 mg protein per gram of cellulase, the glucose concentration increased by 14, 11, and 12%, respectively. When the cellulase dosage was in the range of 15–25 mg protein

per gram of cellulose, the glucose concentrations only increased by 7 and 10% by CTec 2.0 and LLC 4, respectively, while # 7 did not show the increase in glucose concentration. The xylose utilization by # 7 was the optimal because of the low glucose concentration provided.

The SSCF assay shows that the ethanol titer and yield from the SSCF by CTec 2.0, LLC 4, and # 7 were close (about 80 g/L) when the cellulase dosage was in the higher range of 10, 15, 25 mg protein per gram of cellulose, indicating the higher overdose of cellulase enzyme did not help in further improvement of ethanol titer and yield. High cellulase dosage led to the high glucose concentration in the pre-hydrolysis step, and the high glucose inhibited the cell growth and conversion rate of glucose and xylose to ethanol. However, the minimum dosage of cellulase (5 mg protein per gram of cellulose) showed the differences in the ethanol titer and yield by CTec 2.0, LLC 4, and # 7 were significantly different: the 96-h SSCF generated 68.1, 68.7, 56.1 g/L of ethanol and 66.6, 69.0, 49.8% of conversion yields by CTec 2.0, LLC 4, and # 7, respectively. The ethanol titer and yield by # 7 were relatively lower at the minimum dosage of cellulose. As shown in Fig. 2c, d, the activities of filter paper unit (FPU) and cellobiose (CBU) of # 7 were much lower than those of CTec 2.0 and LLC 4 at the same cellulase dosage, especially in the minimum range, which inevitably led to lower hydrolysis yield, ethanol titer, and ethanol yield by # 7.

Based on the SSCF results at the 10 mg protein per gram of cellulose, for producing 1 kg of cellulosic ethanol, 21.6 g of cellulase protein of CTec 2.0, equivalent to 247 g of the liquid enzyme; or 20.9 g cellulase protein of LLC 4, equivalent to 275 g of the liquid enzyme; or 21.3 g of cellulase protein of # 7, equivalent to 457 g of the solid enzyme, is needed. If the cellulase enzyme is based on the same price per kilogram enzyme by volume or weight, then CTec 2.0 is the least expensive. However, if the enzyme is sold based on the total proteins of the enzyme product, the cost of the enzymes is similar for the three enzymes.

Conclusion

The industrial cellulase enzymes shows significantly different performances in activity and cellulose hydrolysis yield, and less significant ethanol titer and yield based

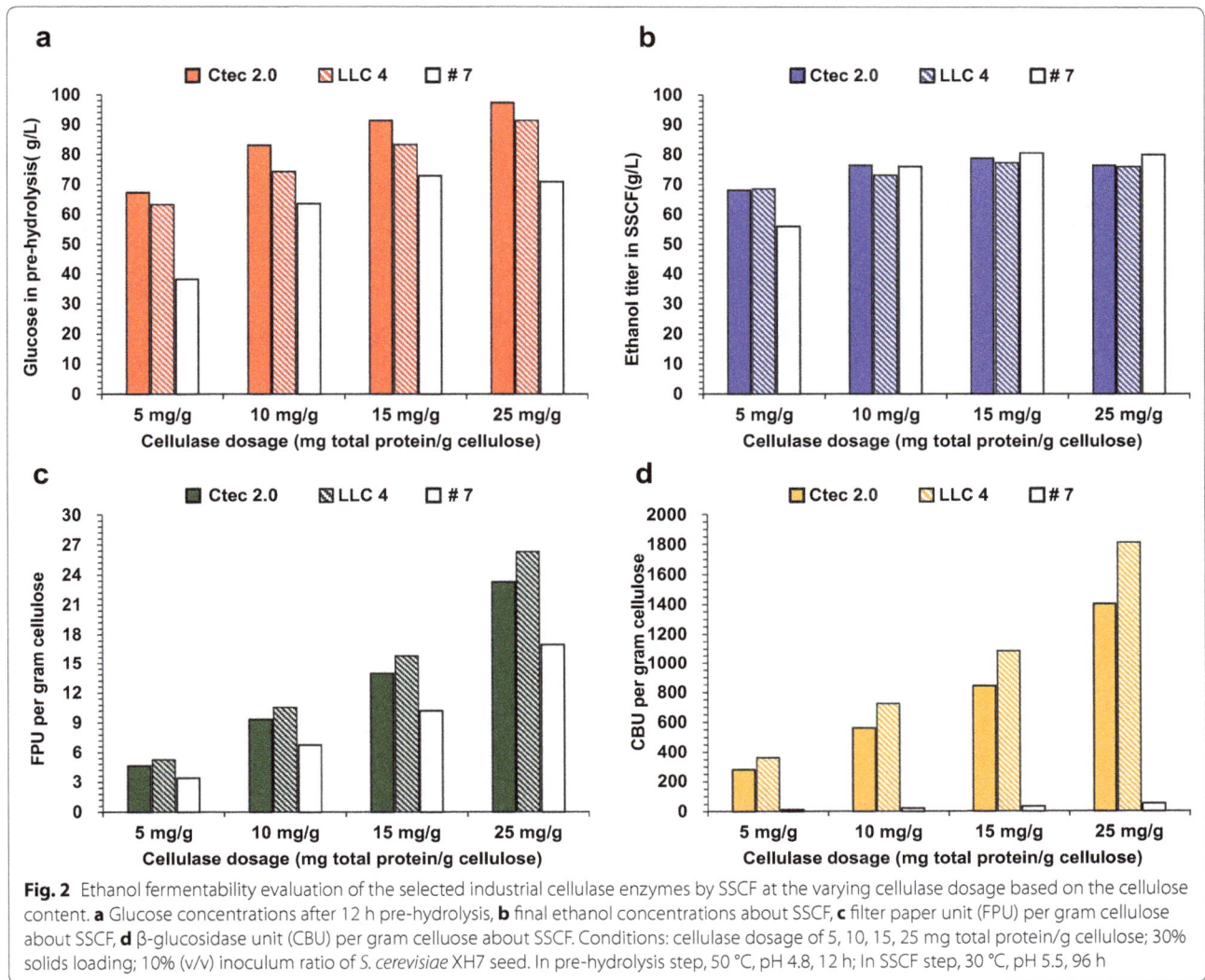

Fig. 2 Ethanol fermentability evaluation of the selected industrial cellulase enzymes by SSCF at the varying cellulase dosage based on the cellulose content. **a** Glucose concentrations after 12 h pre-hydrolysis, **b** final ethanol concentrations about SSCF, **c** filter paper unit (FPU) per gram cellulose about SSCF, **d** β-glucosidase unit (CBU) per gram celluose about SSCF. Conditions: cellulase dosage of 5, 10, 15, 25 mg total protein/g cellulose; 30% solids loading; 10% (v/v) inoculum ratio of *S. cerevisiae* XH7 seed. In pre-hydrolysis step, 50 °C, pH 4.8, 12 h; In SSCF step, 30 °C, pH 5.5, 96 h

on the total protein dosage per gram of cellulose of corn stover. In general, the industrial enzymes available as cellulase products in the market give satisfactory performance and can be applied for the practical cellulosic ethanol production on a commercial scale.

Authors' contributions
QZ and JB designed and wrote the experiment; QZ conducted the experiment; JB conceived the research. Both authors read and approved the final manuscript.

Competing interests
The authors declare that they have no competing interests.

Funding
This research was supported by the funding granted to JB by the National High-Tech Program of China (2014AA021901).

References
Adney B, Baker J (1996) Measurement of cellulase activities. AP-006. NREL. Analytical Procedure. National Renewable Energy Laboratory, Golden CO
Bradford MM (1976) A rapid and sensitive method for the quantitation of microgram quantities of protein utilizing the principle of protein-dye binding. Anal Biochem 72:248–254
Chen XW, Kuhn E, Jennings EW, Nelson R, Tao L, Zhang M (2016) DMR (deacetylation and mechanical refining) processing of corn stover achieves high monomeric sugar concentrations (230 g/L) during enzymatic hydrolysis and high ethanol concentrations (>10% v/v) during fermentation without hydrolysate purification or concentration. Energy Environ Sci 9:1237–1245
Chiaramonti D, Balan V, Kumar S (2013) Review of US and EU initiatives toward development, demonstration, and commercialization of lignocellulosic biofuels. Biofuels Bioprod Bioref 7:732–759
Gang L, Jian Z, Jie B (2016) Cost evaluation of cellulase enzyme for industrial-scale cellulosic ethanol production based on rigorous aspen plus modeling. Bioprocess Biosyst Eng 39:133–140
Ghose TK (1987) Measurement of cellulase activities. Pure Appl Chem 59:257–268
Gu HQ, Zhang J, Bao J (2015) High tolerance and physiological mechanism of zymomonas mobilis to phenolic inhibitors in ethanol fermentation of corncob residue. Biotechnol Bioeng 112:1770–1782

He Y, Zhang J, Bao J (2014) Dry dilute acid pretreatment by co-currently feeding of corn stover feedstock and dilute acid solution without impregnation. Bioresour Technol 158:360–364

He Y, Zhang J, Bao J (2016) Acceleration of biodetoxification on dilute acid pretreated lignocellulose feedstock by aeration and the consequent ethanol fermentation evaluation. Biotechnol Biofuels 9:19

Klein-Marcuschamer D, Oleskowicz-Popiel P, Simmons BA, Blanch HW (2012) The challenge of enzyme cost in the production of lignocellulosic biofuels. Biotechnol Bioeng 109:1083–1087

Lei C, Zhang J, Xiao L, Bao J (2014) An alternative feedstock of corn meal for industrial fuel ethanol production: delignified corncob residue. Bioresour Technol 167:555–559

Li H, Wu M, Xu L, Jin H, Guo T, Bao X (2015) Evaluation of industrial saccharomyces cerevisiae strains as the chassis cell for second-generation bioethanol production. Microb Biotechnol 82:66

Liu S, Wang Q (2014) Response surface optimization of enzymatic hydrolysis process of wet oxidation pretreated wood pulp waste. Cellul Chem Technol 50:803–809

Marcos M, González-Benito G, Coca M, Bolado S, Lucas S (2013) Optimization of the enzymatic hydrolysis conditions of steam-exploded wheat straw for maximum glucose and xylose recovery. J Chem Technol Biotechnol 88:237–246

Qureshi AS, Zhang J, Bao J (2015) High ethanol fermentation performance of the dry dilute acid pretreated corn stover by an evolutionarily adapted saccharomyces cerevisiae, strain. Bioresour Technol 189:399–404

Ramesh D (2013) World's first commercial scale cellulosic biofuels plant opens. Chem Week 175:39

Sluiter A, Hames B, Ruiz R Scarlata C, Sluiter J Templeton D (2008) Determination of sugars, byproducts, and degradation products in liquid fraction process samples. NREL/TP-510-42623. National Renewable Energy Laboratory, Golden CO

Sluiter A, Hames B, Ruiz R, Scarlata C, Sluiter J, Templeton D, Crocker D (2012) Determination of structural carbohydrates and lignin in biomass. NREL/TP-510-42618. National Renewable Energy Laboratory, Golden CO

Zhang J, Bao J (2012) A modified method for calculating practical ethanol yield at high lignocellulosic solids content and high ethanol titer. Bioresour Technol 116:74–79

Zhang Y, Schell D, Mcmillan J (2007) Methodological analysis for determination of enzymatic digestibility of cellulosic materials. Biotechnol Bioeng 96:188–194

Zhang J, Zhu Z, Wang X, Wang N, Wang W, Bao J (2010a) Biodetoxification of toxins generated from lignocellulose pretreatment using a newly isolated fungus Amorphotheca resinae ZN1 and the consequent ethanol fermentation. Biotechnol Biofuels 3:26

Zhang J, Chu D, Huang J, Yu Z, Dai G, Bao J (2010b) Simultaneous saccharification and ethanol fermentation at high corn stover solids loading in a helical stirring bioreactor. Biotechnol Bioeng 105:718–728

Zhang J, Wang X, Chu D, He Y, Bao J (2011) Dry pretreatment of lignocellulose with extremely low steam and water usage for bioethanol production. Bioresour Technol 102:4480–4488

Zhang J, Shao S, Bao J (2015) Long term storage of dilute acid pretreated corn stover feedstock and ethanol fermentability evaluation. Bioresour Technol 201:355–359

Zhang H, Han X, Wei C, Bao J (2016) Oxidative production of xylonic acid using xylose in distillation stillage of cellulosic ethanol fermentation broth by gluconobacter oxydans. Bioresour Technol 224:573–580

8

Food waste: a potential bioresource for extraction of nutraceuticals and bioactive compounds

Krishan Kumar[1*†] ⓘ, Ajar Nath Yadav[2†], Vinod Kumar[2†], Pritesh Vyas[2†] and Harcharan Singh Dhaliwal[2†]

Abstract

Food waste, a by-product of various industrial, agricultural, household and other food sector activities, is rising continuously due to increase in such activities. Various studies have indicated that different kind of food wastes obtained from fruits, vegetables, cereal and other food processing industries can be used as potential source of bioactive compounds and nutraceuticals which has significant application in treating various ailments. Different secondary metabolites, minerals and vitamins have been extracted from food waste, using various extraction approaches. In the next few years these approaches could provide an innovative approach to increase the production of specific compounds for use as nutraceuticals or as ingredients in the design of functional foods. In this review a comprehensive study of various techniques for extraction of bioactive components citing successful research work have been discussed. Further, their efficient utilization in development of nutraceutical products, health benefits, bioprocess development and value addition of food waste resources has also been discussed.

Keywords: Bioactive compounds, Food waste, Nutraceuticals, Extraction, Diseases

Background

Food waste is produced in all the phases of food life cycle, i.e. during agricultural production, industrial manufacturing, processing and distribution. Up to 42% of food waste is produced by household activities, 39% losses occurring in the food manufacturing industry and 14% in food service sector (ready to eat food, catering and restaurants), while 5% is lost during distribution. Food waste is expected to rise to about 126 Mt by 2020, if any prevention policy or activities are not undertaken (Mirabella et al. 2014). It can be achieved through the extraction of high-value components such as proteins, polysaccharides, fibres, flavour compounds, and phytochemicals, which can be re-used as nutraceuticals and functional ingredients (Baiano 2014).

Attempts have been made broadly for the past few decades to develop methods and find different ways to utilise fruit and vegetable wastes therapeutically. Generally, agro-industrial wastes have been used extensively as animal feeds or fertilisers. Recent reports shows development of high value products (such as cosmetics, foods and medicines) from agro-industrial by-products (Rudra et al. 2015).

Natural bioactive compounds are being searched for the treatment and prevention of human diseases. These compounds efficiently interact with proteins, DNA, and other biological molecules to produce desired results, which can then be used for designing natural therapeutic agents (Ajikumar et al. 2008). There is growing interest of consumers towards food bioactives that provide beneficial effects to humans in terms of health promotion and disease risk reduction. Detailed information about food bioactives is required in order to obtain appropriate functional food products (Kumar 2015).

Nutraceuticals are medicinal foods that play a role in enhancing health, maintaining well being, improving immunity and thereby preventing as well as treating

*Correspondence: krishankumar02007@gmail.com
†Krishan Kumar, Ajar Nath Yadav, Vinod Kumar, Pritesh Vyas and Harcharan Singh Dhaliwal contributed equally to this work
[1] Department of Food Technology, Akal College of Agriculture, Eternal University, Sirmour 173001, India
Full list of author information is available at the end of the article

specific diseases. Phytochemicals have specific role and can be used in different forms, e.g. as antioxidants and have a positive effect on human health. Recently, lot of attention has been given to phytochemicals that possess cancer preventive properties (Kumar and Kumar 2015). Nowadays, there is growing trend in the food industry toward the development and manufacture of functional and nutraceutical products. This new class of food products have got huge attention in food market due to the increased consumer interest for "healthy" food. Hence, pharmaceutical and food domains have common interest to obtain new natural bioactive components which can be used as drugs, functional food ingredients, or nutraceuticals (Joana Gil-Chávez et al. 2013). Bioactives from food waste can be extracted and utilized for development of nutraceuticals and functional foods. This review describes the utilization of different extraction techniques for extraction of bioactives and nutraceuticals from food waste and their uses in prevention of chronic and lifestyle diseases.

Food waste as a source of bioactive compounds

Bioactive compounds comprise an excellent pool of molecules for the production of nutraceuticals, functional foods, and food additives (Joana Gil-Chávez et al. 2013). Fruits and vegetables represent the simplest form of functional foods because they are rich in several bioactive components. Fruits containing polyphenols and carotenoids have been shown to have antioxidant activity and diminish the risk of developing certain types of cancer (Day et al. 2009). The vegetable waste includes trimmings, peelings, stems, seeds, shells, bran and residues remaining after extraction of juice, oil, starch and sugar. The animal-derived waste includes waste from dairy processing and seafood industry. The recovered biomolecules and by-products can be used to produce functional foods in food processing or in medicinal and pharmaceutical preparations (Baiano 2014). Bioactive phytochemicals like sterols, tocopherols, carotenes, terpenes and polyphenols extracted from tomato by-products contain significant amounts of antioxidant activities. Therefore, these value adding components isolated from such waste can be used as natural antioxidants for the formulation of functional foods or can serve as additives in food products to extend their shelf-life (Kalogeropoulos et al. 2012).

The bioactive compounds present in mango peel are phenolic compounds, carotenoids, vitamin C and dietary fibre. It has been well recognized that these compounds contribute to lower the risk of cancer, cataracts, Alzheimer's disease and Parkinson's disease (Ayala-Zavala et al. 2010). Wastes from wine making industry include biodegradable solids namely stems, skins, and seeds. Bioactive compounds from winery by-products have been shown

to improve health promoting activities both in vitro and in vivo. These compounds act as effective agents for prevention of degenerative processes through their incorporation into functional foods, nutraceuticals, and cosmetics (Teixeira et al. 2014). These are commonly utilized for the production of pharmaceuticals and as food additives to increase the functionality of foods (Ayala-Zavala et al. 2010). Citrus is the most abundant fruit crop in the world. Its one-third of the crop is processed. Oranges, lemons, grapefruits and mandarins represent approximately 98% of the entire industrialized crop. Citrus fruits are processed, not only to obtain juice, but also, in the canning industry to produce jam and segments of mandarin.

Lemes et al. (2016) reported bioactive peptides as the new generation of biologically active regulators that can prevent oxidation and microbial degradation in foods, and might be helpful in treatment of various diseases. These can be extracted from residual waste and incorporated into value added products. Their encapsulated form may be utilized in a controlled manner for efficient use in human body. Development of suitable techniques for large-scale recovery and purification of peptides will increase their applications in pharmaceutical and food industries.

Pujol et al. (2013) investigated the chemical composition of exhausted coffee waste generated in a soluble coffee industry and found that total polyphenols and tannins represent <6 and <4% of the exhausted coffee wastes, respectively. Zuorro and Lavecchia (2012) extracted total phenolic content of 17.75 mg gallic acid equivalent GAE (gallic acid equivalents)/g from spent coffee grounds (SCG) collected from coffee bars and 21.56 mg GAE/g from coffee capsules unloaded from an automatic espresso machine. Mussatto et al. (2011) optimised the extraction of antioxidant phenolic compounds from SCG and found that extraction using 60% methanol in a solvent/solid ratio of 40 ml/g SCG, for 90 min, was the most appropriate condition to produce an extract with 16 mg GAE/g SCG of phenolic compounds having high antioxidant activity, i.e. Ferric reducing antioxidant power (FRAP) of 0.10 mM Fe(II)/g). Rebecca et al. (2014) extracted the amount of caffeine from used tea leaves of black, white, green and red tea using dichloromethane as solvent and found that caffeine content was maximum (60 mg/100 g) in green tea and minimum in red tea (3 mg/100 g). Some of the bioactive components found in different food waste residues are summarized in Table 1.

Extraction technologies for bioactive compounds from food waste

Bioactive components present in agro-industrial waste can be recovered using various techniques. Availability of these techniques provides an opportunity for optimal use

Table 1 Bioactive components in different industrial food waste residues

S. no.	Source	Residue	Bioactive components	References
Fruits				
1.	Apple	Peel and pomace	Epicatechin, catechins, anthocyanins, quercitin glycosides, chlorogenic acid, hydroxycinnamates, phloretin glycosides, procyanidins	Wolfe and Liu (2003), Foo and Lu (1999), Lu and Foo (1997)
2.	Avocado	Peel and seeds	Epicatechin, catechin, gallic acid, chlorogenic acid, cyanidin 3-glucoside, homogentisic acid	Deng et al. (2012)
3.	Banana	Peel	Gallocatechin, anthocyanins, delphindin, cyaniding, catecholamine	Someya et al. (2002), Kanazawa and Sakakibara (2000), González-Montelongo et al. (2010)
4.	Citrus fruits	Peel	Hesperidin, naringin, eriocitrin, narirutin	Coll et al. (1998)
5.	Grapes	Seed and skin	Coumaric acid, caffeic acid, ferulic acid, chlorogenic acid, cinnamic acid, neochlorogenic acid, p-hydroxybenzoic acid, protocatechuic acid, vanillic acid, gallic acid, proanthocyanidins, quercetin 3-o-gluuronide, quercetin, resvaratrol	Shrikhande (2000), Negro et al. (2003), Maier et al. (2009)
6.	Guava	Skin and seeds	Catechin, cyanidin 3-glucoside, galangin, gallic acid, homogentisic acid, kaempferol	Deng et al. (2012)
7.	Litchi	Pericarp, seeds	Cyanidin-3-glucoside, cyanidin-3-rutonoside, malvidin-3-glucoside, gallic acid, epicatechin-3-gallate	Lee and Wicker (1991), Duan et al. (2007)
8.	Mango	Kernel	Gallic acid, ellagic acid, gallates, gallotannins, condensed tannins	Arogba (2000), Puravankara et al. (2000)
9.	Palm	By-products of palm oil milling	Tocopherols, tocotrienols, sterols, and squalene, phenolic antioxidants	Tan et al. (2007), Choo et al. (1996)
10.	Pomegranate	Peel and pericarp	Gallic acid, cyanidin-3,5-diglucoside, cyanidin-3-diglucoside, delphinidin-3,5-diglucoside	Noda et al. (2002), Gil et al. (2000)
Vegetables				
11.	Carrot	Peel	Phenols, beta-carotene	Chantaro et al. (2008)
12.	Cucumber	Peel	Chlorophyll, pheophytin, phellandrene, caryophyllene	Zeyada et al. (2008)
13.	Potato	Peel	Gallicacid, caffeic acid vanillic acid	Zeyada et al. (2008)
14.	Tomato	Skin and pomace	Carotenoids	Strati and Oreopoulou (2011)
Cereal crops				
15.	Barley	Bran	β-Glucan	Sainvitu et al. (2012)
16.	Rice	bran	γ-Oryzanol, bran oil	Perretti et al. (2003), Oliveira et al. (2012)
17.	Wheat	Bran and germs	Phenolic acids, antioxidants	Wang et al. (2008)

of any of these for recovery of specific compounds. Based on literature survey, the extraction techniques for bioactive compounds are mainly based on solvent extraction (SE), supercritical fluid extraction (SFE), subcritical water extraction (SCW), use of enzymes, ultrasounds and microwaves. In the following sections, these techniques have been discussed independently in reference to recent studies.

Solvent extraction technique

In this extraction approach, the suitably sized raw material is exposed to different organic solvents, which takes up soluble components of interest and also other flavouring and colouring agents such as anthocyanins which are anti-cancerous and anti-inflammatory (Vyas et al. 2009, 2014) (Fig. 1). Samples are usually centrifuged and filtered to remove solid residue, and the extract could be used as additive, food supplement or for the preparation of functional foods (Zulkifli et al. 2012). Solvent Extraction is beneficial compared to other methods due to low processing cost and ease of operation. However, this method uses toxic solvents, requires an evaporation/concentration step for recovery, and usually calls for large amounts of solvent and extended time to be carried out. Moreover, the possibility of thermal degradation of natural bioactive components cannot be ignored due to the high temperatures of the solvents during the long times of extraction. Solvent extraction has been improved by other methods such as Soxhlet's, ultrasound, or microwave extraction and SFE in order to obtain better yields (Szentmihályi et al. 2002).

Baysal et al. (2000) utilized ethanol for extraction of Lycopene and β-carotene from tomato pomace containing dried and crushed skins (rich in lycopene and carotenes) and seeds of the fruit along with supercritical CO_2 for resulting in recoveries of up to 50%. Gan and Latiff (2011) studied the extraction of polyphenolic compounds from *Parkia speciosa* pod powders using 50% acetone solution. They concluded that that 50% acetone yielded the highest content of polyphenols compared to methanol, ethanol, ethyl-acetate and hexane. Safdar et al. (2016) extracted and quantified polyphenols from kinnow (*Citrus reticulate* L.) peel. Maximum polyphenols were extracted with 80% methanol (32.48 mg GAE/g extract) using ultrasound assisted extraction, whereas, minimum phenolics (8.64 mg GAE/g extract) were obtained with 80% ethyl acetate through the maceration technique.

Bandar et al. (2013) found that out of the various organic solvents used in their study, ethanol was the most efficient one, producing the highest extraction yield and hexane gave the lowest yield in extracting bioactive compounds by these methods. Further they found that there was an increase in the yield of extracted compounds with increasing extraction time.

Supercritical fluid extraction

Supercritical fluid extraction is an environment friendly technology and is commonly used for extraction of bioactive compounds from natural sources such as plants, food by-products, algae and microalgae. Supercritical carbon dioxide (SC-CO_2) is an attractive alternative to organic solvents as it is non explosive, non-toxic and inexpensive. It possesses the ability to solubilise lipophilic substances, and can be removed easily from the final products (Wang and Weller 2006).

During the process of extraction, raw material is placed in an extraction container equipped with temperature and pressure controllers to maintain the required conditions. Following this, the extraction container is pressurized with the fluid by a pump. Once the fluid and dissolved compounds are transported to separators, the products are collected through a tap located in the lower part of the separators. Finally, the fluid is regenerated and cycled or released to the environment. Selection of supercritical fluids is very important for proper functioning of this process and a wide range of compounds can be used as solvents in this technique (Sihvonen et al. 1999).

Giannuzzo et al. (2003) found that SC-CO_2 modified with ethanol gave higher extraction yields of naringin (flavonoid) from citrus waste than pure SC-CO_2 at 9.5 MPa and 58.6 °C. Ashraf-Khorassani and Taylor (2004) extracted polyphenols and procyanidins from grape seeds using SFE, where, methanol was used as modifier and methanol modified CO_2 (40%) released more than 79% of catechin and epicatechin from grape

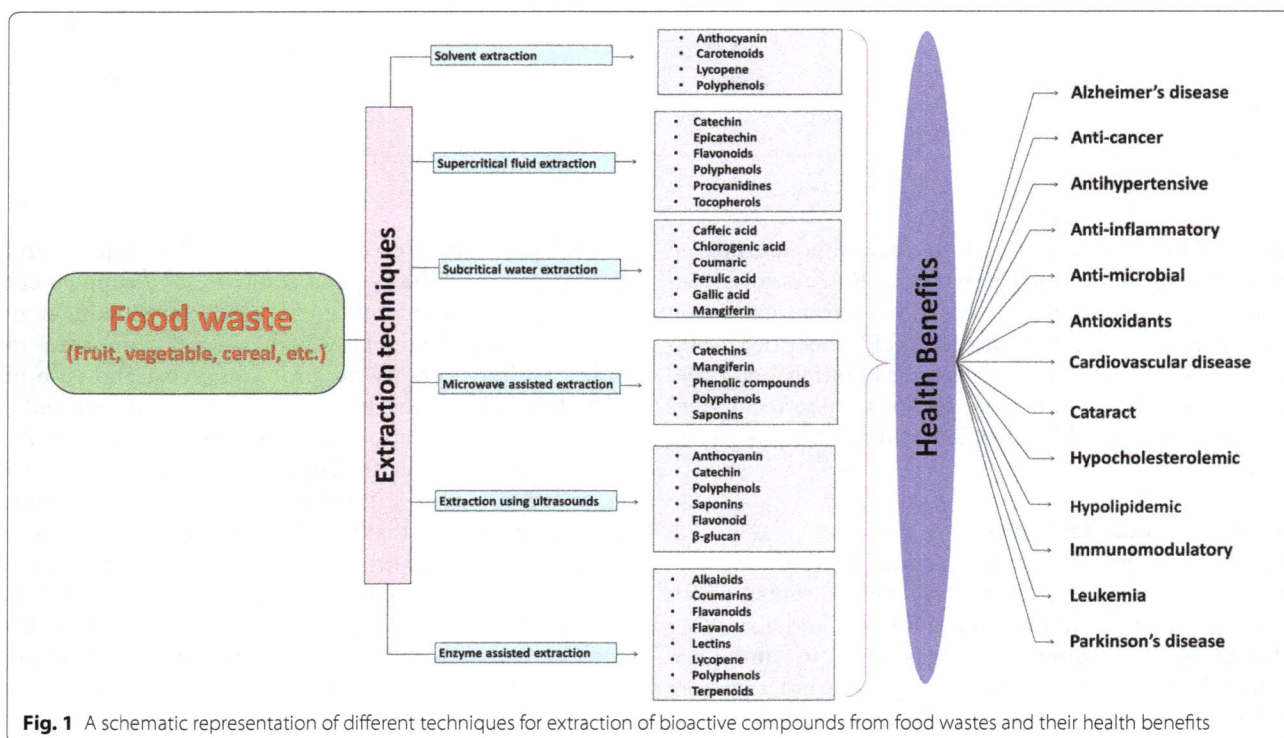

Fig. 1 A schematic representation of different techniques for extraction of bioactive compounds from food wastes and their health benefits

seed. Liza et al. (2010) studied the feasibility of the SFE method to extract lipophilic compounds such as tocopherols, phytosterols, policosanols and free fatty acids from sorghum and the preventive role of these compounds in many diseases (skin, cardiovascular, coronary heart diseases, and cancer).

Farías-Campomanes et al. (2015) extracted polyphenols (gallic, protocatechuic, vanillic, syringic, ferulic derivatives and p-coumaric derivatives) and flavonoids (quercetin and its derivatives) from Lees generated from pisco-making process with SFE at 20 MPa and 313 K. Jung et al. (2012) carried out extraction of oil from wheat bran which is a rich source of antioxidants using SC-CO$_2$ and Soxhlet extraction. Pressure and temperature ranged from 10 to 30 MPa and 313.15–333.15 K, respectively during SC-CO$_2$ extraction. It was observed that oil obtained by SC-CO$_2$ extraction had higher resistance against oxidation and higher radical scavenging activity compared to hexane extracted oil. Wenzel et al. (2016) extracted phenolic compounds from black walnut (*Juglans nigra*) husks using SC-CO$_2$ with an ethanol modifier. The optimal extraction conditions were 68 °C and 20% ethanol in SC-CO$_2$.

Ahmadian-Kouchaksaraie and Niazmand (2017) used the SC-CO$_2$ for the extraction of antioxiant compounds from *Crocus sativus* petals at 62 °C for 47 min and 164 bar pressure. Extraction using these optimized conditions resulted in recovery of 1423 mg/100 g total phenolics, 180 mg/100 g total flavonoid and 103.4 mg/100 g total anthocyanin content. Wang and Weller (2006) described supercritical method as a significant substitute to conventional extraction methods using organic solvents for extracting biologically active compounds.

Some of the conditions used for the extraction, recovery and characterization of bioactive compounds from food and plants using SC-CO$_2$ are summarized in Table 2.

Subcritical water extraction

Subcritical water extraction is a growing alternative technology for extraction of phenolic compounds from different foods. Subcritical water refers to water at temperature between 100 and 374 °C and a pressure which is high enough to maintain the liquid state (below the critical pressure of 22 MPa). Main advantages of SCW over conventional extraction techniques are shorter extraction time, lower solvent cost, higher quality of the extraction and environment-friendly (Herrero et al. 2006). SCW is the most promising engineering approach that offers an environmentally friendly technique for extracting various compounds from plants and algae (Zakaria and Kamal 2016).

Tunchaiyaphum et al. (2013) extracted phenolic compounds from mango peels using SCW. The amount of phenolic compounds from mango peels using SCW extraction was higher than that using Soxhlet extraction technique. Therefore, SCW extraction is an alternative green technology for phenolic compounds extraction from agricultural wastes, which substitute conventional method using organic solvents.

Rangsriwong et al. (2009) studied the use of SCW for extraction of polyphenolic compounds from *Terminalia chebula Retz.* fruits and it was found that the amounts of extracted gallic acid and ellagic acid increased with an increasing in subcritical water temperature up to 180 °C, while the highest amount of corilagin was recovered at 120 °C. Kim et al. (2010) extracted mangiferin, a pharmacological active component from *Mahkota Dewa* using subcritical water extractiont at temperatures range of 323–423 K and pressures 0.7–4.0 MPa with extraction times ranging from 1 to 7 h.

Singh and Saldaña (2011) extracted eight phenolic compounds (gallic acid, chlorogenic acid, caffeic acid, protocatechuic acid, syringic acid, p-hydroxyl benzoic acid, ferulic acid, and coumaric acid) from potato peel using subcritical water. Phenolic compounds were recovered highest (81.83 mg/100 g fresh wt.) at 180 °C and extraction time of 30 min. Chlorogenic acid (14.59 mg/100 g) and gallic acid (29.56 mg/100 g) were the main phenolic compounds obtained from potato peel at 180 °C. It was concluded that subcritical water at 160–180 °C, 6 MPa and 60 min might be a good substitute to organic solvents (such as methanol and ethanol) to obtain phenolic compounds from potato peel.

Ahmadian-Kouchaksaraie et al. (2016) investigated subcritical water extraction as a green technology for the extraction of phenolic compounds from *Crocus sativus* petals. The optimum conditions of extraction were of 36 ml/g (water to solid ratio) 159 °C temperature and an extraction time of 54 min. Subcritical water extraction using these optimized conditions leads to extraction of 1616 mg/100 g total phenolics, 239 mg/100 g total flavonol content and 86.05% 2,2-diphenyl-1-picrylhydrazyl (DPPH).

Ko et al. (2016) enhanced production of individual phenolic compounds by subcritical water hydrolysis in pumpkin leaves by varying temperatures from 100 to 220 °C at 20 min and also by varying reaction times from 10 to 50 min at 160 °C. Caffeic acid, p-coumaric acid, ferulic acid, and gentisic acid were the major phenolic compounds in the hydrolysate of pumpkin leaves. Mayanga-Torres et al. (2017) utilized two abundant coffee waste residues (powder and defatted cake) for extraction of total phenolic compounds using subcritical water under semi-continuous flow conditions. The highest total phenolic compounds (26.64 mg GAE/g coffee powder) was recovered at 200 °C and 22.5 MPa.

Table 2 Extraction, recovery and characterization of bioactive compounds using supercritical fluid extraction

S. no.	Sources	Temperature (°C)	Pressure (Bar)	Co-solvent	Bioactive compounds	References
Fruits						
1.	Blueberry residue	40	150–300		Anthocyanins	Paes et al. (2013)
2.	Apricot pomace	39.85–59.85	304–507	Dimethoxy propane	Carotenoids	Sanal et al. (2004)
3.	Red grape residue	45	100–250	Methanol	Pro-anthocyanidins	Louli et al. (2004)
4.	Citrus peel	58.6	95	Ethanol	Naraingin	Giannuzzo et al. (2003)
5.	Grape by products	35	400	Ethanol	Resveratrol (19.2 mg/100 g)	Casas et al. (2010)
6.	Banana peel	40–50	100–300		Essential oils	Comim et al. (2010)
7.	Grape peel	37–46	137–167	Ethanol	Phenolic, anti-oxidants, anthocyanins	Ghafoor et al. (2010)
8.	Orange peel	19.85–49.85	80–280		Limonene and linalool	Mira et al. (1999)
9.	Guava seeds	40–60	100–300	Ethyl acetate,	Phenolic compounds	Castro-Vargas et al. (2010)
10.	Apricot by products	59	310	Ethanol	β-Carotene	Sanal et al. (2005)
11.	Pistachio hull	45	355	Methanol	Polyphenols (7810 mg GAE/100 g)	Goli et al. (2005)
Vegetable						
12.	Tomato waste	40–80	200–300		Trans-lycopene	Nobre et al. (2009)
13.	Tomato skin	75	350	Ethanol	Carotenoids	Shi et al. (2009)
14.	Sweet potato waste	40–80	350		Beta-carotene and alpha tocopherol	Okuno et al. (2002)
15.	Carrot press cake	55	345	Ethanol	β-Carotene	Vega et al. (1996)
Others						
16.	Green tea leaves	60	310	Ethanol	Catechins	Chang et al. (2000)
17.	Tea seed cake	80	200	Ethanol	Kaempferol glycosides (11.4 mg/g)	Li et al. (2010)
18.	Spearmint leaves	40–60	100–300	Ethanol	Flavonoids	Bimakr et al. (2009)

There are number of advantages of SCW extraction over traditional extraction techniques used, such as higher quality of the extracts, lower extraction times, lower costs of the extracting agent, and an environment friendly technique (Joana Gil-Chávez et al. 2013).

Enzyme assisted extraction

There is wide use of enzymes for extraction of bioactive components from food wastes. The main sources for extraction of antioxidants are plant tissues. Plant cell walls contain polysaccharides such as cellulose, hemicellulose, and pectins which act as barriers to the release of intracellular substances. Enzymes such as cellulase, β-glucosidase, xylanase, β-gluconase, and pectinase help to degrade cell wall structure and depolymerize plant cell wall polysaccharides, facilitating the release of linked compounds (Moore et al. 2006; Singh et al. 2016). Because of using water as a solvent instead of organic chemicals, the enzyme assisted extraction is recognized as more eco-friendly technology for extraction of bioactive compounds and oil (Puri et al. 2012).

Zuorro et al. (2011) studied the enzyme-assisted extraction of lycopene from the peel fraction of tomato processing waste and found that the recovery of lycopene could be greatly improved by the use of mixed enzyme preparations with cellulolytic and pectinolytic activities, and the comparatively low cost of commercial food-grade enzyme preparations, having possible implementation on industrial scale. Puri et al. (2012) studied the enzyme-assisted extraction of bioactive compounds stevioside from *Stevia rebaudiana* from plant sources particularly for food and nutraceutical purposes. Reshmitha et al. (2017) prepared lycopene rich extracts by enzyme assisted extraction of tomato peel using cellulase (20 units/g) and pectinase (30 units/g) at 50 °C for 60 min.

These studies suggested that the release of bioactive compounds from plant cells by cell disruption and extraction can be optimized using enzyme preparations either alone or in mixtures. Enzyme-assisted extraction is a promising alternative to conventional solvent-based extraction methods. It is based on the ability of enzymes to catalyze reactions, under mild processing conditions, in aqueous solutions (Gardossi et al. 2010).

Extraction using ultrasounds

Ultrasound-assisted extraction is considered as a simpler and more effective technique compared to traditional extraction methods for the extraction of bioactive compounds from natural products. Ultrasound induces a greater diffusion of solvent into cellular materials, thus improving mass transfer and also disrupts cell walls, thus facilitating the release of bioactive components. Extraction yield is greatly influenced by ultrasound frequency, depending on the nature of the plant material to be extracted (Wang et al. 2008).

Wang et al. (2011) used ultrasound-assisted extraction to extract three dibenzylbutyrolactone lignans (including tracheloside, hemislienoside, and arctiin) from *Hemistepta lyrata*. High-performance liquid chromatography was used for simultaneous determination of the target compounds in the corresponding extracts.

Rostagno et al. (2003) studied extraction efficiency of four isoflavone derivatives, i.e. glycitin, daidzin, genistin, and malonyl genistin from soybean with mix-stirring method using different extraction times and solvents. Use of ultrasound was found to improve the extraction yield depending on solvent use. Ghafoor et al. (2011) extracted anthocyanins and phenolic compounds from grape peel using ultrasound-assisted extraction technique. Bimakr et al. (2013) also applied the ultrasound-assisted extraction technique for the extraction of bioactive valuable compounds from winter melon (*Benincasa hispida*) seeds.

Piñeiro et al. (2016) optimised and validated ultrasound-assisted extraction for rapid extraction of stilbenes from grape canes. By this method, stilbenes in grape canes was extracted 10 min only using extraction temperature of 75 °C and ethanol (60%) as the extraction solvent. It was concluded that grape cane by-products were potential sources of bioactive compounds of interest for pharmaceutical and food industries.

Aguiló-Aguayo et al. (2017) studied the effect of ultrasound technology in extract of water soluble polysaccharides from dried and milled by-products generated from *Agaricus bisporus*. β-Glucan were obtained in amounts of 1.01 and 0.98 g/100 g dry mass in particle sizes of 355–250 and 150–125 μm, respectively, from the mushroom by-products. The highest extraction yield of 4.7% was achieved with an extraction time of 15 min, amplitude of 100 μm with 1 h of precipitation in 80% ethanol.

Microwave assisted extraction

Microwave-assisted extraction (MAE) is a new extraction technique that combines microwave and traditional solvent extraction. It is an advantageous technique due to shorter extraction time, higher extraction rate, less requirement of solvent and lower cost over traditional method of extraction of compounds (Delazar et al. 2012). The main advantage of MAE over ultrasonic assisted extraction and Soxhlet extraction is that, it can

be used to extract plant metabolites at a shorter time interval (Afoakwah et al. 2012). Padmapriya et al. (2012) extracted mangiferin present in *Curcuma amada* with the help of MAE using ethanol as a solvent. The mangiferin content was reportedly increased until 500 W, but decreased as the microwave power was increased further. An optimal mangiferin yield of 41 μg/ml was obtained from an extraction time of 15.32 s for a microwave power of 500 W. Kerem et al. (2005) extracted saponins from chickpea (*Cicer arietinum*) using MAE and found this method superior over Soxhlet extraction with regard to amounts of solvents required, time and energy expended. The pure chickpea saponin exhibited significant inhibitory activity against *Penicillium digitatum* and additional filamentous fungi (Fig. 1).

Kulkarni and Rathod (2016) extracted mangiferin from the *Mangifera indica* leaves by microwave assisted extraction conditions using water as a solvent. The maximum extraction yield of 55 mg/g was obtained at extraction time 5 min, solid to solvent ratio 1:20 and microwave power of 272 W. In comparison to the sequential batch extraction and Soxhlet extraction, MAE increased the yield of extraction in a short span of time and also reduced the solvent requirement as compared to the conventional methods.

Smiderle et al. (2017) studied MAE and pressurized liquid extraction (PLE) as advanced techniques to obtain polysaccharides (particularly biologically active β-glucans) from *Pleurotus ostreatus* and *Ganoderma lucidum* fruiting bodies and detected β- and α-glucans and heteropolysaccharides in all extracts. In an interesting study, Filip et al. (2017) optimized MAE by response surface methodology in order to enhance the extraction of polyphenols from basil (*Ocimum basilicum* L). Optimal conditions for extraction were 50% ethanol, microwave power of 442 W, and an extraction time of 15 min. Under these conditions, obtained basil liquid extract contained 4.299 g GAE/100 g of total polyphenols and 0.849 g catechin equivalents/100 g DW of total flavonoids.

Therefore, it can be concluded that microwave assisted method has many advantages compared with other methods due to its higher extraction efficiency, reduced extraction time, less labor and high extraction selectivity which makes it a favourable method in extraction of bioactive compounds (Bandar et al. 2013).

Comparative evaluation of different extraction technologies for recovery of bioactive compounds

Zhang et al. (2005), compared a number of extraction methods for the recovery of alkaloids from fruit of *Macleaya cordata* (Willd) R. Br. The techniques used include maceration, ultrasound-assisted extraction (UAE), MAE and percolation. The method of MAE was found to be the most effective method capable of yielding 17.10 ± 0.4 mg/g sanguinarine and 7.04 ± 0.14 mg/g chelerythrine with 5 min of extraction time. They further concluded that the alkaloid content of the fruit shell was much greater than that of seed.

Corrales et al. (2008) extracted bioactive substances from grape by-products such as anthocyanins which can be used as natural antioxidants or colouring agents. They studied the effect of heat treatment at 70 °C combined with the effect of different novel technologies such as high hydrostatic pressure (600 MPa) (HHP), ultrasonics (35 kHz) and pulsed electric fields (3 kV cm^{-1}) (PEF). After 1 h extraction, the total phenolic content of samples subjected to novel technologies was 50% higher than in the control samples. They further concluded that the application of novel technologies increased the antioxidant activity of the extracts with PEF fourfold, with HHP three-fold and with ultrasonics two-fold higher than the control extraction carried out in a water bath incubated at a temperature of 70 °C for 1 h.

Plaza et al. (2011) studied the effect of different extraction technologies on extraction of bioactive compounds of orange juice. Juice was treated by high pressure (HP) (400 MPa/40 °C/1 min), pulsed electric fields (PEF) (35 kV cm^1/750 ms) and low pasteurization (LPT) (70 °C/30 s). They extracted various bioactive compounds such as lutein, zeaxantin, α and β-cryptoxanthin, α and β-carotene, naringenin and hesperetin from grape juice. It was concluded that HPT was the most effective treatment for extraction of bioactive components from orange juice with highest recovery of bioactive components from orange juice followed by PEF and LPT.

Drosou et al. (2015) compared the extraction yield of air dried Agiorgitico red grape pomace by-products by three different extraction methods using water, water: ethanol (1:1) and ethanol as solvents. The methods included the conventional Soxhlet extraction, MAE and ultrasound assisted extraction (UAE). They concluded that UAE water: ethanol extracts were found to be rich in phenolic compounds (up to 438,984 ppm GAE in dry extract).

Jayathunge et al. (2017), investigated the influence of moderate intensity pulsed electric field pre-processing on increasing the lycopene bioaccessibility of tomato fruit, and the combined effect of blanching, ultrasonic and high intensity pulsed electric field processing on further enhancement of the lycopene bioaccessibility after juicing. They concluded that only the treatment of blanching followed by high intensity pulsed electric field showed a significant release of *trans*-(4.01 ± 0.48) and *cis*-(5.04 ± 0.26 lg/g) Lycopene. They further concluded that processing of pre-blanched juice using high intensity

pulsed electric field, derived from pre-processed tomato was the most excellent approach to achieve the highest nutritive value.

Kehili et al. (2017) extracted lycopene and carotene as oleoresin from a Tunisian industrial tomato peels by-product using supercritical CO_2 and solvent extraction using hexane, ethyl acetate and ethanol. Supercritical CO_2 extraction resulted in a lycopene extraction of 728.98 mg/kg of dry tomato peels under processing conditions of 400 bar, 80 °C and 4 g CO_2/min for 105 min. Solvent extraction of lycopene using overnight maceration with hexane, ethyl acetate and ethanol yielded 608.94 ± 10.05 mg/kg, 320.35 and 284.53 mg/kg of dry tomato peels, respectively. They further concluded that SC-CO_2 extraction method resulted in a higher lycopene production as compared to solvent extraction under the above mentioned processing conditions.

Espinosa-Pardo et al. (2017) extracted total phenolic contents (TPC) from the pomace generated in the industrial processing of orange (*Citrus sinensis*) juice in Brazil by SFE method. Process was carried out at pressures of 15, 25 and 35 MPa and temperatures of 40, 50 and 60°C, using pure ethanol and ethanol: water (9:1 v/v) as co-solvents. They observed that high pressures improved the recovery of TPC (18–21.8 mg GAE/g dry extract) from pomace. They further observed that the use of ethanol 90% as co-solvent enhanced the extraction of antioxidant compounds. Finally, it was concluded that biotransformation process improved the TPC and provided extracts with higher antioxidant activities.

Some of the methods, optimum conditions and yield of some bioactive compounds from different food wastes are summarized in Table 3.

Use of bioactive compounds as nutraceuticals and functional foods for human health

Being health related compounds, bioactive compounds are known to lower the risk of developing various diseases like cancer, alzheimer, cataracts and parkinson, among others. These beneficial effects have been attributed mainly to their antioxidant and radical scavenging activities which can delay or inhibit the oxidation of DNA, proteins and lipids. Indeed, these compounds have also shown antimicrobial effects, playing an important role in fruits' protection against pathogenic agents, penetrating the cell membrane of microorganisms, causing lysis (Ayala-Zavala and González-Aguilar 2011).

An imbalance between the production of reactive oxygen species (ROS), and their eradication by defensive mechanisms in our body creates oxidative stress. Antioxidant systems of our body detoxify the reactive intermediates and results in reduction of oxidative stress

(Al-Dalaen and Al-Qtaitat 2014). ROS can be divided into free radicals and non-radicals. Molecules containing one or more unpaired electrons are called free radicals whereas non-radical forms are created when two free radicals share their unpaired electrons. The three major ROS of physiological importance are superoxide anion (O_2^-.), hydroxyl radical (.OH), and hydrogen peroxide (H_2O_2) (Birben et al. 2012). There should be interaction between free radicals, antioxidants and co-factors for maintaining health and prevention from aging and age-related diseases. Oxidative stress caused by free radicals is balanced by the endogenous antioxidant systems of our body which get strengthened by the intake of exogenous antioxidants with an input from co-factors. Production of free radicals in excess of the defensive effects of antioxidants and some co-factors causes oxidative damage which gets accumulated during life cycle resulting in aging, and chronic diseases such as cancer, cardiovascular diseases, neurodegenerative disorders, and other life style diseases (Rahman 2007).

Free radicals generated in the body during normal metabolic functions affect the vital cellular structures and functions resulting in various degenerative diseases. These free radicals are deactivated by antioxidant enzymes that catalyze oxidation/reduction reactions and serve as redox biomarkers in various human diseases along with controlling the redox state of functional proteins. Redox regulators with antioxidant properties related to active intermediates, cell organelles, and the neighbouring environments are involved in diseases related to redox imbalance including neurodegenerative diseases, aging cancer, ischemia/reperfusion injury and other lifestyle diseases (Yang and Lee 2015).

Nutraceuticals are usually consumed in pharmaceutical preparations such as pills, capsules, tablets, powder, and vials (Espín et al. 2007). Núñez Selles et al. (2016) repor-tred that Mangiferin (1,3,6,7-tetrahydroxyxanthone-C2-β-D-glucoside), a natural bioactive xanthonoid found in many plant species such as mango tree (*Mangifera indica* L) has attracted the attention of research groups around the World for cancer treatment. Single administration of mangiferin or in combination with known anticancer chemicals has shown the potential benefits of this molecule in brain, lung, cervix, breast and prostate cancers, and leukemia besides its antioxidant and anti-inflammatory properties.

Meat industry by-products such as brains, nervous systems and spinal cords are a source of cholesterol, which after extraction are used for the synthesis of vitamin D3 (Ejike and Emmanuel 2009). Chávez-Santoscoy et al. (2016) studied the the health promoting benefits of flavonoids and saponins from black bean seed coats. The effect of adding flavonoids and saponins from black

bean seed coat to whole wheat bread formulation was resulted in retention of more than 90% of added flavonoids and saponins, and 80% of anthocyanins in bread after baking. Use of such breads rich in these health promoting compounds might have significant health consequences.

In the production of rolled oats, phenolic compounds derived from natural sources such as benzoin, catechin, chlorogenic acid, and ferulic acid, mixed with the other ingredients prior to extrusion might obtain products more resistant to oxidation (retardation of hexanal formation). Although processing resulted in a 24–26% reduction of the amount of the phenolic compounds added (Viscidi et al. 2004). Lozano-Sánchez et al. (2017) reported that olive by-product, so called "pâté," generated during a modern two-phase centrifugal processing technique can be used as a natural source of bioactive compounds. It was characterized by the presence of hydroxytyrosol, β-hydroxyverbascoside, oleoside derivative, luteolin etc., as potential ingredients for nutraceuticals preparations or feed industry.

Conclusion and future prospects

As an indication, various reports of diverse array of bioactive compounds from specific food residues and availability of highly sensitive measurement tools provide a great opportunity to quantify metabolites in different range of food waste materials. Based on higher quantity of specific bioactive components, a food waste by-product could be utilized for its extraction using any of approaches discussed above. Utility of extraction methods is evident based on various reports and supercritical fluid extraction technology was proved to be very useful. A suitable extraction method could be adopted based on outcome of optimization process. Development of a bioprocess with better efficiency of bioactive component recovery will not only add value to the food waste but also be useful in reducing cost of formulated products and decreasing the use of synthetic chemicals in such formulations. With increasing setup of food processing industries and post harvest losses of fruits and vegetables, the increasing amount of food and agriculture waste is available and its utilization as a source of bioactive compounds will increase the financial status of farmers and decrease the burden of waste management. Improvement in extraction technology with lesser or no use of solvent will be of great significance towards a sustainable bioprocess.

Moreover, in India, the discarded portion of industrial waste is very high and it creates a serious waste disposal problem. Organic wastes generated from industries are hazardous to the environment and can be used as a potential bioresource for extraction of bioactive

Table 3 Comparative evaluation of different extraction techniques for extraction of bioactive compounds

S. no.	Bioactive component	Sources	Method	Extraction solvent used	Optimum conditions	Yield	References
1.	Alkaloid (sanguinarine)	*Macleaya cordata*	Maceration	Hydrochloric acid	100 °C/30 min	16.87[a]	Zhang et al. (2005)
			MAE	Hydrochloric acid	280 W/5 min	17.10	
			UAE	Hydrochloric acid	250 W/30 min	10.74	
			Percolation	Hydrochloric acid and sodium hydroxide	–	6.14	
2.	Anthocyanin	Grape	WE	Water	70 °C	7.93[b]	Corrales et al. (2008)
			Ultrasonics	Water and ethanol	600 MPa	7.76	
			HHP	Water and glycol	35 kHz	11.21	
			PEF	Water and ethanol	3 kV cm^{-1}	14.05	
3.	Hesperetin	Orange	LPT	–	70 °C/30 s	11.56[f]	Plaza et al. (2011)
			HPT	–	400 MPa/40 °C/1 min	13.34	
			PEF	–	35 kV cm^1/750 ms	11.09	
4.	Lutein	Orange	LPT	–	70 °C/30 s	226.42[e]	Plaza et al. (2011)
			HPT	–	400 MPa/40 °C/1 min	361.17	
			PEF	–	35 kV cm^1/750 ms	260.86	
5.	Lycopene	Tomato waste	SFE	Liquid CO_2	400 bar/80 °C/4 g CO_2/min/105 min	728.98[c]	Kehili et al. (2017)
			SE	Hexane	–	608.94	
				Ethyl acetate	–	320.35	
				Ethanol	–	284.53	
6.	Naringenin	Orange	LPT	–	70 °C/30 s	3.87[e]	Plaza et al. (2011)
			HPT	–	400 MPa/40 °C/1 min	4.43	
			PEF	–	35 kV cm^1/750 ms	3.42	
7.	Total phenolic	Red grape pomace	SE	Water	Refluxing for 2–3 h	96,386[d]	Drosou et al. (2015)
				Ethanol	Refluxing for 5–6 h	102,995	
			UAE	Water	25 kHz/300 W/20 °C/60 min	50,959	
				Water and ethanol	25 kHz/300 W/20 °C/60 min	438,984	
			MAE	Water	50 °C/200 W/60 min	52,645	
				Water and ethanol	50 °C/200 W/60 min	200,025	
8.	Total phenolic content	Orange pomace (dry)	SFE	Pure ethanol	25 MPa and 60 °C	21.2[g]	Espinosa-Pardo et al. (2017)
				Ethanol:water-9:1	25 MPa and 60 °C	20.7	
		Orange pomace (fermented)	SFE	Pure ethanol	25 MPa and 60 °C	19.0	
				Ethanol:water-9:1	25 MPa and 60 °C	47.0	
9.	Zeaxantin	Orange	LPT	–	70 °C/30 s	259.95[g]	Plaza et al. (2011)
			HPT	–	400 MPa/40 °C/1 min	408.56	
			PEF	–	35 kV cm^1/750 ms	278.70	
10.	α-Carotene	Orange	LPT	–	70 °C/30 s)	25.04[g]	Plaza et al. (2011)
			HPT	–	400 MPa/40 °C/1 min	38.06	
			PEF	–	35 kV cm^1/750 ms	26.64	
11.	α-Cryptoxanthin	Orange	LPT	–	70 °C/30 s	93.99	Plaza et al. (2011)
			HPT	–	400 MPa/40 °C/1 min	167.26	
			PEF	–	35 kV cm^1/750 ms	101.52	

components. The present review ascertains how the use of different technologies can result into the extraction of bioactive compounds which can be used as nutraceuticals and dietary supplements. The replacement of environmentally troublesome organic solvents in such extraction techniques, with green and safe solvents such as CO_2, ethanol, and water is the main objective of this review. Steps should be taken to help build a more rational use of our natural resources. A detailed economic analysis of these extraction techniques will help setting up

Table 3 continued

S. no.	Bioactive component	Sources	Method	Extraction solvent used	Optimum conditions	Yield	References
12.	β-Carotene	Orange	LPT	–	70 °C/30 s	32.72[g]	Plaza et al. (2011)
			HPT	–	400 MPa/40°C/1 min	53.78	
			PEF	–	35 kV cm[1]/750 ms	33.74	
13.	β-Cryptoxanthin	Orange	LPT	–	70 °C/30 s	235.21[g]	Plaza et al. (2011)
			HPT	–	400 MPa/40 °C/1 min	330.07	
			PEF	–	35 kV cm[1]/750 ms	230.53	

HHP high hydrostatic pressure, *HPT* high-pressure treatment, *LPT* low pasteurization treatment, *MAE* microwave assisted extraction, *PEF* pulsed electric field, *SE* solvent extraction, *WE* water extraction

[a] mg/g sample; [b] mg Cy-3-glu eq. g^{-1} dry matter; [c] mg/kg'; [d] ppm GAE in dry extract of air dried grape pomace; [e] μg/100ml; [f] mg/100ml; [g] GAE/g of extract

commercial units, thereby establishing a commercial use for such residues. This will help in complete utilization of the industrial waste thereby providing extra compensation to the industries by sale of residues and will also help in eradicating environmental pollution caused by the poor dumping of industrial food waste.

Authors' contributions
All authors read and approved the final manuscript.

Author details
[1] Department of Food Technology, Akal College of Agriculture, Eternal University, Sirmour 173001, India. [2] Department of Biotechnology, Akal College of Agriculture, Eternal University, Sirmour 173001, India.

Acknowledgements
The authors duly acknowledge the Department of Biotechnology, Govt. of India for the financial support provided (Grant No. BT/AGR/BIOFORTI/PHII/NIN/2011), Ministry of Food Processing Industries (MoFPI) Govt. of India grant for infrastructural facility development (F. No. 5-11/2010-HRD) and Vice Chancellor, Eternal University for providing the motivation and research infrastructure.

Competing interests
The authors declare that they have no competing interests.

Funding
Department of Biotechnology, Govt. of India for the financial support provided (Grant No. BT/AGR/BIOFORTI/PHII/NIN/2011) and Ministry of Food Processing Industries (MoFPI) Govt. of India grant for infrastructural facility development (F. No. 5-11/2010-HRD).

References
Afoakwah A, Owusu J, Adomako C, Teye E (2012) Microwave assisted extraction (MAE) of antioxidant constituents in plant materials. Glob J Biosci Biotechnol 1:132–140
Aguiló-Aguayo I, Walton J, Viñas I, Tiwari BK (2017) Ultrasound assisted extraction of polysaccharides from mushroom by-products. LWT Food Sci Technol 77:92–99
Ahmadian-Kouchaksaraie Z, Niazmand R (2017) Supercritical carbon dioxide extraction of antioxidants from *Crocus sativus* petals of saffron industry residues: optimization using response surface methodology. J Supercrit Fluids 121:19–31
Ahmadian-Kouchaksaraie Z, Niazmand R, Najafi MN (2016) Optimization of the subcritical water extraction of phenolic antioxidants from *Crocus sativus* petals of saffron industry residues: Box-Behnken design and principal component analysis. Innov Food Sci Emerg Technol 36:234–244
Ajikumar PK, Tyo K, Carlsen S, Mucha O, Phon TH, Stephanopoulos G (2008) Terpenoids: opportunities for biosynthesis of natural product drugs using engineered microorganisms. Mol Pharm 5:167–190
Al-Dalaen S, Al-Qtaitat A (2014) Review article: oxidative stress versus antioxidants. Am J Biosci Bioeng 2:60–71
Arogba SS (2000) Mango (*Mangifera indica*) kernel: chromatographic analysis of the tannin, and stability study of the associated polyphenol oxidase activity. J Food Compos Anal 13:149–156
Ashraf-Khorassani M, Taylor LT (2004) Sequential fractionation of grape seeds into oils, polyphenols, and procyanidins via a single system employing CO$_2$-based fluids. J Agric Food Chem 52:2440–2444
Ayala-Zavala JF, González-Aguilar GA (2011) Use of additives to preserve the quality of fresh-cut fruits and vegetables. CRC Press, Boca Raton, pp 231–254
Ayala-Zavala J, Rosas-Domínguez C, Vega-Vega V, González-Aguilar G (2010) Antioxidant enrichment and antimicrobial protection of fresh-cut fruits using their own by products: looking for integral exploitation. J Food Sci 75:R175–R181
Baiano A (2014) Recovery of biomolecules from food wastes-a review. Molecules 19:14821–14842
Bandar H, Hijazi A, Rammal H, Hachem A, Saad Z, Badran B (2013) Techniques for the extraction of bioactive compounds from Lebanese Urtica Dioica. Am J Phytomed Clin Ther 6:507–513
Baysal T, Ersus S, Starmans D (2000) Supercritical CO$_2$ extraction of β-carotene and lycopene from tomato paste waste. J Agric Food Chem 48:5507–5511
Bimakr M, Rahman RA, Taip FS, Chuan L, Ganjloo A, Selamat J, Hamid A (2009) Supercritical carbon dioxide (SC-CO$_2$) extraction of bioactive flavonoid compounds from spearmint (*Mentha spicata* L.) leaves. Eur J Sci Res 33:679–690
Bimakr M, Abdul Rahman R, Taip FS, Mohd Adzahan N, Sarker ZI, Ganjloo A (2013) Ultrasound-assisted extraction of valuable compounds from winter melon (*Benincasa hispida*) seeds. Int Food Res J 20:331–338
Birben E, Sahiner UM, Sackesen C, Erzurum S, Kalayci O (2012) Oxidative stress and antioxidant defense. World Allergy Org J 5:9–19
Casas L, Mantell C, Rodríguez M, de la Ossa EM, Roldán A, De Ory I, Caro I, Blandino A (2010) Extraction of resveratrol from the pomace of *Palomino fino* grapes by supercritical carbon dioxide. J Food Eng 96:304–308
Castro-Vargas HI, Rodríguez-Varela LI, Ferreira SR, Parada-Alfonso F (2010) Extraction of phenolic fraction from guava seeds (*Psidium guajava* L.) using supercritical carbon dioxide and co-solvents. J Supercrit Fluids 51:319–324
Chang CJ, Chiu K-L, Chen Y-L, Chang C-Y (2000) Separation of catechins from green tea using carbon dioxide extraction. Food Chem 68:109–113
Chantaro P, Devahastin S, Chiewchan N (2008) Production of antioxidant high dietary fiber powder from carrot peels. LWT Food Sci Technol 41:1987–1994
Chávez-Santoscoy RA, Lazo-Vélez MA, Serna-Sáldivar SO, Gutiérrez-Uribe JA (2016) Delivery of flavonoids and saponins from black bean (*Phaseolus vulgaris*) seed coats incorporated into whole wheat bread. Int J Mol Sci 17:222

Choo Y-M, Yap S-C, Ooi C-K, Ma A-N, Goh S-H, Ong AS-H (1996) Recovered oil from palm-pressed fiber: a good source of natural carotenoids, vitamin E, and sterols. J Am Oil Chem Soc 73:599–602

Coll M, Coll L, Laencina J, Tomas-Barberan F (1998) Recovery of flavanones from wastes of industrially processed lemons. Eur Food Res Technol 206:404–407

Comim SR, Madella K, Oliveira J, Ferreira S (2010) Supercritical fluid extraction from dried banana peel (Musa spp., genomic group AAB): extraction yield, mathematical modeling, economical analysis and phase equilibria. J Supercrit Fluids 54:30–37

Corrales M, Toepfl S, Butz P, Knorr D, Tauscher B (2008) Extraction of anthocyanins from grape by-products assisted by ultrasonics, high hydrostatic pressure or pulsed electric fields: a comparison. Inn Food Sci Emerg Technol 9:85–91

Day L, Seymour RB, Pitts KF, Konczak I, Lundin L (2009) Incorporation of functional ingredients into foods. Trends Food Sci Technol 20:388–395

Delazar A, Nahar L, Hamedeyazdan S, Sarker SD (2012) Microwave-assisted extraction in natural products isolation. Methods Mol Biol 864:89–115

Deng G-F, Shen C, Xu X-R, Kuang R-D, Guo Y-J, Zeng L-S, Gao L-L, Lin X, Xie J-F, Xia E-Q (2012) Potential of fruit wastes as natural resources of bioactive compounds. Int J Mol Sci 13:8308–8323

Drosou C, Kyriakopoulou K, Bimpilas A, Tsimogiannis D, Krokida M (2015) A comparative study on different extraction techniques to recover red grape pomace polyphenols from vinification byproducts. Ind Crops Prod 75:141–149

Duan X, Jiang Y, Su X, Zhang Z, Shi J (2007) Antioxidant properties of anthocyanins extracted from litchi (Litchi chinenesis Sonn.) fruit pericarp tissues in relation to their role in the pericarp browning. Food Chem 101:1365–1371

Ejike CE, Emmanuel TN (2009) Cholesterol concentration in different parts of bovine meat sold in Nsukka, Nigeria: implications for cardiovascular disease risk. Afr J Biochem Res 3:095–097

Espín JC, García-Conesa MT, Tomás-Barberán FA (2007) Nutraceuticals: facts and fiction. Phytochemistry 68:2986–3008

Espinosa-Pardo FA, Nakajima VM, Macedo GA, Macedo JA, Martínez J (2017) Extraction of phenolic compounds from dry and fermented orange pomace using supercritical CO_2 and cosolvents. Food Bioprod Process 101:1–10

Farías-Campomanes AM, Rostagno MA, Coaquira-Quispe JJ, Meireles MAA (2015) Supercritical fluid extraction of polyphenols from lees: overall extraction curve, kinetic data and composition of the extracts. Bioresour Bioprocess 2:45

Filip S, Pavlić B, Vidović S, Vladić J, Zeković Z (2017) Optimization of microwave-assisted extraction of polyphenolic compounds from Ocimum basilicum by response surface methodology. Food Anal Method. doi:10.1007/s12161-017-0792-7

Foo LY, Lu Y (1999) Isolation and identification of procyanidins in apple pomace. Food Chem 64:511–518

Gan C-Y, Latiff AA (2011) Optimisation of the solvent extraction of bioactive compounds from Parkia speciosa pod using response surface methodology. Food Chem 124:1277–1283

Gardossi L, Poulsen PB, Ballesteros A, Hult K, Švedas VK, Vasić-Rački Đ, Carrea G, Magnusson A, Schmid A, Wohlgemuth R (2010) Guidelines for reporting of biocatalytic reactions. Trends Biotechnol 28:171–180

Ghafoor K, Park J, Choi Y-H (2010) Optimization of supercritical fluid extraction of bioactive compounds from grape (Vitis labrusca B.) peel by using response surface methodology. Innov Food Sci Emerg Technol 11:485–490

Ghafoor K, Hui T, Choi YH (2011) Optimization of ultrasonic-assisted extraction of total anthocyanins from grape peel using response surface methodolog. J Food Biochem 35:735–746

Giannuzzo AN, Boggetti HJ, Nazareno MA, Mishima HT (2003) Supercritical fluid extraction of naringin from the peel of Citrus paradisi. Phytochem Anal 14:221–223

Gil MI, Tomás-Barberán FA, Hess-Pierce B, Holcroft DM, Kader AA (2000) Antioxidant activity of pomegranate juice and its relationship with phenolic composition and processing. J Agric Food Chem 48:4581–4589

Goli AH, Barzegar M, Sahari MA (2005) Antioxidant activity and total phenolic compounds of pistachio (Pistachia vera) hull extracts. Food Chem 92:521–525

González-Montelongo R, Lobo MG, González M (2010) Antioxidant activity in banana peel extracts: testing extraction conditions and related bioactive compounds. Food Chem 119:1030–1039

Herrero M, Cifuentes A, Ibañez E (2006) Sub-and supercritical fluid extraction of functional ingredients from different natural sources: plants, food-by-products, algae and microalgae: a review. Food Chem 98:136–148

Jayathunge K, Stratakos AC, Cregenzán-Albertia O, Grant IR, Lyng J, Koidis A (2017) Enhancing the lycopene in vitro bioaccessibility of tomato juice synergistically applying thermal and non-thermal processing technologies. Food Chem 221:698–705

Joana Gil-Chávez G, Villa JA, Fernando Ayala-Zavala J, Basilio Heredia J, Sepulveda D, Yahia EM, González-Aguilar GA (2013) Technologies for extraction and production of bioactive compounds to be used as nutraceuticals and food ingredients: an overview. Compr Rev Food Sci Food Saf 12:5–23

Jung G-W, Kang H-M, Chun B-S (2012) Characterization of wheat bran oil obtained by supercritical carbon dioxide and hexane extraction. J Ind Eng Chem 18:360–363

Kalogeropoulos N, Chiou A, Pyriochou V, Peristeraki A, Karathanos VT (2012) Bioactive phytochemicals in industrial tomatoes and their processing byproducts. LWT Food Sci Technol 49:213–216

Kanazawa K, Sakakibara H (2000) High content of dopamine, a strong antioxidant, in cavendish banana. J Agric Food Chem 48:844–848

Kehili M, Kammlott M, Choura S, Zammel A, Zetzl C, Smirnova I, Allouche N, Sayadi S (2017) Supercritical CO_2 extraction and antioxidant activity of lycopene and β-carotene-enriched oleoresin from tomato (Lycopersicum esculentum L.) peels by-product of a Tunisian industry. Food Bioprod Process 102:340–349

Kerem Z, German-Shashoua H, Yarden O (2005) Microwave-assisted extraction of bioactive saponins from chickpea (Cicer arietinum L). J Sci Food Agric 85:406–412

Kim W-J, Veriansyah B, Lee Y-W, Kim J, Kim J-D (2010) Extraction of mangiferin from Mahkota Dewa (Phaleria macrocarpa) using subcritical water. J Ind Eng Chem 16:425–430

Ko J, Ko M, Kim D, Lim S (2016) Enhanced production of phenolic compounds from pumpkin leaves by subcritical water hydrolysis. Prev Nutr Food Sci 21:132

Kulkarni V, Rathod V (2016) Green process for extraction of mangiferin from Mangifera indica leaves. J Biol Act Prod Nat 6:406–411

Kumar K (2015) Role of edible mushrooms as functional foods—a review. South Asian J Food Technol Environ 1:211–218

Kumar K, Kumar S (2015) Role of nutraceuticals in health and disease prevention: a review. South Asian J Food Technol Environ 1:116–121

Lee H, Wicker L (1991) Anthocyanin pigments in the skin of lychee fruit. J Food Sci 56:466–468

Lemes AC, Sala L, Ores JDC, Braga ARC, Egea MB, Fernandes KF (2016) A review of the latest advances in encrypted bioactive peptides from protein-rich waste. Int J Mol Sci 17:950

Li B, Xu Y, Jin Y-X, Wu Y-Y, Tu Y-Y (2010) Response surface optimization of supercritical fluid extraction of kaempferol glycosides from tea seed cake. Ind Crops Prod 32:123–128

Liza M, Rahman RA, Mandana B, Jinap S, Rahmat A, Zaidul I, Hamid A (2010) Supercritical carbon dioxide extraction of bioactive flavonoid from Strobilanthes crispus (Pecah Kaca). Food Bioprod Process 88:319–326

Louli V, Ragoussis N, Magoulas K (2004) Recovery of phenolic antioxidants from wine industry by-products. Bioresour Technol 92:201–208

Lozano-Sánchez J, Bendini A, Di Lecce G, Valli E, Gallina Toschi T, Segura-Carretero A (2017) Macro and micro functional components of a spreadable olive by-product (pâté) generated by new concept of two-phase decanter. Eur J Lipid Sci. doi:10.1002/ejlt.201600096

Lu Y, Foo LY (1997) Identification and quantification of major polyphenols in apple pomace. Food Chem 59:187–194

Maier T, Schieber A, Kammerer DR, Carle R (2009) Residues of grape (Vitis vinifera L.) seed oil production as a valuable source of phenolic antioxidants. Food Chem 112:551–559

Mayanga-Torres P, Lachos-Perez D, Rezende C, Prado J, Ma Z, Tompsett G, Timko M, Forster-Carneiro T (2017) Valorization of coffee industry residues by subcritical water hydrolysis: recovery of sugars and phenolic compounds. J Supercrit Fluids 120:75–85

Mira B, Blasco M, Berna A, Subirats S (1999) Supercritical CO_2 extraction of essential oil from orange peel. Effect of operation conditions on the extract composition. J Supercrit Fluids 14:95–104

Mirabella N, Castellani V, Sala S (2014) Current options for the valorization of food manufacturing waste: a review. J Clean Prod 65:28–41

Moore J, Cheng Z, Su L, Yu L (2006) Effects of solid-state enzymatic treatments on the antioxidant properties of wheat bran. J Agric Food Chem 54:9032–9045

Mussatto SI, Ballesteros LF, Martins S, Teixeira JA (2011) Extraction of antioxidant phenolic compounds from spent coffee grounds. Sep Purif Technol 83:173–179

Negro C, Tommasi L, Miceli A (2003) Phenolic compounds and antioxidant activity from red grape marc extracts. Bioresour Technol 87:41–44

Nobre BP, Palavra AF, Pessoa FL, Mendes RL (2009) Supercritical CO_2 extraction of trans-lycopene from Portuguese tomato industrial waste. Food Chem 116:680–685

Noda Y, Kaneyuki T, Mori A, Packer L (2002) Antioxidant activities of pomegranate fruit extract and its anthocyanidins: delphinidin, cyanidin, and pelargonidin. J Agric Food Chem 50:166–171

Núñez Selles AJ, Daglia M, Rastrelli L (2016) The potential role of mangiferin in cancer treatment through its immunomodulatory, anti-angiogenic, apoptotic, and gene regulatory effects. BioFactors 42:475–491

Okuno S, Yoshinaga M, Nakatani M, Ishiguro K, Yoshimoto M, Morishita T, Uehara T, Kawano M (2002) Extraction of antioxidants in sweetpotato waste powder with supercritical carbon dioxide. Food Sci Technol Res 8:154–157

Oliveira R, Oliveira V, Aracava KK, da Costa Rodrigues CE (2012) Effects of the extraction conditions on the yield and composition of rice bran oil extracted with ethanol—a response surface approach. Food Bioprod Process 90:22–31

Padmapriya K, Dutta A, Chaudhuri S, Dutta D (2012) Microwave assisted extraction of mangiferin from Curcuma amada. 3 Biotech 2:27–30

Paes J, Dotta R, Martínez J (2013) Extraction of phenolic compounds from blueberry (Vaccinium myrtillus L.) residues using supercritical CO_2 and pressurized water. J Supercritic Fluids. doi:10.1016/j.supflu.2014.07.025

Perretti G, Miniati E, Montanari L, Fantozzi P (2003) Improving the value of rice by-products by SFE. J Supercrit Fluids 26:63–71

Piñeiro Z, Marrufo-Curtido A, Serrano MJ, Palma M (2016) Ultrasound-assisted extraction of stilbenes from grape canes. Molecules 21:784

Plaza L, Sánchez-Moreno C, De Ancos B, Elez-Martínez P, Martín-Belloso O, Cano MP (2011) Carotenoid and flavanone content during refrigerated storage of orange juice processed by high-pressure, pulsed electric fields and low pasteurization. LWT - Food Sci Technol 44(4):834–839

Pujol D, Liu C, Gominho J, Olivella M, Fiol N, Villaescusa I, Pereira H (2013) The chemical composition of exhausted coffee waste. Ind Crops Prod 50:423–429

Puravankara D, Boghra V, Sharma RS (2000) Effect of antioxidant principles isolated from mango (Mangifera indica L) seed kernels on oxidative stability of buffalo ghee (butter-fat). J Sci Food Agric 80:522–526

Puri M, Sharma D, Barrow CJ (2012) Enzyme-assisted extraction of bioactives from plants. Trends Biotechnol 30:37–44

Rahman K (2007) Studies on free radicals, antioxidants, and co-factors. Clin Interv Aging 2:219–236

Rangsriwong P, Rangkadilok N, Satayavivad J, Goto M, Shotipruk A (2009) Subcritical water extraction of polyphenolic compounds from Terminalia chebula Retz. fruits. Sep Purif Technol 66:51–56

Rebecca LJ, Seshiah C, Tissopi T (2014) Extraction of caffeine from used tea leaves. Ann Valahia Univ Targ pp 19–22

Reshmitha T, Thomas S, Geethanjali S, Arun K, Nisha P (2017) DNA and mitochondrial protective effect of lycopene rich tomato (Solanum lycopersicum L.) peel extract prepared by enzyme assisted extraction against H_2O_2 induced oxidative damage in L6 myoblasts. J Funct Foods 28:147–156

Rostagno MA, Palma M, Barroso CG (2003) Ultrasound-assisted extraction of soy isoflavones. J Chromatogr 1012:119–128

Rudra SG, Nishad J, Jakhar N, Kaur C (2015) Food industry waste: mine of nutraceuticals. Int J Sci Environ Technol 4:205–229

Safdar MN, Kausar T, Jabbar S, Mumtaz A, Ahad K, Saddozai AA (2016) Extraction and quantification of polyphenols from kinnow (Citrus reticulate L.) peel using ultrasound and maceration techniques. J Food Drug Anal. doi:10.1016/j.jfda.2016.07.010

Sainvitu P, Nott K, Richard G, Blecker C, Jérôme C, Wathelet J-P, Paquot M, Deleu M (2012) Structure, properties and obtention routes of flaxseed lignan secoisolariciresinol: a review. Biotechnol Agron Soc Environ 16:115

Şanal İ, Güvenç A, Salgın U, Mehmetoğlu Ü, Çalımlı A (2004) Recycling of apricot pomace by supercritical CO_2 extraction. J Supercrit Fluids 32:221–230

Şanal İ, Bayraktar E, Mehmetoğlu Ü, Çalımlı A (2005) Determination of optimum conditions for SC-(CO2 + ethanol) extraction of β-carotene from apricot pomace using response surface methodology. J Supercrit Fluids 34:331–338

Shi J, Khatri M, Xue SJ, Mittal GS, Ma Y, Li D (2009) Solubility of lycopene in supercritical CO_2 fluid as affected by temperature and pressure. Sep Purif Technol 66:322–328

Shrikhande AJ (2000) Wine by-products with health benefits. Food Res Int 33:469–474

Sihvonen M, Järvenpää E, Hietaniemi V, Huopalahti R (1999) Advances in supercritical carbon dioxide technologies. Trends Food Sci Tech 10:217–222

Singh PP, Saldaña MD (2011) Subcritical water extraction of phenolic compounds from potato peel. Food Res Int 44:2452–2458

Singh G, Verma A, Kumar V (2016) Catalytic properties, functional attributes and industrial applications of β-glucosidases. 3 Biotech 6:3

Smiderle FR, Morales D, Gil-Ramírez A, de Jesus LI, Gilbert-López B, Iacomini M, Soler-Rivas C (2017) Evaluation of microwave-assisted and pressurized liquid extractions to obtain β-D-glucans from mushrooms. Carbohydr Polym 156:165–174

Someya S, Yoshiki Y, Okubo K (2002) Antioxidant compounds from bananas (Musa cavendish). Food Chem 79:351–354

Strati IF, Oreopoulou V (2011) Effect of extraction parameters on the carotenoid recovery from tomato waste. Int J Food Sci Technol 46:23–29

Szentmihályi K, Vinkler P, Lakatos B, Illés V, Then M (2002) Rose hip (Rosa canina L.) oil obtained from waste hip seeds by different extraction methods. Bioresour Technol 82:195–201

Tan YA, Sambanthamurthi R, Sundram K, Wahid MB (2007) Valorisation of palm by-products as functional components. Eur J Lipid Sci Technol 109:380–393

Teixeira A, Baenas N, Dominguez-Perles R, Barros A, Rosa E, Moreno DA, Garcia-Viguera C (2014) Natural bioactive compounds from winery by-products as health promoters: a review. Int J Mol Sci 15:15638–15678

Tunchaiyaphum S, Eshtiaghi M, Yoswathana N (2013) Extraction of bioactive compounds from mango peels using green technology. Int J Chem Eng Appl 4:194

Vega PJ, Balaban M, Sims C, O'keefe S, Cornell J (1996) Supercritical carbon dioxide extraction efficiency for carotenes from carrots by RSM. J Food Sci 61:757–759

Viscidi KA, Dougherty MP, Briggs J, Camire ME (2004) Complex phenolic compounds reduce lipid oxidation in extruded oat cereals. LWT Food Sci Technol 37:789–796

Vyas P, Chaudhary B, Mukhopadhyay K, Bandopadhyay R (2009) Anthocyanins: looking beyond colors. In: Bhowmik PK, Basu SK, Goyal A (eds) Advances in biotechnology. Bentham Science Publishers Ltd., Oak Park, pp 152–184

Vyas P, Haque I, Kumar M, Mukhopadhyay K (2014) Photocontrol of differential gene expression and alterations in foliar anthocyanin accumulation: a comparative study using red and green forma Ocimum tenuiflorum. Acta Physiol Plant 36:2091–2102

Wang L, Weller CL (2006) Recent advances in extraction of nutraceuticals from plants. Trends Food Sci Technol 17:300–312

Wang J, Sun B, Cao Y, Tian Y, Li X (2008) Optimisation of ultrasound-assisted extraction of phenolic compounds from wheat bran. Food Chem 106:804–810

Wang W, Wu X, Han Y, Zhang Y, Sun Dong F (2011) Investigation on ultrasound-assisted extraction of three dibenzylbutyrolactone lignans from Hemistepta lyrata. J Appl Pharm Sci 1:24

Wenzel J, Storer Samaniego C, Wang L, Burrows L, Tucker E, Dwarshuis N, Ammerman M, Zand A (2016) Antioxidant potential of Juglans nigra, black walnut, husks extracted using supercritical carbon dioxide with an ethanol modifier. Food Sci Nutr. doi:10.1002/fsn3.385

Wolfe KL, Liu RH (2003) Apple peels as a value-added food ingredient. J Agric Food Chem 51:1676–1683

Yang H-Y, Lee T-H (2015) Antioxidant enzymes as redox-based biomarkers: a brief review. BMB Rep 48:200

Zakaria SM, Kamal SMM (2016) Subcritical water extraction of bioactive compounds from plants and algae: applications in pharmaceutical and food ingredients. Food Eng Rev 1:23–34

Zeyada NN, Zeitoum M, Barbary O (2008) Utilization of some vegetables and fruit waste as natural antioxidants. Alex J Food Sci Technol 5:1–11

Zhang F, Chen B, Xiao S, S-z Yao (2005) Optimization and comparison of different extraction techniques for sanguinarine and chelerythrine in fruits of *Macleaya cordata* (Willd) R. Br. Sep Purif Technol 42:283–290

Zulkifli KS, Abdullah N, Abdullah A, Aziman N, Kamarudin W (2012) Bioactive phenolic compounds and antioxidant activity of selected fruit peels. In: 2012 international conference on environment, chemistry and biology, vol 49. IACSIT Press, Singapore, pp 66–70

Zuorro A, Lavecchia R (2012) Spent coffee grounds as a valuable source of phenolic compounds and bioenergy. J Clean Prod 34:49–56

Zuorro A, Fidaleo M, Lavecchia R (2011) Enzyme-assisted extraction of lycopene from tomato processing waste. Enzyme Microb Technol 49:567–573

9

Utilization of corncob xylan as a sole carbon source for the biosynthesis of endo-1,4-β xylanase from *Aspergillus niger* KIBGE-IB36

Urooj Javed[1], Afsheen Aman[1] and Shah Ali Ul Qader[2*]

Abstract

Background: Xylan is a hemicellulose polysaccharide which is composed of β-1,4-linked D-xylosyl residues. Endo-1,4-β xylanase has the ability to cleave xylan back bone chains to release xylose residues. They are produced by a number of prokaryotic and eukaryotic organisms. Among them, filamentous fungi are attracting great attention due to high secretion of xylanolytic enzymes. Endo-1,4-β xylanase has wide industrial applications such as in animal feed, bread making, food and beverages, textile, bleaching of wood pulp, and biofuel production.

Results: In this study, different *Aspergillus* species were screened for the production of endo-1,4-β xylanase, and *Aspergillus niger* KIBGE-IB36 was selected for optimum production of enzyme in submerged fermentation technique. Influence of various fermentation conditions was investigated to produce high titer of endo-1,4-β xylanase. The results indicated that *A. niger* KIBGE-IB36 showed optimum production of endo-1,4-β xylanase at 30 °C, pH 8 after 6 days of incubation. Different macro- and micronutrients were also amalgamated in the fermentation medium to increase the enzyme production. The parametric optimization of endo-1,4-β xylanase resulted in tenfold increase after hydrolysis of 20 g L^{-1} corncob xylan.

Conclusions: The use of low-cost substrate approach for high production of endo-1,4-β xylanase has been developed successfully that can be consumed in different industrial applications especially in paper and pulp industry.

Keywords: *Aspergillus* species, Corncob, Fermentation, Hemicellulose, Endo-1,4-β xylanase

Background

Substantial consideration has been given for the use of microorganisms in industrial processes particularly for the production of enzymes. Amid different microorganisms, fungi, bacteria, and actinomycetes are the abundant producers of endo-1,4-β xylanase (Lu et al. 2008). Filamentous fungi such as *Aspergillus* and *Trichoderma* species have immense significance over bacteria due to their efficient ability to degrade plant cell wall (Kaushik and Malik 2009). *Aspergillus niger* is a filamentous fungus that has been used extensively in different biotechnological applications. According to Food and Drug Administration (FDA), *A. niger* can be "generally regarded as safe" (GRAS) under good manufacturing* practices for industrial products and they can be isolated easily from soil, compost, and plant-decaying materials (Klich 2002; Schuster et al. 2002). *Aspergillus niger* has the ability to produce high yield of broad range of enzymes under both submerged and solid-state fermentation conditions, and approximately 80–90% endo-1,4-β xylanases are produced using submerged fermentation technique (Polizeli et al. 2005; Pel et al. 2007).

*Correspondence: ali_kibge@yahoo.com
[2] Department of Biochemistry, University of Karachi, Karachi, Pakistan
Full list of author information is available at the end of the article

Among different industrially important enzymes, xylanolytic enzymes have been used extensively in food and pharmaceutical industries. This complex enzyme includes endo-1,4-β- xylanase [EC 3.2.1.8], β-xylosidase [EC 3.2.1.37], α-arabinofuranosidase [EC 3.2.1.55], and acetyl xylan esterase [EC 3.1.1.72] (Biely 1993). Xylan is the second most abundant resource after cellulose and is the main constituent of hemicellulose which consists of long chain of 1,4-β-D-xylose monomers (Izidoro and Knob 2014). Endo-1,4-β xylanase is an enzyme which has the ability to hydrolyze β-1,4 glycosidic bonds in xylan into small series of xylooligosaccharides (Chanwicha et al. 2015). In the presence of β-xylosidase, these oligosaccharides are further hydrolyzed into xylose molecules. Alone exo-xylanase will not be able to hydrolyze the complex xylan structure. After this synergistic effect, more xylose is produced as a by-product which confirms the presence of endo-1,4-β xylanase. Being an industrially important enzyme, endo-1,4-β xylanase has several applications: in baking and in food industries, it is utilized as a taste and texture enhancer; in poultry, it is used as a food additive; and in beverages, it acts as a juice clarifying agent. Commonly, it is also used in pre-bleaching process of kraft and pulp to diminish the use of harmful chemicals (Arulanandham and Palaniswamy 2014).

To achieve entire enzymatic degradation of xylan into its monosaccharide components, a group of synergistic xylanolytic enzymes is required due to the presence of differences in xylan structure from different sources (Latif et al. 2006). Previously, corncob is reported as one of the valuable by-products of food industry which can be utilized as growth-inducing substrate for bacteria and fungi. In addition, it is also used to synthesize xylose, alcohol, xylitol, and xylooligosaccharides (Chapla et al. 2012). In this study, commercial corncob xylan was used for the synthesis of endo-1,4-β xylanase by A. niger KIBGE-IB36 under submerged fermentation conditions. Different physiological and chemical factors were also optimized to enhance the production of endo-1,4-β xylanase.

Results and discussion
Screening of fungal species for endo-1,4-β xylanase production
To screen the production of endo-1,4-β xylanase, four different species of Aspergillus were used. It was observed that A. niger KIBGE-IB36 is a hyper producer endo-1,4-β xylanase (837 U mg^{-1}) as compared to other species (Fig. 1a). In addition, it was also confirmed by Congo red dye method in which A. niger KIBGE-IB36 expressed a clear xylanolytic zone around the growing colony on medium containing corncob xylan as a sole carbon source (Fig. 1b). Further experimental studies were carried out using A. niger KIBGE-IB36 for the production of endo-1,4-β xylanase.

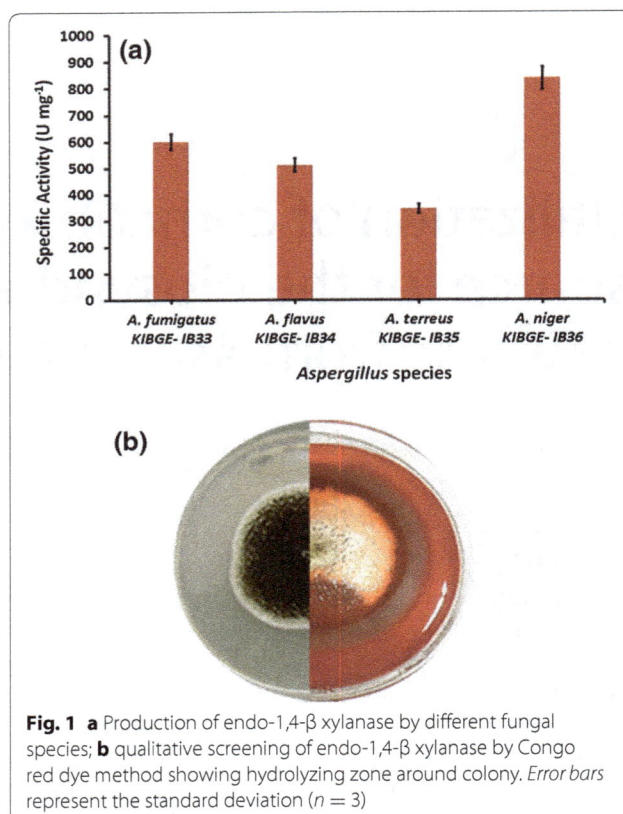

Fig. 1 a Production of endo-1,4-β xylanase by different fungal species; b qualitative screening of endo-1,4-β xylanase by Congo red dye method showing hydrolyzing zone around colony. Error bars represent the standard deviation (n = 3)

Selection of fermentation medium
To produce high titers of endo-1,4-β xylanase, five different reported media were analyzed. Among these media, maximum endo-1,4-β xylanase was synthesized in Czapek medium (1239 U mg^{-1}) as compared to other media (Fig. 2).

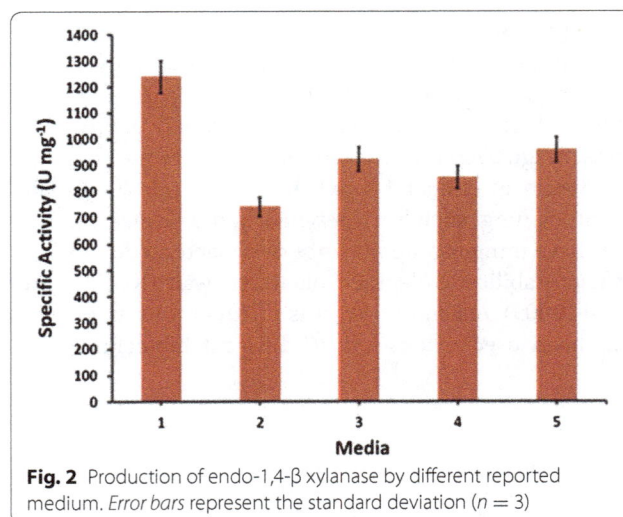

Fig. 2 Production of endo-1,4-β xylanase by different reported medium. Error bars represent the standard deviation (n = 3)

Selection of fermentation temperature

The fermentation temperature not only influences the growth curve of an organism but it also has an impact on the production of endo-1,4-β xylanase (Senthilkumar et al. 2005). Most of the filamentous fungi grow in between 25 and 35 °C, while some thermophilic fungal species can also grow at high temperature with maximum at or above 50 °C (Suresh and Chandrasekaran 1999; Maheshwari et al. 2000). Mostly *Aspergillus niger* is reported to show growth pattern in between 25 and 30 °C (Adinarayana et al. 2003). In current study, optimum temperature for the production of enzyme was recorded at 30 °C with 1485 U mg^{-1} of endo-1,4-β xylanase. A gradual decline in enzyme titer was noted at 50 °C (50 U mg^{-1}) and this decline is due to the lower growth rate of this fungi at high temperature (Fig. 3a). The high- and low incubation temperatures cause the inhibition of fungal growth that ultimately leads to the decline in enzyme synthesis (Lenartovicz et al. 2003). This investigation synchronizes with those of the previously reported data where activity of endo-1,4-β xylanase was optimum at 30 °C (Subbulakshmi and Iyer 2014; Kanimozhi and Nagalakshmi 2014).

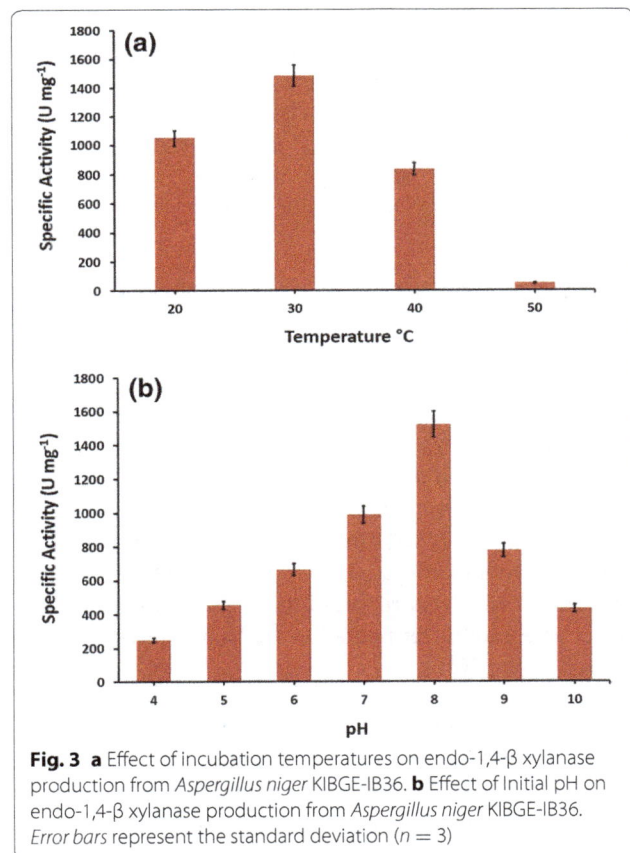

Fig. 3 a Effect of incubation temperatures on endo-1,4-β xylanase production from *Aspergillus niger* KIBGE-IB36. **b** Effect of Initial pH on endo-1,4-β xylanase production from *Aspergillus niger* KIBGE-IB36. *Error bars* represent the standard deviation (n = 3)

Selection of fermentation pH

The pH of the medium plays a significant role in enzyme production. It can either lower the enzyme production by effecting the growth of microorganism or by creating unsuitable toxic environment that leads to the denaturation or inactivation of enzyme produced (Bajaj and Abbass 2011). In the present study, the optimum pH for the production of endo-1,4-β xylanase was achieved at pH 8.0 with specific activity of 1523 U mg^{-1}, whereas minimum activity was observed at pH 4.0 (249 U mg^{-1}) (Fig. 3b). Most of the researchers reported maximum endo-1,4-β xylanase production by filamentous fungi in acidic pH ranging from 5.0 to 6.5 and also near pH 8.0 (Murthy and Naidu 2012; Bajaj and Abbass 2011). Some other investigators reported enzyme production at pH 9.0 and 10.0 (Kapilan and Arasaratnam 2012; Nair et al. 2008). In the present study, *A. niger* KIBGE-IB36 showed effective tolerance and potential to grow and produce endo-1,4-β xylanase at both acidic and alkaline pH values (pH 6.0–10.0).

Selection of fermentation time period

In this experiment, the production of endo-1,4-β xylanase by *A. niger* KIBGE-IB36 was determined during different intervals of time (01–10 days) along with fungal biomass estimation. The exponential phase was observed from days 01 to 06, entering the stationary phase up to day 08, and afterwards unstable decline was observed at day 10. The enzyme synthesis increased in the exponential phase (days 03 to 06) and after reaching its maxima both the enzyme activity and fungal biomass then started to decline gradually (Fig. 4a). It has been suggested that prolonged incubation period stimulates the secretion of nonspecific proteases that degrade the synthesized enzyme in the medium. Therefore, it is suggested to control the end point of fermentation period (Pal and Khanum 2010). It has been proposed previously that the optimum fermentation period relies on the nature of substrate, organism, macro- and micronutrients and many other fermentation events (Dekker 1983). This obtained result coincides with the previous studies in which maximum endo-1,4-β xylanase was achieved in 6 days of fermentation period (Sharma et al. 2015; Pal and Khanum 2010).

Effect of substrate concentration

Specific substrate plays an important role for any enzyme production. In this study, endo-1,4-β xylanase was synthesized using different concentrations (5–20 g L^{-1}) of corncob xylan which was studied in many past studies (Ahmad et al. 2012). In present study, the efficiency of corncob xylan in the maximum induction of endo-1,4-β xylanase production was established in 20 g L^{-1} of concentration, while 25g L^{-1} of corncob xylan created inhibitory effect on

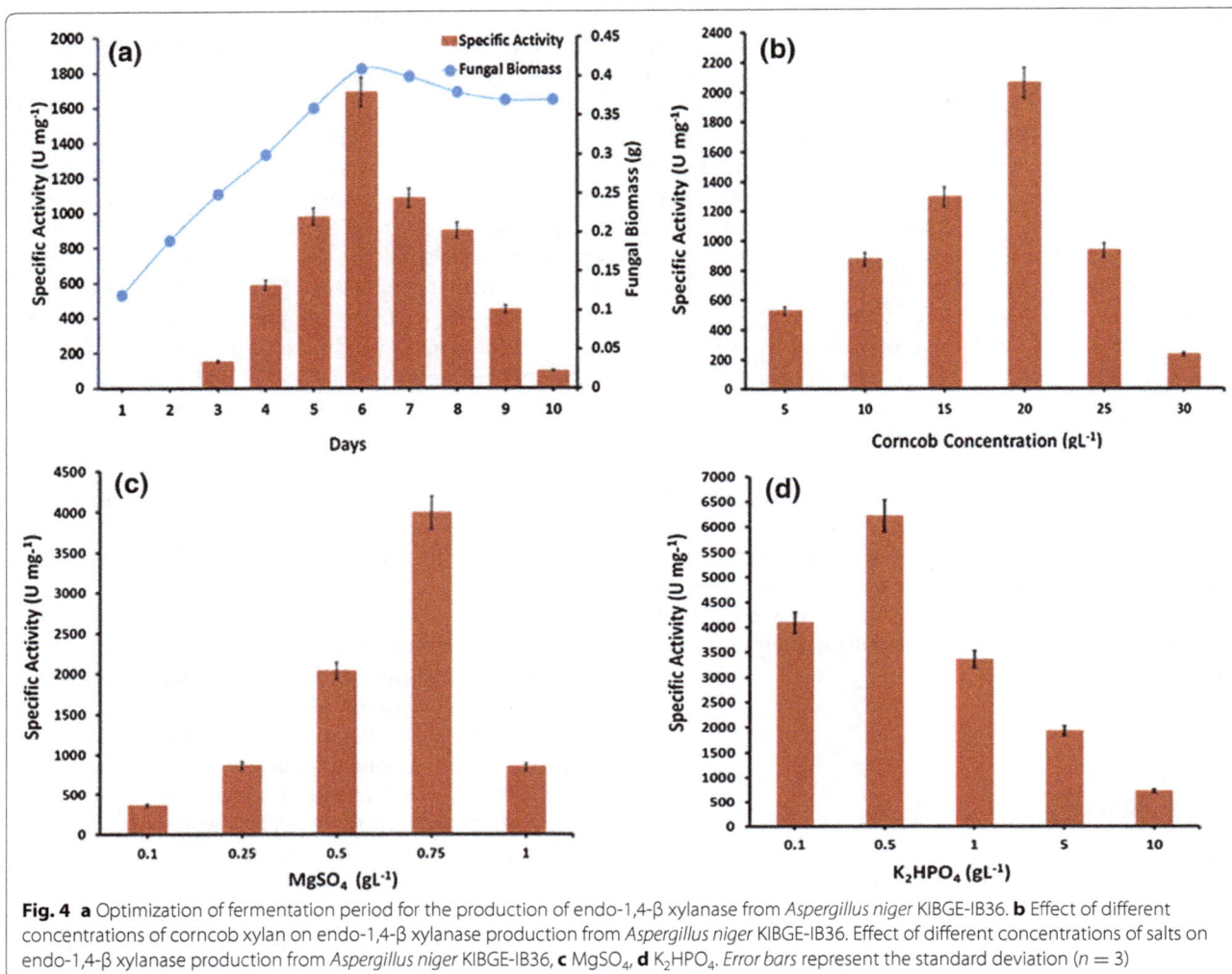

Fig. 4 a Optimization of fermentation period for the production of endo-1,4-β xylanase from *Aspergillus niger* KIBGE-IB36. **b** Effect of different concentrations of corncob xylan on endo-1,4-β xylanase production from *Aspergillus niger* KIBGE-IB36. Effect of different concentrations of salts on endo-1,4-β xylanase production from *Aspergillus niger* KIBGE-IB36, **c** $MgSO_4$, **d** K_2HPO_4. *Error bars* represent the standard deviation ($n = 3$)

the production of endo-1,4-β xylanase (Fig. 4b). It might be due to the increased viscosity of the medium that ultimately leads to feedback inhibition of enzyme (Karim et al. 2014). Our results are in line with other research in which they used 20 g L^{-1} of xylan for the induction of endo-1,4-β xylanase (Shah and Madamwar 2005).

Effect of different nitrogen sources

In the present study, different nitrogen sources (organic/in organic) were studied (Table 1). The result showed that organic nitrogen sources have a profound effect on the production of endo-1,4-β xylanase as compared to inorganic nitrogen sources. Among different organic sources, peptone proved was the inducer for maximum endo-1,4-β xylanase production (3069 U mg^{-1}). Previously, it is also reported that endo-1,4-β xylanase yield was enhanced by the supplementation of peptone (Qinnghe et al. 2004). Other organic nitrogen sources such as tryptone, meat

extract, and yeast extract also showed endo-1,4-β xylanase production but to a lower extent as compared to peptone. When different nitrogen sources were combined, a very unique expression was observed to the production of xylanase suggesting that *A. niger* requires different types of nitrogen sources for its growth and enzyme production. Among five different combinations, meat extract and peptone produce high titers of endo-1,4-β xylanase (4219 U mg^{-1}) which was 1.37 times higher when compared with the medium incorporated solely with peptone. Various researchers have also reported the augmentation of nitrogen source into the fermentation media (Gomes et al. 1994; Bansod et al. 1993).

Effect of K_2HPO_4 and $MgSO_4$ concentrations

Microbial metabolism and regulation of enzyme production are reciprocal to the supplementation of salts in the medium (Maciel et al. 2008). In this study, the medium

Table 1 Effect of different nitrogen sources on endo-1,4-β xylanase production from *Aspergillus niger* KIBGE-IB36

Types of nitrogen source	Nitrogen source (5 g L^{-1})	Specific activity (U mg^{-1})
Organic nitrogen sources	Yeast extract	1713 ± 9.24
	Meat extract	2166 ± 11.4
	Tryptone	2028 ± 7.75
	Peptone	3069 ± 18.49
Inorganic nitrogen sources	Urea	1090 ± 8.22
	NH$_4$Cl	642 ± 8.86
	KNO$_3$	492 ± 10.43
	NH$_4$NO$_3$	360 ± 20.85
Combination of organic nitrogen sources	Yeast extract + tryptone	3116 ± 13.54
	Yeast extract + peptone	2539 ± 18.90
	yeast extract + meat extract	1215 ± 24.37
	Meat extract + peptone	4219 ± 8.65
	Meat extract + tryptone	3056 ± 9.46

All the experiments were performed in triplicate and the results expressed are the mean values of all experimental setup

was optimized using different concentrations of K$_2$HPO$_4$ and MgSO$_4$·7H$_2$O. Magnesium ions have a significant effect on the equalization of the ribosomes and cellular membrane (Cui and Zhao 2012). In this study, with the increase of MgSO$_4$·7H$_2$O, a gradual increase was noticed in endo-1,4-β xylanase production up to 0.75 g L^{-1} (3987 U mg^{-1}), while at 1 g L^{-1} of magnesium salt the production of endo-1,4-β xylanase was declined (850 U mg^{-1}) (Fig. 4c). In contrast, Naveen and Siddalingeshwara (2015) reported 0.1 g L^{-1} of MgSO$_4$·7H$_2$O as a finest inducer of endo-1,4-β xylanase.

On the other hand, the presence of K$_2$HPO$_4$ in the growth medium also showed a positive effect on the yield of endo-1,4-β xylanase. At 0.5 g L^{-1} of K$_2$HPO$_4$, maximum endo-1,4-β xylanase production was observed (6214 U mg^{-1}) but as the concentration of salt increased, it leads towards a decline in endo-1,4-β xylanase production (729 U mg^{-1}) (Fig. 4d). According to the previous investigation, 0.3 g L^{-1} of K$_2$HPO$_4$ was found to be appropriate for the production of endo-1,4-β xylanase (Salihu et al. 2015). It is reported that K$_2$HPO$_4$ has the potential to maintain the ideal osmotic pressure for high endo-1,4-β xylanase production (Berk 2000). Hence, after optimizing all the medium formulation and physical parameters, a tenfold increase in endo-1,4-β xylanase synthesis was observed (Fig. 5).

The result obtained in the present study indicates that among different *Aspergillus* species, *A. niger* KIBGE-IB36 has more potential to saccharify corncob xylan efficiently into ample amount of endo-1,4-β xylanase through submerged fermentation. The production at low temperature (30 °C) indicates the mesophilic nature of *A. niger* KIBGE-IB36 and the production of endo-1,4-β xylanase at alkaline pH makes this enzyme a promising

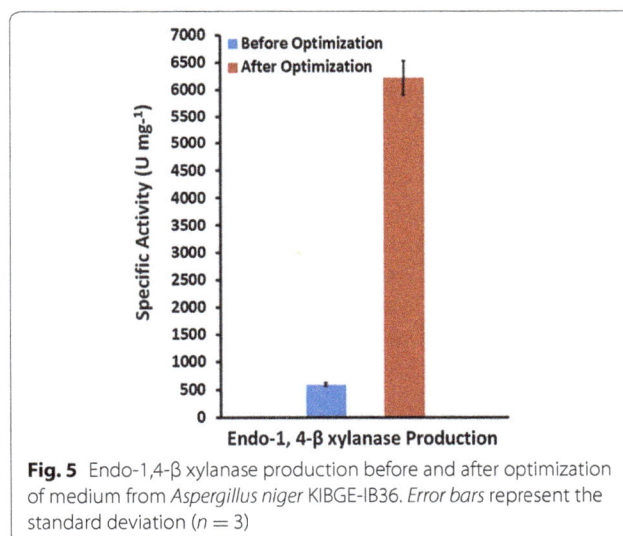

Fig. 5 Endo-1,4-β xylanase production before and after optimization of medium from *Aspergillus niger* KIBGE-IB36. *Error bars* represent the standard deviation (n = 3)

candidate for bio-bleaching processes and other industrial applications. Hence, the current investigation provides direct comparison of enhanced production of endo-1,4-β xylanase up to tenfold after optimization of different physico-chemical parameters of fermentation medium such as pH, temperature, fermentation period, substrate concentration, suitable nitrogen source, and salt concentration.

Methods

Microorganisms

The initial screening was carried out using four strains of *Aspergillus* species namely *Aspergillus fumigatus* KIBGE-IB33 [GenBank: KF905648], *Aspergillus flavus*

KIBGE-IB34 [GenBank: KF905649], *Aspergillus terreus* KIBGE-IB35 [GenBank: KF905649], and *A. niger* KIBGE-IB36 [GenBank: KF905650] which were isolated previously (Pervez et al. 2015).

Screening of endo-1,4-β xylanase production

All fungal isolates were grown on xylan-containing medium containing; g L^{-1}: (corncob xylan (Carbosynth, UK) 5.0; Nutrient broth 13.0; K_2HPO_4 2.5; KH_2PO_4 0.5; $CaCl_2$ 0.1; and NH_2SO_4 0.5). After incubation at 30 °C for 05 days, culture broth was centrifuged at 4000 rpm at 4 °C for 30 min and filtered through Whatman filter paper No. 1. The supernatant was used for the estimation of endo-1,4-β xylanase production. For qualitative confirmation of endo-1,4-β xylanase production, the selected strain was grown on corncob xylan agar medium for 03 days maintaining the condition same as used for the fermentation. The clear hydrolytic zone around the fungal colony was observed after flooding with Congo red (Teather and Wood 1982).

Endo-1,4-β xylanase assay

The enzyme activity of endo-1,4-β xylanase was estimated by evaluation of reducing sugars released from 10 g L^{-1} of xylan in 10 mM citrate phosphate buffer (pH 5.0) by 3,5,dinitrosalicylic acid method using xylose as a standard (Miller 1959). One unit of enzyme of enzyme activity is defined as the amount of enzyme required to release 1 μmol of xylose per minute of reaction under standard assay conditions. Specific activities were expressed as unit of enzyme per milligram of protein.

Total protein estimation

The total protein was determined in the supernatant by Lowry's method (1951) using BSA (bovine serum albumin) as a standard.

Selection of fermentation medium

Five different reported media were initially used for the production of endo-1,4-β xylanase and these reported media were designated as media 1 (Kulkarni and Gupta 2013), media 2 (Adhyaru et al. 2014), media 3 (Kocabas and Ozben 2014), media 4 (Yuan et al. 2005), and media 5 (Bibi et al. 2014).

After selection of suitable medium, fermentation conditions were optimized by varying different physicochemical parameters using *A. niger* KIBGE-IB36 for the production of maximum endo-1,4-β xylanase.

Optimization of fermentation temperature, pH, and time

For the selection of appropriate temperature for the production of endo-1,4-β xylanase, different temperatures were tested ranging from 20 to 50 °C.

In the next step, the culture was also grown in a different pH medium ranging from 4.0 to 10.0.

The time course and fungal biomass for the production of endo-1,4-β xylanase were also estimated by incubating *A. niger* KIBGE-IB36 for different time intervals ranging from 03 to 10 days.

Optimization of macro- and micronutrients

In the present study, corncob xylan was used as a substrate for endo-1,4-β xylanase production with different concentrations of xylan ranging from 5 to 30 g L^{-1}.

To determine the effect of nitrogen source, different organic and inorganic nitrogen sources were assimilated in the production medium. Furthermore, the effect of different nitrogen sources in combination was also investigated.

To analyze the influence of $MgSO_4$ and K_2HPO_4 on endo-1,4-β xylanase production, different concentrations of salts were incorporated in the production medium ranging from 0.1–1 to 0.1 to 10 g L^{-1}, respectively.

Authors' contributions

All authors discussed the results and proofread the manuscript. All authors read and approved the final manuscript.

Author details

[1] The Karachi Institute of Biotechnology and Genetic Engineering, University of Karachi, Karachi, Pakistan. [2] Department of Biochemistry, University of Karachi, Karachi, Pakistan.

Acknowledgements

This research was funded by The Karachi Institute of Biotechnology and Genetic Engineering (KIBGE), University of Karachi, Pakistan.

Competing interests

The authors declare that they have no competing interests.

References

Adhyaru DN, Bhatt NS, Modi HA (2014) Enhanced production of cellulase-free, thermo-alkali-solvent-stable xylanase from *Bacillus altitudinis* DHN8, its characterization and application in sorghum straw saccharification. Biocatal Agric Biotechnol 3:182–190. doi:10.1016/j.bcab.2013.10.003

Adinarayana K, Prabhakar T, Srinivasulu V, Rao MA, Lakshmi PJ, Ellaiah P (2003) Optimization of process parameters for cephalosporin C production under solid state fermentation from *Acremonium chrysogenum*. Process Biochem 39:171–177. doi:10.1016/S0032-9592(03)00049-9

Ahmad Z, Butt MS, Anjum FM, Awan MS, Rathore HA, Nadeem MT, Ahmad A, Khaliq A (2012) Effect of corn cobs concentration on xylanase biosynthesis by *Aspergillus niger*. Afr J Biotechnol 11:1674–1682. doi:10.5897/ajb11.1769

Arulanandham TV, Palaniswamy M (2014) Production of xylanase by *Aspergillus nidulans* isolated from litter soil using rice bran as substrate by solid state fermentation. World J Pharm Sci 3: 1805–13. http://www.wjpps.com

Bajaj BK, Abbass M (2011) Studies on an alkali-thermostable xylanase from *Aspergillus fumigatus* MA28. 3 Biotech 1:161–171. doi:10.1007/s13205-011-0020-x

Bansod SM, Dutta-Choudhary M, Srinivasan MC, Rele MV (1993) Xylanase active at high pH from an alkalotolerant *Cephalosporium* species. Biotechnol Lett 15:965–970. doi:10.1007/BF00131765

Beily P (1993) Biochemical aspects of the production of microbial hemicellulase. In: Coughlan MP, Hazlewood GP (eds) Hemicellulose and hemicellulases. Portland Press, London, pp 29–51

Berk A (2000) Molecular cell biology, vol 4. WH Freeman, New York

Bibi Z, Ansari A, Zohra RR, Aman A, Qader SA (2014) Production of xylan degrading endo-1,4-β-xylanase from thermophilic *Geobacillus stearothermophilus* KIBGE-IB29. J Radiat Res Appl Sci 7:478–485. doi:10.1016/j.jrras.2014.08.001

Chanwicha N, Katekaew S, Aimi T, Boonlue S (2015) Purification and characterization of alkaline xylanase from *Thermoascus aurantiacus var. levisporus* KKU-PN-I2-1 cultivated by solid-state fermentation. Mycoscience 56:309–318. doi:10.1016/j.myc.2014.09.003

Chapla D, Pandit P, Shah A (2012) Production of xylooligosaccharides from corncob xylan by fungal xylanase and their utilization by probiotics. Bioresour Technol 115:215–221. doi:10.1016/j.biortech.2011.10.083

Cui F, Zhao L (2012) Optimization of xylanase production from *Penicillium* sp. WX-Z1 by a two-step statistical strategy: Plackett–Burman and Box–Behnken experimental design. Int J Mol Sci 13:10630–10646. doi:10.3390/ijms130810630

Dekker RF (1983) Bioconversion of hemicellulose: aspects of hemicellulase production by *Trichoderma reesei* QM 9414 and enzymic saccharification of hemicellulose. Biotechnol Bioeng 25:1127–1146. doi:10.1002/bit.260250419

Gomes DJ, Gomes J, Steiner W (1994) Factors influencing the induction of endo-xylanase by *Thermoascus aurantiacus*. J Biotechnol 33:87–94. doi:10.1016/01681656(94)90101-5

Izidoro SC, Knob A (2014) Production of xylanases by an *Aspergillus niger* strain in wastes grain. Acta Sci Biol Sci 36:313–319. doi:10.4025/actascibiolsci.v36i3.20567

Kanimozhi K, Nagalakshmi PK (2014) Xylanase production from *Aspergillus niger* by solid state fermentation using agricultural waste as substrate. Int J Curr Microbiol App Sci 3: 437–446. http://www.ijcmas.com

Kapilan R, Arasaratnam V (2012) Comparison of the kinetic properties of crude and purified xylanase from *Bacillus pumilus* with commercial xylanase from *Aspergillus niger*. Vingnanam J Sci 10:1–7. doi:10.4038/vingnanam.v10i1.4072

Karim A, Nawaz MA, Aman A, Ul Qader SA (2014) Hyper production of cellulose degrading endo (1,4) β-D-glucanase from *Bacillus licheniformis* KIBGE-IB2. J Radiat Res Appl Sci 8:160–165. doi:10.1016/j.jrras.2014.06.004

Kaushik P, Malik A (2009) Fungal dye decolourization: recent advances and future potential. Environ Int 35:127–141. doi:10.1016/j.envint.2008.05.010

Klich MA (2002) Biogeography of *Aspergillus* species in soil and litter. Mycologia 94: 21–27. http://www.mycologia.org/content/94/1/21.short

Kocabas DS, Ozben N (2014) Co-production of xylanase and xylooligosaccharides from lignocellulosic agricultural wastes. RSC Adv 4:26129–26139. doi:10.1039/C4RA02508C

Kulkarni P, Gupta N (2013) Screening and evaluation of soil fungal isolates for xylanase production. Recent Res Sci Technol 5. http://recent-science.com/

Latif F, Asgher M, Saleem R, Akrem A, Legge RL (2006) Purification and characterization of a xylanase produced by *Chaetomium thermophile* NIBGE. World J Microbiol Biotechnol 22:45–50. doi:10.1007/s11274-005-5745-4

Lenartovicz V, Marques De Souza CG, Guillen Moreira F, Peralta RM (2003) Temperature and carbon source affect the production and secretion of a thermostable β-xylosidase by *Aspergillus fumigatus*. Process Biochem 38:1775–1780

Lowry OH, Rosebrough NJ, Farr AL, Randall RJ (1951) Protein measurement with the Folin phenol reagent. J Biol Chem 193: 265–275. http://www.jbc.org/content/193/1/265.citation.full.html#ref-list-1

Lu F, Lu M, Lu Z, Bie X, Zhao H, Wang Y (2008) Purification and characterization of xylanase from *Aspergillus ficuum* AF-98. Bioresour Technol 99:5938–5941. doi:10.1016/j.biortech.2007.10.051

Maciel GM, de Souza Vandenberghe LP, Haminiuk CW, Fendrich RC, Della Bianca BE, da Silva Brandalize TQ, Pandey A, Soccol CR (2008) Xylanase production by *Aspergillus niger* LPB 326 in solid-state fermentation using statistical experimental designs. Food Technol Biotechnol 46:183–189. http://www.ftb.com.hr/index.php/component/content/article/69-volume-46-issue-no-2/281-xylanase-production-by-aspergillus-niger-lpb-326-in-solid-state-fermentation-using-statistical-experimental-designs

Maheshwari R, Bharadwaj G, Bhat MK (2000) Thermophilic fungi: their physiology and enzymes. Microbiol Mol Biol Rev 64:461–488

Miller GL (1959) Use of dinitrosalicylic acid reagent for determination of reducing sugar. Anal Chem 31:426–428. doi:10.1021/ac60147a030

Murthy PS, Naidu MM (2012) Production and application of xylanase from *Penicillium* sp. utilizing coffee by-products. Food Bioprocess Technol 5:657–664. doi:10.1007/s11947-010-0331-7

Nair SG, Sindhu R, Shashidhar S (2008) Purification and biochemical characterization of two xylanases from *Aspergillus sydowii* SBS 45. Appl Biochem Biotechnol 149:229–243. doi:10.1007/s12010-007-8108-9

Naveen M, Siddalingeshwara KG (2015) Influence of metal source for the production of xylanase from *Penicillium citrinum*. Int J Curr Microbiol App Sci 4: 815–819. http://www.ijcmas.com

Pal A, Khanum F (2010) Production and extraction optimization of xylanase from *Aspergillus niger* DFR-5 through solid-state-fermentation. Bioresour Technol 101:7563–7569. doi:10.1016/j.biortech.2010.04.033

Pel HJ, de Winde JH, Archer DB, Dyer PS, Hofmann G, Schaap PJ, Turner G, de Vries RP, Albang R, Albermann K, Andersen MR (2007) Genome sequencing and analysis of the versatile cell factory *Aspergillus niger* CBS 513.88. Nat Biotechnol 25:221–231. doi:10.1038/nbt1282

Pervez S, Siddiqui NN, Ansari A, Aman A, Qader SAU (2015) Phenotypic and molecular characterization of *Aspergillus* species for the production of starch-saccharifying amyloglucosidase. Ann Microbiol 65:2287–2291. doi:10.1007/s13213

Polizeli ML, Rizzatti AC, Monti R, Terenzi HF, Jorge JA, Amorim DS (2005) Xylanases from fungi: properties and industrial applications. Appl Microbiol Biotechnol 67:577–591. doi:10.1007/s00253-005-1904-7

Qinnghe C, Xiaoyu Y, Tiangui N, Cheng J, Qiugang M (2004) The screening of culture condition and properties of xylanase by white-rot fungus *Pleurotus ostreatus*. Process Biochem 39:1561–1566. doi:10.1016/S0032-9592(03)00290-5

Salihu A, Bala SM, Olagunju A (2015) A statistical design approach for xylanase production by *Aspergillus niger* using soybean hulls: optimization and determining the synergistic effects of medium components on the enzyme production. Jordan J Biol Sci 8:319–323. doi:10.1016/S0032-9592(03)00290-5

Schuster E, Dunn-Coleman N, Frisvad JC, Van Dijck P (2002) On the safety of *Aspergillus niger*—a review. Appl Microbiol Biotechnol 59:426–435. doi:10.1007/s00253-002-1032-6

Senthilkumar SR, Ashokkumar B, Raj KC, Gunasekaran P (2005) Optimization of medium composition for alkali-stable xylanase production by *Aspergillus fischeri* Fxn 1 in solid-state fermentation using central composite rotary design. Bioresour Technol 96:1380–1386. doi:10.1016/j.biortech.2004.11.005

Shah AR, Madamwar D (2005) Xylanase production by a newly isolated *Aspergillus foetidus* strain and its characterization. Process Biochem 40:1763–1771. doi:10.1016/j.procbio.2004.06.041

Sharma S, Vaid S, Bajaj BJ (2015) Screening of thermo-alkali stable fungal xylanases for potential industrial applications. Curr Res Microbiol Biotechnol 3:536–541. doi:10.1016/j.procbio.2004.06.041

Subbulakshmi S, Iyer PR (2014) Production and purification of enzyme xylanase by *Aspergillus niger*. Int J Curr Microbiol App Sci 3: 664–668. http://www.ijcmas.com

Suresh PV, Chandrasekaran M (1999) Impact of process parameters on chitinase production by an alkalophilic marine *Beauveria bassiana* in solid state fermentation. Process Biochem 34:257–267. doi:10.1016/S0032-9592(98)00092-2

Teather RM, Wood PJ (1982) Use of Congo red-polysaccharide interactions in enumeration and characterization of cellulolytic bacteria from the bovine rumen. Appl Environ Microbiol 43:777–780

Yuan QP, Wang JD, Zhang H, Qian ZM (2005) Effect of temperature shift on production of xylanase by *Aspergillus niger*. Process Biochem 40:3255–3257. doi:10.1016/j.procbio.2005.03.020

10

Heterologous expression of an acidophilic multicopper oxidase in *Escherichia coli* and its applications in biorecovery of gold

Shih-I Tan[1], I-Son Ng[1,2*] and You-Jin Yu[1]

Abstract

Background: Copper oxidase is a promising enzyme for detection of oxidation, which can function as a biosensor and in bioremediation. Previous reports have revealed that the activity of the multicopper oxidase (MCO, EC 1.10.3.2) from the *Proteus hauseri* ZMd44 is induced by copper ions, and has evolved to participate in the mechanism of copper transfer.

Results: From *P. hauseri* ZMd44, a full-length, 1497-base-pair gene, *lacB*, encoding 499 amino acids without signal peptide, was cloned into *Escherichia coli* (*E. coli*) to obtain high amounts of MCO. The use of the pET28a vector yielded better enzyme activity, which was approximately 400 and 500 U/L for the whole cell and soluble enzyme extracts, respectively. The crude enzyme showed activity at an optimal temperature of 55 °C and it remained highly active in the range of 50–65 °C. The optimal pH was 2.2 but the activity was significantly inhibited by chloride ions. This MCO has great potential for Au adsorption (i.e., 38% w/w) and the Au@NPs were directly adsorbed on enzyme's surface.

Conclusion: An acidophilic MCO from bioelectricity generating bacterium, *P. hauseri*, is first cloned and heterologously expressed in *E. coli* with high amounts and activity. This MCO has great potential for Au adsorption and can be used as a biosensor or applied to bioremediation of electronic waste.

Keywords: Multicopper oxidase, *Proteus hauseri*, Recombinant protein, Copper effect, Au adsorption

Background

The enzyme multicopper oxidase (MCO) is a type of laccase (EC 1.10.3.2) that has important industrial applications owing to its oxidizing and degrading activities on a wide variety of aromatic compounds (Mayer and Staples 2002; Claus 2003; Rodgers et al. 2010). For instance, laccases are used in the process of paper-pulp bleaching to degrade lignin (Larsson et al. 2001). Furthermore, MCO, as one of laccases, has a high demand in different manufacturing processes such as wine production, medical analysis, electrochemical detection (Li et al. 2015), bioremediation (Santhanam et al. 2011), and gold nanoparticle preparation (Guo et al. 2015).

Proteus hauseri strain ZMd44 is a gram-negative bacterium with outstanding performance in biodecolorization of azo-dyes (Ng et al. 2013), and has been used in the microbial fuel cell (MFC) system (Chen et al. 2010, 2012). In addition, its MCO-laccase activity of 357 U/L at optimal cultivation condition was induced by copper (Zheng et al. 2013). Furthermore, this MCO-laccase participates in the transport of copper ions from the medium to the cells (Grass and Rensing 2001). By whole genome sequencing, it has been found that *P. hauseri* ZMd44 possesses LacA and LacB (Wang et al. 2014). However, genetic engineering of any laccase or multicopper oxidase from the *Proteus* genus has not been illustrated in the literature thus far. Owing to the insufficient enzyme production by naturally occurring microbes, heterologous expression of the laccase genes via *E. coli* cloning is urgently required.

*Correspondence: yswu@mail.ncku.edu.tw
[1] Department of Chemical Engineering, National Cheng Kung University, Tainan 70101, Taiwan
Full list of author information is available at the end of the article

Genetic heterologous expression may be influenced by different factors, including protein structure (Gopal and Kumar 2013), effect of bacterial strain (Moreira et al. 2014), and the use of different expression vectors (Rosano and Ceccarelli 2014). In general, the most common plasmid used in recombinant engineering of *E. coli* is the pET system, owing to its strong expression levels. In this study, three different vectors, pET22b, pET28a, and pET32a, are used to produce recombinant MCO. All the vectors contain the same T7 promoter, and the resulting proteins are fused to a 6-Histidine tag for protein affinity purification. While pET28a is a vector in which recombinant protein would be overexpressed intracellularly, the pET22b vector includes a signal peptide, *pel*B, so that the recombinant protein is secreted to the periplasam. Finally, vector pET32a contains a fusion protein, TrxA, which works as a chaperone to assist in protein folding and to enhance soluble protein expression (Baneyx and Mujacic 2004).

Over the past decades, the number of consumers using electronic devices manufactured by the semiconductor industry is rapidly increasing, resulting in the accumulation of a huge amount of electronic waste (Natarajan and Ting 2014, 2015). Routine recovery of gold (Au) from industrial metal waste is expensive. Further, biotechnology industries are taking eco-friendly approaches to recover metals from waste, which have significantly contributed to control the pollution in the environment. Previous studies showed *Proteus* spp. (Chen et al. 2010, 2012) and *Shewanella* spp. (Ng et al. 2014, 2015) with good performance in biodecolorization and metal absorption, but they never been used of genetic approach. Therefore, the use of genetically engineered MCO for bioremediation is a promising and cost-effective solution.

Until now, only few research studies have explored the heterologous expression of MCO in *E. coli*. This is the first attempt to determine the optimal vector to express

MCO. Additionally, the optimal reaction conditions, including temperature, pH, and ion effect, were assessed. Finally, the use of the recombinant MCO in Au adsorption and gold nanoparticle (Au@NPs) preparation, and its catalytic activity, were also explored.

Methods

Cloning and construction of recombinant LacB in *E. coli*

All cloning was performed in *E. coli* strain DH5α. Reagents used in polymerase chain reaction (PCR), including long and accurate LA-*Taq* DNA Polymerase, PCR-grade dNTPs, restriction enzymes, T4 DNA ligase, and DNA ladder marker, were obtained from Takara (Dalian, China). PCR products and restriction-digested DNA were purified by DNA gel extraction and PCR cleanup kits (Axygen). Genomic DNA was isolated from 5 mL of overnight cultures at 37 °C in LB broth, using bacterial genomic DNA miniprep Kit (Axygen). The entire open reading frames of *lacB* were amplified with primers shown in Table 1, using genomic DNA from the ZMd44 as DNA template. The resulting PCR products were inserted between the T7 promoter and the terminator in the vectors pET32a, pET28a, and pET22b, and introduced into *E. coli* BL21(DE3) cells, as shown in Fig. 1a–c. Recombinant colonies grown on LB plates with corresponding antibiotics (i.e., kanamycin or ampicillin) were verified by colony PCR and double enzymatic digestion with *Nco*I and *Xho*I. All strains, plasmids, and primers used in this study are shown in Table 1.

Culture and heterologous expression of recombinant MCO

The expression host BL21 (DE3) harboring the recombinant ZMd44-*lacB* vector was cultivated in LB medium with corresponding antibiotics and agitation (200 rpm) at 37 °C. Once the cultures reached a biomass with OD_{600} between 0.6 and 0.8, the isopropyl β-D-1-thiogalactopyranoside (IPTG) inducer and the

Table 1 List of strains, plasmids, and oligonucleotide primers in this study

Strains, plasmid, or primer	Description
E. coli strains	
DH5α	F^- *recA1 endA1 hsdR17(r_k^- m_k^+) supE44 thi-1 gyrA relA1*
BL21(DE3)	F^- *ompT hsdS$_B$ (r_B^- m_B^-) gal dcm* (DE3)
Plasmids	
pET-32a	Amp^r T7 promoter *trxA*-tag His-tag T7 terminator lacI f1 pBR322 origin *E. coli* expression vector
pET-28a	Kan^r T7 promoter His-tag T7 terminator lacI f1 pBR322 origin *E. coli* expression vector
pET-22b	Amp^r T7 promoter *pel*B His-tag T7 terminator lacI f1 pBR322 origin *E. coli* expression vector
Primers[a]	
ZMd-lacB-XhoI-r	5′-at<u>CTCGAG</u>TTTACTCACAGTAAAACCCG
ZMd-lacB-NcoI-f	5′-ta<u>CCATGG</u>ATCAAAGTAACACGCCTTC

[a] Nucleotides shown in underline represented the restriction sites

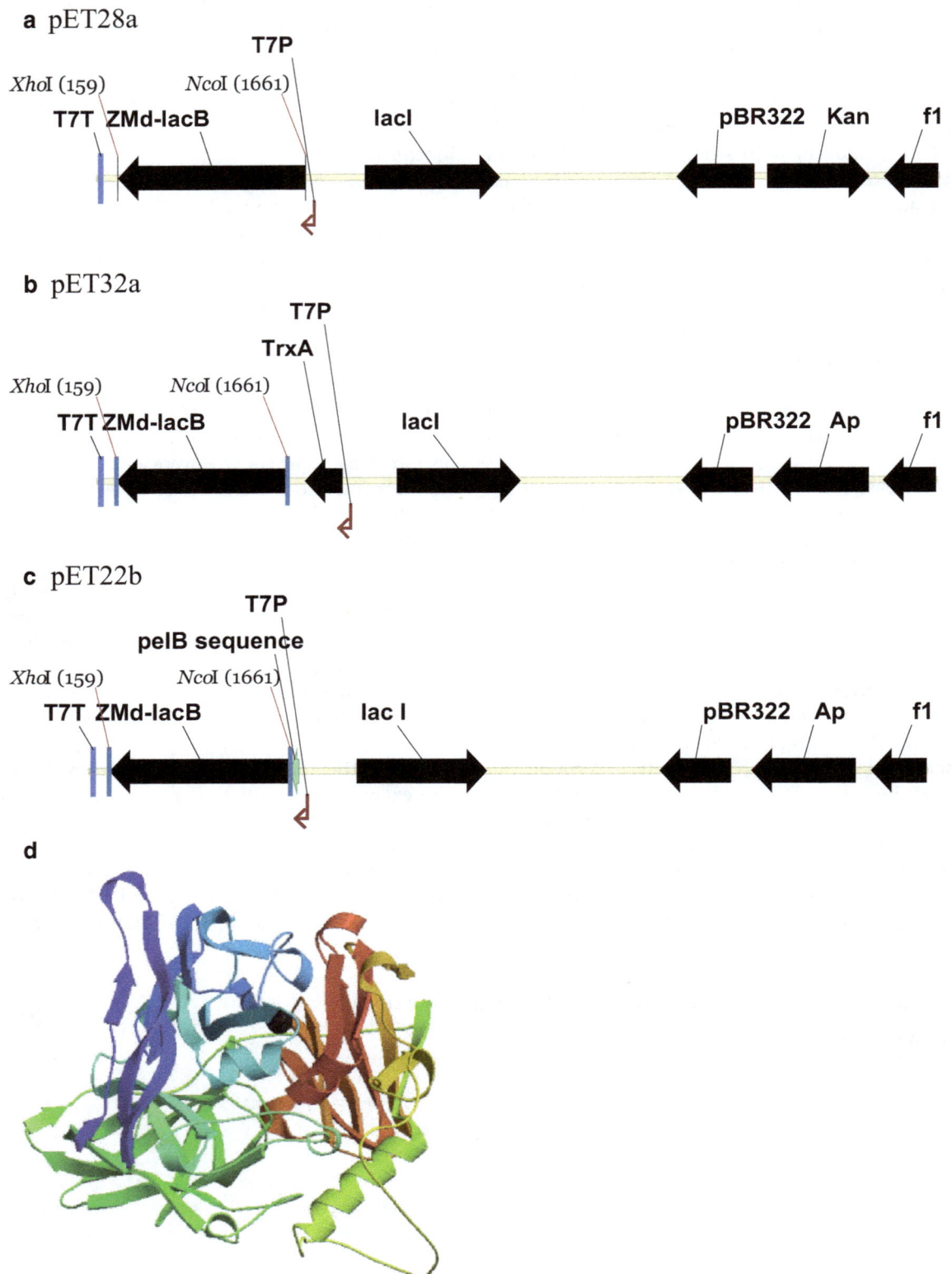

Fig. 1 Construction map for *lacB* in different vectors as **a** pET28a, **b** pET32a, and **c** pET22b and **d** molecular structure simulation of LacB via SWISS-MODEL

key factor $CuSO_4$ were added to the cultures at a final concentration of 0.1 mM and 0.5 mM, respectively. At this point, the incubation temperature was changed to 22 °C with the same agitation speed. After 12 h in culture, the recombinant bacteria were placed in an incubator at 22 °C without agitation for an additional 12 h. For cell density analysis, sample was taken out from the broth to measure the optical density by a spectrophotometer at a wavelength of 600 nm (VersaMaxTM microplate reader, Molecular Devices, CA). The OD_{600} values were converted to biomass in terms of g/L via a calibration between optical density and dry cell weight. All the experiments were run in triplicate and designated as (A) +IPTG and Cu, (B) +IPTG, and (C) without induction, taking into consideration different cell fractions (i.e., W = whole cell, S = supernatant, P = pellet, and M = periplasm).

Determination of MCO activity and protein concentration

MCO activity was determined by a spectrophotometric method based on the use of ABTS [2,2-azinobis(3-ethylbenzothiazoline-6-sulfonic acid)] as substrate. To this end, a 200 µL reaction mixture containing 100 µL of 2 mM ABTS, 95 µL of 50 mM reaction buffer (pH 3.0), and 5 µL of enzyme solution was prepared. Enzyme activity was monitored with a spectrophotometer (VersaMax™ microplate reader, Molecular Devices, CA) set up at a wavelength of 420 nm (i.e., OD_{420}) for the ABTS substrate. Reaction rates were calculated using a molar extinction coefficient of 36 mM/cm. One unit was defined as the amount of enzyme that oxidized 1 µmol of substrate per minute (min). Protein concentrations were determined by the Bradford method (Bradford 1976), using bovine serum albumin as the standard.

Protein expression determined by SDS-PAGE

Proteins were run in gels prepared with 0.1% SDS, using 12% separating gel and 4% stacking gel. Tris–glycine buffer (pH 8.3) containing 0.1% SDS was used as electrode buffer. Samples at same concentration in terms of OD = 5 were treated with buffer and heated at 100 °C for 5 min before loading onto the gel. Electrophoresis was run from the cathode to the anode at a constant current of 20 mA per slab at room temperature in a Biorad mini gel electrophoresis unit. Proteins were visualized by staining with Coomassie blue R-250. Stained SDS gels were scanned on the Image scanner Labscan 6.0 (GE Healthcare). Subsequently, band intensities were quantified by densitometry, using the Quantity One 4.6.2 analysis software (BioRad).

Biochemical characterization of LacB

LacB activity was analyzed using 2 mM ABTS in 50 mM sodium citrate buffer (pH 2.2), and the reaction was incubated at a temperature range of 37–65 °C. The crude enzyme of MCO was analyzed in 50 mM buffer at variable pHs. Different concentrations of $CuSO_4$, $CuCl_2$, and NaCl were included in the ABTS substrate at a range of 0–2 mM.

Purification of recombinant LacB

The supernatant of the recombinant LacB solution produced from the pET28a-lacB vector in BL21(DE3) cells was loaded onto the AKTA system (GE, USA), using a His-trap affinity column for purification. The purified enzymes were put into a micro-centrifuge tube at 30-kDa molecular weight cut-off (GE, Millipore), to remove the imidazole and concentrate the solution for Au adsorption.

Application of recombinant LacB on Au adsorption
Adsorption of free Au ions

Solutions containing 100 ppm of $AuHCl_4$ and 50 ppm of purified recombinant LacB were prepared. Two milliliters of both solutions were mixed and agitated for 1 h at 70 rpm. Following agitation, 4 mL of the solution was transferred into the Amicon® device equipped with a 3 kDa molecular cut-off filter and centrifuged for 25 min. The resulting filtrate was used to measure the Au concentration by inductively coupled plasma with atomic emission spectroscopy (ICP-AES) (Thomas 2001).

Adsorption of Au@NPs

A solution containing 1 mM $AuHCl_4$ and 38.8 mM sodium citrate was prepared as follows: 60 mL of $AuHCl_4$ solution was heated until boiling, followed by the addition of 6 mL of sodium citrate solution. Once the color of the solution changed from yellow to purple, the solution was left to cool at room temperature. This solution was the Au nanoparticle solution. Next, 300 µL of Au nanoparticle solution and 200 ppm of purified recombinant LacB were mixed together and agitated for 1 h at 70 rpm. Following agitation, 200 µL of the solution was loaded into a 96-well plate to monitor the optical density with a spectrophotometer set up at a wavelength of 350–750 nm (VersaMax™ microplate reader, Molecular Devices, CA).

Characterization of MCO-Au

Zeta potential (Malvern Zetasizer Nano ZS, UK) was used to measure the isoelectric points of MCO or MCO with Cu (MCO^{+Cu}) following Au adsorption. The zeta potential was measured in 100 mM phosphate buffer.

Each sample was analyzed at least by triplicate. MCO^{+Cu} and MCO^{+Cu} after Au adsorption were further analyzed for their kinetic parameters against ABTS. The ABTS concentrations ranged from 0.01 to 2 mM at pH 2.2. The Michaelis–Menten kinetics was assumed and fitted by Lineweaver–Burk plot, using Sigmaplot 10.0 software.

Nucleotide sequence accession number
The sequences of *lacB* of *P. hauseri* ZMd44 have been deposited in the GenBank database under accession number JF718783.

Results and discussion
Cloning Mco-lacB from P. hauseri ZMd44
The full-length 1578 bp gene sequence encoding LacB, which included a signal peptide of 27 amino acids, was obtained from genomic annotation of the biodecolorizing bacterium, *Proteus hauseri* Zmd44 (Ng et al. 2013; Chen et al. 2010). Although the native multicopper laccase activity has been reported in *P. hauseri* (Zheng et al. 2013) and *P. mirabilis* (Olukanni et al. 2010), the heterologous expression of such kind of enzyme in *E. coli* has never been explored. Therefore, we cloned and overexpressed *lacB* in the *E. coli* for the first time. LacB exhibited activity after expression in the pET22b vector but it is not satisfying. LacB was shown to be a simple monomer by SWISS-MODEL simulation (Fig. 1d); therefore, this study focussed on the heterologous expression of LacB in different vectors.

Mco-lacB expression in different vectors
The constructions of recombinant protein were confirmed by enzymatic digestion with the restriction enzymes *Nco*I and *Xho*I. The triplicate results from the growth curves shown in Fig. 2 reveal that the addition of IPTG and CuSO$_4$ into the cultures negatively affected cell growth. This could be due to two reasons: (i) the IPTG and copper ion may have a toxic impact on cell growth; (ii) the recombinant pET22b-lacB may have higher toxicity in the cells owing to the signal peptide, *pel*B, which is secreted to the periplasm and accumulates copper, leading to cell death. Next, protein expression patterns detected by SDS-PAGE were different from those of the selected vectors, as shown in Fig. 3a. For pET28a, most of the recombinant LacB was soluble and had an estimated yield of 1330 mg/L, as quantified by in-gel densitometry (Fig. 3a). However, LacB expression from the pET32a and pET22b vectors was found in the pellet (Fig. 3a). We speculate that the fusion protein TrxA expressed in the pET32a vector could have increased the load of the cells, resulting in the wrong folding of the MCO, which may become an inclusion body. On the other hand, in the pET22b vector, the recombinant MCO was expressed in the periplasm and

also may have become an inclusion body. The time course of pET22b-lacB biomass shown in Fig. 2c indicates considerable cell death at 24 h, whereas no cell lysis occurred in case of pET28a-lacB and pET32a-lacB. As shown in the activity analysis of Fig. 3b, the recombinant MCO expressed by pET28a had the highest activity among the three expression vectors. This activity was about 400 and 500 U/L at an optimal pH of 2.2 for whole cell and supernatant, respectively. This result was consistent with the color of the cell pellet obtained by centrifugation (Fig. 2d), where the chelation of copper ion represents a key event for MCO activity (Grass and Rensing 2001). In the case of the pET28a-lacB vector, copper ions can be transferred into the cell and combined with the MCO, generating a blue color (Fig. 2d) that translates into the highest MCO activity. Recently, a cell-free system to synthesize an active multicopper oxidase has been reported (Li et al. 2016). In such publication, the yield was almost 1.2 mg/mL, which is as similar to the heterologous expression in the pET system. The alternative method to improve heterologous protein in *E. coli* can be accomplished by shifting to lower temperature and applied experimental design (Wu et al. 2017). In our strategy based on vector's property, we found that the best protein production came from the pET28a vector, which included neither the original signal peptide nor the *pel*B signal peptide from pET22b or fusion TrxA from pET32a in *E. coli*.

MCO-lacB characterization
The pH effect on crude MCO is shown in Fig. 3b. Increasing the pH value caused the MCO activity to drop dramatically. The best pH value was 2.2, where MCO showed a 10-fold increase in activity when compared to its activity at pH 3.5. In the presence of ABTS, only single electron transfer takes place in the reaction instead of a proton transfer (Bertrand et al. 2002). Consequently, the optimal pH would be lower. The optimal temperature ranged from 55 to 65 °C, as shown in Fig. 4a. In addition, the recombinant MCO showed relatively high activity in the range of 50–65 °C. Compared to the native MCO of *P. hauseri* ZMd44 (Zheng et al. 2013), the recombinant MCO possessed a higher enzymatic activity and a wider thermo-resistance range. Therefore, the catalytic performance of MCO can be enhanced through heterologous expression, which increases its applicability in industry. Regarding the effect of the use of different ions (Fig. 4b), our results showed that the addition of copper ion (CuSO$_4$) at the optimal concentration of 0.5–1 mM could increase MCO activity. However, addition of the copper ion at concentrations higher than 1 mM resulted in an unexpected decrease of MCO activity. This suggests that the appropriate addition of the copper ion may help electrons to be transferred to the substrate, whereas an excess

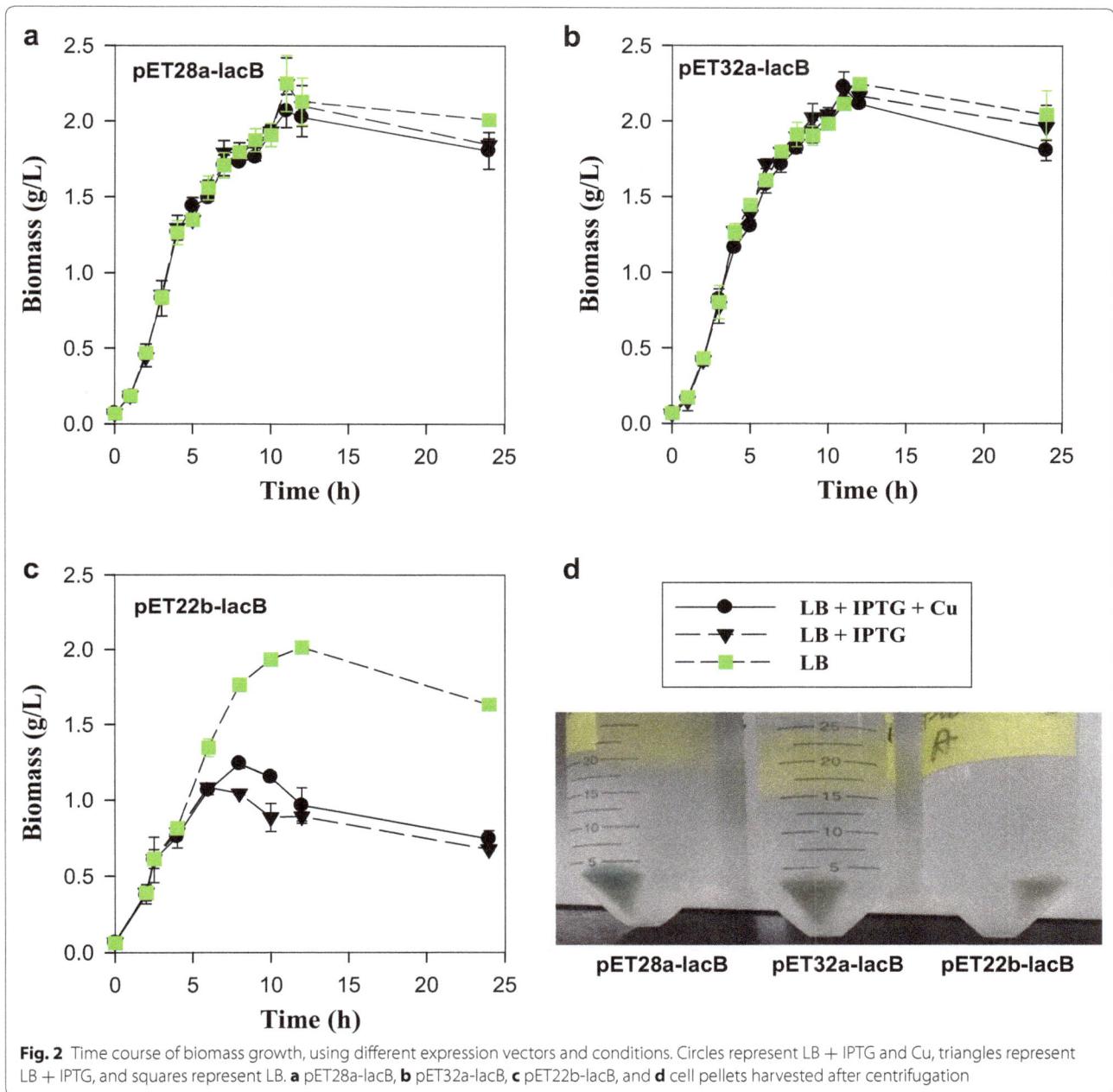

Fig. 2 Time course of biomass growth, using different expression vectors and conditions. Circles represent LB + IPTG and Cu, triangles represent LB + IPTG, and squares represent LB. **a** pET28a-lacB, **b** pET32a-lacB, **c** pET22b-lacB, and **d** cell pellets harvested after centrifugation

of copper ion may become an obstacle and may slow down the rate of electron transfer. Although the catalytic activity of MCOs can be enhanced by copper addition, the inhibition of such activity by chloride ion became more evident upon addition of copper (II) chloride and sodium chloride. As the chloride ion was added in the reaction system, the MCO activity dropped dramatically in spite of the presence of copper ion. Other groups have reported similar results in which MCO activity could be

strongly inhibited by the chloride ion (Naqui and Varfolomeev 1980; Kepp 2015).

Recently, Sondhi et al. reported that the strain *Bacillus tequilensis* SN4 had the highest catalytic activity of MCO (i.e., k_{cat} is 4020/min) at pH 5.5 and 85 °C (Sondhi et al. 2014). As shown in Table 2, the reaction conditions of MCO activity are very diverse among different microbes (Ye et al. 2010; Ausec et al. 2015; Yang et al. 2016; Safary et al. 2016). However, rare forms of MCO

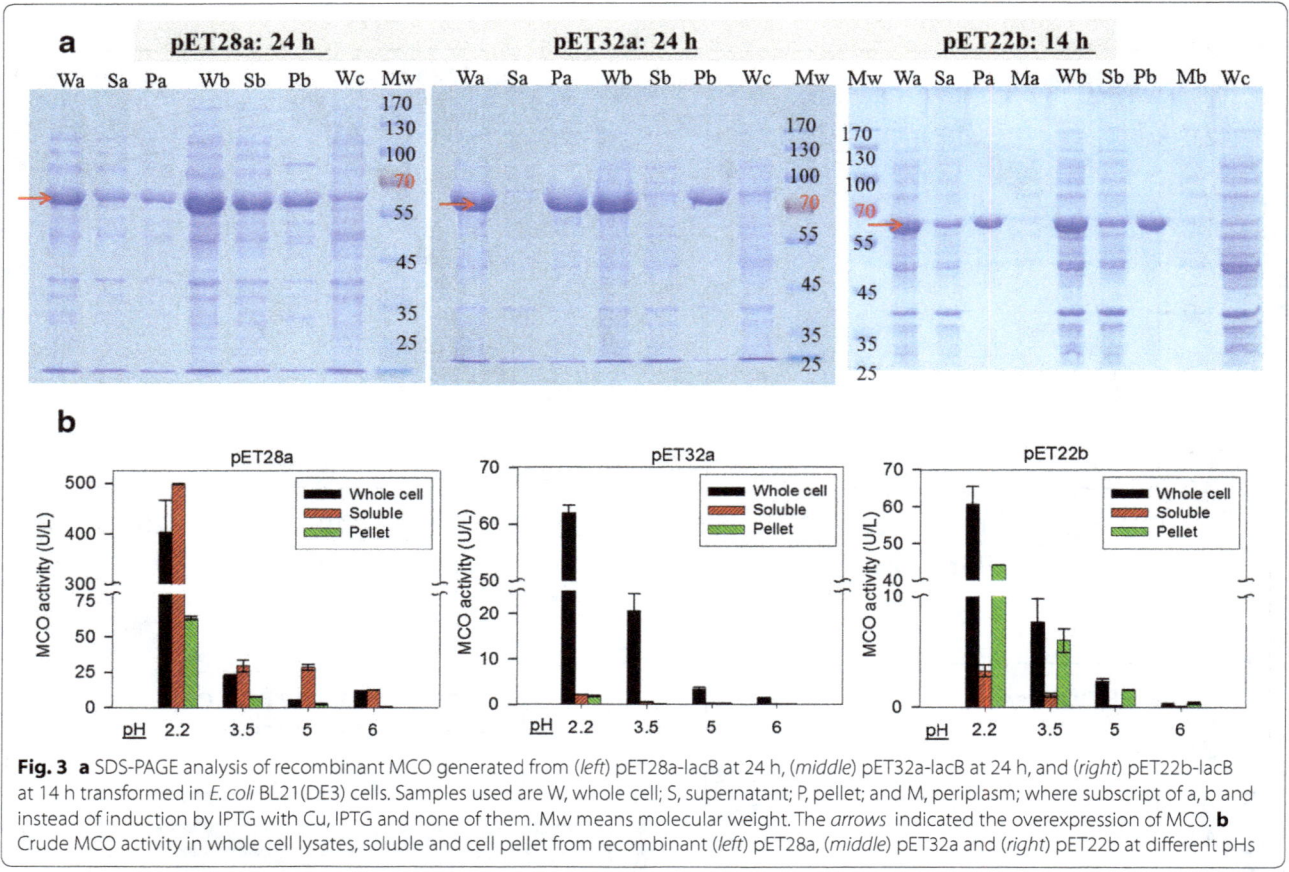

Fig. 3 **a** SDS-PAGE analysis of recombinant MCO generated from (*left*) pET28a-lacB at 24 h, (*middle*) pET32a-lacB at 24 h, and (*right*) pET22b-lacB at 14 h transformed in *E. coli* BL21(DE3) cells. Samples used are W, whole cell; S, supernatant; P, pellet; and M, periplasm; where subscript of a, b and instead of induction by IPTG with Cu, IPTG and none of them. Mw means molecular weight. The *arrows* indicated the overexpression of MCO. **b** Crude MCO activity in whole cell lysates, soluble and cell pellet from recombinant (*left*) pET28a, (*middle*) pET32a and (*right*) pET22b at different pHs

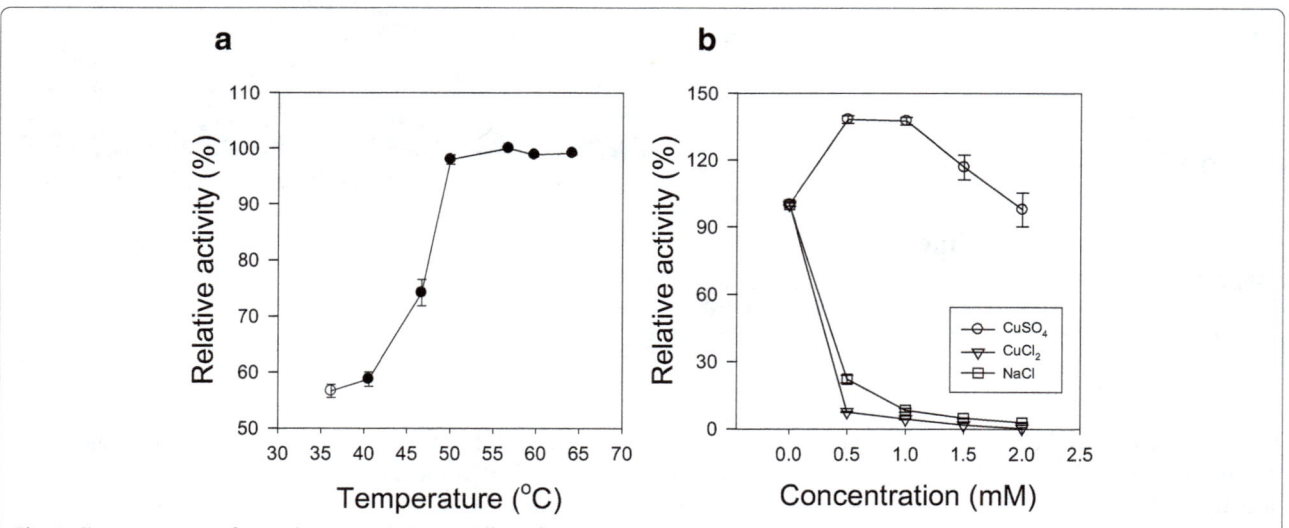

Fig. 4 Characterization of recombinant crude LacB. **a** Effect of incubation temperature on MCO activity. **b** Effect of metal ion addition on MCO activity

Table 2 Comparison of the catalytic kinetic parameters of MCO on ABTS from different microorganisms

Sources	Gene	V_{max} µmol/min	k_{cat} 1/min	K_m mM	k_{cat}/K_m 1/mM[a]/min	Reaction condition	Reference
Bacillus tequilensis SN4	SN4LAC	ND	4020	1.4	2871	pH 5.5 at 85 °C	Sondhi et al. (2014)
Mangrove soil	Lac591[a]	ND	2.52	0.09	28.0	pH 7.4 at 55 °C	Ye et al. (2010)
Thioalkalivibrio sp. ALRh	lacc	1500	129	4.6	28.0	pH 5 at 50 °C	Ausec et al. (2015)
Pseudomonas sp. *593*	cumA	ND	23.58	0.1	235.8	pH 5 at 55 °C	Yang et al. (2016)
Bacillus sp. *SL-1*	cotA	ND	1380	0.046	30,000	pH 4.5 at 50 °C	Safary et al. (2016)
Proteus hauseri	LacB[a]	58	1328	0.303	4382	pH 2.2 at 50 °C	This study
Proteus hauseri	LacB-Au[b]	5.59	52.7	0.129	408.5	pH 2.2 at 50 °C	This study

[a] The yields of LacB of *Proteus hauseri* and Lac591 of Mangrove soil expressed in *E. coli* are 1330 and 1381 mg/L, respectively

[b] LacB-Au means with gold binding and followed by activity analysis

show acidophilic properties and accordingly our results of *P. hauseri* MCO had optimal activity at pH 2.2. More importantly, this is the first attempt to detect the gold-binding properties of this MCO. The enzymatic activity was highly inhibited after addition of $AuHCl_4$ because of chloride inhibition. Finally, the Au ion was from $AuHCl_4$; thus, the V*max* of LacB is tenfold than that of LacB-Au (Table 2), which represented the chloride inhibition was dominating. This mechanism when gold binding to enzyme will be discussed in the next section.

Application and characterization of recombinant MCO-lacB for Au adsorption

It has been reported that Au nanoparticle could be adsorbed on MCO and function as a switch on ABTS reaction (Guo et al. 2015). We attempted to evaluate the ability of recombinant MCOs to adsorb Au following two approaches: free ion and nanoparticle. Purified recombinant MCO was used for Au ion adsorption experiments (Fig. 5a). This showed that 37 and 39% Au (mg) was adsorbed per milligram of MCO with (Sa) and without (Sb) copper ion in significant level (*P* value < 0.05). MCOs supplemented with copper ion possessed a very similar ratio of Au adsorption. This adsorption percentage is relatively higher than other protein for gold adsorption. Maruyama et al. applied three proteins to adsorb the gold ion, which showed that the adsorption percentage is 3.6, 4, and 1.6% for ovalbumin, BSA, and lysozyme, respectively (Tatsuo et al. 2007). Actually, in the Au biorecovery research, some reports revealed that a specific peptide bond on the gold surface (1,1,1) but not adsorbed of gold ions (Brown 1997, 2000). Alternatively, the adsorption of Au nanoparticles (Au@NPs) on MCO was similar in the presence or absence of copper ion (Fig. 5b). The Au@NPs were directly adsorbed on the MCO's surface, as the adsorption ability is totally determined by the surface property. As shown in Fig. 5b, the adsorption of Au@NPs showed a peak at a wavelength of

530 nm, and the optical density increased as the recombinant MCO was added to the culture. Adsorption of Au@NPs on the MCO could have caused a blockade of the light pathway, and as a result, there would be an increase in optical density.

The initial working conditions of MCO and MCO^{+Cu} (i.e., cultured with Cu) for adsorbing Au were determined at pH 7.75 and 7.73, respectively. However, the zeta potential of both samples following Au adsorption was determined at different pHs. These results are shown in Fig. 5c. The isoelectric points of MCO and MCO^{+Cu} were determined at pH 5.29 and pH 5.51, respectively. This showed that the recombinant proteins were anionic and the Au adsorption on the protein was through electrostatic interactions. The isoelectric points of MCO and MCO^{+Cu} upon Au adsorption were determined at pH 6.21 and 6.15, respectively. At this point, the proteins were still anionic, but the isoelectric point shifted to a higher pH, especially for MCO. We suggest that Au ions compete for the same binding site within the protein, affecting the chelation of the copper ion by the MCO and causing structural changes. Higher isoelectric points showed that, at the same pH, there were fewer negative charges of the protein after Au adsorption. This is consistent with the fact that Au possesses protons combined to the proteins, which partially neutralize the protein's negative charges.

Conclusion

We optimized the heterologously expression of MCO and investigated the extent of its applications. Among the three different vectors analyzed in this study (i.e., pET28a, pET22b, and pET32a), pET28a, carrying the lacB gene, showed the best results regarding expression levels, cell growth, and enzyme activity at different pHs. The optimal pH of the recombinant MCO was 2.2 at an incubation temperature of 55 °C with the addition of 0.5 mM of copper. We also found that the addition of chloride ion strongly inhibited MCO activity. This is

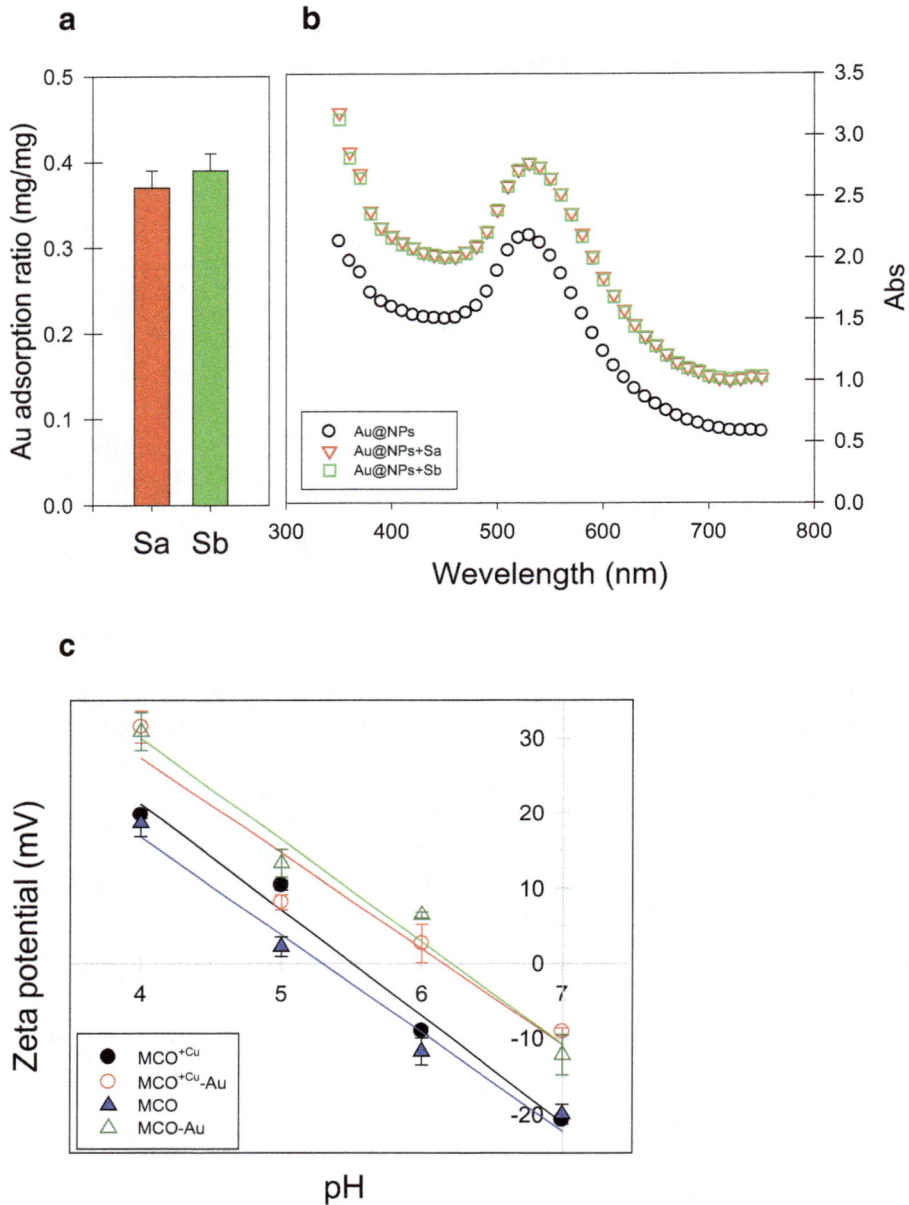

Fig. 5 Application of recombinant LacB in Au adsorption. **a** Supernatant fractions of MCO produced by BL21 cells harboring the pET28a-lacB vector under IPTG and copper (Sa), or IPTG only (Sb). **b** Spectrum scanning of free Au@NPs or Au@NPs adsorption by Sa and Sb. **c** Zeta potential analysis of MCO and MCO+Cu before and after adsorption of Au

the first attempt to explore the ability of MCO from *P. hauseri* to adsorb Au or Au@NPs, which may become a novel application for bioremediation in the future.

Authors' contributions
IS and SI designed the experiment and analyzed the data, SI performed most of experiments in genetic section, and YJ did the major part of gold adsorption. IS and SI wrote the manuscript. All authors read and approved the final manuscript.

Author details
[1] Department of Chemical Engineering, National Cheng Kung University, Tainan 70101, Taiwan. [2] Research Center for Energy Technology and Strategy, National Cheng Kung University, Tainan 70101, Taiwan.

Acknowledgements
Not applicable.

Competing interests
The authors declare that they have no competing interests.

Funding
This work was support by the Ministry of Science and Technology (MOST 105-2221-E-006-225-MY3 and MOST 105-2621-M-006-012-MY3) in Taiwan.

References

Ausec L, Črnigoj M, Šnajder M, Ulrih NP (2015) Characterization of a novel high-pH-tolerant laccase-like multicopper oxidase and its sequence diversity in *Thioalkalivibrio* sp. Appl microbiol biotechnol 99(23):9987–9999

Baneyx F, Mujacic M (2004) Recombinant protein folding and misfolding in *Escherichia coli*. Nat Biotechol 22(11):1399–1408

Bertrand T, Jolivalt C, Briozzo P, Caminade E, Joly N, Madzak C, Mougin C (2002) Crystal structure of a four-copper laccase complexed with an arylamine: insights into substrate recognition and correlation with kinetics. Biochemistry 41(23):7325–7333

Bradford MM (1976) A rapid and sensitive method for the quantitation of protein utilizing the principle of protein–dye binding. Anal Biochem 72(1–2):248–254

Brown S (1997) Metal-recognition by repeating polypeptides. Nat Biotechnol 15:269–272

Brown S, Sarikaya M, Johnson E (2000) A genetic analysis of crystal growth. J Mol Biol 299(3):725–735

Chen BY, Zhang MM, Chang CT, Ding Y, Lin KL, Chiou CS, Hsueh CC, Xu H (2010) Assessment upon azo dye decolorization and bioelectricity generation by *Proteus hauseri*. Bioresour Technol 101(12):4737–4741

Chen BY, Wang YM, Ng IS, Liu SQ, Hung JY (2012) Deciphering simultaneous bioelectricity generation and dye decolorization using *Proteus hauseri*. J Biosci Bioeng 113(4):502–507

Claus H (2003) Laccases and their occurrence in prokaryotes. Arch Microbiol 179(3):145–150

Gopal GJ, Kumar A (2013) Strategies for the production of recombinant protein in *Escherichia coli*. Protein J 32(6):419–425

Grass G, Rensing C (2001) CueO Is a multi-copper oxidase that confers copper tolerance in *Escherichia coli*. Biochem Biophys Res Commun 286(5):902–908

Guo S, Li H, Liu J, Yang Y, Kong W, Qiao S, Huang H, Liu Y, Kang Z (2015) Visible-light-induced effects of Au nanoparticle on laccase catalytic activity. ACS appl mater inter 7(37):20937–20944

Kepp KP (2015) Halide binding and inhibition of laccase copper clusters: the role of reorganization energy. Inorg Chem 54(2):476–483

Larsson S, Cassland P, Jonsson LJ (2001) Development of a *Saccharomyces cerevisiae* strain with enhanced resistance to phenolic fermentation inhibitors in lignocellulose hydrolysates by heterologous expression of laccase. Appl Environ Microb 67(3):1163–1170

Li D, Lv P, Zhu J, Lu Y, Chen C, Zhang X, Wei Q (2015) NiCu alloy nanoparticle-loaded carbon nanofibers for phenolic biosensor applications. Sensors 15(11):29419–29433

Li J, Lawton TJ, Kostecki JS, Nisthal A, Fang J, Mayo SL, Rosenzweig AC, Jewett MC (2016) Cell-free protein synthesis enables high yielding synthesis of an active multicopper oxidase. Biotechnol J 11(2):212–218

Mayer AM, Staples RC (2002) Laccase: new functions for an old enzyme. Phytochemistry 60(5):551–565

Moreira GM, Cunha CE, Salvarani FM, Goncalves LA, Pires PS, Conceicao FR, Labato FC (2014) Production of recombinant botulism antigens: a review of expression systems. Anaerobe 28:130–136

Naqui A, Varfolomeev SD (1980) Inhibition mechanism of Polyporus laccase by fluoride ion. FEBS Lett 113(2):157–160

Natarajan G, Ting YP (2014) Pretreatment of e-waste and mutation of alkali-tolerant cyanogenic bacteria promote gold biorecovery. Bioresour Technol 152:80–85

Natarajan G, Ting YP (2015) Gold biorecovery from e-waste: an improved strategy through spent medium leaching with pH modification. Chemosphere 136:232–238

Ng IS, Zheng X, Chen BY, Chi X, Lu Y, Chang C (2013) Proteomics approach to decipher novel genes and enzymes characterization of a bioelectricity-generating and dye-decolorizing bacterium *Proteus hauseri* ZMd44. Biotechnol Bioproc Eng 18:8–17

Ng IS, Chen T, Lin R, Zhang X, Ni C, Shun D (2014) Decolorization of textile azo dye and congo red by an isolated strain of the dissimilatory manganese-reducing bacterium *Shewanella xiamenensis* BC01. Appl Microbiol Biotechnol 98(5):2297–2308

Ng IS, Isaac NC, Zhou Y, Wu X (2015) Cultural optimization and metal effects of *Shewanella xiamenensis* BC01 growth and swarming motility. Bioresour Bioprocess 2:28–37

Olukanni OD, Osuntoki AA, Kalyani DC, Gbenle GO (2010) Decolorization and biodegradation of Reactive Blue 13 by *Proteus mirabilis* LAG. J Hazard Mater 184(1–3):290–298

Rodgers CJ, Blanford CF, Giddens SR, Skamnioti P, Armstrong FA, Gurr SJ (2010) Designer laccases: a vogue for high-potential fungal enzymes? Trends Biotechnol 28(2):63–72

Rosano GL, Ceccarelli EA (2014) Recombinant protein expression in *Escherichia coli*: advances and challenges. Front Microbiol 5:172

Safary A, Moniri R, Hamzeh-Mivehroud M, Dastmalchi S (2016) A strategy for soluble overexpression and biochemical characterization of halo-thermo-tolerant *Bacillus* laccase in modified *E. coli*. J Biotechnol 227(227):56–63

Santhanam N, Vivanco JM, Decker SR, Reardon KF (2011) Expression of industrially relevant laccases: prokaryotic style. Trends Biotechnol 29(10):480–489

Sondhi S, Sharma P, Saini S, Puri N, Gupta N (2014) Purification and characterization of an extracellular, thermo-alkali-stable, metal tolerant laccase from *Bacillus tequilensis* SN4. PLoS ONE 9:e96951

Tatsuo M, Hironari M, Yukiko S, Ichiro K, Misa H, Saori S, Noriho K, Masahiro G (2007) Proteins and protein-rich biomass as environmentally friendly adsorbents selective for precious metal ions. Environ Sci Technol 41(4):1359–1364

Thomas R (2001) A beginner's guide to ICP-MS. Spectroscopy 16:38–42

Wang N, Ng IS, Chen PT, Li Y, Chen YC, Chen BY, Lu Y (2014) Draft genome sequence of the bioelectricity-generating and dye-decolorizing bacterium *Proteus hauseri* strain ZMd44. Genome Announc 2(1):e00992–e009913

Wu Z, Hong H, Zhao X, Wang X (2017) Efficient expression of sortase A from *Staphylococcus aureus* in *Escherichia coli* and its enzymatic characterization. Bioresour Bioprocess 4:13–22

Yang S, Long Y, Yan H, Cai H, Li Y, Wang X (2016) Gene cloning, identification and characterization of the multicopper oxidase CumA from *Pseudomonas* sp. 593. Biotechnol Appl Bioc. doi:10.1002/bab.1501

Ye M, Li G, Liang WQ, Liu YH (2010) Molecular cloning and characterization of a novel metagenome-derived multicopper oxidase with alkaline laccase activity and highly soluble expression. Appl Microbiol Biot 87(3):1023–1031

Zheng X, Ng IS, Ye C, Chen BY, Lu Y (2013) Copper ion-stimulated McoA-laccase production and enzyme characterization in *Proteus hauseri* ZMd44. J Biosci Bioeng 115(4):388–393

11

Detoxification of parthenium (*Parthenium hysterophorus*) and its metamorphosis into an organic fertilizer and biopesticide

Naseer Hussain[1], Tasneem Abbasi[1,2] and Shahid Abbas Abbasi[1*]

Abstract

Background: Vermicompost of the toxic and allelopathic weed parthenium (*Parthenium hysterophorus*) was explored for its possible use as an organic fertilizer. Replicated plant growth trials were conducted using four levels of parthenium vermicompost (0, 2.5, 3.75, and 5 t/ha) to assess their effects on the germination, growth, and fruition of a typical food plant ladies finger (*Abelmoschus esculentus*). Additionally the role of vermicompost in reducing plant pests and disease was evaluated.

Results: Vermicompost encouraged the germination and growth of ladies finger at all levels of vermicompost application, with best results obtained in 5 t/ha treatments. The positive impact extended up to the fruit yield. Vermicompost application also improved the quality of fruits in terms of mineral, protein, and carbohydrate contents, and reduced the disease incidence and pest attacks.

Conclusions: The studies establish the fact that parthenium acquires all the qualities of a good organic fertilizer with concomitant loss of its toxic and allelopathic properties after it gets vermicomposted. The findings raise the prospects of economical and eco-friendly utilization of billions of tons of parthenium biomass which is generated annually but goes to waste at present.

Keywords: Invasive weed, Allelopathy, Organic fertilizer, Biopesticide, Vermicomposting

Background

Parthenium hysterophorus, commonly called as congress grass, is among the world's seven most noxious and devastating weeds (Patel 2011). It is an annual flowering, erect, and severally branched ubiquitous herb, which grows aggressively in a wide range of habitats (Akter and Zuberi 2009). Due to the absence of any effective natural enemies and due to its allelopathic nature, large seed bank, and fast growth rate, parthenium grows luxuriantly all through the year, infesting millions of hectares of land masses including agricultural fields, parks, orchards, railway tracks, and other open areas (Wiesner et al. 2007; Nigatu et al. 2010; Qureshi

et al. 2014). This proves disastrous in terms of monopolizing of space and nutrients by parthenium at the expense of other vegetation, consequent loss of biodiversity, and associated ecological imbalances (Hussain et al. 2016a, b). Parthenium's dominance over other vegetation is fostered by the presence of allelopathic compounds in parthenium, especially parthenin, hysterin, ambrosin, and flavonoids (Maishi et al. 1998; Khan and Abbasi 1998; Knox et al. 2011; Patel 2011; Kaur et al. 2014). These compounds are leached when dew or rain falls on parthenium (Abbasi and Abbasi 2011), and reach the underlying soil. There they cause toxicity, discouraging the growth of other vegetation in the vicinity of parthenium, thereby aiding and abetting the spread of parthenium monocultures (Hussain et al. 2016a, b).

Parthenium has mammalian toxicity as well. It causes dermatitis, eczema, asthma, allergic rhinitis, hay fever,

*Correspondence: abbasi.cpee@gmail.com; prof.s.a.abbasi@gmail.com
[1] Centre for Pollution Control & Environmental Engineering, Pondicherry University, Chinakalapet, Puducherry 605 014, India
Full list of author information is available at the end of the article

black spots, burning, and blisters around eyes in mammals, including humans (Gunaseelan 1987; Towers and Rao 1992; Maishi et al. 1998; Morin et al. 2009; Akhtar et al. 2010). Exposure to parthenium also causes systemic toxicity including loss of skin pigmentation, dermatitis, diarrhea, and degenerative changes in liver and kidneys in livestock who accidentally graze upon parthenium (Gunaseelan 1987; Rajkumar et al. 1988; Lakshmi and Srinivas 2007).

The eradication of parthenium is a major challenge, primarily because of its epidemic proliferation, strong reproductive potential, hardiness, and competitiveness, apart from its wide ecological adoptability. Efforts made across the world to find persistent methods of controlling parthenium by way of mechanical, chemical, or biological means, have at best achieved only partial and temporary success (Manoj 2014; Hussain et al. 2016a). The weed has never been eradicated from any country and its spread in all the tropical and sub-tropical regions of the world is only increasing with time. When viewed as a resource, for the generation of green manure, biogas production, biopesticides, and drugs (Abbasi et al. 1990; Kishor et al. 2010; Kumar et al. 2012; Gunaseelan 1987; Patel 2011; Tauseef et al. 2013; Singh and Garg 2014; Kumar et al. 2014; Anwar et al. 2015; Hussain et al. 2016b), parthenium has proved uneconomical, expensive, and unsustainable. Thus finding an ecologically sound and economically viable means by which parthenium can be gainfully utilized in large quantities appears to be the only recourse which can make it profitable to regularly harvest the weed, thereby keeping it under some control.

One such option is conversion of parthenium biomass into organic fertilizer through vermicomposting. When a substrate is vermicomposted, it converts the latter into fine peat-like material and transforms some of its nutrients into more bioavailable forms (Hussain et al. 2015, 2016a, b). The vermicompost acquires several species of microflora, besides hormones and enzymes as it passes though the earthworms' gut (Pramanik et al. 2007; Ievinsh 2011). Past studies have reported that vermicompost derived from animal manure stimulated seed germination (Atiyeh et al. 2000; Zaller 2007; Lazcano et al. 2010), enhanced plant growth (Edwards Clive 2004; Lazcano et al. 2009; Samrot et al. 2015), and the yield and the quality of fruits (Singh et al. 2008; Doan et al. 2015) of several plant species. They are also believed to induce resistance in plants against pests and disease (Yardim et al. 2006; Edwards et al. 2010; Serfoji et al. 2010; Carr and Nelson 2014). But it is not yet established whether vermicompost derived from plants, more so from toxic and allelopathic plants like parthenium, can be as benign and effective an organic fertilizer as manure-derived vermicomposts are. In our recent studies (Gajalakshmi and Abbasi 2002; Hussain et al. 2016a, b) it was seen that soil augmented with parthenium vermicompost had enabled better germination success of four

species of food plants compared to the control soil. But will the beneficial effect extend to plant growth, fruit yield, and quality of the fruit? Will the parthenium vermicompost also help the fertilized plants to repel pathogens the way manure-derived vermicomposts are known to do? In order to find definite answers to these questions, the present field-scale study has been carried out on the effect of parthenium vermicompost on a common vegetable ladies finger (*Abelmoschus esculentus*) from germination stage right up to the quality and yield of fruit.

Experimental

The experiment was carried out at the Pondicherry University, Puducherry, India, which is located along the eastern coast of the South India (11° 56′ N, 79° 53′ E). This region experiences hot summers, with maximum day temperature 35–38 °C, during March–July and mild winters during December–February (maximum day temperature 29–32 °C). The average annual rainfall is about 1300 mm, concentrated mainly during October–December but with a few rainy days occurring in July–August and January as well. The study was conducted during February–May which is ideal for growing ladies finger in the study area (ICAR 2011). The vermicompost used in the experiment was produced from parthenium leaves, which were collected from the vicinity of Pondicherry University campus. The leaves were washed to remove the adhering soil and subjected to the earthworm species *Eisenia foetida* in pulse-fed, high-rate vermireactors, as detailed by Nayeem-Shah (2014), and Nayeem-Shah et al. (2015). There was no pre-composting or any manure supplementation. The vermicast was periodically harvested in each pulse and this precisely quantifiable product of the earthworm action was deemed as vermicompost (Abbasi et al. 2009, 2015). The characteristics of the vermicompost are detailed earlier by Hussain et al. (2016a, b).

To study the effect of parthenium vermicompost on several stages of plant growth, an outdoor experiment was conducted using low-density polyethylene (LDPE) bags of 50-l capacity as containers of the soil. In separate treatments, vermicompost was supplemented in bags to the extent of 0, 2.5, 3.75, and 5 t/ha. For each treatment 35 bags were set and in each bag 5 ladies finger (*Abelmoschus esculentus*) seeds were sown. Soil used in the study was not used for cultivation in the past and had not received any anthropogenic input of fertilizers. Germination success was assessed up to 8 days and has been presented as germination percentage. After recording the germination success, seedlings in each bag were thinned to single while discarding the other four.

The growth experiment with daily monitoring was continued for 15 weeks, during which all the bags were irrigated with tap water. After 100 days of growth, five plants from each treatment were randomly harvested for the determination of shoot length, root length, plant biomass, number of leaves, stem diameter, and number of branches. The harvested plants were washed with tap water to remove the soil adhering to their roots, wiped, and weighed. They were then oven dried at 105 °C to a constant weight, to calculate their dry weight. Flowering was assessed in terms of number of days to the appearance of the first flower, and the total flowers emerging per plant. The fruits (pods) were harvested at each alternative days and the yield was assessed in terms of number and weight of pods harvested per plant. Further the average length (cm) and diameter (mm) of the pods were also recorded. The chlorophyll and carotenoid contents of the leaves were estimated by following the procedure of Moran and Porath (1980) using N, N-dimethyl formamide (DMF) as an extractant. The optical density of the extract was read at 470, 647, and 664 nm in a UV–Visible spectrophotometer, and the concentration of pigments was determined as detailed by Wellburn (1994). The fruits (pods) of the ladies finger were analyzed for protein, carbohydrate, and mineral content by Kjeldahl, Anthrone, and dry ashing methods, respectively (Nielsen 2010). The total solid content was determined by heating the pods at 105 °C to a constant weight.

In the course of the experiment the plants were infested with leaf miners and leaf spot disease. The leaf miner infection, traced to *Liriomyza* spp. was seen in the symptoms of feeding punctures and leaf mines appearing as white speckles on the upper leaf surface (Ahmed 2000). Plants were considered infected by the fungus, *Alternaria alternate* when there were light brown spots on leaves, which later turned into concentric dark brown spots (Cho and Moon 1980; Werner 1987; Tohyama et al. 2005; Arain et al. 2012). When the intensity of the infection was particularly severe, the infected leaves become brown, eventually dying and falling off (Canihos et al. 1999; Amenduini et al. 2003; Antonijevic et al. 2007). There was also borer infestation in the fruits due to the *Earias vittella* (Sharma et al. 2010; Halder et al. 2015); it was quantified as weight percentage of the infected fruits to the total weight of fruits per treatment.

The data were statistically analyzed for assessing the extent of significance in the observed variations—especially by one-way analysis of variance and least significant difference (LSD)—as per standardized protocols (Alan and David 2001; Field 2009).

Results and discussion

The substitution of the soil with parthenium vermicompost enabled significantly greater germination success of ladies finger seeds in comparison to controls (Table 1). In comparison with the controls (62.29%), a germination success of 85.71% was achieved with vermicompost treatment 5 t/ha followed by 81.71 and 77.14% with 3.75 and 2.5 t/ha, respectively. The increase in seed germination in vermicompost-amended soils may be due to the increased concentrations of nitrate and ammonium in them, relative to the control soils. It is now beyond the dispute that nitrate and ammonium are efficient breakers of seed dormancy, facilitating germination (Bewley and Black 1982; Hilhorst and Karssen 2000). In recent studies, Hussain et al. 2016a reported that parthenium vermicompost significantly enhanced the relative concentrations of nitrate and ammonium in soil compared with the controls.

There was also better growth of ladies finger in terms of all the variables studied compared with the controls (Table 1). An increase in growth was observed with increasing concentration of vermicompost in the soil, the trend being control <2.5 < 3.75 < 5 t/ha. The growth of ladies finger had shown significant enhancement even with a relatively small concentration of parthenium vermicompost (2.5 t/ha) in the container medium. Maximum shoot length 122.6 ± 10.21 cm, root length 53.0 ± 5.24 cm, shoot diameter 13.38 ± 1.30 mm, shoot dry weight 30.67 ± 3.06 g, root dry weight 9.38 ± 0.93 g, number of leaves 30.6 ± 2.97, and number of branches 5.2 ± 0.84 were recorded in plants grown in soil amended with 5 t/ha VC treatments. Parthenium vermicompost also induced early flowering and significantly higher number of flowers in ladies finger plants, in comparison to controls. The yield, in terms of number and weight of pods per plant, and the length and diameter of the pods, was also significantly higher in VC treatments than the controls (Fig. 1). Past studies have demonstrated that vermicompost derived from animal manure increased the growth and yield of several plant species (Doan et al. 2013; Joshi et al. 2015; Ayyobi et al. 2014; Xu et al. 2014; Akhzari et al. 2015; Kumar et al. 2015; Saxena et al. 2015).

Vermicompost derived from different substrates especially from animal manures are known to contain all the necessary plant nutrients in more bioavailable form than is present in the parent substrate (Edwards et al. 2011). It also contains diverse microflora, which are beneficial for soil health and the plant growth. A number of studies also reported the presence of plant growth regulators especially humic and fulvic acids, and phytohormones in manure-based vermicompost (Muscolo et al. 1999; Atiyeh et al. 2000, 2001, 2002; Arancon et al. 2005, 2008; Ievinsh 2011). It is the combined action of bioavailable nutrients, plant growth regulators, and soil microflora in the vermicompost that is responsible for enhancing the plant growth and yield (Chan and Griffiths 1988; Edwards and Burrows 1988; Wilson and Carlile 1989; Atiyeh et al. 1999; Ayyobi et al. 2014; Xu

Table 1 Seed germination, plant growth, flowering, and disease incidence in ladies finger plants grown in soil fortified with different concentrations of parthenium vermicompost

Parameters, average value	Vermicompost concentration				F value
	0 t/ha (control)	2.5 t/ha	3.75 t/ha	5 t/ha	
Germination					
Germination percentage	62.29 ± 6.46^a	77.14 ± 7.10^b	81.71 ± 7.47^c	85.71 ± 9.17^d	63.253*
Growth					
Shoot length (cm)	23.6 ± 2.97^a	86.0 ± 6.86^b	98.8 ± 7.98^c	122.6 ± 10.21^d	159.515*
Root length (cm)	34.4 ± 3.21^a	41.4 ± 3.85^{bc}	46.0 ± 4.58^c	53.0 ± 5.24^d	16.627*
Shoot diameter (mm)	4.48 ± 0.38^a	10.22 ± 1.00^{bc}	10.78 ± 0.89^c	13.38 ± 1.30^d	77.381*
Shoot dry weight (g)	1.96 ± 0.31^a	17.26 ± 0.83^b	26.00 ± 2.48^c	30.67 ± 3.06^d	195.511*
Root dry weight (g)	1.39 ± 0.13^a	4.94 ± 0.47^b	6.12 ± 0.59^c	9.38 ± 0.93^d	149.649*
No of leaves	11.8 ± 1.10^a	24.8 ± 2.49^b	28.4 ± 2.41^{cd}	30.6 ± 2.97^d	64.352*
No of branches	0.0 ± 0.00^a	2.4 ± 0.55^b	4.4 ± 0.89^{cd}	5.2 ± 0.84^d	59.852*
Flowering					
Days to first flowering	52.70 ± 4.85^a	41.00 ± 2.91^{bc}	39.10 ± 3.31^{cd}	36.80 ± 1.93^d	42.882*
Number of flowers per plant	2.90 ± 0.32^a	7.80 ± 1.14^b	11.80 ± 1.14^c	12.90 ± 0.99^d	224.036*
Diseases incidence					
Diseases incidence percentage	21.43 ± 4.95^a	13.57 ± 4.29^{bcd}	12.14 ± 4.88^{cd}	12.14 ± 6.34^d	$2.976^{n.s}$

Results which do not differ significantly (LSD test; $p < 0.05$) carry at least one character in the superscript which is common

n.s not significant

* $p < 0.05$

et al. 2014; Akhzari et al. 2015; Kumar et al. 2015; Saxena et al. 2015). In the present study we suggest that parthenium vermicompost may have also imbibed with similar attributes to that of the manure-based vermicomposts that has resulted in greater germination success, better plant growth, and yield of ladies finger plants. Recently (Hussain et al. 2016a, b) have reported that parthenium vermicompost contains a number of fatty acids, alcohols, alkanes, alkenes, and nitrogenous compounds in it and enhanced the microbial biomass carbon of the soil. Beside these factors, parthenium vermicompost, as like manure-based vermicomposts, is also known to induce positive impact on soil physical properties and hence may also have contributed to the better plant growth (Hussain et al. 2016a).

The levels of pigments in the leaves of the ladies finger plants were significantly influenced by the vermicompost application (Table 2). Maximum chlorophyll (1.43 ± 0.09 mg/g) and carotenoid (0.90 ± 0.07 mg/g) content was recorded in 5 t/ha vermicompost treatments. The total solids and ash (mineral) content of vermicompost-treated plants were also significantly higher compared to the controls. An increase in the protein and carbohydrate concentrations were also recorded in the plants grown in vermicompost-amended soils (Fig. 2). All these gains are perhaps due to the greater bioavailability of nutrients in vermicompost treatments compared to the controls, as has been earlier seen with manure-based vermicompost (Abduli et al. 2013; Ayyobi et al. 2014; Akhzari et al. 2015; Yadav et al. 2015).

Table 2 Chlorophyll and carotenoid content of ladies finger plants grown in soil fortified with different concentrations of parthenium vermicompost

Parameters, average value	Vermicompost concentration				F value
	0 t/ha (control)	2.5 t/ha	3.75 t/ha	5 t/ha	
Chlorophyll 'a' (mg/g)	0.47 ± 0.04^a	0.62 ± 0.04^{bc}	0.63 ± 0.04^c	0.91 ± 0.07^d	68.604*
Chlorophyll 'b' (mg/g)	0.25 ± 0.02^{ac}	0.29 ± 0.02^{bc}	0.26 ± 0.01^c	0.52 ± 0.04^d	120.380*
Total Chlorophyll (mg/g)	0.72 ± 0.04^a	0.90 ± 0.03^{bc}	0.89 ± 0.04^c	1.43 ± 0.09^d	149.604*
Carotenoid (mg/g)	0.30 ± 0.03^a	0.43 ± 0.04^{bc}	0.40 ± 0.04^c	0.90 ± 0.07^d	154.545*

Results which do not differ significantly (LSD test; $p < 0.05$) carry at least one character in the superscript which is common

* $p < 0$

Parthenium vermicompost was effective in inducing resistance in the ladies finger plants against pests and pathogens. A significant reduction in the infestation of leaf miners, leaf spot disease, and fruit borers was observed (Table 1; Fig. 1E), the trend being 0 t/ha (control) >2.5 > 3.75 > 5 t/ha. At the early stages of growth the

seedlings were seen severely infested by leaf liner; however, as the growth increased the number of incidents decreased in proportion. Previous studies have reported that plants grown in soil amended with manure-based vermicompost have shown a reduction in the pest and disease attack (Edwards et al. 2010; Cardoza and Buhler

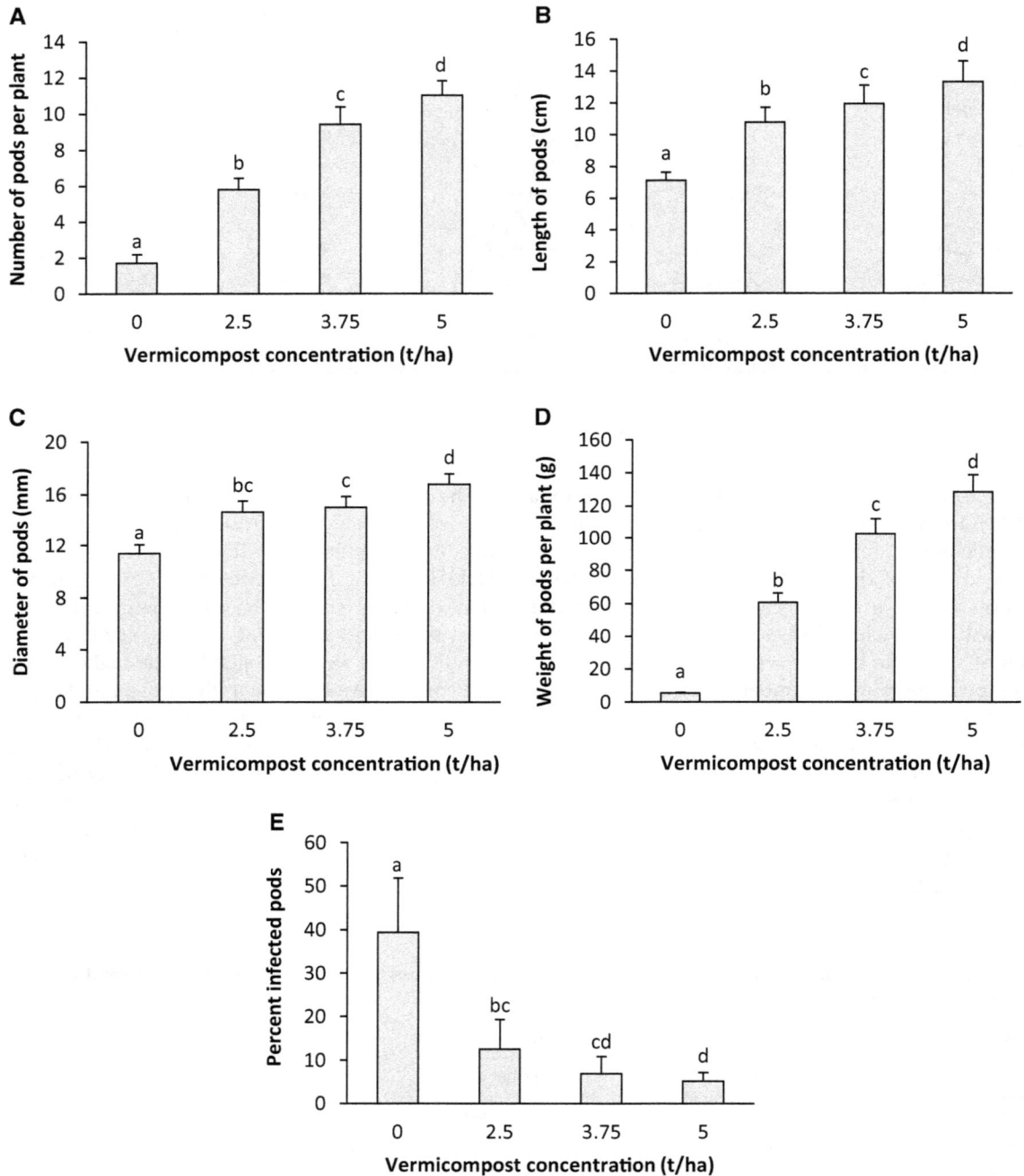

Fig. 1 Effect of parthenium vermicompost on **A** number of pods, **B** length of pods, **C** diameter of pods, **D** weight of pods per plant, and **E** weight of infected pods of ladies finger. The standard deviation is indicated on the chart. Results which do not differ significantly (LSD test; $p < 0.05$) carry at least one character in the superscript which is common

2012). Different authors have provided different explanations for the pesticidal properties of the vermicompost which basically revolve around two conjectures: better nutrient availability hence greater vitality in warding off infection, and presence of pathogen-destroying microorganisms (Arancon et al. 2005; 2008; Yardim et al. 2006; Cardoza 2011; Singh et al. 2013; Xiao et al. 2016). The present studies indicate that the vermicompost of parthenium is also imbibed with a similar attribute. Past studies on manure-based vermicompost have indicated that better nutrient availability and presence of antimicrobial compounds such as flavonoids, phenolics, and humic acids in the vermicompost may have induced the resistance to pathogens in the plants (Graham and Webb 1991; Hill et al. 1999; Haviola et al. 2007; Sahni et al. 2008; Edwards et al. 2010). Similarly beneficial attributes seem to be present in parthenium's vermicompost as well.

Conclusion

In a field study, effect of vermicompost produced solely from an allelopathic weed parthenium has been investigated on germination, growth, yield, and quality of ladies finger (*Abelmoschus esculentus*). The effect of the vermicompost in inducing resistance in ladies finger against disease was also assessed. In general vermicompost application increased germination success, plant growth, and yield—the positive effect increased in prominence as the extent of vermicompost application was enhanced from 2.5 to 5 t/ha. Parthenium vermicompost also induced beneficial changes in the biochemical and mineral content of the ladies finger. Additionally, ipomoea vermicompost induced resistance in ladies finger towards disease and pest attacks. Overall, contrary to the toxic and allelopathic nature of parthenium, its vermicompost manifests the attributes of highly plant-friendly organic fertilizer that vermicomposts derived from animal manure are known to possess.

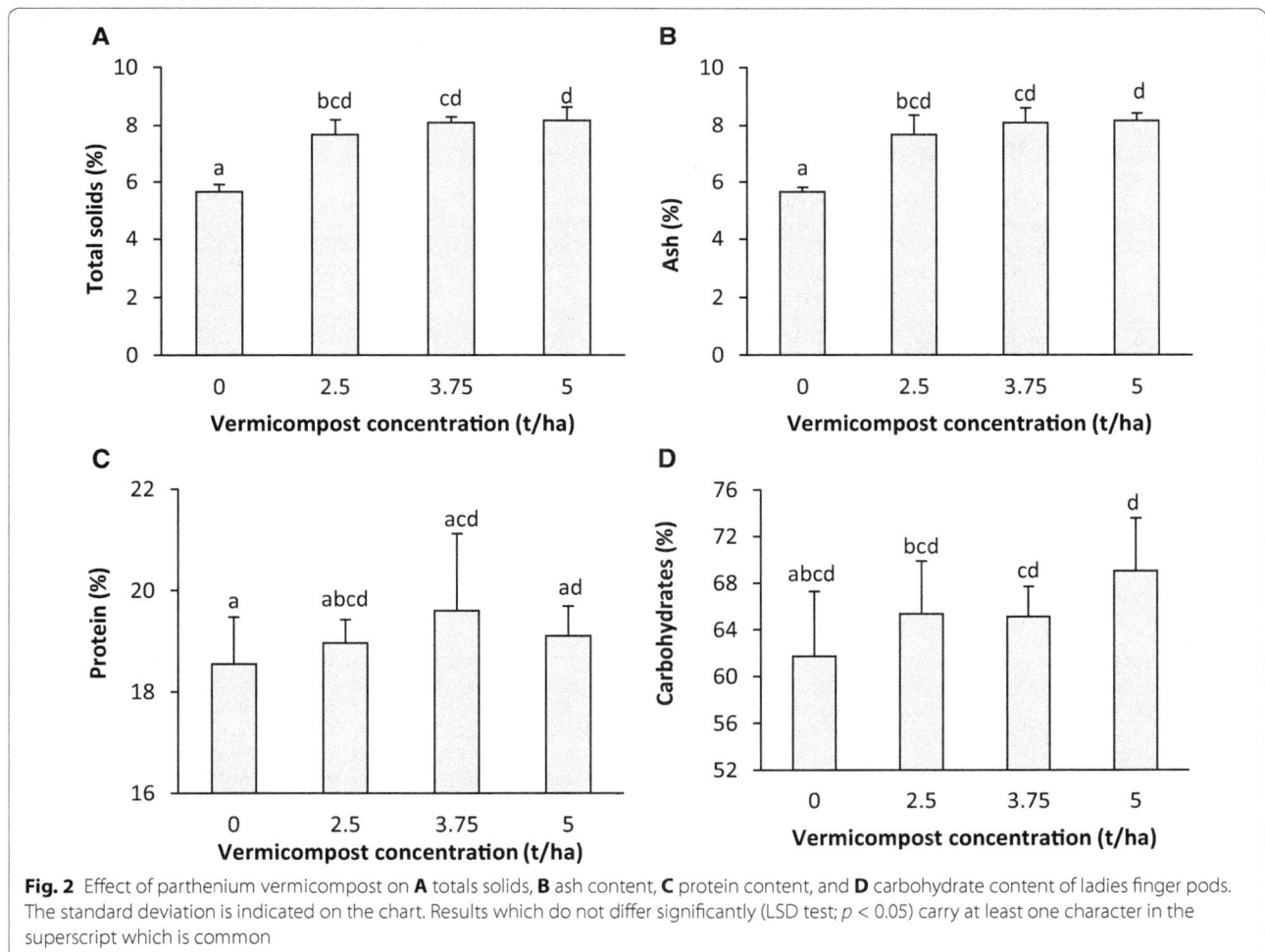

Fig. 2 Effect of parthenium vermicompost on **A** totals solids, **B** ash content, **C** protein content, and **D** carbohydrate content of ladies finger pods. The standard deviation is indicated on the chart. Results which do not differ significantly (LSD test; $p < 0.05$) carry at least one character in the superscript which is common

Authors' contributions

The experiments were planned and designed by NH, TA, and SAA who also jointly interpreted the findings. The experiments were conducted by NH. All authors read and approved the final manuscript.

Author details

[1] Centre for Pollution Control & Environmental Engineering, Pondicherry University, Chinakalapet, Puducherry 605 014, India. [2] Department of Fire Protection Engineering, Worcester Polytechnic Institute, Worcester, MA 01609, USA.

Acknowledgements

Authors thank the University Gants Commission, New Delhi, for support.

Competing interests

The authors declare that they have no competing interests.

References

Abbasi T, Abbasi SA (2011) Sources of pollution in rooftop rainwater harvesting systems and their control. Crit Rev Environ Sci Technol (Taylor and Francis) 41(23):2097–2167

Abbasi SA, Nipaney PC, Schaumberg GD (1990) Bioenergy potential of 8 common aquatic weeds. Biolo Wastes (Elsevier) 34(4):359–366

Abbasi T, Gajalakshmi S, Abbasi SA (2009) Towards modeling and design of vermicomposting systems: mechanisms of composting/vermicomposting and their implications. Indian J Biotechnol 8:177–182

Abbasi SA, Nayeem-Shah M, Abbasi T (2015) Vermicomposting of phytomass: limitations of the past approaches and the emerging directions. J Clean Prod 93:103–114

Abduli MA, Amiri L, Madadian E, Gitipour S, Sedighian S (2013) Efficiency of vermicompost on quantitative and qualitative growth of tomato plants. Int J Environ Res 7(2):467–472

Ahmed MMM (2000) Studies on the control of insect pests in vegetables (okra, tomato, and onion) in Sudan with special reference to neem-preparations Doctoral dissertation, University of Göttingen, Göttingen

Akhtar N, Satyam A, Anand V, Verma KK, Khatri R, Sharma A (2010) Dysregulation of T_H type cytokines in the patients of Parthenium induced contact dermatitis. Clin Chimica Acta 411:2024–2028

Akhzari D, Attaeian B, Arami A, Mahmoodi F, Aslani F (2015) Effects of vermicompost and arbuscular mycorrhizal fungi on soil properties and growth of Medicago polymorpha L. Compos Sci Util 23(3):142–153

Akter A, Zuberi MI (2009) Invasive alien species in Northern Bangladesh: identification, inventory and impacts. Int J Biodivers Conserv 15:129–134

Alan Clewer G, David Scarisbrick H (2001) Practical statistics and experimental design for plant and crop sciences. Wiley, England

Amenduini M, D'Amico M, Colella C, Cirulli M (2003) Severe outbreaks of alternaria leaf spot on kiwi in Southern Italy. Ferguson-Basilicata 53:39–43

Antonijevic D, Fakultet P, Zemun B, Mitrovic P (2007) Leaf spot of oilseed rape Biljni lekar (Serbia). Plant Doctor 35:443–449

Anwar MF, Yadav D, Kapoor S, Chander J, Samim M (2015) Comparison of antibacterial activity of Ag nanoparticles synthesized from leaf extract of Parthenium hystrophorus L in aqueous media and gentamicin sulphate, In-vitro. Drug Dev Ind Pharm 41:43–50

Arain AR, Jiskani MM, Wagan KH, Khuhro SN, Khaskheli MI (2012) Incidence and chemical control of okra leaf spot disease. Pak J Bot 44(5):1769–1774

Arancon NQ, Galvis PA, Edwards CA (2005) Suppression of insect pest populations and damage to plants by vermicomposts. Bioresource Tech 96(10):1137–1142

Arancon NQ, Edwards CA, Babenko A, Cannon J, Galvis P, Metzger JD (2008) Influences of vermicomposts, produced by earthworms and microorganisms from cattle manure, food waste and paper waste, on the germination, growth and flowering of petunias in the greenhouse. Appl Soil Ecol 39(1):91–99

Atiyeh RM, Subler S, Edwards CA, Metzger J (1999) Growth of tomato plants in horticultural media amended with vermicompost. Pedobiologia 43:724–728

Atiyeh RM, Arancon N, Edwards CA, Metzger JD (2000) Influence of earthworm-processed pig manure on the growth and yield of greenhouse tomatoes. Biores Technol 75:175–180

Atiyeh RM, Edwards CA, Subler S, Metzger JD (2001) Pig manure vermicompost as a component of a horticultural bedding plant medium: effects on physicochemical properties and plant growth. Biores Technol 78:11–20

Atiyeh RM, Lee S, Edwards CA, Arancon NQ, Metzger JD (2002) The influence of humic acids derived from earthworms-processed organic wastes on plant growth. Biores Technol 84:7–14

Ayyobi H, Hassanpour E, Alaqemand S, Fathi S, Olfati JA, Peyvast G (2014) Vermicompost leachate and vermiwash enhance French dwarf bean yield. Int J Veg Sci 20(1):21–27

Bewley JD, Black M (1982) Physiology and biochemistry of seeds in relation to germination. Springer, Berlin

Canihos Y, Peever TL, Timmer LW (1999) Temperature, leaf wetness and isolate effects on infection of Minneola tangelo leaves by Alternaria sp. Plant Diseases 83(5):429–433

Cardoza YJ (2011) Arabidopsis thaliana resistance to insects, mediated by an earthworm-produced organic soil amendment. Pest Manag Sci 67(2):233–238

Cardoza YJ, Buhler WG (2012) Soil organic amendment impacts on corn resistance to Helicoverpa zea: Constitutive or induced?. Pedobiologia 55:343–347

Carr EA, Nelson EB (2014) Disease-suppressive vermicompost induces a shift in germination mode of Pythium aphanidermatum zoosporangia. Plant Dis 98:361–367

Chan PLS, Griffiths DA (1988) The vermicomposting of pre-treated pig manure. Biol Wastes 24:57–69

Cho JT, Moon BJ (1980) The occurrence of strawberry black leaf spot caused by Alternaria alternata (Fr.) Keissler in Korea. Korean J Plant Prot 19(4):221–226

Doan TT, Ngo PT, Rumpel C, Nguyen BV, Jouquet P (2013) Interactions between compost, vermicompost and earthworms influence plant growth and yield: a 1 year greenhouse experiment. Sci Hortic 160:148–154

Doan TT, Henry-Des-Tureaux T, Rumpel C, Janeau J-L, Jouquet P (2015) Impact of compost, vermicompost and biochar on soil fertility, maize yield and soil erosion in Northern Vietnam: a 3 year mesocosm experiment. Sci Total Environ 514:147–154

Edwards CA, Burrows I (1988) The potential of earthworm composts as plant growth media. In: Edwards CA, Neuhauser E (eds) Earthworms in waste and environmental management. SPB Academic Press, The Hague, pp 21–32

Edwards Clive A (2004) Earthworm ecology, 2nd edn. CRC Press, Washington

Edwards CA, Arancon NQ, Bennett MV, Askar A, Keeney G, Little B (2010) Suppression of green peach aphid (Myzus persicae) (Sulz.), citrus mealybug (Planococcus citri), and two spotted spider mite (Tetranychus urticae) (Koch.) attacks on tomatoes and cucumbers by aqueous extracts from vermicomposts. Crop Prot 29:80–93

Edwards CA, Norman QA, Sherman R (2011) Vermiculture technology, earthworms, organic waste and environmental management. CRC Press, Washington, pp 17–19

Field A (2009) Discovering statistics using SPSS. In SAGE Publications, London

Gajalakshmi S, Abbasi SA (2002) Effect of the application of water hyacinth compost/vermicompost on the growth and flowering of Crossandra undulaefolia, and on several vegetables. Bioresour Technol 85(2):197–199

Graham RD, Webb MJ (1991) Micronutrients and disease resistance and tolerance in plants. In: Mortvedt JJ, Cox FR, Shuman LM, Welch RM (eds) Micronutrients in agriculture, 2nd edn. Soil Science Society of America, Fitchburg, pp 329–370

Gunaseelan VN (1987) Parthenium as an additive with cattle manure in biogas production. Biol Wastes 21:195–202

Halder J, Sanwal SK, Rai AK, Rai AB, Singh B, Singh BK (2015) Role of physico-morphic and biochemical characters of different okra genotypes in relation to population of okra shoot and fruit borer, Earias vittella (Noctuidae: Lepidoptera). Indian J Agric Sci 85(2):278–282

Haviola S, Kapari L, Ossipov V, Rantala MJ, Ruuhola T, Haukioja E (2007) Foliar phenolics are differently associated with Epirrata autumnata growth and immune competence. J Chem Ecol 33:1013–1023

Hilhorst HWM, Karssen CM (2000) Effect of chemical environment on seed germination. In: Fenner M (ed) Seeds: the ecology of regeneration in plant communities. CABI publishing, Wallingford, pp 293–310

Hill WJ, Clarke BB, Murphy JA (1999) Take-all suppression in creeping bentgrass with manganese and copper. Hortscience 34:891–892

Hussain N, Abbasi T, Abbasi SA (2015) Vermicomposting eliminates the toxicity of Lantana (Lantana camara) and turns it into a plant friendly organic fertilizer. J Hazard Mater 298:46–57

Hussain N, Abbasi T, Abbasi SA (2016a) Vermicomposting transforms allelopathic parthenium into a benign organic fertilizer. J Environ Manage 180:180–189

Hussain N, Abbasi T, Abbasi SA (2016b) Transformation of a highly pernicious and toxic weed parthenium into an eco-friendly organic fertilizer by vermicomposting. Int J Environ Stud 73:731–745

ICAR (2011) Handbook of agriculture. Indian Council of agricultural Research, Pusa

Ievinsh G (2011) Vermicompost treatment differentially affects seed germination, seedling growth and physiological status of vegetable crop species. Plant Growth Regul 65:169–181

Joshi R, Singh J, Vig AP (2015) Vermicompost as an effective organic fertilizer and biocontrol agent: effect on growth, yield and quality of plants. Rev Environ Sci Bio/Technol 14(1):137–159

Kaur M, Aggarwal NK, Kumar V, Dhiman R (2014) Effects and management of Parthenium hysterophorus: a weed of global significance, international scholarly research notices, p 1–12

Khan FI, Abbasi SA (1998) Techniques and methodologies for risk analysis in chemical process industries. Discovery Publishing House, New Delhi

Kishor P, Maurya BR, Ghosh AK (2010) Use of uprooted Parthenium before flowering as compost: a way to reduce its hazards worldwide. Int J Soil Sci 5:73–81

Knox J, Jaggi D, Paul MS (2011) Population dynamics of Parthenium hysterophorus (Asteraceae) and its biological suppression through Cassia occidentalis (Caesalpiniaceae). Turk J Bot 35:111–119

Kumar A, Kumar A, Kumar V, Kumar M, Singh B, Cjouhan P (2012) Recycling of harmful weeds through NADEP composting. Vegetos 25:315–318

Kumar S, Pandey S, Pandey AK (2014) In vitro antibacterial, antioxidant, and cytotoxic activities of Parthenium hysterophorus and characterization of extracts by LC-MS analysis. Biomed Res Int 49:51–54

Kumar R, Singh MK, Kumar V, Verma RK, Kushwah JK, Pal M (2015) Effect of nutrient supplementation through organic sources on growth, yield and quality of coriander (Coriandrum sativum L.). Indian J Agric Res 49:278–281

Lakshmi C, Srinivas CR (2007) Parthenium: A wide angle view. Ind J Dermatol Venereol Leprol 73:296–306

Lazcano C, Arnold J, Tato A, Zalle JG, Domínguez J (2009) Compost and vermicompost as nursery pot components: effects on tomato plant growth and morphology. Span J Agric Res 7:944–951

Lazcano C, Sampedro L, Zas R, Domínguez J (2010) Assessment of plant growth promotion by vermicompost in different progenies of maritime pine (Pinus pinasterAit.). Compost Sci Util 18:111–118

Maishi AI, Ali PKS, Chaghtai SA, Khan GA (1998) A proving of Parthenium hysterophorus L. Br Homoeopath J 87:17–21

Manoj EM (2014) In vasive plants a threat to wild life, The Hindu, http://www.thehindu.com/news/national/kerala/invasive-plants-a-threat-to-wildlife/article6432731.ece, p3

Moran R, Porath D (1980) Chlorophyll determination in intact tissues using N, N-di-Methyl formamide. Plant Physiol 65:478–479

Morin L, Reid AM, Sims-Chilton NM, Buckley YM, Dhileepan K, Hastwell GT, Nordblom TL, Raghu S (2009) Review of approaches to evaluate the effectiveness of weed biological control agents. Biol Control 5:1–15

Muscolo A, Bovalo F, Gionfriddo F, Nardi F (1999) Earthworm humic matter produces auxin-like effects on Daucus carota cell growth and nitrate metabolism. Soil Biol Biochem 31:1303–1311

Nayeem-Shah M (2014) Exploration of methods for gainful utilization of phytomass-based biowaste, Ph.D. thesis, Pondicherry University, Kalapet, pp. 92

Nayeem-Shah M, Gajalakshmi S, Abbasi SA (2015) Direct, rapid and sustainable vermicomposting of the leaf litter of neem (Azadirachta indica). Appl Biochem Biotechnol 175:792–801

Nielsen SS (ed) (2010) Food analysis. Springer, New York, p 550

Nigatu L, Hassen A, Sharma J, Adkins SW (2010) Impact of Parthenium hysterophorus on grazing land communities in North-Eastern Ethiopia. Weed Biol Manag 10:143–152

Patel S (2011) Harmful and beneficial aspects of Parthenium hysterophorus: an update. 3 Biotech 1:1–9

Pramanik P, Ghosh GK, Ghosal PK, Banik P (2007) Changes in organic-C, N, P and K and enzyme activities in vermicompost of biodegradable organic wastes under liming and microbial inoculants. Biores Technol 98:2485–2494

Rajkumar EDM, Kumar NVN, Haranm NVH, Ram NVS (1988) Antagonistic effect of P. hysterophorus on succinate dehydrogenase of sheep liver. J Environ Biol 9:231–237

Sahni S, Sarma BK, Singh DP, Singh HB, Singh KP (2008) Vermicompost enhances performance of plant growth-promoting rhizobacteria in Cicer arietinum rhizosphere against Sclerotium rolfsii. Crop Prot 27:369–376

Samrot AV, Vignesh MP, Vignesh V (2015) Utilization of vermicompost of organic waste for plant growth. Int J Pharm Bio Sci 6:B326–B332

Saxena J, Choudhary S, Pareek S, Choudhary AK, Iquebal MA (2015) Recycling of organic waste through four different composts for disease suppression and growth enhancement in mung beans. Clean Soil Air Water 43:1066–1071

Serfoji P, Rajeshkumar S, Selvaraj T (2010) Management of root-knot nematode, Meloidogyne incognita on tomato cv Pusa Ruby by using vermicompost, AM fungus, Glomus aggregatum and mycorrhiza helper bacterium, Bacillus coagulans. J Agric Technol 6:37–45

Sharma RP, Swaminathan R, Bhati KK (2010) Seasonal incidence of fruit and shoot borer of okra along with climatic factors in Udaipur region of India. Asian J Agric Res 4(4):232–236

Singh RK, Garg Arti (2014) Parthenium hysterophorus L.—neither noxious nor an obnoxious weed. Indian For 140:1260–1262

Singh R, Sharma RR, Kumar S, Gupta RK, Patil RT (2008) Vermicompost substitution influences growth, physiological disorders, fruit yield and quality of strawberry (Fragaria X ananassa Duch.). Biores Technol 99:8507–8511

Singh R, Singh R, Soni SK, Singh SP, Chauhan UK, Kalra A (2013) Vermicompost from biodegraded distillation waste improves soil properties and essential oil yield of Pogostemon cablin (patchouli) Benth. Appl Soil Ecol 70:48–56

Tauseef SM, Premalatha M, Tasneem Abbasi, Abbasi SA (2013) Methane capture from livestock manure. J Environ Manag 117:187–207

Tohyama A, Hayashi K, Taniguchi N, Naruse C, Ozawa Y, Shishiyama J, Tsuda M (2005) A new post-harvest disease of okra pods caused by Alternaria alternate. Ann Phytopathol Soc Jpn 61(4):340–345

Towers GHN, Subba Rao PV (1992) Impact of the pantropical weed, Parthenium hysterophorus L. on human aff airs. In: Richardson RG (ed) Proceedings of the 1st international weed control congress Australia. Weed Sci Soc Victoria, Melbourne, pp 135–138

Wellburn AR (1994) The spectral determination of chlorophylls a and b, as well as total carotenoids, using various solvents with spectrophotometers of different resolutions. J Plant Physiol 144:307–313

Werner M (1987) Necrotic leaf spot of apple caused by fungi of the genus Alternaria. Ochrona Roslin (Poland) 31:6–7

Wiesner M, Tessema T, Hoffmann A, Wilfried P, Buettner C, Mewis I, Ulrichs C (2007) Impact of the pan-tropical weed Parthenium hysterophorus L on human health in Ethiopia. Institute of Horticultural Science, Urban Horticulture

Wilson DP, Carlile WR (1989) Plant growth in potting media containingworm-worked duck waste. Acta Hortic 238:205–220

Xiao Z, Liu M, Jiang L, Chen X, Griffiths BS, Li H, Hu F (2016) Vermicompost increases defense against root-knot nematode (Meloidogyne incognita) in tomato plants. Appl Soil Ecol 105:177–186

Xu Y, Wang C, Bi Y, Zhang Y, Cheng W, Sun Z, Zhang J, Lv Z, Guo X (2014) Influence of cow manure vermicompost soil mixtures on two flowers seedling cultivation. Acta Hort 1018:583–588

Yadav A, Suthar S, Garg VK (2015) Dynamics of microbiological parameters, enzymatic activities and worm biomass production during vermicomposting of effluent treatment plant sludge of bakery industry. Environ Sci Pollut Res 22(19):14702–14709

Yardim EN, Arancon NQ, Edwards CA, Oliver TJ (2006) Suppression of tomato hornworm (Manduca quinquemaculata) and cucumber beetles (Acalymma vittatum and Diabotrica undecimpunctata) populations and damage by vermicomposts. Pedobiologia 50:23–29

Zaller JG (2007) Vermicompost as a substitute for peat in potting media: effects on germination, biomass allocation, yields and fruit quality of three tomato varieties. Sci Hortic 112:191–199

Gajalakshmi S, Abbasi SA, (2002) Effect of the application of water hyacinth compost/vermicompost on the growth and flowering of Crossandra undu-

12

Lobster processing by-products as valuable bioresource of marine functional ingredients, nutraceuticals, and pharmaceuticals

Trung T. Nguyen[1,2,3*], Andrew R. Barber[1,2], Kendall Corbin[1,2,4] and Wei Zhang[1,2]

Abstract

The worldwide annual production of lobster was 165,367 tons valued over $3.32 billion in 2004, but this figure rose up to 304,000 tons in 2012. Over half the volume of the worldwide lobster production has been processed to meet the rising global demand in diversified lobster products. Lobster processing generates a large amount of by-products (heads, shells, livers, and eggs) which account for 50–70% of the starting material. Continued production of these lobster processing by-products (LPBs) without corresponding process development for efficient utilization has led to disposal issues associated with costs and pollutions. This review presents the promising opportunities to maximize the utilization of LPBs by economic recovery of their valuable components to produce high value-added products. More than 50,000 tons of LPBs are globally generated, which costs lobster processing companies upward of about $7.5 million/year for disposal. This not only presents financial and environmental burdens to the lobster processors but also wastes a valuable bioresource. LPBs are rich in a range of high-value compounds such as proteins, chitin, lipids, minerals, and pigments. Extracts recovered from LPBs have been demonstrated to possess several functionalities and bioactivities, which are useful for numerous applications in water treatment, agriculture, food, nutraceutical, pharmaceutical products, and biomedicine. Although LPBs have been studied for recovery of valuable components, utilization of these materials for the large-scale production is still very limited. Extraction of lobster components using microwave, ultrasonic, and supercritical fluid extraction were found to be promising techniques that could be used for large-scale production. LPBs are rich in high-value compounds that are currently being underutilized. These compounds can be extracted for being used as functional ingredients, nutraceuticals, and pharmaceuticals in a wide range of commercial applications. The efficient utilization of LPBs would not only generate significant economic benefits but also reduce the problems of waste management associated with the lobster industry. This comprehensive review highlights the availability of the global LPBs, the key components in LPBs and their current applications, the limitations to the extraction techniques used, and the suggested emerging techniques which may be promising on an industrial scale for the maximized utilization of LPBs.

Keywords: Lobster processing by-products, Marine functional ingredients and nutraceuticals, Chitin and chitosan, Astaxanthin, Lobster flavors, Lobster lipids, Lobster protein

Global lobster processing industry generates a large amount of by-products

In 2004, the global production of lobster yielded 165,367 tons (Holmyard and Franz 2006) which had an estimated value of $3.32 billion. Over the last decade, these figures have been rising to reach 304,000 tons (captures and aquaculture) in 2012 (Sabatini 2015). Lobster production can be found across the world; however, the majority of production is concentrated in only three countries: Canada (34%), America (29%), and Australia (11%) (Fig. 1) (Annie and McCarron 2006). The four main commercial lobster species produced are the American lobster (*Homarus americanus*), Tropical or Spiny lobster

*Correspondence: nguy0514@flinders.edu.au
[1] Centre for Marine Bioproducts Development, Flinders University, Adelaide, Australia
Full list of author information is available at the end of the article

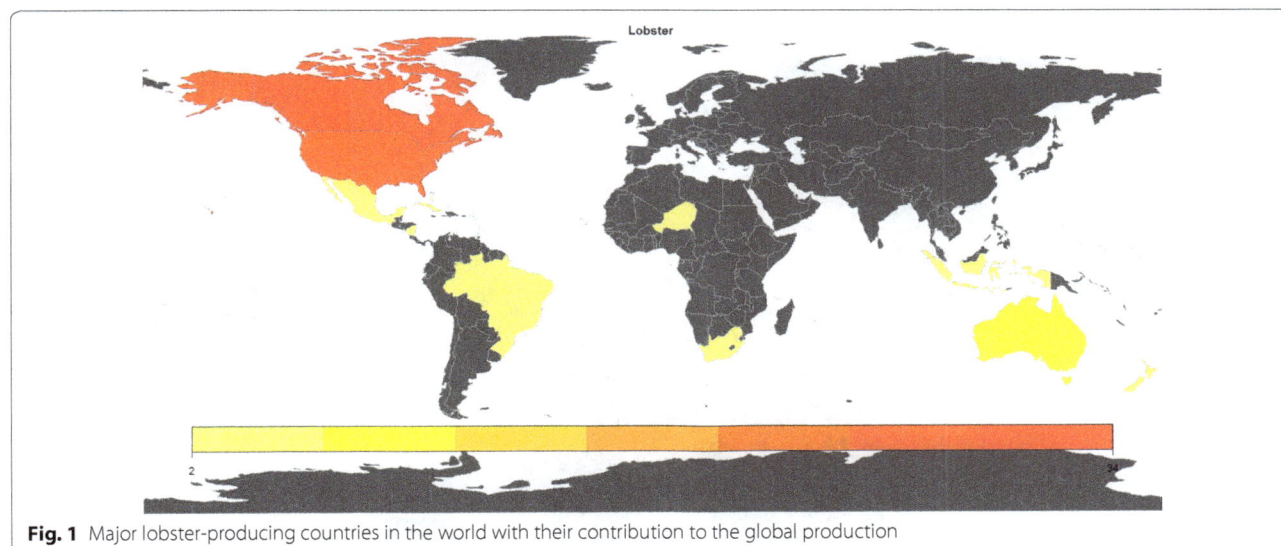

Fig. 1 Major lobster-producing countries in the world with their contribution to the global production

(*Panulirus sp*), Rock lobster (*Jasus sp*), and European lobster (*Homarus gammarus*).

The most abundant species produced in the world is the American lobster which is mainly harvested in Canada and America (Fig. 2) (Annie and McCarron 2006). In 2012, both Canada and America produced 140,000 tons of lobsters (Thériault et al. 2013) with the majority (74,790 tons, valued at $662.8 million) originating from Canada (Ilangumaran 2014). The second-most readily available commercial species is Spiny lobster accounting for 38% of the global production, while the contribution of the Rock lobster is 6%. This latter species is predominantly harvested from Australia, which includes four main commercial species: Western rock lobster (60%), Southern rock lobster (30%), Tropical rock lobster (8%), and eastern rock lobster (2%) with the total yield about 9650 tons annually (Gary 2012). The term 'Rock lobster' has been used to describe lobster species such as *Jasus* and *Panulirus* which are caught by Australian lobster fishery (Holmyard and Franz 2006).

As lobster are consumed globally but are predominantly produced in a few countries, there is a rapidly growing export market of lobsters. Although live lobsters are preferred by consumers around the world, the export of live lobsters is limited due to its high cost, complexity, and high rates of mortality and loss during shipment. In contrast, processed lobsters has several advantages such as ease of handling in transport and storage, extended shelf life, availability of the products, convenience in food preparation, and higher potential to adding value to raw products. This ease in handling and increase in profits have resulted in over half of the landed lobster in the major lobster-producing countries being processed (Barker and Rossbach 2013; Denise and Jason 2012; Ilangumaran 2014).

Lobsters are commercially processed into various products such as fresh lobster meat, picked lobster meat, canned lobster, lobster medallion, whole cooked lobsters, and frozen lobsters (Holmyard and Franz 2006). During processing, the inedible parts are removed including heads, shells, roe, and livers (Fig. 3), and are traditionally discarded. The types and proportion of lobster processing by-products (LPBs) generated vary depending on the processing process but on average accounts for around 75% (w/w) of the starting material (Table 1). As a result of this, the annual estimate of LPBs produced from the major lobster processing countries (Canada, America, and Australia) is about 50,000 tons.

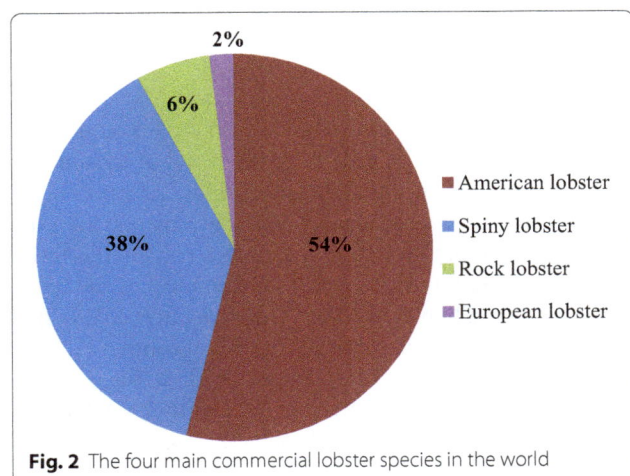

Fig. 2 The four main commercial lobster species in the world

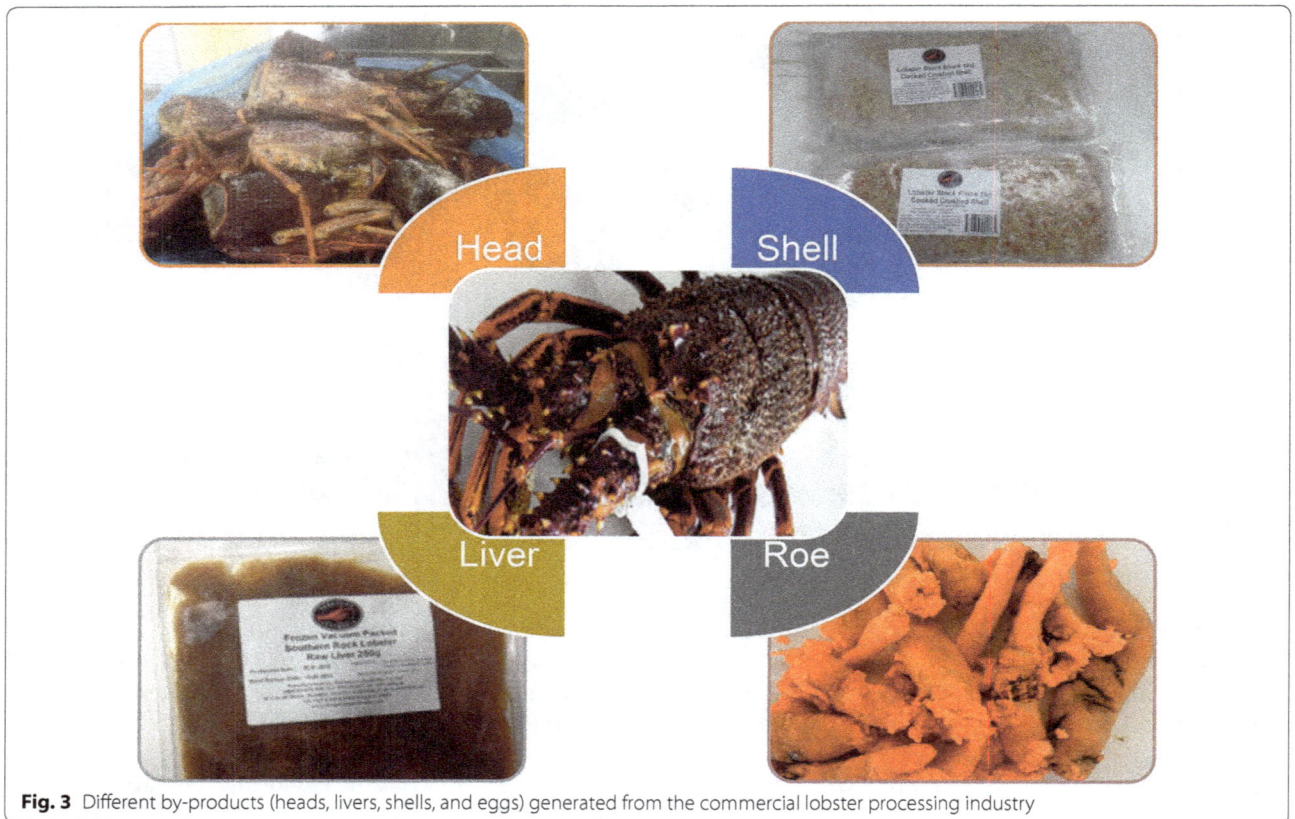

Fig. 3 Different by-products (heads, livers, shells, and eggs) generated from the commercial lobster processing industry

Table 1 The amount of by-products generated from the different lobster processing industries

Lobster processing industries	Types of by-products	Percentage of by-products based on starting material (%)	References
Canning of Canadian lobsters	Lobster body	45	Ross (1927)
Canadian lobster meat	Lobster head, hard carapace, viscera, mandibles, and gills	>75	Tu (1991)
Brazilian lobster tails	Lobster head (cephalothorax)	75	Vieira et al. (1995)
Fresh meat picked from Australian rock lobster	Lobster head, shell, and viscera	60	Lien (2004)
High hydrostatic pressure production of American lobster meat	Lobster shells, viscera, residual meat	75–80	Denise and Jason (2012)

To maximize the production yield and profit, lobster processors have begun to utilize some LPBs to produce several products such as lobster tomalley, lobster roe, lobster concentrate, and lobster meat paste (Holmyard and Franz 2006). However, the amount of LPBs being utilized compared to the tons generated is still very limited. The slow uptake and growth of industries using this waste material may be attributed to the current lack of efficient and standardized techniques to transform these materials into a marketable form. Thus, the vast majority of LPBs is discarded at a cost incurred by lobster processors. In some countries such as Australia, this fee can be

in excess of $150 per ton (Knuckey 2004; Yan and Chen 2015). In addition, not only is the disposal of lobster by-products and waste management a financial burden for lobster processing companies globally costing an estimated $7.5 million per year, but is also considered to be environmentally unfriendly due to dumping in the landfill or the sea (Chen et al. 2016; Yan and Chen 2015). This could create disposal problems and environmental pollutions (Hamed et al. 2016; Sayari et al. 2016). Moreover, this underutilizes a marine bioresource that could be mined to produce several valuable ingredients for a wide range of commercial industries. Shell waste, for instance, has been considered as a source of useful chemicals for many commercial applications (Chen et al. 2016; Yan and Chen 2015). Crude crustacean shells or chitin derived from chitin shells could be used for production of low molecular weight chitosan and chitin oligomers with numerous applications employing simple processes (Chen et al. 2017; Zhang and Yan 2016). As a result, the utilization of these LPBs for recovery of marine functional ingredients, nutraceuticals, and pharmaceuticals for incorporation into highly value-added products could bring significantly both economic and environmental benefits to regions where lobsters are processed.

LPBs containing several valuable components
Proteins

By-products generated from lobster processing industry are protein-rich sources for recovery. For example, the lobster liver (green) contains up to 41% protein on a dry basis (Nguyen et al. 2015), while the lobster head containing residual meats (body, breast, and leg) up to 20% of the lobster weight (Vieira et al. 1995) would be another major high-quality protein source. In addition, lobster shells (carapace) constituted by a large amount of proteins (about 25%) are also another potential source of protein for mining (Nguyen et al. 2016).

The amino acid profile of crustacean protein such as lobster is comparable to that of red meat protein, but it contains more nonprotein nitrogen (amino acids, small peptides, trimethylamine oxide (TMAO), trimethylamine, creatine, creatinine, and nucleotides) ranging from 10–40%. Thus, crustacean protein is more palatable than meat proteins (Venugopal 2009). As compared with other marine species such as finfish, proteins derived from crustacean generally contain larger amounts of arginine, glutamic acid, glycine, and alanine; this makes crustacean proteins more palatable than finfish proteins. Due to its ideal essential amino acid pattern, moreover, the nutritional value of crustacean protein is equal to or better than that of milk protein (casein) and red meat proteins (Venugopal 2009) or soya-bean proteins (Yan and Chen 2015). Lobster proteins are rich in all the essential amino acids (EAAs) with its proportion approximately reaching 41.2% for protein in lobster head meat (Vieira et al. 1995) and 34% for lobster shell protein (Nguyen et al. 2016). Especially, the nutritional value of lobster protein is fortified significantly by its natural combination with a large amount of astaxanthin (295 µg/g) as a powerful antioxidant to form a protein complex known as carotenoprotein. This protein was found in lobster shells with high proportion (16%) (Tu et al. 1991).

Apart from its high delicacy, palatability, and nutritional values, lobster proteins have also been shown to have excellent functional properties. For example, protein derived

from lobster head meats has been shown to have excellent wettability, high solubility, and emulsification (Vieira et al. 1995). Solubility of lobster shell protein (LSP) recovered either by aqueous extraction or enzymatic digestion was over 93%, which is independent of the pH value and the ionic strength of the solution used (Nguyen et al. 2016; Oviedo et al. 1982). High water binding of LSP was also reported by the fact that beef mince when added with 2% of LSP resulted in its water-binding capacity being 2.5 times higher than that when added with egg white protein (Nguyen et al. 2016). In addition, lobster protein hydrolysate (LPH) performed an excellent emulsifying property (69.7 vs 50.3 m^2/g of cow gelatin) (He et al. 2016).

Chitin and chitin derivatives

Chitin is a cationic linear polysaccharide composed of β-(1–4)-linked N-acetyl-D-glucosamine monomers (Fig. 4a) and is the second-most abundant biopolymer only after cellulose in the biosphere (Kumar 2000; Kurita 2006). Chitin is present in lobster shells with contents of 16–23% in the form of α-chitin (Lien 2004; Rinaudo 2006; Tu 1991). Chito san represents a family of N-deacetylated chitin with various degrees of deacetylation (Fig. 4b), while chito-oligosaccharides (COS) are derived from chitin and chitosan by either chemical or enzymatic hydrolysis.

LPBs such as lobster shell waste are abundant and rich in chitin attributing to its importance as a potential source of commercial chitin. The utilization of LPBs for recovery of chitin has received a great interest since chitin is a natural biopolymer with high biodegradability, biocompatibility, and nontoxicity (Karagozlu and Kim 2015). Derivatives generated from chitin such as chitosan and COS have significant potential economic values with more than 200 applications in water treatment, food, agriculture, healthcare products, environmental sector, pharmaceuticals, and biomedicine (Kaur and Dhillon 2013; Muzzarelli 1989; Sandford 1989; Synowiecki and Al-Khateeb 1997). There is an increasing global demand for chitin and chitin derivatives with the estimated quantities of 11,400 tons for chitin and 33,400 ton for chitin derivatives (Hayes 2012). Economic value of chitin and chitin derivatives was $2.0 billion in 2016, but this figure is estimated to increase significantly to $4.2 billion by 2021 (Pathak 2017).

Lipids

Lipid is another valuable constituent of LPBs due to its nutraceutical-rich composition (polyunsaturated fatty acids (PUFAs); ω-3 fatty acids; and lipid-soluble vitamins). Lobster body contains less than 2% of lipids (Shahidi 2006), but concentration of this constituent is significantly higher in some other parts of the lobsters. Cephalothorax, a by-product generated from lobster processing industry,

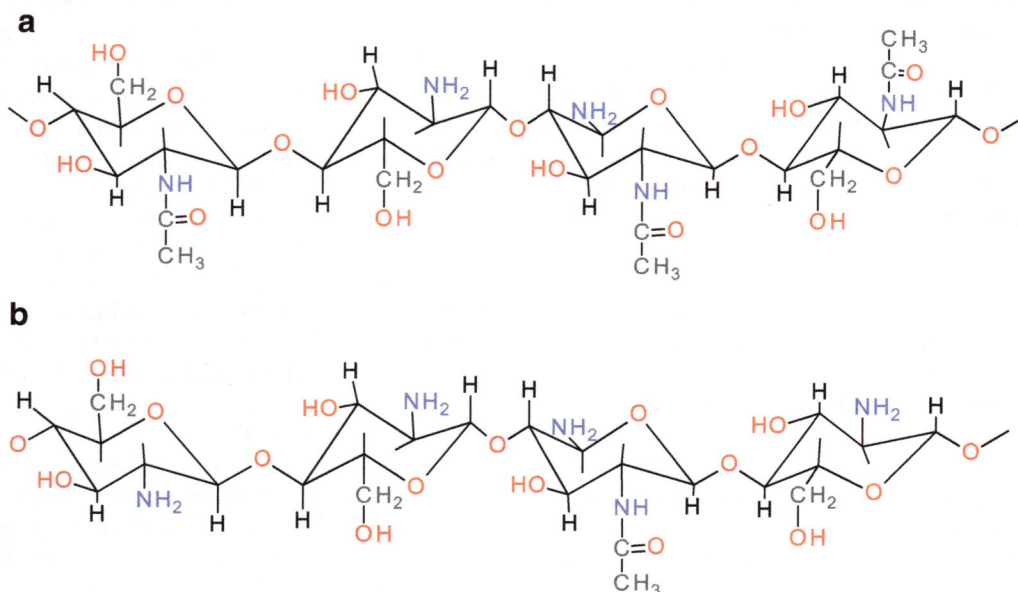

Fig. 4 The chemical structure of the main polymer present in lobster shell: chitin (**a**), and the *N*-deacetylation form of chitin, chitosan (**b**)

is one of the lipid-rich parts of lobster with various lipid contents depending on lobster habitat, season, and species. In summer season, Norway lobster cephalothorax has the highest lipid content with a value of 11.5% (Albalat et al. 2016). Cephalothorax of Norway lobster contains lower lipid content than that of Australian lobster which accounts for up to 19.4% (Tsvetnenko et al. 1996). Lobster liver is another lipid-rich part of lobster with the lipid content of which is up to 24.3% (Nguyen et al. 2015). Among crustaceans, LPBs contain more lipids than crab and shrimp by-products, while this value is comparable to that of krill being used for lipid production (Albalat et al. 2016). Lobster lipids are rich in PUFAs and ω-3 fatty acids as reported in the study of Tsvetnenko et al. (1996): lipids extracted from lobster cephalothorax contained 22.6% of fatty acids in which ω-3 fatty acids accounted for 50.8%. However, the amounts of PUFAs and ω-3 fatty acids were significantly higher (22.6 vs 31.3%, 50.8 vs 58%) when lipids were recovered from lobster livers using supercritical carbon dioxide (SC-CO_2) extraction technique (Nguyen et al. 2015). Richness of PUFAs and ω-3 fatty acids in lobster lipids is comparable to that of menhaden lipids (Tsvetnenko et al. 1996) or even higher than that of krill lipids (Albalat et al. 2016). Particularly, lobster lipids also contain carotenoids and astaxanthin with concentrations of 70.4 and 41.6 μg/mL, respectively (Nguyen 2017).

Astaxanthin

Astaxanthin is the oxygenated derivatives of carotenoids occurring widely and naturally in marine organisms

including crustaceans (lobster, shrimp, crab, and krill) and fish (salmon, sea bream). Astaxanthin is one of the first pigments isolated and characterized from lobster (Kuhn and Soerensen 1938). By-products generated from lobsters, shrimps, crabs, crayfish, and krill are a vital source of natural carotenoids, mainly astaxanthin (Sachindra et al. 2007). Apart from containing a large proportion of mineral salts (15–35%), proteins (25–50%), chitin (25–35%), and lipids (19.4–24.3%) (Lee and Peniston 1982; Nguyen et al. 2015), LPBs are also constitutive of a certain amount of carotenoids. Astaxanthin content in crustaceans varies depending on species, season, and environmental grown conditions, but lobster by-products contain double the amount of astaxanthin compared with shrimp (Table 2). Astaxanthin exists in a free form and/or in a complex form known as carotenoprotein. Due to their containing a high proportion of astaxanthin, LPBs have been utilized for the recovery of astaxanthin

Table 2 Total astaxanthin in by-products of lobsters compared with other crustacean species

Source	Total astaxanthin (mg/100 g)	References
Shrimp (*P. borealis*)	4.97	Torrissen et al. (1981)
Crawfish (*P. clarkii*)	15.30	Meyers and Bligh (1981)
Backs snow crab (*Ch. opilio*)	11.96	Shahidi and Synowiecki (1991)
Lobster (*Homarus americanus*)	9.80	Tu (1991)

(Auerswald and Gäde 2008; Gäde and Auerswald 2005). Two processes were patented for astaxanthin extraction from lobster heads (Kozo 1997; Sunda et al. 2012), while food-grade astaxanthin was recovered from the shells of American lobster generated from the high hydrostatic pressure process (Denise and Jason 2012).

Various applications of extracts derived from LPBs

The most common traditional utilization of LPBs is their use as a source of nutrients for soil amendment considered as an informal way of disposal (Cousins 1997), but it brings no economic benefits for lobster producers. Recently, LPBs have been studied as an important bioresource in the recovery of marine functional ingredients, nutraceuticals, and pharmaceuticals for numerous applications (Table 3).

Lobster protein: dietary protein supplement, food functional ingredients, or flavorings

The recovery of edible meat from lobster by-products is not novel. However, the use of these by-products as a

source of sustainable proteins for food products has been less explored. In one study, the meats (body meat, breast meat, leg meat, roe, and liver) recovered from Canadian lobster by-products were used to create a canned food product and lobster paste (Ross 1927). Although the final products prepared from these protein sources were flavor-rich, nutritious, and palatable, the feasibility for commercialization was found impractical as the methods used for residual meat recovery were inefficient. To address this key problem, a process was developed for the recovery of food-grade lobster meats from spiny lobster head by-product by freezing the heads and later cutting them for meat picking. The recovered lobster meat was sold as high-value, gourmet food products (Meyers and Machada 1978).

The actual flavor of lobster itself is considered a highly valued product, which may be extracted and sold, creating an additional processing stream for lobster wastes. The practice of converting LPBs into natural lobster flavors has been standardized and has now become an established industrial practice. The cephalothorax of Brazilian lobster by-products

Table 3 The different types of lobster by-products generated with their valuable components for potential areas of applications

Lobster by-products	Functional ingredients, nutraceuticals, pharmaceuticals	Suggested application areas	References
Lobster shells (carapace)	Chitin, chitosan	Water treatment	Gustavo et al. (2005); Pathiraja (2014)
	Chitin, chitosan, chitin-oligosaccharides, chitosan-oligosaccharide	Agriculture	Borges et al. (2000); Falcón et al. (2002); Cabrera and Van Cutsem (2005); Falcón Rodríguez et al. (2010); Ilangumaran (2014)
	Chitosan, chitosan film	Food processing and preservative	Defang et al. (2001); García et al. (2015)
	Water-soluble chitosan, chitosan particles	Pharmacy	Safitri et al. (2014); De la Paz et al. (2015)
	Chitosan film	Biomedicine	Malho et al. (2014); Qi (2015)
	Carotenoprotein	Aquafeed	Dauphin (1991); Simpson et al. (1993); Tu (1991); Tu et al. (1991)
	Astaxanthin	Food, nutraceutical, pharmaceutical; feed additive	Auerswald and Gäde (2008); Denise and Jason (2012); Gäde and Auerswald (2005)
	Proteins	Food and nutraceutical	Nguyen et al. (2016); Oviedo et al. (1982)
	Flavors and nutrient broth	Crackers, biscuits	Lien (2004)
Lobster heads (cephalothorax)	Body meat, breast meat, and leg meat	Lobster paste, canned products	Ross (1927)
	Lobster meat	Gourmet food products	Meyers and Machada (1978)
		Feed additive	Daniel (2008)
	Lobster protein hydrolysate	Flavor enhancer, protein supplement	Vieira et al. (1995)
Lobster roes	Raw roe	Lobster paste, canned products	Ross (1927)
Lobster livers (hepatopancreas)	Raw liver	Lobster paste, canned product	Ross (1927)
	ω-3 rich lipids	Lobster oils, infused oils	Nguyen et al. (2015); Tsvetnenko et al. (1996)
Lobster blood (hemolymph)	Phenol oxidase	Anti-microbial proteins	Fredrick and Ravichandran (2012)
	Crustin	Anti-microbial proteins	Battison et al. (2008); Pisuttharachai et al. (2009)
	Bioactive fragment	Pharmaceutical and/or cosmetic treatment of viral and other neoplastic or pre-neoplastic mammalian tissue lesions	Bayer (2015)

(*Panulirus* spp.) has been utilized for lobster flavor production by enzymatic hydrolysis of lobster head meats (Vieira et al. 1995). Hydrolyzed lobster protein could be used as flavor enhancers for various formulated food products. The key aromatic components derived from cooked tail meat of American lobster (*Homarus americanus*) was investigated by Lee et al. (2001). In this study, 3-methylbutanal, 2,3-butanedione, (Z)-heptenal, 3-(methylthio)propanal, 1-octadien-3-one, and (E,Z)-2,6-nonadienal were identified as dominant aroma components of cooked American lobster tail meat with high odor intensities. By this reason, flavorants extracted from lobster shells by either frying with edible oils or cooking for nutrient extraction were used for production of infused lobster oil and lobster cracker biscuit (Lien 2004). Due to possessing several functional properties that are favorable for use in food industry, LSP has also been characterized and trialed for various applications including food-functional ingredients or as protein supplements (Nguyen et al. 2016; Oviedo et al. 1982), enhancing water-binding or reducing lipidemic effects of meat protein (Nguyen et al. 2016), emulsifier (He et al. 2016).

Lobsters chitin, chitosan, and their derivatives as natural biopolymers for multiple applications

In recent decades, a greater knowledge of chitin chemistry accompanied by the increased availability of crustacean shells as by-products of the crustacean-processing industry have led to significant development and wide applications of chitin and its derivatives produced by several pathways (Fig. 5). In particular, derivatives from chitin such as chitosan and COS are highly valued compounds as they have more than 200 commercial applications (Kaur and Dhillon 2013; Muzzarelli 1989; Sandford 1989; Synowiecki and Al-Khateeb 1997). In addition, the biopolymers generated from marine crustaceans, lobster chitin, chitosan, and their derivatives have been used in multitude of industries including water treatment, agriculture, food production, pharmaceuticals, and biomedicine (Table 4).

Water treatment

One of common applications of chitin and its derivatives is for the treatment of water. Chitin and chitosan with their high absorbance, chelating, and affinity properties have been used as coagulation agents, chelating polymers, or bio-absorbents in water treatment for decades. In particular, chitin and chitosan prepared from lobster have shown high affinity for metal chelating. Peniche-Covas et al. (1992) reported lobster chitosan could effectively remove mercuric ions from solutions. Lobster chitosan was later used successfully for separation of heavy metals (Cu, Hg) (Gustavo et al. 2005). The heavy metal removal efficiency of lobster chitosan was comparable to that of commercial resin. Lobster chitin/chitosan has been used for the removal of reactive dyes (vinyl sulfone and chlorotriazine) from aqueous solution (Juang et al. 1997). Lobster chitin/chitosan produced from this study had significantly high adsorption capacity of reactive dyes (50–500 mg/L) compared with a conventional absorbent (activated carbon). The result of this study has been recently amplified by a study of

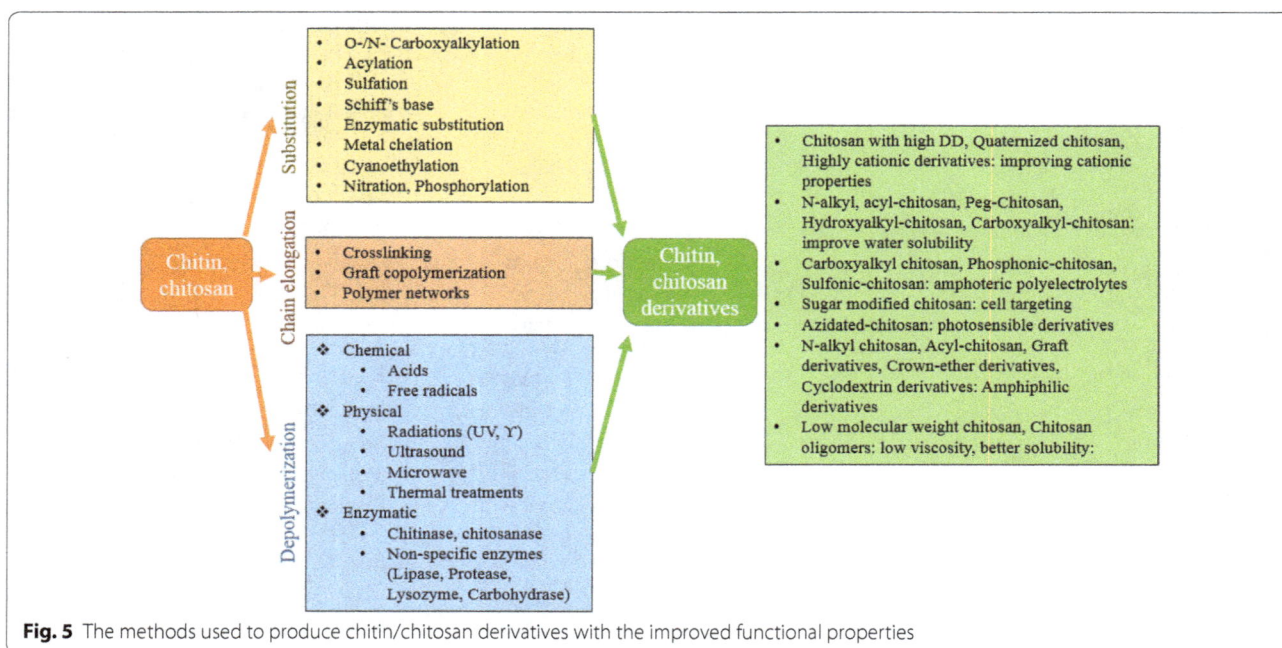

Fig. 5 The methods used to produce chitin/chitosan derivatives with the improved functional properties

Table 4 Applications of lobster chitin, chitosan, and their derivatives

Industry	Applications	References
Water treatment	Removal of mercuric ion	Peniche-Covas et al. (1992)
	Removal of reactive dyes from aqueous solution	Juang et al. (1996, 1997)
	Removal of heavy metals (Cu, Hg)	Gustavo et al. (2005)
	Dye absorbents for treatment of industrial effluent	Pathiraja (2014)
Agriculture	Seedling growth and antimycorrhizal in tomato crop	Iglesias et al. (1994)
	Fungicides on plant fungal diseases	Pombo (1995)
	Amending media and seeds for inhibition of pathogen fungus growth	Borges et al. (2000)
	Inducing systemic resistance agents in tobacco plants	Falcón et al. (2002)
	Elicitors of plants defense reactions	Cabrera and Van Cutsem (2005)
	Plant growth regulators	Hipulan (2005)
	Inducing defensive agents	Cabrera et al. (2010); Falcón Rodríguez et al. (2010)
	Plant protection	Ilangumaran (2014)
Food	Organic polymer flocculants	Defang et al. (2001)
	Edible or biodegradable films	Casariego et al. (2008, 2009)
	Biodegradable packages	Hudson et al. (2015)
	Fiber and nutrient supplement	Harikrishnan et al. (2012)
	Antioxidants, antimicrobials	Garcia et al. (2015); Sayari et al. (2016)
Pharmaceuticals	Direct compression excipients for pharmaceutical application	Mir et al. (2008)
	Co-diluent in direct compression of tablets	Mir et al. (2010)
	Water-soluble lobster chitosan salt as materials for drug carriers	Cervera et al. (2011)
	Targeted drug delivery film	Bamgbose et al. (2012)
	Bio-mouth spray for anti-halitosis	Safitri et al. (2014)
	The stability and safety of lobster chitosan salts	De la Paz et al. (2015); Lagarto et al. (2015)
	Natural additive for pharmaceuticals	Sayari et al. (2016)
Biomedicine	Immobilization of multi-enzyme extract	Osinga et al. (1999)
	Novel hybrid biomaterials for medical application	Malho et al. (2014)
	Biomimetic functional materials or epithelial treatment	Qi (2015); Da Silveira et al. (2013)

Pathiraja (2014) where lobster chitin was used as dye absorbents for effective treatment of industrial effluent.

Agriculture

Chitin and its derivatives have been shown to have profound beneficial effects when used for agriculture applications. For example, these polymers stimulate plant growth and improve crop yields (Badawy and Rabea 2011; Deepmala et al. 2014; El Hadrami et al. 2010; Sharp 2013). In addition, when applied to crops, they exhibit toxicity to plant pests and pathogens, induce plant defenses, have high anti-fungal activity, and stimulate the growth and activity of beneficial microbes (Pombo 1995; Sharp 2013). Based on these findings, cheap sources rich in chitin such as lobster shells have been trialed as fungicides, biocides, and bio-stimulants.

Other biological effects induced by lobster chitosan when applied to plant seeds were reported by Borges et al. (2000). Those authors used chitosan prepared from lobster chitin to amend agar and coat tomato seeds together with menadione sodium bisulfite for inhibition of pathogenic fungus and disease development on tomato seeds. The results showed that lobster chitosan significantly inhibited the fungal growth and diminished disease occurrence in the roots from seeds, which had been treated. In another study, lobster chitin was shown to have positive effects on seedling growth and mycorrhizal infection of tomato crop (Iglesias et al. 1994). Furthermore, chitin extracted from Cuban lobster shells has been shown to have biological functions. Researchers at the University of Havana prepared chitosan from lobster chitin to coat tomato seeds as well as encapsulate somatic embryos to produce artificial seeds for accelerating yields. Under laboratory conditions, the coated seeds exhibited considerably higher rates of germination and growth compared with the noncoated seeds. From this study, it was concluded that the lobster-shell-derived chitosan served as a biological stimulant yielding better seed germination, increased plant height, enhanced stem thickness, and dry biomass yield (Hareyan 2007).

Chitin, chitosan, and COS recovered from lobster shells have also been studied for their biocide and biostimulant activities for agricultural applications. A study carried out by Falcón et al. (2002) investigated the use of lobster chitosan and chitosan oligomers for crop protection. Both lobster chitosan and its enzymatic hydrolysate exhibited high pathogen-resistant activity against *Phytophthora parasitica* on tobacco plants at low concentrations ranging from 5–500 mg/L. With this success, the study was extended by preparing various chitosan and chitosanoligomers from Cuban lobster chitin. Both lobster chitosan and its derivatives were tested for antimicrobial activity (versus fungus and oomycetes) and their ability to induce defensive and protective responses in tobacco and rice plants. Some of the lobster chitosan derivatives were found to be active protectants against infection for both cultivars at field scale (Falcón Rodríguez et al. 2010).

In another study, Peters et al. (2006) used raw lobster shells and compost as a soil amendment to supply nutrients for plants and as a biological control method for soil-borne fungi pathogenic to potatoes. The results showed that lobster shells served as a nutrient source for plant growth, enhanced beneficial soil microbial communities, suppressed soil-borne diseases, and was an organic production process. Following this success, Ilangumaran (2014) studied isolation of soil microbes for bioconversion of lobster shells into chitin and chitin-oligomers for plant disease management. The extracted bioproducts showed significant induction of disease resistance in *Arabidopsis* confirming the findings in previous studies.

Food industry

The use of chitin, chitosan, and their derivatives in food industry has aroused a great interest in recent years because these biopolymers possess several interesting biological activities and functional properties. Chitin recovered from Spiny lobster shells was used to produce low molecular weight chitosan for food applications using gamma irradiation. Lobster chitosan produced from this irradiation process had high antioxidant activity and was proposed to be utilized as a food preservative (García et al. 2015). Apart from that, chitosan originated from Norway lobster shells had excellent antimicrobial activity against several food-poisoning bacteria and fungus (Sayari et al. 2016). This makes lobster chitosan has a great potential to be used for preservation of foods from microbial deterioration. To recover soluble proteins from food-processing processes, several compounds could be used, but chitosan and chitosan complexes have a great potential (Lu et al. 2011; Wibowo et al. 2007); besides, chitosan prepared from lobster showed promising results for recovery of solids in food-processing plants (Defang et al. 2001). Moreover, chitin and chitosan could be used

for producing functional biomaterials (membranes, films, and packages) used for applications in the food industry (e.g., processing, packaging, or storage). Lobster chitosan films were used as the film matrix in combination with clay micro/nanoparticles for the preparation of chitosan/clay films (Casariego et al. 2008, 2009). With its significant improvements in physical properties (water solubility, water vapor, oxygen and carbon dioxide permeability, and optical, mechanical, and thermal properties), the lobster chitosan/clay film was proposed for use in coating in order to extend the shelf life of food products (Casariego et al. 2008, 2009). Apart from retarding moisture migration and the loss of volatile compounds, reducing the respiration rate, and delaying changes in textural properties, another advantage of chitosan films is that it is biodegradable. Thus, chitosan films are environmentally friendly alternatives to synthetic, nonbiodegradable films, which may be further modified to create biodegradable food packages (Hudson et al. 2015).

Pharmaceuticals and biomedicine

Recently, the applications of chitin, chitosan, and their derivatives in pharmaceuticals and biomedicines have received more attention not only because they are biocompatible, biodegradable, and nontoxic (Jeon and Kim 2000), but also because they exhibit several biological and physiological characteristics with known medical benefits. For example, these polymers and their derivatives are antioxidant, antimicrobial, anticancer, immune-stimulant, hypocholesterolemic, hypoglycemic, angiotensin-I-converting enzyme (ACE) inhibitors, and anticoagulant (Sayari et al. 2016; Wijesekara and Kim 2010).

Apart from the above, chitin and chitin derivatives also possess several physical properties that are favorable for their applications in pharmaceutical and biomedical sectors. Chitin and chitosan prepared from lobster were studied on their deformation and compaction properties for use as pharmaceutical direct compression excipients (Mir et al. 2008). In comparison with other established direct compression excipients (microcrystalline cellulose), lobster chitin/chitosan performed better at both tendencies of plastic deformation and compression behavior. This result indicates that lobster chitin and chitosan have a potential use as co-excipients for direct compression applications.

The most commonly used chitosan derivatives used in drug delivery are the water-soluble lobster chitosan acid salts (Cervera et al. 2011). Lobster chitosan salts prepared by spray-drying have a higher tendency toward sphericity, which are good excipients for pharmaceutical applications. Moreover, lobster chitosan acid salts maintain their physical, chemical, and microbiological characteristics for a period of 12 months when stored correctly at room temperature in a dry place (De la Paz et al. 2015). Apart from being stable for extended periods (1 year), the toxicity levels of lobster chitosan acid salts are negligible. This was indicated by a study investigating the single- and repeated-dose toxicity of chitosan and its salts (lactate and acetate) on rats. At oral doses of 2000 mg/kg, no fatalities or changes in the general behavior of the rats in both the acute- and repeated-dose toxicity studies were observed (Lagarto et al. 2015). This led us to a conclusion that chitosan obtained from lobster shells may be safe for use in the pharmaceutical industry.

The welling potential of lobster chitosan film in various solvents was determined for use as targeted drug delivery or drug film (Bamgbose et al. 2012). Since it can generate a membrane with structure porous and stability in several organic solvents, lobster chitosan film could be incorporated in devices for development of targeted drug delivery. Recently, novel hybrid biomaterials for medical application was also prepared from lobster chitosan (Malho et al. 2014). A bifunctional protein was successfully attached to lobster chitin to generate biosynthetic materials with advanced functional properties. Bio-inspired chitin/protein nanocomposites were developed using lobster chitin nanofibers and recombinant chitin-binding resilin (Qi 2015). Furthermore, chitosan nanoparticles with its antioxidant, antimicrobes in combination with high absorption have been used as pharmaceutical ingredients. Recently, Safitri et al. (2014) reported chitosan nanoparticles produced from lobster had positive results in prevention and treatment of halitosis. Based on these findings, the lobster chitosan nanoparticles were used as an active ingredient in a anti-halitosis bio-mouth spray. Ongoing advances such as these could considerably expand the use of lobster chitin nanofiber-based composites and functional materials.

Lobster lipids: as source of nutraceuticals, pharmaceuticals, and flavorants

Although lipids have often been condemned, the use of lipid and its products has drawn a dramatic interest in recent years due to findings related to their effects on human health. Apart from enhancing flavor, texture, and mouthfeel to foods, lipids also provide essential fatty acids [eicosapentaenoic acid (EPA), docosahexaenoic acid (DHA), arachidonic acid (AA), and γ-linolenic acid], fat-soluble vitamins (A, D, E, K), and other minor components (phospholipids, tocopherols, tocotrienols, carotenoids, sterols, and phenolic compounds) (Rizliya and Mendis 2014; Shahidi 2006). Particularly, the roles of carotenoids, EPA, and/or DHA in heart health, mental health, brain, and retina development, have been well documented (Alabdulkarim et al. 2012; Guerin et al. 2003; Swanson et al. 2012; Yamashita 2013), and such

lipid constituents have been regconized as nutraceuticals and pharmaceuticals for improving human health. By this reason, fish oils have been used for enriching DHA and EPA of many food products such as powder milk formulate, salad oil, fruit beverage, vegetable juice (Kolanowski and Berger 1999), dairy products (Kolanowski and Weißbrodt 2007), soft goat cheese (Hughes et al. 2012), or cookies (Jeyakumari et al. 2016). Lobster oils have a potential use as a source of natural ω-3 fatty acids for many fortified products since they have been demonstrated to contain PUFAs and ω-3 fatty acids as high as that of fish oil (menhaden) or krill oils (Albalat et al. 2016; Tsvetnenko et al. 1996). Moreover, higher bioavailability of fatty acids derived from crustacean oils compared to those of fish oils (Köhler et al. 2015) together with the antioxidant superior of crustacean oils provided by carotenoids make them ideal for application as a novel and beneficial food ingredient (Tetens 2009) or as oil supplement (Köhler et al. 2015). With a significant richness in astaxanthin, PUFAs, and ω-3 fatty acids, lobster oil was suggested for use as a dietary supplement (Nguyen 2017) since oils produced from fish livers are often considered as an important source of vitamins A and D with several therapeutic properties (Gunstone 2006; Rizliya and Mendis 2014). In addition, lobster liver oil contains very strong specific flavors, which combined with the unique and inherent ability of oils for absorbing and preserving flavors, would be a promising application for flavor industry. For this reason, lobster lipids were investigated for production of infused lobster oil, salt plated with lobster flavors, and lobster seasoning, and they obtained promising results (Nguyen 2017).

Astaxanthin as a powerful antioxidant

Oxidative molecules or free radicals such as hydroxyls, peroxides, and reactive oxygen species generated during normal aerobic metabolism are necessary for sustaining life processes. However, under certain conditions or periods of exposure such as physiological stress, air pollution, smoking, chemical inhalation, or exposure to UV light, the increased production of these free radicals can be detrimental. This threat arises due to the highly reactive nature of free radicals with essential cellular components such as proteins, lipids, carbohydrates, and DNA (Di Mascio et al. 1991). As a result of oxidative damage through a chain reaction known as oxidative stress, proteins and lipids are oxidized, while DNA is severely damaged. It has been suggested that diseases such as macular degeneration, retinopathy, carcinogenesis, arteriosclerosis, and Alzheimer may be induced by such damages (Maher 2000).

The human body controls and reduces oxidation by self-producing enzymatic antioxidants including catalase, peroxidase, super oxide dismutase, and other antioxidant activity molecules. However, the levels of these compounds in many cases are not sufficient to protect the body against oxidative stress, and an additional supplement of water-soluble antioxidants (vitamin C) and lipophilic antioxidants (vitamin E, carotenoids: beta-carotene and astaxanthin) are required. The use of astaxanthin as an antioxidant has been receiving significant attention because it possesses superior antioxidant activity. Antioxidant activity of astaxanthin was found to be 10 times higher than that of zeaxanthin, lutein, canthaxanthin, and β-caroten, and 100 times higher compared with vitamin E (Miki 1991). With its superior antioxidant activity, astaxanthin has been used as a natural antioxidant in edible oil (Rao et al. 2007), nutraceuticals (Guerin et al. 2003), and cosmetics (Tominaga et al. 2012). Particularly, astaxanthin has shown great potential for promoting human health and in the prevention/treatment of various diseases (Table 5). Its efficiency has been proven in over 65 clinical studies and featured in over 300 peer-reviewed publications (Yamashita 2013). It should be noted that other carotenoids can act as a prooxidant under specific conditions such as high oxygen and partial pressure, while there is currently no information available

regarding astaxanthin. Therefore, astaxanthin is considered as a high-value product due to its being increasingly marketed as a functional food ingredient with prices ranging between 3000 and $12,000 per kg (Lordan et al. 2011).

Industrially applicable techniques for efficient recovery of functional and bioactive nutraceuticals from lobster processing by-products

Extraction of functional and nutritional proteins by isoelectric solubilization/precipitation and ultrasound-assisted extraction

Commercial lobster processing for fresh lobster meat, picked lobster meat, or canned lobster generates large amounts of lobster by-products containing residual meat, which is not frequently recovered by hand, or using mechanical equipment. This underutilization of the lobster results in a waste with highly valuable protein, which must be disposed of accordingly, frequently at a cost to the producers. Several possibly suitable techniques for the recovery of proteins from crustacean-, fish-, or meat-processing by-products have been considered. One of the most conventional methods used to transform fishery by-products into a marketable and consumer-friendly

Table 5 Several beneficial effects of astaxanthin on promoting human health

Human health	Health benefits	References
Neurovascular protection	Decreases oxidation of red blood cells; decreases the chances of ischemic stroke; and improves memory and learning	Yook et al. (2015); Zhang et al. (2014a, b)
Eye fatigue relief	Reduces eye fatigue relieve in subjects suffering from visual display syndrome	Kajita et al. (2009); Kidd (2011); Nagaki et al. (2006); Serrano and Narducci (2014); Seya et al. (2009)
Immune system booster	Has an immunomodulating effect, strong immune system stimulator, anti-tumor, very effective for autoimmune conditions such as rheumatoid arthritis	Chew et al. (2010); Chew and Park (2004); Jyonouchi et al. (1995); Nir and Spiller (2002); Park et al. (2010)
Cardiovascular health	Improves blood lipid profiles, decreases blood pressure, offers protection from hypertension and stroke, reduces the consequences of a heart attack and vascular inflammation, reduces the area of infarction and the damage, reduces the area of infarction and the damage	Fassett and Coombes (2012); Fassett and Coombes (2011); Gross and Lockwood (2005); Guerin et al. (2003); Hussein et al. (2006); Hussein et al. (2005); Iwamoto et al. (2000); Miyawaki et al. (2008)
Liver health and metabolic syndrome	Improves blood lipids and increases adiponectin, prevents fatty liver disease, reduces the risk of atherosclerotic plaque, inhibits progression of fatty liver disease, restores insulin–glucose balance, increases fat burning, and decreases inflammatory markers	Kindlund and BioReal (2011); Kishimoto et al. (2016); Shen et al. (2014); Yilmaz et al. (2015)
Diabetes and Kidneys	Reduces glucose toxicity and kidney inflammation; improves pancreatic function, insulin resistance, and insulin sensitivity	Naito et al. (2004); Ni et al. (2015); Savini et al. (2013); Uchiyama et al. (2002)
Fertility	Improves sperm parameters and fertility	Comhaire et al. (2005); Donà et al. (2013); Mina et al. (2014)
Muscle resilience	Enhances power output, endurance, and recovery after exercise; prevents muscle damage and muscle atrophy	Earnest et al. (2011); Malmstena and Lignellb (2008); Yamashita (2011)
Capillary circulation	Improves blood flow and capillary integrity; reduces blood cell oxidation and risk of thrombosis	Kanazashi et al. (2013)
Anti-aging (skin cells)	Prevents UV-induced wrinkle formation, skin sagging, and age-spots; improves skin elasticity and skin dryness	Seki et al. (2001); Tominaga et al. (2012); Yamashita (2005)

products is the use of endogenous or added proteolytic enzymes (Tong-Xun and Mou-Ming 2010; Venugopal and Shahidi 1995). However, the slow rate of hydrolysis, generation of short-chain peptides, loss of functionality of the native proteins, and the absence of homogeneous hydrolysates are other major limitations of this process (Kristinsson and Rasco 2000). Moreover, enzymes used often require an inactivation step and thus cannot be recycled for subsequent reactions leading to the rise of processing costs (Kristinsson and Rasco 2000). In addition, low yield, taste defects, and the overall economic feasibility are still major issues for using enzymatic hydrolysis on industrial scales. Other approaches, which have been explored, include the use of chemicals. Using acidic or alkaline solutions for degrading proteins into peptides of varying sizes is a nonselective and rapid method. However, the severe conditions (HCl 6 N, 118 °C, 18 h, or pH 12.5, 95 °C, 20 min) used in chemical hydrolysis can have negative consequences such as racemization, bitter taste, reduced nutritional quality, and poor functionality, resulting in products of lower value, i.e., fertilizer (Chobert et al. 1996).

A more promising technique is isoelectric solubilization and precipitation (ISP). The shifting in pH of the solutions used during this processing induces solubility of residual proteins while simultaneously separating of lipids and the inedible parts such as shells, membranes, bones, scales, and skin, not intended for human consumption (Gehring et al. 2011). Apart from its generating high yield of protein recovery, this process also produces high-quality proteins, which still maintain their functional properties and nutritional value (Chen et al. 2007b; Gigliotti et al. 2008; Nolsoe and Undeland 2009; Taskaya et al. 2009a; Taskaya et al. 2009b). Since the ISP process is simple and quick, it has been used for the recovery of fish proteins at both laboratory and pilot scales using batch mode (Choi and Park 2002; Kim et al. 2003; Kristinsson and Hultin 2003; Mireles Dewitt et al. 2002; Undeland et al. 2002). Furthermore, the presence of dioxin and polychlorinated biphenyls (PCBs), one of the most predominant bio-toxic compounds in fish protein, has been found to be reduced significantly in the ISP-recovered proteins (Marmon et al. 2009). When applied to other biological waste materials such as beef-processing by-products (Chen et al. 2007a; Mireles Dewitt et al. 2002) or chicken-processing by-products (Tahergorabi et al. 2011; Tahergorabi et al. 2012), high yields of protein recovery were still achieved. With its significant advantages over conventional methods, the ISP process could have a great potential for application on protein recovery from LPBs.

Recently, ultrasound-based extractions have been shown to be an effective technique for improving the rates of various extraction processes (Lebovka et al. 2011; Majid et al. 2015; Vilkhu et al. 2008). Ultrasound processing disrupts cells and creates microcavities in the tissue, which enhances the surface area and thus in the penetration of the solvent into the material, mass transfer, and improves protein release. The use of ultrasound in protein extraction from fish, meat, and beef by-products resulted in higher extraction yields with reduced processing time and solvent consumption compared with the use of ISP alone (Chemat and Khan 2011; Saleem et al. 2015; Vardanega et al. 2014; Vilkhu et al. 2008). Therefore, recovery of protein by ultrasound-assisted extraction has been scaled up to an industrial level due to their high economic feasibility (Álvarez and Tiwari 2015; Tu et al. 2015).

Supercritical fluid extraction for recovery of rich-ω-3 lipids and astaxanthin

Solvent extraction is the most common method for lipid or astaxanthin extraction (Sindhu and Sherief 2011; Tsvetnenko et al. 1996). In this method, organic solvents including acetone, ethyl acetate, hexane, isopropanol, methanol, methyl ethyl ketone, ethanol, dichloromethane, dimethyl sulfoxide, or chloroform may be used for the extraction. Although some of these solvents can be used to extract lipids for food applications, others such as dichloromethane, dimethyl sulfoxide, and chloroform cannot be used due to their toxicity (FDA 2010). Regardless of the solvent used, there is increasing public awareness of the hazards related to the use of any organic solvents in the extraction of compounds for food or medical applications due to the possibility of solvent contamination in the final extracts. Apart from this, high demands for natural astaxanthin and bioactive lipid components such as ω-3 fatty acids, physterols, tocopherols, and tocotrineols have stimulated the search for green and sustainable extraction methods (Delgado Vargas and Paredes-Lopez 2003). One such approach that is being considered is the use of supercritical fluid extraction techniques.

In recent years, supercritical fluid extraction (SFE) has become an important technology for extracting high-quality lipids from fishery-processing by-products (Letisse et al. 2006; Rubio-Rodríguez et al. 2008). Furthermore, it is an effective separation technique in the production of nutraceutical supplements and functional foods (Parajó et al. 2008; Reverchon and De Marco 2006). The operational conditions of SFE are also favorable from an environmental and industrial processing viewpoint. SFE can extract bioactive nutraceuticals at moderate temperatures in a non-oxygen environment with very low lipid oxidation, selectively extracts low polar lipid compounds, and does not co-extract polar impurities such as some organic derivatives containing heavy metals (Rubio-Rodríguez et al. 2012).

Supercritical carbon dioxide (SC-CO$_2$) extraction is presently being evaluated as a promising technology compared with conventional methods (López-Cervantes et al. 2006) due to its ability to extract the heat-sensitive, easily oxidized compounds (PUFAs, ω-3 fatty acids, and astaxanthin) without the use of toxic solvents. Moreover, CO$_2$ is a generally recognized as safe (GRAS), relatively cheap, and easy to evaporate from the matrix and extracts (Mercadante 2008; Reverchon and De Marco 2006; Sahena et al. 2009). The high extraction yields achieved using this technique is attributable to the high diffusivity and solubility but low viscosity of SC-CO$_2$. In contrast to the products extracted from the conventional methods, the SC-CO$_2$ extracts are rich in nutraceuticals with high purity. SC-CO$_2$ have been used to extract lipids and carotenoids from vegetables (Filho et al. 2008; Hardardottir and Kinsella 1988; Mendes et al. 1995; Silva et al. 2008) and animal matrices (Froning et al. 1990; Hardardottir and Kinsella 1988; Letisse et al. 2006; Tanaka and Ohkubo, 2003). The SC-CO$_2$ extraction technique has also been studied for extraction of lipids and astaxanthin from crustacean-processing waste such as shrimp by-products (Charest et al. 2001; Felix-Valenzuela et al. 2001; Kamaguchi et al. 1986; Lopez et al. 2004). Recently, ω-3-rich lipids have been recovered with high yield (94%) from Rock lobster livers by SC-CO$_2$ extraction (Nguyen et al. 2015).

Microwave-intensified production of chitin and chitin derivatives

Chitin is found in lobster shells and is closely associated with proteins, minerals, and pigments, which need to be completely removed and separated from chitin during the extraction process. Although conventional methods removes nearly all these compounds from the shells, the use of strong chemicals and high temperatures during the process can cause deacetylation and depolymerisation leading to inconsistent physical properties of the extracted chitin (Jung et al. 2007; Kjartansson et al. 2006; Percot et al. 2003). In addition, under these harsh conditions, undesirable secondary reactions between amino acids and the alkaline medium as well as racemization occur rending the proteins and minerals unusable (Synowiecki and Al-Khateeb 2003). Moreover, the conventional method requires large volumes of water for washing steps, generating a huge amount of waste water (Wang and Chio 1998). To circumvent these issues, various biological processes have been employed for the production of chitin from crustacean shells such as fermentation or using commercially available enzymes (Giyose et al. 2010; Jung et al. 2007; Manni et al. 2010; Oh et al. 2007; Sini et al. 2007; Sorokulova et al. 2009; Xu et al. 2008). However, these proposed new methods require a longer

production time (8–72 h), while their removal degrees (deproteinization, demineralization) are relatively low.

More recently, microwave has emerged as a promising nonconventional energy source for performing organic synthesis. Heat generated from microwave irradiation accelerates chemical reactions and enhances the rate of enzyme-catalyzed reactions so spectacularly that it cannot be explained by the effect of rapid heating alone (De La Hoz et al. 2005). Apart from thermal effects, microwave irradiation is accompanied with several nonthermal effects such as overheating, hot spots, selective heating, highly polarizing field, and mobility and diffusion. Microwave-assisted extraction has been proven to be an efficient technique for extracting small molecular weight from various biological samples due to its high ability to intensify processes, low usage of extraction chemicals, and shorter extraction time (Teo et al. 2013).

Microwave technology has been used as an environmentally friendly and cost-effective method for chitin production by assisting demineralization of deproteinized shrimp shells with lactic acid (Valdez-Pena et al. 2010). This process produced high yields of chitin with low residual minerals (0.2%); thus, demineralization of lobster shells by the microwave process was also optimized for chitin production obtaining promising results such as high degree of demineralization, low residues, and recovery of lobster minerals (Nguyen et al. 2017). The rate and yield of extraction can be further exploited by combining multiple processing approaches such as using microwave technology to increase the rate of enzymatic hydrolysis of lobster shells (Nguyen et al. 2016). Microwave irradiation has also been demonstrated as highly efficient in chemical deacetylation of chitin into chitosan. The degree of deacetylation in chitosan production within 5.5 min of microwave irradiation was as high as it was deacetylated at 121 °C, 15 psi for 4 h (Sahu et al. 2009). To generate chitosan with high solubility or desired functional properties, microwave irradiation has been extensively utilized for the chemical modification of chitosan (Ge and Luo 2005; Huacai et al. 2006; Liu et al. 2004). Particularly, the microwave has been demonstrated as a green and sustainable technology for degradation of chitin (Ajavakom et al. 2012; Roy et al. 2003) and chitosan (García et al. 2015; Li et al. 2012; Petit et al. 2015; Wasikiewicz and Yeates 2013) into low molecular weight chitosan or chito-oligomers in a wide range of applications.

Conclusion

The global lobster processing industry produces a large amount of by-products with an estimated yield of 50,000 tons, which are currently being underutilized or discarded annually costing lobster companies in excess of $7.5 million/year for disposal. Finding alternative uses for this waste material could result in more environmentally friendly processes and ultimately result in an economic advantage for the lobster industry. This comprehensive review discusses the global availability of LPBs, the composition of the by-products, and the high-value compounds that may be extracted from them. It also discussed the key areas of applications—that these extracted compounds show potential to be used in—like water treatment, agriculture, food, nutraceutical, pharmaceutical products, and biomedicine. Furthermore, it addresses the limitations to current techniques used for the recovery of these valuable components and suggests emerging innovative techniques (i.e., microwave, ultrasonic, and supercritical fluid extraction) which may be more promising at the industrial scale. Although the potential value of such lobster ingredients is currently being ignored, several recent studies on biorefinery have shown that recovery of these valuable ingredients for value-added products is very promising due to their richness, highly commercial value, and numerous applications. Developing the simplified processes combined with using industrially applicable technologies for economic recovery of these valuable ingredients would be a practical solution for maximizing the utilization of these by-products. In this way, LPBs could be economically turned into a highly profitable source rather than a traditional pollution and costing source.

Abbreviations
LPBs: lobster processing by-products; LSP: lobster shell protein; LPH: lobster protein hydrolysate; COS: chito-oligosaccharides; PUFAs: polyunsaturated fatty acids; SC-CO$_2$: supercritical carbon dioxide; SFE: supercritical fluid extraction; EAA: essential amino acids.

Authors' contributions
All authors were involved in the drafting and revision of the manuscript. All authors read and approved the final manuscript.

Author details
[1] Centre for Marine Bioproducts Development, Flinders University, Adelaide, Australia. [2] Department of Medical Biotechnology, School of Medicine, Flinders University, Adelaide, Australia. [3] Department of Food Science and Technology, Agricultural and Natural Resources Faculty, An Giang University, Long Xuyen, Vietnam. [4] Centre for NanoScale Science Technology (CNST), Chemical and Physical Sciences, Flinders University, Adelaide, Australia.

Acknowledgements
The authors wish to thank the Australian Government for offering Trung Nguyen a Ph.D. scholarship. Thanks to the South Australian Government, the Ferguson Lobster Co. Ltd, and the Premier's Research and Industry Fund—Innovation Voucher Project (PRIF-IVP) for financial support. Thanks to the Flinders Medical Biotechnology Department and Centre for Marine Bio-products Development for technical supports.

Competing interests
The authors declare that they have no competing interests.

Funding
This manuscript is a part of the South Australia Premier's Research and Industry Fund—Innovation Voucher Project (PRIF-IVP37) which was co-funded by the South Australia government and Ferguson Australia Pty Ltd.

References
Ajavakom A, Supsvetson S, Somboot A, Sukwattanasinitt M (2012) Products from microwave and ultrasonic wave assisted acid hydrolysis of chitin. Carbohydr Polym 90(1):73–77

Alabdulkarim B, Bakeet ZAN, Arzoo S (2012) Role of some functional lipids in preventing diseases and promoting health. J King Saud Univ-Sci 24(4):319–329

Albalat A, Nadler LE, Foo N, Dick JR, Watts AJ, Philp H, Neil DM, Monroig O (2016) Lipid composition of oil extracted from wasted Norway lobster (Nephrops norvegicus) heads and comparison with oil extracted from Antarctic krill (Euphasia superba). Mar Drugs 14(12):219

Álvarez C, Tiwari BK (2015) Ultrasound assisted extraction of proteins from fish processing by-products. In: institute of food technologist. Chicago, US

Annie T, McCarron P (2006) Lobster market overview. Maine Lobstermen's Association

Auerswald L, Gäde G (2008) Simultaneous extraction of chitin and astaxanthin from waste of lobsters Jasus lalandii, and use of astaxanthin as an aquacultural feed additive. Afr J Mar Sci 30(1):35–44

Badawy MEI, Rabea EI (2011) A biopolymer chitosan and its derivatives as promising antimicrobial agents against plant pathogens and their applications in crop protection. Int J Carbohydr Chem 2011:1–29

Bamgbose JT, Bamigbade AA, Adewuyi S, Dare EO, Lasisi AA, Njah AN (2012) Equilibrium swelling and kinetic studies of highly swollen chitosan film. J Chem Chem Eng 6(3):272–283

Barker E, Rossbach M (2013) Western rock lobster fishery—2013/2014 season. Commercial fisheries production bulletin, vol 48. Department of fisheries, Government of western Australia, pp 1–8

Battison AL, Summerfield R, Patrzykat A (2008) Isolation and characterisation of two antimicrobial peptides from haemocytes of the American lobster Homarus americanus. Fish Shellfish Immunol 25(1):181–187

Bayer RC (2015) Lobster hemolymph as a utility for treatment of mammalian tissue lesions, Google Patents

Borges AA, Borges-Pérez A, Gutiérrez A, Paz-Lago D, Cabrera G, Fernández M, Ramírez MA, Acosta A (2000) Tomato-Fusarium oxysporum interactions: l-chitosan and MSB effectively inhibits fungal growth. Cultivos Trop 21(4):13–16

Cabrera JC, Van Cutsem P (2005) Preparation of chitooligosaccharides with degree of polymerization higher than 6 by acid or enzymatic degradation of chitosan. Biochem Eng J 25(2):165–172

Cabrera J-C, Boland A, Cambier P, Frettinger P, Van Cutsem P (2010) Chitosan oligosaccharides modulate the supramolecular conformation and the biological activity of oligogalacturonides in Arabidopsis. Glycobiology 20(6):775–786

Casariego A, Souza B, Cruz L, Díaz R, Teixeira J, Vicente A (2008) Chitosan coating and films: evaluation of surface, permeation, mechanical and thermal propertiess. Valnatura-A Europe-Latin América post-graduate research network in the valorization of natural resources, institute for biotechnology and bioengineering

Casariego A, Souza B, Cerqueira M, Teixeira J, Cruz L, Díaz R, Vicente A (2009) Chitosan/clay films 'properties as affected by biopolymer and clay micro/nanoparticles' concentrations. Food Hydrocoll 23(7):1895–1902

Cervera MF, Heinämäki J, de la Paz N, López O, Maunu SL, Virtanen T, Hatanpää T, Antikainen O, Nogueira A, Fundora J (2011) Effects of spray drying on physicochemical properties of chitosan acid salts. AAPS Pharm Sci Tech 12(2):637–649

Charest DJ, Balaban MO, Marshall MA, Cornell JA (2001) Astaxanthin extraction from crawfish shells by supercritical CO$_2$ with ethanol as cosolvent. J Aquat Food Prod Technol 3:81–96

Chemat F, Khan MK (2011) Applications of ultrasound in food technology: processing, preservation and extraction. Ultrason Sonochem 18(4):813–835

Chen YC, Nguyen J, Semmens K, Meamer S, Jaczynski J (2007a) Physicochemical changes in omega-3-enhanced farmed rainbow trout (Oncorhynchus mykiss) muscle during refrigerated storage. Food Chem 104(3):1143–1152

Chen YC, Tou JC, Jaczynski J (2007b) Amino acid, fatty acid and mineral profiles of materials recovered from rainbow trout (Oncorhynchus mykiss)

Chen X, Yang H, Yan N (2016) Shell biorefinery: dream or reality? Chem-A Eur J 22:13402–13421

Chen X, Yang H, Zhong Z, Yan N (2017) Base-catalysed, one-step mechano-chemical conversion of chitin and shrimp shells into low molecular weight chitosan. Green Chem 19(12):2783–2792

Chew BP, Park JS (2004) Carotenoid action on the immune response. J Nutr 134(1):257S–261S

Chew B, Park J, Chyun J, Mahoney M, Line L (2010) Astaxanthin decreased oxidative stress and inflammation and enhanced immune response in humans. Nutr Metab 7(1):18

Chobert JM, Briand L, Gueguen J, Popineau Y, Larre C, Haertle T (1996) Recent advances in enzymatic modifications of food proteins for improving their functional properties. Nahr-Food 40(4):177–182

Choi YJ, Park JW (2002) Acid-aided protein recovery from enzyme-rich Pacific whiting. J Food Sci 67:2962–2967

Comhaire F, Garem YE, Mahmoud A, Eertmans F, Schoonjans F (2005) Combined conventional/antioxidant "Astaxanthin" treatment for male infertility: a double blind, randomized trial. Asian J Androl 7(3):257–262

Cousins JA (1997) Primary resource industry waste on Prince Edward Island. The PEI department of fishery and environment

Da Silveira FF, De Souza KL, Nunes Filho A, de Lima Mendes JU, Ladchuma-nanandasivan R (2013) Synthesis and characterization of chitosan used in the treatment of epithelial lesions. In: 22nd international congress of mechanical engineering. Brazil

Daniel M (2008) Evaluating by-products of the Atlantic shellfish industry as alternative feed ingredients for laying hens. Poult Sci 91:2189–2200

Dauphin L (1991) Enhancing value of lobster waste by enzymatic methods. In: food and agriculture chemistry, vol. Master, Mc Gill. Canada

De La Hoz A, Diaz-Ortiz A, Moreno A (2005) Microwaves in organic synthesis. Thermal and non-thermal microwave effects. Chem Soc Rev 34(2):164–178

De la Paz N, García C, Fernández M, García L, Martínez V, López OD, Nogueira A (2015) Stability of spray-dried chitosan salts derived from lobster chitin as a raw material. Ars Pharm 56(4):217–224

Deepmala K, Hemantaranjan A, Bharti S, Nishant Bhanu A (2014) A future perspective in crop protection: chitosan and its oligosaccharides. Adv Plantsand Agric Res 1(1):1–8

Defang Z, Gang Y, Pengyi Z, Zhiwei F (2001) The modified process for preparing natural organic polymer flocculant chitosan. Chin J Environ Sci 22(3):123–125

Delgado Vargas F, Paredes-Lopez (2003) Carotenoids, CRC Press. LLC NW

Denise S, Jason B (2012) Food grade astaxanthin from lobster shell discards. Maine agricultural center

Di Mascio P, Murphy ME, Sies H (1991) Antioxidant defense systems: the role of carotenoids, tocopherols, and thiols. Am J Clin Nutr 53(1):194S–200S

Donà G, Kožuh I, Brunati AM, Andrisani A, Ambrosini G, Bonanni G, Ragazzi E, Armanini D, Clari G, Bordin L (2013) Effect of astaxanthin on human sperm capacitation. Mar Drugs 11(6):1909–1919

Earnest CP, Lupo M, White K, Church T (2011) Effect of astaxanthin on cycling time trial performance. Int J Sports Med 32(11):882–888

El Hadrami A, Adam LR, El Hadrami I, Daayf F (2010) Chitosan in plant protection. Mar Drugs 8(4):968–987

Falcón Rodríguez A, Rodríguez AT, Ramírez MA, Rivero D, Martínez B, Cabrera JC, Costales D, Cruz A, González LG, Jiménez MC (2010) Chitosans as bioactive macromolecules to protect conomically relevant crops from their main pathogens. Biotecnol Apl 27(4):305–309

Falcón A, Ramírez M, Márquez R, Hernández M (2002) Chitosan and its hydrolysate at tobacco-Phytophthora parasitica interaction. Cultivos Trop 23(1):61–66

Fassett RG, Coombes JS (2011) Astaxanthin: a potential therapeutic agent in cardiovascular disease. Mar Drugs 9(3):447–465

Fassett RG, Coombes JS (2012) Astaxanthin in cardiovascular health and disease. Molecules 17(2):2030–2048

FDA (2010) Listing of food additives status, vol 2014, FDA. EUA

Felix-Valenzuela L, Higuera-Ciapara I, Goycoolea FM, Arguelles-Monal W (2001) Supercritical CO2/ethanol extraction of astaxanthin from blue crab (Calinectes sapidus) shell waste. J Food Eng 24:101–112

Filho GL, De Rosso VV, Meireles MAA, Rosa PTV, Oliveira AL, Mercadante AZ, Cabral FA (2008) Supercritical CO2 extraction of carotenoids from pitanga fruits (Eugenia uniflora L.). J Supercrit Fluids 46:33–39

Fredrick WS, Ravichandran S (2012) Hemolymph proteins in marine crustaceans. Asian Pac J Trop Biomed 2(6):496–502

Froning GW, Wheling RL, Cuppett SL, Pierce MM, Niemann L, Seikman DK (1990) Extraction of cholesterol and other lipids from dried egg yolk using supercritical carbon dioxide. J Food Sci 55:95–98

Gäde G, Auerswald L (2005) The West Coast rock lobster Jasus lalandii as a valuable source for chitin and astaxanthin. Afr J Mar Sci 27(1):257–264

Garcia MA, de la Paz N, Castro C, Rodriguez JL, Rapado M, Zuluaga R, Ganán P, Casariego A (2015) Effect of molecular weight reduction by gamma irradiation on the antioxidant capacity of chitosan from lobster shells. J Radiat Res Appl Sci 8(2):1–11

Gary M (2012) Southern rock lobster strategic plan 2011–2016. Southern Rock Lobster Ltd

Ge H-C, Luo D-K (2005) Preparation of carboxymethyl chitosan in aqueous solution under microwave irradiation. Carbohydr Res 340(7):1351–1356

Gehring CK, Gigliotti JC, Moritz JS, Tou JC, Jaczynski J (2011) Functional and nutritional characteristics of proteins and lipids recovered by isoelectric processing of fish by-products and low value-fish: a review. Food Chem 124:422–431

Gigliotti JC, Jaczynski J, Tou JC (2008) Determination of nutritional value, protein quality and safety of krill protein concentrate isolated using an isoelectric solubilization/precipitation technique. Food Chem 111(1):209–214

Giyose N, Mazomba N, Mabinya L (2010) Evaluation of proteases produced by Erwinia chrysanthemi for the deproteinization of crustacean waste in a chitin production process. Afr J Biotech 9(5):707–710

Gross GJ, Lockwood SF (2005) Acute and chronic administration of disodium disuccinate astaxanthin (CardaxTM) produces marked cardioprotection in dog hearts. Mol Cell Biochem 272(1–2):221–227

Guerin M, Huntley ME, Olaizola M (2003) Haematococcus astaxanthin: applications for human health and nutrition. Trends Biotechnol 21(5):210–216

Gunstone FD (2006) Modifying lipids for use in food, 1st edn. Woodhead Publishing, Cambridge

Gustavo C, Galo CT, Alexei TVE (2005) Retention capacity of chitosan for copper and mercury ions. J Chil Chem Soc 48(1):1–11

Hamed I, Özogul F, Regenstein JM (2016) Industrial applications of crustacean by-products (chitin,chitosan, and chitooligosaccharides): a review. Trends Food Sci Technol 48:40–50

Hardardottir I, Kinsella JE (1988) Extraction of lipids and cholesterol from fish muscle with supercritical fluid. J Food Sci 53:1656–1661

Hareyan A (2007) Chitin from lobster shell shows great healing, bio-stimulant properties, vol 2014. Universidad de Granada, Cuba

Harikrishnan R, Kim J-S, Balasundaram C, Heo M-S (2012) Dietary supplementation with chitin and chitosan on haematology and innate immune response in Epinephelus bruneus against Philasterides dicentrarchi. Exp Parasitol 131(1):116–124

He S, Nguyen T, Zhang W, Peng S (2016) Protein hydrolysates produced from Rock lobster (Jasus edwardsii) head: emulsifying capacity and food safety. Food Sci Nutr 4(6):869–877

Hipulan JRAM (2005) Chitosan from waste shrimp and lobster shells as plant growth regulator. In: intel Philippines science fair. Cagayan De Oro City

Holmyard N, Franz N (2006) Lobster markets. FAO

Huacai G, Wan P, Dengke L (2006) Graft copolymerization of chitosan with acrylic acid under microwave irradiation and its water absorbency. Carbohydr Polym 66(3):372–378

Hudson R, Glaisher S, Bishop A, Katz JL (2015) From lobster shells to plastic objects: a bioplastics activity. J Chem Educ 92(11):1882–1885

Hughes BH, Brian Perkins L, Calder BL, Skonberg DI (2012) Fish oil fortification of soft goat cheese. J Food Sci 77(2):S128–S133

Hussein G, Nakamura M, Zhao Q, Iguchi T, Goto H, Sankawa U, Watanabe H (2005) Antihypertensive and neuroprotective effects of astaxanthin in experimental animals. Biol Pharm Bull 28(1):47–52

Hussein G, Goto H, Oda S, Sankawa U, Matsumoto K, Watanabe H (2006) Antihypertensive potential and mechanism of action of astaxanthin: III. Antioxidant and histopathological effects in spontaneously hypertensive rats. Biol Pharm Bull 29(4):684–688

Iglesias R, Gutierrez A, Fernandez F (1994) The influence of chitin from lobster exoskeleton seedling growth and mycorrhizal infection in tomato crop (Lycopersion esculentum mill). Cultivos Trop 15(2):48–49

Ilangumaran G (2014) Microbial degradation of lobster shells to extract chitin derivatives for plant disease management, vol. Master of Science, Dalhousie. Halifax, Nova Scotia

Iwamoto T, Hosoda K, Hirano R, Kurata H, Matsumoto A, Miki W, Kamiyama M, Itakura H, Yamamoto S, Kondo K (2000) Inhibition of low-density lipoprotein oxidation by astaxanthin. J Atheroscler Thromb 7(4):216–222

Jeon YJ, Kim SK (2000) Production of chitooligosaccharides using an ultrafiltration membrane reactor and their antibacteria activity. Carbohydr Polym 41:133–141

Jeyakumari A, Janarthanan G, Chouksey M, Venkateshwarlu G (2016) Effect of fish oil encapsulates incorporation on the physico-chemical and sensory properties of cookies. J Food Sci Technol 53(1):856–863

Juang RS, Tseng RL, Wu FC, Lin SJ (1996) Use of chitin and chitosan in lobster shell wastes for color removal from aqueous solutions. J Environ Sci Health, Part A 31(2):325–338

Juang RS, Tseng RL, Wu FC, Lee SH (1997) Adsorption behavior of reactive dyes from aqueous solutions on chitosan. J Chem Technol Biotechnol 70(4):391–399

Jung W, Jo G, Kuk J, Kim Y, Oh K, Park R (2007) Production of chitin from red crab shell waste by successive fermentation with Lactobacillus paracasei KCTC-3074 and Serratia marcescens FS-3. Carbohydr Polym 68(4):746–750

Jyonouchi H, Sun S, Tomita Y, Gross MD (1995) Astaxanthin, a carotenoid without vitamin A activity, augments antibody response in cultures including T-helper cell clones and suboptimal doses of antigen. J Nutr 125(10):2483–2492

Kajita M, Tsukahara H, Kato M (2009) The effects of a dietary supplement containing astaxanthin on the accommodation function of the eye in middle-aged and older people. Med Consult New Rem 46:89–93

Kamaguchi K, Murakami M, Nakano H, Konosu S, Kokura T, Yamamoto H, Kosaka M, Hata K (1986) Supercritical carbon dioxide extraction of oils from Antartic krill. J Agric Food Chem 34:904–907

Kanazashi M, Okumura Y, Al-Nassan S, Murakami S, Kondo H, Nagatomo F, Fujita N, Ishihara A, Roy R, Fujino H (2013) Protective effects of astaxanthin on capillary regression in atrophied soleus muscle of rats. Acta Physiol 207(2):405–415

Karagozlu MZ, Kim S-K (2015) Anti-cancer effects of chitin and chitosan derivatives. In: Kim S-K (ed) Handbook of anticancer drugs from marine origin. Springer, Heidelberg, pp 413–421

Kaur S, Dhillon GS (2013) Recent trends in biological extraction of chitin from marine shell wastes: a review. Crit Rev Biotechnol 35(1):1–18

Kidd P (2011) Astaxanthin, cell membrane nutrient with diverse clinical benefits and anti-aging potential. Altern Med Rev 16(4):355–364

Kim YS, Park JW, Choi YJ (2003) New approaches for the effective recovery of fish proteins and their physicochemical characteristics. Fish Sci 69(6):1231–1239

Kindlund P, BioReal A (2011) AstaREAL®, natural astaxanthin–nature's way to fight the metabolic syndrome. Wellness Foods 1:8–13

Kishimoto Y, Yoshida H, Kondo K (2016) Potential anti-atherosclerotic properties of astaxanthin. Mar Drugs 14(2):1–13

Kjartansson GT, Zivanovic S, Kristbergsson K, Weiss J (2006) Sonication-assisted extraction of chitin from shells of fresh water prawns (Macrobrachium rosenbergii). J Agric Food Chem 54:3317–3323

Knuckey I (2004) Utilisation of seafood processing waste—challenges and opportunites. In: Australian New Zealand soils conference. University of Sydney, Australia

Köhler A, Sarkkinen E, Tapola N, Niskanen T, Bruheim I (2015) Bioavailability of fatty acids from krill oil, krill meal and fish oil in healthy subjects–a randomized, single-dose, cross-over trial. Lipids Health Dis 14(19):1–10

Kolanowski W, Berger S (1999) Possibilities of fish oil application for food products enrichment with omega-3 PUFA. Int J Food Sci Nutr 50(1):39–49

Kolanowski W, Weißbrodt J (2007) Sensory quality of dairy products fortified with fish oil. Int Dairy J 17(10):1248–1253

Kozo F (1997) Extraction of astaxanthin from shell of lobster or shrimp or crab and apparatus therefore. In: Patent J (ed) Japanese patent, vol JP A9301950. Japan

Kristinsson HG, Hultin HO (2003) Changes in conformation and subunit assembly of cod myosin at low and high pH after subsequent refolding. J Agric Food Chem 51(24):7187–7196

Kristinsson HG, Rasco BA (2000) Fish protein hydrolysates: production, biochemical and functional properties. Crit Rev Food Sci Nutr 40:43–81

Kuhn R, Soerensen N (1938) The coloring matters of the lobster (Astacus gammarus L.). Z Angew Chem 51:465–466

Kumar MNR (2000) A review of chitin and chitosan applications. React Funct Polym 46(1):1–27

Kurita K (2006) Chitin and chitosan: functional biopolymers from marine crustaceans. Mar Biotechnol 8(3):203–226

Lagarto A, Merino N, Valdes O, Dominguez J, Spencer E, de la Paz N, Aparicio G (2015) Safety evaluation of chitosan and chitosan acid salts from Panurilus argus lobster. Int J Biol Macromol 72:1343–1350

Lebovka N, Vorobiev E, Chemat F (2011) Enhancing extraction processes in the food industry. CRC Press, Boca Raton

Lee JE, Peniston Q (1982) Utilization of shellfish waste for chitin and chitosan production. In: Roy EM (ed) Chemistry and biochemistry of marine food products, vol 2. Academic Press, Cambridge, pp 1–5

Lee GH, Suriyaphan O, Cadwallader KR (2001) Aroma components of cooked tail meat of American lobster (Homarus americanus). J Agricand Food Chem 49(9):4324–4332

Letisse M, Rozieres M, Hiol A, Sergent M, Comeau L (2006) Enrichment of EPA and DHA from sardine by supercritical fluid extraction without organic modifier: I. Optimization of extraction conditions. J Supercrit Fluids 38(1):27–36

Li K, Xing R, Liu S, Qin Y, Meng X, Li P (2012) Microwave-assisted degradation of chitosan for a possible use in inhibiting crop pathogenic fungi. Int J Biol Macromol 51(5):767–773

Lien LJ (2004) Development of value added foods from low value lobster portions. In: school of agriculture and wine, vol. The bachelor of food technology and management, The University of Adelaide. Adelaide

Liu L, Li Y, Li Y, Fang Y-E (2004) Rapid N-phthaloylation of chitosan by microwave irradiation. Carbohydr Polym 57(1):97–100

Lopez M, Arce L, Garrido J, Rios A, Valcarcel M (2004) Selective extraction of astaxanthin from crustaceans by use of supercritical carbon dioxide. Talanta 64:726–731

López-Cervantes J, Sánchez-Machado DI, Rosas-Rodríguez JA (2006) Analysis of free amino acids in fermented shrimp waste by high-performance liquid chromatography. J Chromatogr A 1105:106–110

Lordan S, Ross RP, Stanton C (2011) Marine bioactives as functional food ingredients: potential to reduce the incidence of chronic diseases. Mar Drugs 9(6):1056–1100

Lu Y, Shang Y, Huang X, Chen A, Yang Z, Jiang Y, Cai J, Gu W, Qian X, Yang H (2011) Preparation of strong cationic chitosan-graft-polyacrylamide flocculants and their flocculating properties. Ind Eng Chem Res 50(12):7141–7149

Maher T (2000) Astaxanthin. Continuing education module 1:1–5

Majid I, Nayik GA, Nanda V (2015) Ultrasonication and food technology: a review. Cogent Food Agric 1(1):1071022

Malho J-M, Heinonen H, Kontro I, Mushi NE, Serimaa R, Hentze H-P, Linder MB, Szilvay GR (2014) Formation of ceramophilic chitin and biohybrid materials enabled by a genetically engineered bifunctional protein. Chem Commun 50(55):7348–7351

Malmstena CL, Lignellb Å (2008) Dietary supplementation with astaxanthin-rich algal meal improves strength endurance–A double blind placebo controlled study on male students. J Carotenoid Sci 13:20–22

Manni L, Ghorbel-Bellaaj O, Jellouli K, Younes I, Nasri M (2010) Extraction and characterization of chitin, chitosan, and protein hydrolysates prepared from shrimp waste by treatment with crude protease from Bacillus cereus SV1. Appl Biochem Biotechnol 162(2):345–357

Marmon SK, Liljelind P, Undeland I (2009) Removal of lipids, dioxin and polychlorinated biphenyls during production of protein isolates from Baltic herring (Clupea harengus) using pH-shift process. J Agric Food Chem 57:7819–7825

Mendes RL, Fernandes HI, Coelho JP, Reis EC, Cabral JM, Novais JM, Palavra AF (1995) Supercritical CO$_2$ extraction of carotenoids and other lipids from Chlorella vulgaris. Food Chem 53:99–103

Mercadante AZ (2008) Analysis of carotenoids. In: Socaciu C, Presss CRC (eds) Food colorants: chemical and functionality. Taylor & Francis. LLC, USA, pp 447–472

Meyers SP, Bligh D (1981) Characterization of astaxanthin pigment from heat processed crawfish waste. J Agric Food Chem 9:509–512

Meyers SP, Machada ZL (1978) Recovery of food-grade "waste" meat from the Spiny lobster *Panulirus Argus*. In: third annual tropical and subtropical fisheries technology conference of the America, Texas A&M university. Texas, America

Miki W (1991) Biological functions and activities of animal carotenoids. Pure Appl Chem 63(1):141–146

Mina F, Mohammad C, Hamid V (2014) The anitoxidant effects of astaxanthin on quantitative and qualitative parameters of bull sperm. Indian J Fundam Appl Life Sci 4(4):425–430

Mir VG, Heinämäki J, Antikainen O, Revoredo OB, Colarte AI, Nieto OM, Yliruusi J (2008) Direct compression properties of chitin and chitosan. Eur J Pharm Biopharm 69(3):964–968

Mir VG, Heinämäki J, Antikainen O, Sandler N, Revoredo OB, Colarte AI, Nieto OM, Yliruusi J (2010) Application of crustacean chitin as a co-diluent in direct compression of tablets. Am Assoc Pharm Sci 11(1):409–415

Mireles Dewitt CA, Gomez G, James JM (2002) Protein extraction from beef heart using acid solubilization. J Food Sci 67:3335–3341

Miyawaki H, Takahashi J, Tsukahara H, Takehara I (2008) Effects of astaxanthin on human blood rheology. J Clin Biochem Nutr 43(2):69–74

Muzzarelli RAA (1989) Amphoteric derivatives of chitosan and their biological significances. In: Skjak-Braek G, Anthonsen T, Sandford P (eds) Chitin and chitosan. Elsevier Applied Science, New York, pp 87–99

Nagaki Y, Mihara M, Tsukahara H, Ono S (2006) The supplementation effect of astaxanthin on accommodation and asthenopia. J Clin Ther Med 22(1):41–54

Naito Y, Uchiyama K, Aoi W, Hasegawa G, Nakamura N, Yoshida N, Maoka T, Takahashi J, Yoshikawa T (2004) Prevention of diabetic nephropathy by treatment with astaxanthin in diabetic db/db mice. BioFactors 20(1):49–59

Nguyen TT (2017) Biorefinery process development for recovery of functional and bioactive compounds from lobster processing by-products for food and nutraceutical applications. In: medical biotechnology, vol. Doctor, Flinders

Nguyen TT, Zhang W, Barber AR, Su P, He S (2015) Significant enrichment of polyunsaturated fatty acids (PUFAs) in the lipids extracted by supercritical CO_2 from the livers of Australian rock lobsters (*Jasus edwardsii*). J Agric Food Chem 63(18):4621–4628

Nguyen TT, Zhang W, Barber AR, Su P, He S (2016) Microwave-intensified enzymatic deproteinization of Australian rock lobster shells (*Jasus edwardsii*) for the efficient recovery of protein hydrolysate as food functional nutrients. Food Bioprocess Technol 9(4):628–636

Nguyen TT, Barber AR, Luo X, Zhang W (2017) Application and optimization of the highly efficient and environmentally-friendly microwave-intensified lactic acid demineralization of deproteinized Rock lobster shells (*Jasusedwardsii*) for chitin production. Food Bioprod Process 102:367–374

Ni Y, Nagashimada M, Zhuge F, Zhan L, Nagata N, Tsutsui A, Nakanuma Y, Kaneko S, Ota T (2015) Astaxanthin prevents and reverses diet-induced insulin resistance and steatohepatitis in mice: a comparison with vitamin E. Sci Rep 5:1–15

Nir Y, Spiller G (2002) BioAstin helps relieve pain and improves performance in patients with rheumatoid arthritis. J Am Coll Nutr 21(5):1–6

Nolsoe H, Undeland I (2009) The acid and alkaline solubilization process for the isolation of muscle proteins: state of art. Food Bioprocess Technol 2:1–27

Oh K-T, Kim Y-J, Jung W-J, Park R-D (2007) Demineralization of crab shell waste by *Pseudomonas aeruginosa* F722. Process Biochem 42(7):1069–1074

Osinga R, Tramper J, Wijffels RH (1999) Stability and stabilization of biocatalysts. In: Meireles AAA (ed) Extracting bioactive compounds for food products: theory and applications. CRC Press, Taylor and Francis Group, LLC, USA

Oviedo D, Garcia I, Mendez A, Henriques R (1982) Functional and nutritional evaluation of lobster shell soluble proteins. Food/Nahrung 26(4):391–395

Parajó JC, Dominguez H, Moure A, Diaz-Reinoso B (2008) Obtaining antioxidants by supercritical fluid extraction applied. In: Meireles AAA (ed)

Park JS, Chyun JH, Kim YK, Line LL, Chew BP (2010) Astaxanthin decreased oxidative stress and inflammation and enhanced immune response in humans. Nutr Metab 7(18):1–10

Pathak A (2017) Chitin and chitosan; technology, applications and global market BCC research

Pathiraja IK (2014) Removal of acid yellow 25 dye onto chitin extracted from waste crab legs and study of adsorption isotherms and kinetics of AY25 dye adsorption. In: Chemistry, vol. Master, Southern Illinois University at Edwardsville

Peniche-Covas C, Alvarez L, Argüelles-Monal W (1992) The adsorption of mercuric ions by chitosan. J Appl Polym Sci 46(7):1147–1150

Percot A, Viton C, Domard A (2003) Optimization of chitin extraction from shrimp shells. Biomacromol 4(1):12–18

Peters RD, Sturz AV, MacLeod JA (2006) The benefits of using lobster processing waste as a soil amendment in organic potato production. In: guelph organic conference, natural sciences symposium Canada

Petit C, Reynaud S, Desbrieres J (2015) Amphiphilic derivatives of chitosan using microwave irradiation. Toward an eco-friendly process to chitosan derivatives. Carbohydr Polym 116:26–33

Pisuttharachai D, Fagutao FF, Yasuike M, Aono H, Yano Y, Murakami K, Kondo H, Aoki T, Hirono I (2009) Characterization of crustin antimicrobial proteins from Japanese spiny lobster *Panulirus japonicus*. Dev Comp Immunol 33(10):1049–1054

Pombo R (1995) Antifungal activity of lobster chitosan on growth of two phythogenic fungi. Congress of cell wall, Santiago De Compostela, Spain

Qi Z (2015) Biomimetic functional materials based on cellulose and chitin nanofibers. 2nd international conference on natural fibers, Portugeese

Rao AR, Sarada R, Ravishankar GA (2007) Stabilization of astaxanthin in edible oils and its use as an antioxidant. J Sci Food Agric 87(6):957–965

Reverchon E, De Marco I (2006) Supercritical fluid extraction and fractionation of natural matter. J Supercrit Fluids 38:146–166

Rinaudo M (2006) Chitin and chitosan: properties and applications. Prog Polym Sci 31(7):603–632

Rizliya V, Mendis E (2014) Biological, physical, and chemical properties of fish oil and industrial applications. In: Kim SK (ed) Seafood processing by-products. Springer, Hiedelberg, pp 285–313

Ross RF (1927) The preparation of lobster paste. The ministry of Canadian marine and fisheries

Roy I, Mondal K, Gupta MN (2003) Accelerating enzymatic hydrolysis of chitin by microwave pretreatment. Biotechnol Prog 19(6):1648–1653

Rubio-Rodríguez N, De-Diego-Rupérez S, Beltrán S, Jaime I, Sanz MT, Rovira J (2008) Supercritical fluid extraction of the omega-3 rich oil contain in hake (*Merluccius capensis–Merluccius paradoxus*) by-products: study of influence of process parameters on the extraction yield and oil quality. J Supercrit Fluids 47:215–226

Rubio-Rodríguez N, Diego SMD, Beltrán S, Jaime I, Sanz MT, Rovira J (2012) Supercritical fluid extraction of fish oil from fish by-products: a comparison with other extraction methods. J Food Eng 109:238–248

Sachindra N, Bhaskar N, Siddegowda G, Sathisha A, Suresh P (2007) Recovery of carotenoids from ensilaged shrimp waste. Biores Technol 98(8):1642–1646

Safitri AU, Fajriah A, Astriandari A, Kartika I (2014) Bio mouth spray anti halitosis (bau mulut) berbasis fine particle chitosan lobster "segar sepanjang hari". Institute of Pertanian Bogor, Indonesia

Sahena F, Zaidul ISM, Jinap S, Saari N, Jahurul HA, Abbas KA, Norulaini NA (2009) PUFAs in fish: extraction, fractionation, importance in health. Compr Rev Food Sce Food Saf 8:59–74

Sahu A, Goswami P, Bora U (2009) Microwave mediated rapid synthesis of chitosan. J Mater Sci Mater Med 20(1):171–175

Saleem R, Hasnain A-U, Ahmad R (2015) Solubilisation of muscle proteins from chicken breast muscle by ultrasonic radiations in physiological ionic medium. Cogent Food Agric 1(1):1–11

Sandford PA (1989) Chitosan: commercial use and potential applications Elsevier Applied Science, London/New York

Savini I, Catani MV, Evangelista D, Gasperi V, Avigliano L (2013) Obesity-associated oxidative stress: strategies finalized to improve redox state. Int J Mol Sci 14(5):10497–10538

Sayari N, Sila A, Abdelmalek BE, Abdallah RB, Ellouz-Chaabouni S, Bougatef A, Balti R (2016) Chitin and chitosan from the Norway lobster by-products:

Seki T, Sueki H, Kono H, Kaoru S, Eiji Y (2001) Effects of astaxanthin from *Haematococcus pluvialis* on human skin. Fr J 12:98–103

Serrano GA, Narducci M (2014) Natural astaxanthin, antioxidant protection power for healthy eyes. Agro Food Ind High Technol 25:11–14

Seya Y, Takahashi J, Imanaka K (2009) Relationship between visual and reaction times—effects of a repetition of a visual task and long-term intake of a supplement food including astaxanthin on reaction time-Japanese. J

Shahidi F (2006) Marine oils: compositional characteristics and health effects. In: nutraceutical and specialty lipids and their co-products, CRC Press

Shahidi F, Synowiecki J (1991) Isolation and characterization of nutrients and value-added products from snow crab (*Chionoecetes opilio*) and shrimp (*Pandalus borealis*) processing discards. J Agric Food Chem 39(8):1527–1532

Sharp RG (2013) A review of the applications of chitin and its derivatives in agriculture to modify plant-microbial interactions and improve crop yields. Agronomy 3(4):757–793

Shen M, Chen K, Lu J, Cheng P, Xu L, Dai W, Wang F, He L, Zhang Y, Chengfen W (2014) Protective effect of astaxanthin on liver fibrosis through modulation of TGF-1 expression and autophagy. Mediat Inflamm 2014:1–15

Silva GF, Gamarra FMC, Oliveira AL, Cabral FA (2008) Extraction of bisin from annatto seeds using supercritical carbon dioxide. Braz J Chem Eng 25:419–426

Simpson BK, Dauphin L, Smith JP (1993) Recovery and characteristics of carotenoprotein from lobster (*Homarus americanus*) waste. J Aquat Food Prod Technol 1(2):129–146

Sindhu S, Sherief P (2011) Extraction, characterization, antioxidant and anti-inflammatory properties of carotenoids from the shell waste of Arabian red shrimp *Aristeus alcocki*. Open Conf Proc J. 2:95–103

Sini TK, Santhosh S, Mathew PT (2007) Study on the production of chitin and chitosan from shrimp shell by using *Bacillus subtilis* fermentation. Carbohydr Res 342(16):2423–2429

Sorokulova I, Krumnow A, Globa L, Vodyanoy V (2009) Efficient decomposition of shrimp shell waste using *Bacillus cereus* and *Exiguobacterium acetylicum*. J Ind Microbiol Biotechnol 36(8):1123–1126

Sunda F, Zhang Q, Zhang H, Zhang C, Qishan L (2012) Extraction method of astaxanthin from lobster heads, Google Patents

Swanson D, Block R, Mousa SA (2012) Omega-3 fatty acids EPA and DHA: health benefits throughout life. Adv Nutr Int Rev J 3(1):1–7

Synowiecki J, Al-Khateeb N (1997) Mycelia of *Mucor rouxii* as a source of chitin and chitosan. Food Chem 60:605–610

Synowiecki J, Al-Khateeb NA (2003) Production, properties, and some new applications of chitin and its derivatives. Crit Rev Food Sci Nutr 43:145–171

Tahergorabi R, Beamer S, Matak KE, Jaczynski J (2011) Effect of isoelectric solubilization/precipation and titanium dioxide on whitening and texture of proteins recovered from dark chicken-meat processing by-products. Food Sci Technol 44(4):896–903

Tahergorabi R, Sivanandan L, Jaczynski J (2012) Dynamic rheology and endothermic transitions of proteins recovered from chicken-meat processing by-products using isolectric solubilization/preicpitation. Food Sci Technol 46(1):148–155

Tanaka Y, Ohkubo T (2003) Extraction of lipids from salmon roes with supercritical carbon dioxide. J Oleo Sci 52:295–301

Taskaya L, Chen YC, Beamer S, Tou JC, Jaczynski J (2009a) Composition characteristics of materials recovered from whole gutted silver carp (*Hypophthalmichthys molitrix*) using isoelectric solubilization/precipitation. J Agric Food Chem 57(10):4259–4266

Taskaya L, Chen YC, Jaczynski J (2009b) Functional properties of proteins recovered from whole gutted silver carp (*Hypophthalmichthys molitrix*) by isoelectric solubilization/precipitation. Food Sci Technol 42(6):1082–1089

Teo CC, Chong WPK, Ho YS (2013) Development and application of microwave-assisted extraction technique in biological sample preparation from small molecule analysis. Metabolomics 9:1109–1128

Tetens I (2009) Safety of 'lipid extract from *Euphausia superba*' as a novel food ingredient. Eur Food Saf Auth 938:1–17

Thériault G, Hanlon J, Creed L (2013) Report of the maritime lobster panel. Report to minister of agriculture, aquaculture and fisheries (New Brunswick), Minister of fisheries and aquaculture (Nova Scotia), and minister of fisheries, aquaculture and rural development (Prince Edward Island), Canada

Tominaga K, Hongo N, Karato M, Yamashita E (2012) Cosmetic benefits of astaxanthin on humans subjects. Acta Biochim Pol 59(1):43

Tong-Xun L, Mou-Ming Z (2010) Thermal pretreatment and chemical modifica tions as a mean to alter hydrolytic characteristics and prevent bitterness in hydrolysates of fishery by-catch (*Decapterus maruadsi*) protein. Int J Food Sci Technol 45:1852–1861

Torrissen O, Tidemann E, Hansen F, Raa J (1981) Ensiling in acid—a method to stabilize astaxanthin in shrimp processing by-products and improve uptake of this pigment by rainbow trout (*Salmo gairdneri*). Aquaculture 26(1–2):77–83

Tsvetnenko E, Kailis S, Evans L, Longmore R (1996) Fatty acid composition of lipids from the contents of rock lobster (*Panulirus cygnus*) cephalotho-rax. J Am Oil Chem Soc 73(2):259–261

Tu Y (1991) Recovery, drying and characterization of carotenoproteins from industrial lobster waste. In: food science and agriculture chemistry, vol. Master of Science, McGill University

Tu Y, Simpson BK, Ramaswamy H, Yaylayan V, Smith JP, Hudon C (1991) Carotenoproteins from lobster waste as a potential feed supplement for cultured salmonids. Food Biotechnol 5(2):87–93

Tu Z-C, Huang T, Wang H, Sha X-M, Shi Y, Huang X-Q, Man Z-Z, Li D-J (2015) Physico-chemical properties of gelatin from bighead carp (*Hypophthalmichthys nobilis*) scales by ultrasound-assisted extraction. J Food Sci Technol 52(4):2166–2174

Uchiyama K, Naito Y, Hasegawa G, Nakamura N, Takahashi J, Yoshikawa T (2002) Astaxanthin protects β-cells against glucose toxicity in diabetic db/db mice. Redox Rep 7(5):290–293

Undeland I, Kelleher SD, Hultin HO (2002) Recovery of functional proteins from herring light muscle by an acid or alkaline solubilization process. J Agric Food Chem 50:7371–7379

Valdez-Pena AU, Espinoza-Perez JD, Sandoval-Fabian GC, Balagurusamy N, Hernandez-Rivera A, De-la-Garza-Rodriguez IM, Contreras-Esquivel JC (2010) Screening of industrial enzymes for deproteinisation of shrimp head for chitin recovery. Food Science Biotechnol 19(2):553–557

Vardanega R, Santos DT, Meireles MAA (2014) Intensification of bioactive compounds extraction from medicinal plants using ultrasonic irradiation. Pharmacogn Rev 8(16):88

Venugopal V (2009) Functional and bioactive nutraceutical compounds from the ocean. In: Mazza G (ed) Marine products for healthcare. CRC Press/ Taylor & Francis, Boca Raton

Venugopal V, Shahidi F (1995) Value-added products from underutilized fish species. Rev Food Sci Nutr 35(5):431–453

Vieira GH, Martin AM, Saker-Sampaiao S, Omar S, Goncalves RC (1995) Studies on the enzymatic hydrolysis of Brazilian lobster (*Panulirus* spp) processing wastes. J Sci Food Agric 69(1):61–65

Vilkhu K, Mawson R, Simons L, Bates D (2008) Applications and opportunities for ultrasound assisted extraction in the food industry—A review. Innov Food Sci Emerg Technol 9(2):161–169

Wang S-L, Chio S-H (1998) Deproteinization of shrimp and crab shell with the protease of *Pseudomonas aeruginosa* K-187. Enzyme Microb Technol 22(7):629–633

Wasikiewicz JM, Yeates SG (2013) "Green" molecular weight degradation of chitosan using microwave irradiation. Polym Degrad Stab 98(4):863–867

Wibowo S, Velazquez G, Savant V, Torres JA (2007) Effect of chitosan type on protein and water recovery efficiency from surimi wash water treated with chitosan–alginate complexes. Biores Technol 98(3):539–545

Wijesekara I, Kim S-K (2010) Angiotensin-I-converting enzyme (ACE) inhibitors from marine resources: prospects in the pharmaceutical industry. Mar Drugs 8(4):1080–1093

Xu Y, Gallert C, Winter J (2008) Chitin purification from shrimp waste by microbial deproteinisation and decalcification. Appl Microbiol Biotechnol 79:687–697

Hayes M (2012) Chitin, chitosan and their derivatives from marine rest raw materials: potential food and pharmaceutical applications. In: Hayes M (ed) Marine bioactive compounds: sources, characterization and applications. Springer, USA

Sabatini P (2015) Lobster—commodity update. Globefish, FAO

Yamashita E (2005) The effects of a dietary supplement containing astaxanthin on skin condition. Food Style 21 9(9):72–80

Yamashita E (2011) Astaxanthin and sports performance. Food Style 21:15–20

Yamashita E (2013) Astaxanthin as a medical food. Funct Foods Health Dis 3(7):254–258

Yan N, Chen X (2015) Don't waste seafood waste. Nature 254:155–157

Yilmaz B, Sahin K, Bilen H, Bahcecioglu IH, Bilir B, Ashraf S, Halazun KJ, Kucuk O (2015) Carotenoids and non-alcoholic fatty liver disease. Hepatobiliary Surg Nutr 4(3):161–171

Yook JS, Okamoto M, Rakwal R, Shibato J, Lee MC, Matsui T, Chang HK, Cho JY, Soya H (2015) Astaxanthin supplementation enhances adult hippocampal neurogenesis and spatial memory in mice. Mol Nutr Food Res 3(60):589–599

Zhang J, Yan N (2016) Formic acid-mediated liquefaction of chitin. Green Chem 18(18):5050–5058

Zhang X-S, Zhang X, Wu Q, Li W, Zhang Q-R, Wang C-X, Zhou X-M, Li H, Shi J-X, Zhou M-L (2014a) Astaxanthin alleviates early brain injury following subarachnoid hemorrhage in rats: possible involvement of Akt/bad signaling. Mar Drugs 12(8):4291–4310

Zhang X-S, Zhang X, Zhou M-L, Zhou X-M, Li N, Li W, Cong Z-X, Sun Q, Zhuang Z, Wang C-X (2014b) Amelioration of oxidative stress and protection against early brain injury by astaxanthin after experimental subarachnoid hemorrhage: laboratory investigation. J Neurosurg 121(1):42–54

13

Biochemical properties of a new thermo- and solvent-stable xylanase recovered using three phase partitioning from the extract of *Bacillus oceanisediminis* strain SJ3

Nawel Boucherba[1], Mohammed Gagaoua[2,3]*, Amel Bouanane-Darenfed[4], Cilia Bouiche[1], Khelifa Bouacem[4], Mohamed Yacine Kerbous[1], Yacine Maafa[1] and Said Benallaoua[1]

Abstract

The present study investigates the production and partial biochemical characterization of an extracellular thermo-stable xylanase from the *Bacillus oceanisediminis* strain SJ3 newly recovered from Algerian soil using three phase partitioning (TPP). The maximum xylanase activity recorded after 2 days of incubation at 37 °C was 20.24 U/ml in the presence of oat spelt xylan. The results indicated that the enzyme recovered in the middle phase of TPP system using the optimum parameters were determined as 50% ammonium sulfate saturation with 1.0:1.5 ratio of crude extract: *t*-butanol at pH and temperature of 8.0 and 10 °C, respectively. The xylanase was recovered with 3.48 purification fold and 107% activity recovery. The enzyme was optimally active at pH 7.0 and was stable over a broad pH range of 5.0–10. The optimum temperature for xylanase activity was 55 °C and the half-life time at this temperature was of 6 h. At this time point the enzyme retained 50% of its activity after incubation for 2 h at 95 °C. The crude enzyme resist to sodium dodecyl sulfate and β-mercaptoethanol, while all the tested ions do not affect the activity of the enzyme. The recovered enzyme is, at least, stable in tested organic solvents except in propanol where a reduction of 46.5% was observed. Further, the stability of the xylanase was higher in hydrophobic solvents where a maximum stability was observed with cyclohexane. These properties make this enzyme to be highly thermostable and may be suggested as a potential candidate for application in some industrial processes. To the best of our knowledge, this is the first report of xylanase activity and recoverey using three phase partitioning from *B. oceanisediminis*.

Keywords: *Bacillus oceanisediminis*, Xylanase, Thermostability, Hydrophobic solvents, Industrial processes, Three phase partitioning

Background

Hemicellulose is the second most abundant renewable biomass after cellulose in nature (Collins et al. 2005). Xylan is the major component of hemicelluloses in wood from angiosperms, where it accounts for 15–30% of the total dry weight. In gymnosperms, however, xylans contribute only 7–12% of the total dry weight. The structure of xylan is complex, and its complete biodegradation requires the concerted action of xylanolytic enzymes (Trajano et al. 2014; Zhang and Viikari 2014). Xylans are heterogeneous polysaccharides with a backbone consisting of β-1,4 linked D-xylosyl residues.

Endo-β-1,4 xylanases (EC 3.2.1.8) are the main enzymes responsible for cleavage of the linkages within the xylan backbone (Collins et al. 2005), to which short side chains of O-acetyl, α-L-arabinofuranosyl, D-α glucuronic, and phenolic acid residues are attached (Collins et al. 2005; Terrasan et al. 2010; Xie et al. 2015). Xylanases have been used in a wide range of industrial applications and processes. They have been applied in

*Correspondence: mohammed.gagaoua@inra.fr; gmber2001@yahoo.fr
[3] UMR1213 Herbivores, INRA, VetAgro Sup, Clermont Université, Université de Lyon, 63122 Saint-Genès-Champanelle, France
Full list of author information is available at the end of the article

the bioconversion of lignocellulosic material and agro-wastes to fermentative products, clarification of juices, improvement in consistency of beer, and the digestibility of animal feed stock (Badhan et al. 2007; Elgharbi et al. 2015a, b; Shameer 2016; Jain and Krishnan 2017). Due to their important activity at alkaline pH (8.0–11) and high temperature (50–90 °C), thermostable alkaline xylanases have attracted special attention in the pulp bio-bleaching industry (Techapun et al. 2003; Bouacem et al. 2014; Boucherba et al. 2014; Bouanane-Darenfed et al. 2016). Xylanase, together with other hydrolytic enzymes, have also proved useful for the generation of bio-fuels, including ethanol, from lignocellulosic biomass. Xylanases are used in pulp pre-bleaching process to remove the hemicelluloses, which bind to the pulp. The hydrolysis of pulp bound hemicelluloses releases the lignin in the pulp, reducing the amount of chlorine required for conventional chemical bleaching and minimizing the toxic, chloroorganic waste. Therefore, xylanases from alkalophilic bacteria and actinomycetes and fungi have been studied widely (Perez-Rodriguez et al. 2014; Wang et al. 2014). However, large scale cultivation of fungi and actinomycetes is often difficult because of their slow generation time, coproduction of highly viscous polymers, and poor oxygen transfer (Wong et al. 1997; Garg et al. 2011). *Bacillus* genus is used more extensively than other bacteria in industrial fermentations, since they produce most of their enzymes. Some *Bacillus* strains have been reported as xylanolytic enzymes producers (Lindner et al. 1994; Seo et al. 2013; Tarayre et al. 2013; Elgharbi et al. 2015a; Zouari et al. 2015).

Bacillus oceanisediminis sp. nov. was first isolated from a marine sediment collected in the South Sea of China (Zhang et al. 2010). Considering the above, the present study was undertaken to described, for the first time, the production of a thermostable xylanase from *B. oceanisediminis* strain SJ3 recently isolated by our laboratory from Algerian soil, an attempt was made to biochemically characterize the xylanase activity secreted by this strain. Also, preliminary investigation using three phase partitioning (TPP) system (Gagaoua et al. 2014; Gagaoua and Hafid 2016) for xylanase purification was performed. In TPP process, firstly an inorganic salt (generally ammonium sulfate) is added to the crude extract containing proteins then mixted with *tert*-butanol in an appropriate amount (Gagaoua et al. 2015, 2016, 2017). When *t*-butanol is added in the presence of ammonium sulfate, it pushes the protein out of the solution. In this process *t*-butanol binds to hydrophobic part of the proteins to reduce the density of the proteins, leading to float above the denser aqueous salt phase. Within approximately an hour, it forms an interfacial (middle) precipitate between the lower aqueous and upper organic phase that usually contains *t*-butanol (Gagaoua and Hafid 2016).

Methods

Substrates, reagents, and chemicals

Birchwood xylan, oat spelt xylan, starch, carboxymethyl cellulose (CMC, low viscosity), *tert*-butanol, ammonium sulfate, and 3,5-dinitrosalicylic acid (DNS) were purchased from Sigma Chemical Company (St. Louis, MO, USA). Unless otherwise specified, all other reagents and chemicals were of the analytical grade or highest level of purity available.

Collection of samples and culture conditions of microorganisms

The garden soil samples were collected from Bejaia north east of Algeria (Kabylia region) in March 2015. The soil was collected from five places and samples were pooled. Sub-samples of approximately 1 g were suspended in 100 ml sterile distilled water. Mixtures were allowed to settle and serial dilutions were prepared. From each dilution, 0.1 ml was taken and spread on agar plates of medium containing in g/l oat spelt xylan 10, yeast extract 2, NaCl 2.5, NH_4Cl 5, KH_2PO_4 15, Na_2HPO_4 30, $MgSO_4 \cdot 7H_2O$ 0.25, and bacteriological agar 15. In this medium, there is a little modification of the main carbon source, the oat spelt xylan was used instead the birchwood xylan (Viet et al. 1991). The plates were incubated at pH 7 and 37 °C for 2 days at 250 rpm. Those colonies that grew well under such conditions and showed an orange zone around the colonies after red Congo were retained for second screening. Colonies with a clear zone formation following the hydrolysis of xylan were evaluated as xylanase producers. Several xylanlolytic strains were isolated and SJ3, which exhibited a large clear zone of hydrolysis, was selected and retained for further experimental study.

Bacterial identification of the isolate SJ3

Analytical profile index (API) strip tests and 16S rRNA gene sequencing were carried out for the identification of the genus to which the strain belong.

API 50 CHB/E and the API 20E strips (bioMérieux, SA, Marcy-l'Etoile, France) were used to investigate the physiological and biochemical characteristics of strain SJ3, as recommended elsewhere (Logan and Berkeley 1984). The growth temperature (4, 10, 15, 20, 25, 30, 35, 40, and 45 °C), pH level values (4, 5, 6, 7, 8, 9, 10, 11, and 12) and sodium chloride regimes were determined.

The 16S rRNA gene was amplified by PCR using forward primer F-d1 5'-AGAGTTTGATCCTGGCTCA G-3', and reverse primer R-d1 5'-AAGGAGGTGATCCAA GCC-3', designed from base positions 8–27 and 1541–1525, respectively, which were the conserved zones within the rRNA operon of *Escherichia coli* (Gurtler and Stanisich 1996). The genomic DNA of strain SJ3 was purified using the Wizard® Genomic DNA Purification Kit (Promega, Madison, WI, USA) and then used as a template for PCR amplification (30 cycles, 94 °C for 45 s denaturation, 60 °C for 45 s primer annealing, and 72 °C for 60 s extension). The amplified ~1.5 kb PCR product was cloned in the pGEM-T Easy vector (Promega, Madison, WI, USA), leading to pSJ3-16S plasmid (this study). The *E. coli* DH5α (F⁻ *supE44 Φ80 δlacZ ΔM15 Δ(lacZYA-argF) U169 endA1 recA1 hsdR17 (r_k^-, m_k^+) deoR thi-1 λ⁻ gyrA96 relA1*) (Invitrogen, Carlsbad, CA, USA) was used as a host strain. All recombinant clones of *E. coli* were grown in Luria–Bertani (LB) broth media with the addition of ampicillin, isopropyl-thio-β-D-galactopyranoside (IPTG), and X-gal for screening. DNA electrophoresis, DNA purification, restriction, ligation, and transformation were all performed according to the method previously described elsewhere (Sambrook et al. 1989).

DNA sequencing and molecular phylogenetic analysis

The nucleotide sequences of the cloned 16S rRNA gene were determined on both strands using BigDye Terminator Cycle Sequencing Ready Reaction kits and the automated DNA sequencer ABI PRISM® 3100-Avant Genetic Analyser (Applied Biosystems, Foster City, CA, USA. The RapidSeq36_POP6 run module was used, and the samples were analyzed using the ABI sequencing analysis software v. 3.7 NT.

The sequences obtained were compared to those present in the public sequence databases and with the EzTaxon-e server (http://eztaxon-e.ezbiocloud.net/), a web-based tool for the identification of prokaryotes based on 16S rRNA gene sequences from type strains (Kim et al. 2012).

Phylogenetic and molecular evolutionary genetic analyses were conducted via the the molecular evolutionary genetics analysis (MEGA) software version 5 (http://www.megasoftware.net). Distances and clustering were calculated using the neighbor-joining method. The tree topology of the neighbor-joining data was evaluated by Bootstrap analysis with 100 re-samplings.

Xylanase assay

Xylanase activity was determined by measuring the release of reducing sugar from soluble xylan using the DNS method (Miller 1959). In brief, 0.9 ml buffer A (10 mg/ml oat spelt xylan in 50 mM sodium-phosphate buffer at pH 7) were mixed with 0.1 ml of the recovered enzyme solution (1 mg/ml). After incubation at 55 °C for 10 min, the reaction was terminated by adding 1.5 ml of the DNS reagent (Maalej et al. 2009). The mixture was

then boiled for 5 min and cooled. Absorption was measured at 540 nm.

One unit of xylanase activity was defined as the amount of enzyme that released 1 μmol of reducing sugar equivalent to xylose per min under the assay conditions.

Xylanase production
Gowth condition of the xylanase activity

To study the properties of the xylanase activity production, the isolates having high xylanase activities were cultivated in 250 ml shake-flasks containing 50 ml basic xylanase production medium at 37 °C. The basic xylanase production medium was prepared at pH 7.0 containing oat spelt xylan. The culture was harvested after 48 h, and centrifuged (10,000 rpm for 10 min). Growth was measured by determining absorbance at 600 nm. The sample was then kept at 4 °C in the refrigerator.

Effect of incubation time on xylanase production

Pre-culture (2%) was used to inoculate 250 ml xylan defined medium at 37 °C for 72 h. culture samples were collected each 4 h during the cultivation period. Immediately after collection, the samples were centrifuged at 4 °C and 10,000*g* for 20 min. Supernatants were analyzed for xylanase activity as described above.

Partial biochemical characterization of the recovered enzyme by TPP
Extraction and partial purification of xylanase by TPP

Aqueous systems such as three phase partitioning (TPP), known as simple, economical and quick methods, were described for the fast recovery of enzymes (Gagaoua and Hafid 2016). This elegant non-chromatographic tool may be performed in a purification process to be used successfully in food or other industries. For its application in this study, the crude extract was first collected after 48 h of batch incubation (Boucherba et al. 2014). The culture supernatant containing secreted xylanases was concentrated using Sartorius membranes (with 10-kDa cutoff membrane; Millipore) after a centrifugation at 10,000 rpm for 10 min. Then, TPP experiments were carried out following the recommendations of Gagaoua et al. (2015). The enzyme exclusively recovered in the interfacial phase was gently separated from the other phases and dissolved in 50 mM Tris–HCl buffer (pH 8.5) and dialyzed overnight at 4–5 °C and used for enzyme characterization.

Effect of temperature and pH on xylanase activity

Optimal temperature was determined by assaying the enzyme activity between 20 and 100 °C, by incubating the enzyme along with the substrate for 10 min at the respective temperature. Relative xylanase activity was determined using 10 mg/ml oat spelt xylan at various pHs. The pH range used varied from 4 to 10. Three different buff-ers (50 mM) were used. Sodium acetate buffer was used for pH 4–6; Sodium-phosphate buffer was used for pH from 6 to 7 and Tris–HCl buffer for pH 7–10.

Effect of temperature on xylanase stability

The thermostability was determined at temperatures of 50, 55, 60, and 95 °C, after incubation with the substrate for different times (from 0.5 to 7 h); remaining xylanase activity was measured under standard assay conditions. The non-heated enzyme, which was left at room temper-ature, was considered as control (100%).

Effect of pH on xylanase stability

For pH stability, the enzyme was incubated with differ-ent buffers viz. 50 mM acetate buffer for pH range 4–6, 50 mM phosphate buffer for pH range 6–7, and 50 mM Tris–HCl buffer for pH range 7–10 at 55 °C for 1 h. Thereafter, enzyme activity was determined using the enzyme assay as described above.

Effect of metal ions and reagents on activity

The effect of metallic ions at concentration of 5 mM, chelating agents, surfactants, and inhibitors on the activ-ity of crude xylanase were determined by preincubating the enzyme in the presence of Na^+, Mg^{2+}, Ca^{2+}, Mn^{2+}, Fe^{2+}, Zn^{2+}, Cu^{2+}, K^+, Hg^{2+}, and Cd^{2+}, EDTA (5 Mm), SDS (1%), β-mercaptoethanol (20 mM), and Triton X-100 (1%) for 30 min at 55 °C before adding the substrate (Ozcan et al. 2011). Subsequently, relative xylanase activi-ties were measured at standard enzyme assay conditions. Relative activity was expressed as the percentage of the activity observed in the absence of any compound.

Activity of crude enzyme on various carbohydrate substrate

The presence of other carbohydrase was analyzed using oat spelt xylan, birchwood xylan, starch, and CMC (10 mg/ml). The reducing sugar released during the assay was quantified by spectroscopy at λ_{540}.

Effect of organic solvents on xylanase activity

Cell free supernatant having maximum xylanase activ-ity was incubated with 30% (v/v) of different organic solvents, namely, acetone, propanol, ethanol, metha-nol, chloroform, heptane, cyclohexane, and toluene for 30 min at 55 °C. The residual xylanase activity was estimated against the control, in which solvent was not present.

Statistical analysis

All determinations were performed at least in three inde-pendent replicates, and the control experiment without xylanase was carried out under the same conditions. The experimental results were expressed as the mean of the rep-licate determinations and standard deviation (mean ± SD). The statistical significance was evaluated using t tests for two-sample comparison and one-way analysis of variance (ANOVA) followed by Duncan test. The results were con-sidered statistically significant for P values of less than 0.05. The statistical analysis was performed using the R package Version 3.1.1 (Vanderbilt University, USA).

Nucleotide sequence accession number

The data reported in this work for the nucleotide sequence of the 16S rRNA (1089 bp) gene of the isolate SJ3 have been deposited in the DDBJ/EMBL/GenBank databases under Accession Number KT222887.

Results and discussion

Screening of xylanase-producing bacteria from Algerain soil and molecular characterization of the target microorganism

In the current study, ten candidates were obtained from the first screening as xylanase producers. Among them, a bacterium called SJ3, displayed the highest extracellu-lar xylanase activity after 2 days incubation in an initial medium (data not shown) and was, therefore, retained for all subsequent studies.

The physiological and biochemical characteristics of the SJ3 isolate presented in this study were investigated according to well-established protocols and criteria described in the Bergey's Manual of Systematic Bacteri-ology as well as the API 50 CHB/E and the API 20E gal-leries for representative strains. The findings indicated that the SJ3 isolate was Gram-stain-positive, motile, rod-shaped, catalase-positive, aerobic, and endospore forming microorganism. Optimal growth temperature was 37 °C; optimal pH was 7.0. According to the results obtained using the API 50 CHB/E medium and the API 20E strips, the characteristics strongly confirmed that the isolate belongs to *Bacillaceae* order and *Bacillus* genus. The physiological and some biochemical properties of the isolate SJ3 are given in Table 1.

The 16S rRNA gene sequence (KT222887) obtained was submitted to GenBank BLAST search analyses, which yielded a strong homology of up to 99% with those of several cultivated strains of *Bacillus*. From the analy-sis of the almost-complete 16S rRNA gene sequence, this strain was found to be similar to *B. oceanisediminis* strain H_2^T (99.16% sequence identity). Through the align-ment of homologous nucleotide sequence of known bac-teria, phylogenetic relationships could be inferred, and the phylogenetic position of the strain and related strains based on the 16S rDNA sequence is shown in Fig. 1. Taken together, the results suggest that this isolate may be assigned as *B. oceanisediminis* strain SJ3.

Table 1 Morphological, physiological, and some biochemical properties of the isolate *Bacillus oceanisediminis* strain SJ3

Characteristics	*Bacillus oceanisediminis* strain SJ3
Isolation source	Soil
Motility	+
Morphology	Spore forming rods
Gram-stain	+
Temperature for growth	37
Temperature optimum range	25–45
pH for growth	7
pH optimum range	6–9
NaCl for growth (%)	0–12
Indole	−
Methyl red	+
Voges-proskauer	−
catalase	+
Glycerol	+
Erythritol	−
D-Arabinose	−
L-Arabinose	−
Ribose	+
D-Xylose	+
Galactose	+
Glucose	+
Fructose	−
D-Mannose	−
Mannitol	−
Sorbitol	−
Cellobiose	−
Maltose	−
Lactose	−
Saccharose	−
Inulin	−
Strach	−
Gelatin	+

Optimization of xylanase production by strain SJ3

In the current study, the bacterial strains were newly isolated from Algerain soil samples (Bejaia north east, Algeria), were screened for their xylanase activities. Using the ratio of the clear zone diameter (onto xylan agar plates) and that of the colony, five isolates exhibiting the highest ratio were tested for xylanase production in liquid culture. Among those strains, a bacterium called strain SJ3, displayed the highest extracellular xylanase activity (20.24 U/ml) after 48 h incubation in an optimized medium (Fig. 2) and was, therefore, retained for all subsequent studies.

Time course of xylanase production showed maximum enzyme activity at 48 h of incubation and thereafter, it remained less constant till 72 h (Fig. 2).

It is the same case with *Bacillus subtilis* strain ASH (Sanghi et al. 2009). The optimum time resulting in maximum enzyme titre is likely to depend on several factors including the microbial strain. A survey of the literature revealed the highest enzyme production from *Bacillus pumilus* strain SV-85S after 36 h (Nagar et al. 2010) and *Bacillus* sp. strain SSP-34 after 96 h (Subramaniyan and Prema 2000) and *B. pumilus* strain VLK-1 after 56 h of incubation (Kumar et al. 2014). In the above reports, the activity of xylanase exhibited a decline after reaching a maximum value, which might be due to proteolysis of the enzyme. However, in the present study, though the incubation period for xylanase production from *B. oceanisediminis* strain SJ3 was shorter than some other *Bacillus* sp. yet it did not decline after attaining the highest level.

Some biochemical properties of the crude enzyme

Xylanase activity from *B. oceanisediminis* strain SJ3 was efficiently recovered using the TPP technique. A purification fold of 3.48 and a recovery yield of 107% were obtained. Using macroaffinity ligand-facilitated TPP, Sharma and Gupta (2002) purified a xylanase from *Aspergillus niger* with a recovery yield of 60% and a 95-fold purification. The authors reported other recovery parameters using the denatured xylanase and the optimal parameters were 93% and a purification factor of 21 (Roy et al. 2004). TPP has been reported to recover different enzyme activities (e.g., xylanase, cellulase, cellobiase, β-glucosidase, and α-chymotrypsin) from their inactivated/denatured forms (Roy et al. 2004, 2005; Sardar et al. 2007). These findings suggest that TPP may be a valuable technique for the simultaneous renaturation/purification of the multiple enzymes present in a protein mixture. Concerning the high yield recovery obtained in this preliminary study several studies reported high recovery yields (>100%) for the purification of enzymes using the TPP system (Gagaoua and Hafid 2016; Gagaoua et al. 2017).

Effect of temperature on xylanase activity

The effect of temperature on the xylanase activity from *B. oceanisediminis* strain SJ3 is shown in Fig. 3a, for 10 min reaction the optimum temperature was 55 °C (assayed in the range 20–100 °C), the xylanase produced by *Bacillus brevis* is also optimally active at the same temperature (Goswami et al. 2013). The optimum temperature of the enzyme is near to that of the xylanases from *B. subtilis* strain CXJZ isolated from the degumming line (60 °C) (Guo et al. 2012) and *Bacillus* sp. strain 41M-1 which

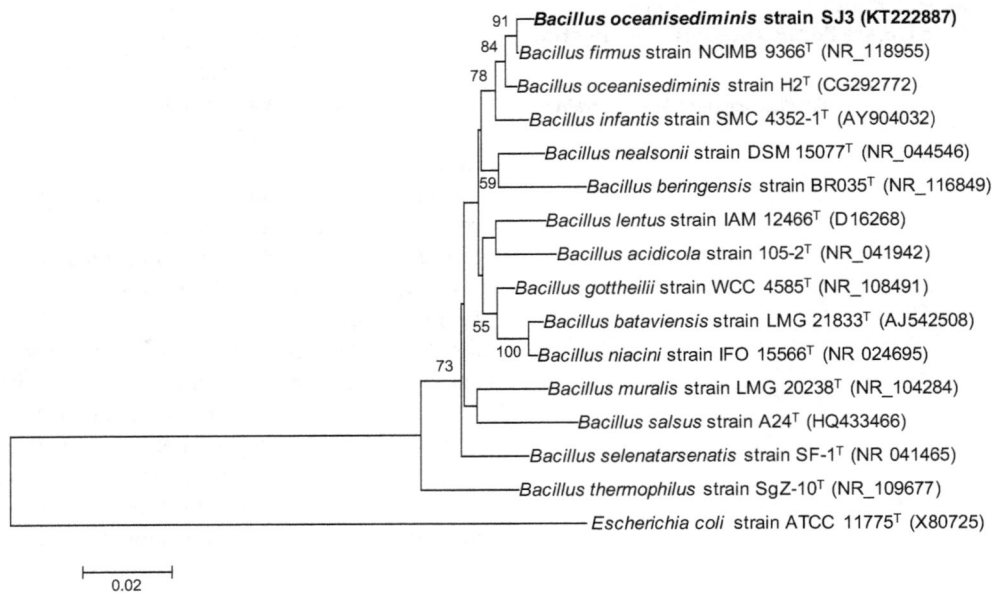

Fig. 1 Phylogenetic tree based on 16S rRNA gene sequences showing the position of strain SJ3 within the radiation of the genus *Bacillus*. The sequence of *E. coli* strain ATCC 11775[T] (Accession No. X80725) was chosen arbitrarily as an outgroup. *Bar* 0.02 nt substitutions per base. Numbers at nodes (>50%) indicate support for the internal branches within the tree obtained by bootstrap analysis (percentages of 100 bootstraps). NCBI accession numbers are presented in *parentheses*

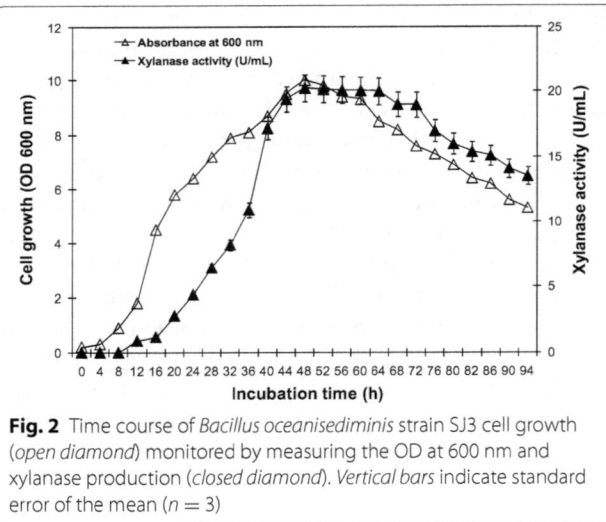

Fig. 2 Time course of *Bacillus oceanisediminis* strain SJ3 cell growth (*open diamond*) monitored by measuring the OD at 600 nm and xylanase production (*closed diamond*). *Vertical bars* indicate standard error of the mean (*n* = 3)

Effect of pH on xylanase activity

The optimum pH of *B. oceanisediminis* strain SJ3 xylanase activity (assayed in the range 4–10) is 7 (Fig. 3b). Other xylanases from *Bacillus* strains so far characterized generally show wide differences in their optimal pH, going from acidic values, such as 4 for the glycosyl hydrolase family 11 xylanase from *B. amyloliquefaciens* strain CH51 (Baek et al. 2012), 5 for the xylanase activity produced by *B. subtilis* strain GN156 (Pratumteep et al. 2010), 5.8 for the xylanase from *B. subtilis* strain CXJZ (Gang et al. 2012), up to 9 in the case of the endoxylanase activity from *B. halodurans* strain TSEV$_1$ (Kumar and Satyanarayana 2013, 2014).

Thermostability profile of the xylanase activity

Thermal stability was carried out by preincubating xylanase up to 7 h at 50, 55, 60, and 95 °C (Fig. 4), at 50 °C there was no significant decrease in xylanase activity during 4 h. The enzyme was stable at 50 °C, with a half-life time of 9 h, a half-life time of 6 and 4.72 h was respectively observed at 55 and 60 °C. *B. brevis* xylanase is less thermostable, it showed a half-life time of 3 h at 55 °C (Goswami et al. 2013).

At 95 °C the profile obtained for thermostability showed that 50% of the original activity was retained after 2 h exposure, the results clearly indicated that the suitable temperature range for industrial application for xylanase

showed maximum activity at 50 °C (Nakamura et al. 1995) and *Bacillus* sp. strain BP-23 (50 °C) (Blanco et al. 1995) but distant from that of the xylanases produced by *Bacillus halodurans* strain TSEV$_1$ (80 °C) (Kumar and Satyanarayana 2014), *Caldicoprobacter algeriensis* strain TH7C1[T] (Bouacem et al. 2014), *B. subtilis* strain GN156 (40 °C) (Pratumteep et al. 2010), and *Bacillus amyloliquefaciens* strain CH51 (25 °C) (Baek et al. 2012).

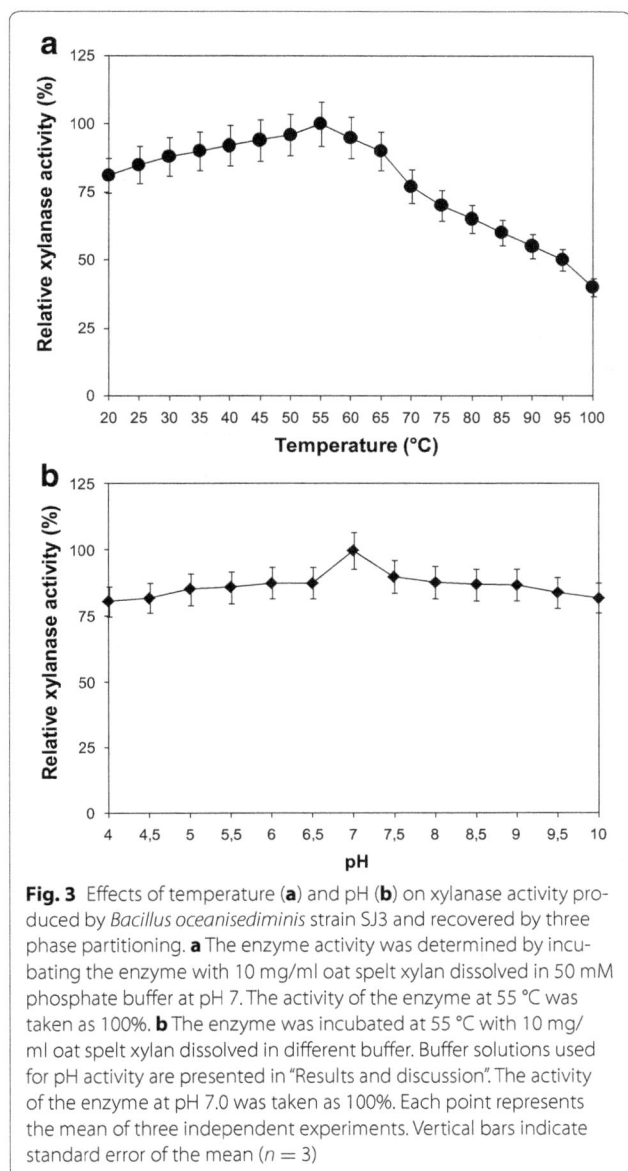

Fig. 3 Effects of temperature (**a**) and pH (**b**) on xylanase activity produced by *Bacillus oceanisediminis* strain SJ3 and recovered by three phase partitioning. **a** The enzyme activity was determined by incubating the enzyme with 10 mg/ml oat spelt xylan dissolved in 50 mM phosphate buffer at pH 7. The activity of the enzyme at 55 °C was taken as 100%. **b** The enzyme was incubated at 55 °C with 10 mg/ml oat spelt xylan dissolved in different buffer. Buffer solutions used for pH activity are presented in "Results and discussion". The activity of the enzyme at pH 7.0 was taken as 100%. Each point represents the mean of three independent experiments. Vertical bars indicate standard error of the mean ($n = 3$)

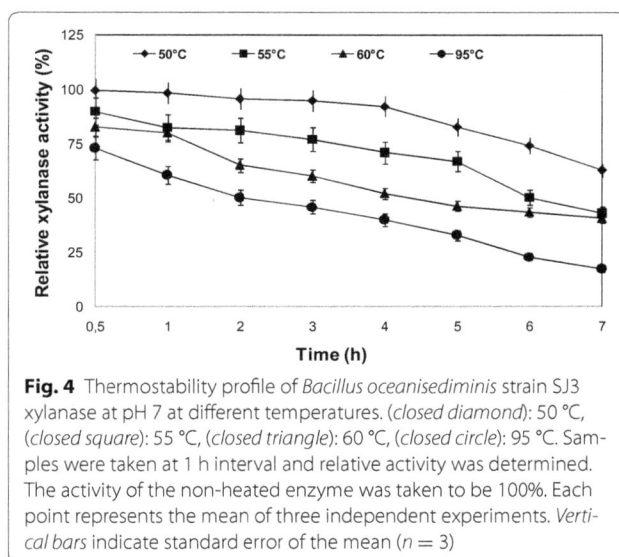

Fig. 4 Thermostability profile of *Bacillus oceanisediminis* strain SJ3 xylanase at pH 7 at different temperatures. (*closed diamond*): 50 °C, (*closed square*): 55 °C, (*closed triangle*): 60 °C, (*closed circle*): 95 °C. Samples were taken at 1 h interval and relative activity was determined. The activity of the non-heated enzyme was taken to be 100%. Each point represents the mean of three independent experiments. *Vertical bars* indicate standard error of the mean ($n = 3$)

from *B. oceanisediminis* strain SJ3 was 50–95 °C. This xylanase is more thermostable than *B. amyloliquefaciens* strain XR44A xylanase activity which showed a half-life time of 5 min at 70 °C, 15 min at both 50 and 60 °C, and 2 h at 40 °C. Interestingly, it retained 90% of activity for at least 2 days at 30 °C, with a half-life time of 7 days. The enzyme immediately loses activity at temperatures higher than 70 °C (Amore et al. 2015), the xylanase produced by *Bacillus aerophilus* strain KGJ2, retained more than 90% activity after incubation at 80–90 °C for 60 min (Gowdhaman et al. 2014). The enzyme produced by *Bacillus* sp. strain DM-15 was stable for 15 min at 60 °C while 95% of the original activity was lost at 90 °C (Ozcan et al. 2011).

The xylanase from *Pseudomonas macquariensis* had half-life time of 2 h at 50 °C whereas it had a half-life time of 1 h at 60 °C. At high temperatures, enzyme gets partly unfolded (Sharma et al. 2008). The xylanase of *B. oceanisediminis* strain SJ3 is highly thermostable, such enzymes with high thermostability and an ability to function at wide pH range are desirable for many industrial processes which take place at very high or low pH and high temperature. With this respect, the strain could be a good source for industrial and biotechnological applications.

pH stability profile of the xylanase activity

It is observed that the highest xylanase activity was established at pH 7.0; on the other hand, it was found to be most stable at pH 7.0–8.0 but it was also stable in a range of pH 5–10 and at pH 10 approximately 80% of its activity was retained (Fig. 5). The enzyme stable in alkaline conditions were characterized by a decreased number of acidic residues and an increased number of arginines (Hakulinen et al. 2003). The similar pattern of pH stability was also found in *Bacillus vallismortis* strain RSPP-15 (Gaur et al. 2015).

Effect of metallic ions, reagents, and inhibitors on xylanase activity

We investigated the effects of metallic ions and other reagents on the activities of the crude xylanase (Table 2). Most of the metallic ions (at concentration of 5 mM) tested had little influence on the activity, the same results were obtained with the xylanases produced by *Bacillus* sp. strain SPS-0 (Bataillon et al. 2000); in this experiment, maximum xylanase production was reported in the presence of Ca^{2+} (138%); some other researchers

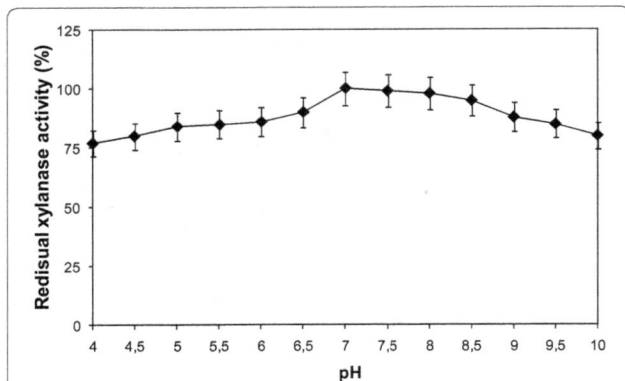

Fig. 5 pH stability of the xylanase activity produced by *Bacillus oceanisediminis* strain SJ3 and recovered by three phase partitioning. The crude enzyme was incubated with 50 mM buffers at 55 °C for 1 h and relative activity was measured under the standard assay conditions. The activity of the enzyme at optimum pH was taken as 100%. Buffer solutions used for pH stability are presented in "Results and discussion". Each point represents the mean of three independent experiments. *Vertical bars* indicate standard error of the mean (*n* = 3)

Table 2 Effect of different metallic ions, surfactants, chelating agents, and inhibitors on xylanase activity

Chemical additives	Concentration	Relative enzyme activity (%)[a]
Control	–	100 ± 2.5
Mg^{2+} (MgCl$_2$)	5 mM	106 ± 2.6
Ca^{2+} (CaCl$_2$)	5 mM	138 ± 4.1
Fe^{2+} (FeSO$_4$)	5 mM	88 ± 2.2
K^+ (KCl)	5 mM	98 ± 2.4
Cu^{2+} (CuCl$_2$)	5 mM	92 ± 2.3
Na^+ (NaCl)	5 mM	94 ± 2.3
Mn^{2+} (MnCl$_2$)	5 mM	99 ± 2.5
Cd^{2+} (CdCl$_2$)	5 mM	83 ± 2.0
Zn^{2+} (ZnCl$_2$)	5 mM	95 ± 2.3
Hg^{2+} (HgCl$_2$)	5 mM	20 ± 0.6
Triton X-100	1%	93 ± 2.3
SDS	1%	87 ± 2.2
EDTA	5 mM	88 ± 2.2
β-Mercaptoethanol	20 mM	86 ± 2.2

Xylanase activity measured in the absence of any chemical additives was taken as control (100%). The non-treated and dialyzed enzyme was considered as 100% for metallic ion assay. Residual activity was measured at pH 7.0 and 55 °C

[a] Values represent means of three independent replicates, and ±standard errors are reported

also reported that Ca^{2+} ion strongly stimulated xylanase activity. Slightly stimulation was also observed by addition of Mg^{2+} (106%) (Mamo et al. 2006; Lv et al. 2008; Ozcan et al. 2011).

On the other hand, the inhibition of xylanases by calcium and magnesium ions have also been reported

(Hmida-Sayari et al. 2012; Chang et al. 2017). Xylanase was strongly inhibited in the presence of Hg^{2+}. Similar results were observed in case of *B. subtilis* (Sanghi et al. 2010) and *Bacillus halodurans* strain PPKS-2 (Prakash et al. 2012), it has been reported that the xylanase activity was inhibited by mercury ion, which might be due to its interaction with sulfhydryl groups of cysteine residue in or close to the active site of the enzyme (Bastawde 1992).

The chelating agent EDTA enveloping metal ions extensively did not change the xylanase activity (Table 2) that means the enzyme did not require metal ions for its catalysis.

Triton X-100 and β-mercaptoethanol had little effect on the xylanase activity (Table 2) whereas the *Bacillus* DM-15 xylanase is sensitive (Ozcan et al. 2011).

Total inactivation due to SDS has already been reported for xylanases of different origins (Fujimoto et al. 1995), in contrast to the resistance to SDS was found in this study, with 87% relative activity after 10 min at 55 °C (Table 2).

Activity of the crude xylanase on various carbohydrate substrates

Activity of the crude enzyme on some carbohydrate was showed at Fig. 6, the crude enzyme mainly contained xylanase as indicated by the highest activity on birchwood xylan (25 U/ml) and oat spelt xylan (20 U/ml). The crude enzyme did not contain amylase but hardly cellulase (1.99 U/ml). Crude enzymes produced by *Bacillus* sp. strain AQ1 not only showed xylanolytic activity but also amylolytic and cellulolytic activity (Wahyuntari et al. 2009). Comparisons to the large literature studies as summarized in Table 3.

Based on the available data from this experiment, the difference in crude enzyme on the different xylan substrate could not be explained yet. It is still needed more

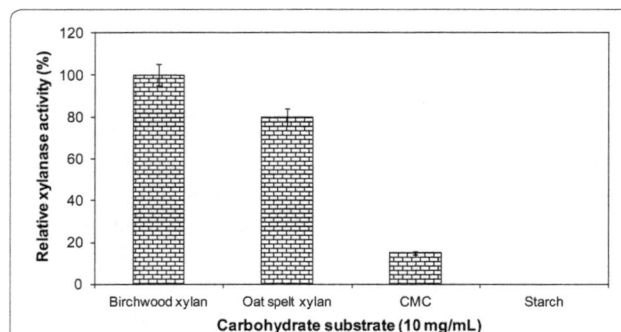

Fig. 6 The effect of the carbohydrate substrate source on the xylanase activity produced by *Bacillus oceanisediminis* strain SJ3 and recovered by three phase partitioning. The enzyme was incubated with 10 mg/ml of substrate at 55 °C and pH 7.0. Each point represents the mean of three independent experiments. *Vertical bars* indicate standard error of the mean (*n* = 3)

Table 3 Production of xylanases from various bacteria, namely *Bacillus* genus and comparisons to our findings

Organism	Substrate	Cultivation conditions (temperature and pH)	Xylanase activity (U/ml or U/mg)	References
Bacillus oceanisediminis SJ3	Oat spelt xylan	55 °C; pH 7.0	20.24 U/ml	Present study
Jonesia denitrificans BN-13	*Birchwood xylan*	50 °C; pH 7.0	77 U/mg	Boucherba et al. (2014)
Bacillus pumilus MTCC 8964	Oat spelt xylan	60 °C; pH 6	241 U/ml	Kumar et al. (2010)
Bacillus brevis	Birchwood xylan	55 °C; pH 7.0	1.52 U/ml	Goswami et al. (2013)
Bacillus brevis	*wheat straw*	55 °C; pH 7.0	4380 U/mg	Goswami et al. (2014)
Bacillus sp. strain BP-23	Birchwood xylan	50 °C; pH 5.5	40.2 U/mg	Blanco et al. (1995)
Bacillus halodurans TSEV$_1$	*Cane molasses*	80 °C; pH 9.0	15 U/ml	Kumar and Satyanarayana (2014)
Bacillus amyloliquefaciens CH51	Birchwood xylan	25 °C; pH 4.0	701.1 U/mg	Baek et al. (2012)
Bacillus subtilis CXJZ	birchwood and oat spelt xylan	60 °C; pH 5.8	36,633 U/mg	Gang et al. (2012)
Bacillus pumilus SSP-34	Oat spelts xylan	50 °C; pH 6.0	1723 U/mg	Subramaniyan (2012)
Bacillus pumilus SV-205	Wheat bran	60 °C; pH 10.0	7382.7 U/ml	Nagar et al. (2012)
Bacillus subtilus BS05	Sugarcane bagasse	50 °C; pH 5.0	17.58 U/ml	Irfan et al. (2012)
Gracilibacillus sp. TSCPVG	Birchwood xylan	pH 7.5	1667 U/mg	Poosarla and Chandra (2014)
Paenibacillus sp. NF1	Oat spelt xylan	60 °C; pH 6.0	3081.05 U/mg	Zheng et al. (2014)
Paenibacillus macerans IIPSP3	Beechwood xylan	60 °C; pH 4.5	4170 U/mg	Dheeran et al. (2012)
Anoxybacillus flavithermus TWXYL3	Oat spelt xylan	65 °C; pH 6.0 and pH 8.0	117.64 U/mg	Ellis and Magnuson (2012)

complete studies to elaborate the type of xylanolytic activities present in the crude enzyme of *B. oceanisediminis* strain SJ3. From preliminary study, it can be observed that the strain SJ3 was able to grow and produce xylanases using commercial xylan. The pH and temperature optima of the preparation were 7 and 55 °C, respectively, and the enzyme was stable in a range of pH 5–10 retained 50% of its activity during 6 h at 55 °C. The enzyme is also resistant to hydrophobic solvents, these properties place this enzyme as promising for industrial and biotechnological applications especially lignocellulose bioconversion and bioethanol production.

Effect of organic solvents of the xylanase activity

The xylanase from *B. oceanisediminis* strain SJ3 is resistant to hydrophobic solvents: heptan, chloroform, toluene, and cyclohexane (the relative activity is 99.2%) but a loss of the enzyme activity was observed by addition of 30% (v/v) of methanol, ethanol, propanol, and acetone (Fig. 7). These alcohols completely inhibited the enzyme from *Termitomyces* sp. and *Macrotermes subhyalinus* at 30% (v/v) and 60% (v/v), respectively. Primary alcohols including methanol, ethanol, and isopropanol as well as polyhydric alcohol containing glycol and glycerol, all showed inhibitory effects on *A. niger* strain C3486 xylanase activity which retained around 90% at the concentration of 2%

(v/v) and less than 60% of its initial activity at 30% (Yang et al. 2010).

In some cases, the presence of solvents enhanced the xylanase activity, for example the xylanase of *B. vallismortis* is extraordinarily stable in the presence of all organic solvents under study. After incubation with *n*-dodecane, isooctane, *n*-decane, xylene, toluene, *n*-hexane, *n*-butanol, and cyclohexane, the xylanase activity increased to 230.8, 137.7, 219.8, 107, 190.5, 194.7, 179.3, and 111.6%, respectively (Gaur and Tiwari 2015).

Conclusion

In conclusion, a new extracellular thermostable xylanase from *B. oceanisediminis* strain SJ3 was produced and characterized in this study. The preliminary results of the use of Three phase partitioning for the recovery of the xylanase were presented. The time course for xylanase accumulation by strain SJ3 in xylan-based medium showed that the highest xylanase activity reached 20.24 U/ml in an optimized medium with oats spelt xylan used as a substrate after 48 h of cultivation. The crude xylanase from strain SJ3 was biochemically characterized. The results revealed that the enzyme was highly stable and active at high temperature (55 °C) and alkaline pH 7.0. Properties of this enzyme such as high specific activity, wide range of pH optimum and stability, and

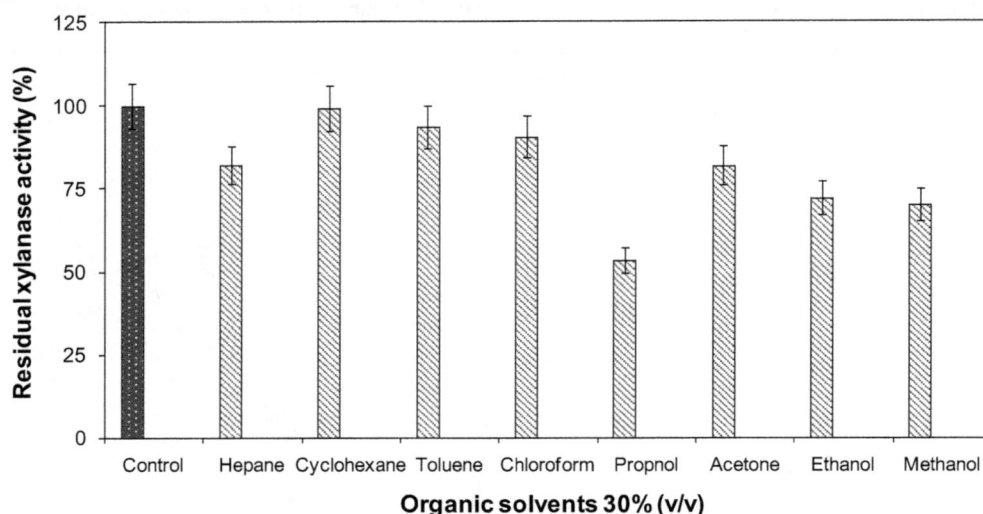

Fig. 7 Effect of organic solvents on xylanase activity produced by *Bacillus oceanisediminis* strain SJ3 and recovered by three phase partitioning. Relative xylanase activity was expressed as a percentage of the control reaction without solvent. Each point represents the mean of three independent experiments. *Vertical bars* indicate standard error of the mean ($n = 3$)

thermostability at elevated temperature as well as organic solvents tolerance, are appropriate for industrial and biotechnological applications. Interestingly, this enzyme presented high xylanolytic activity with oats spelt xylan, and was very effective in the pulp bleaching industry, thus offering a potential promising candidate for application in biotechnological bioprocesses. Accordingly, further studies, some of which are currently underway, are needed to investigate the purification to homogeneity and encoding gene, perform site-directed mutagenesis, and determine its structure–function relationships.

Authors' contributions
NB, MG, and SB designed this research plan and discussed with ABD. NB, MG, CB, KB, MYC performed all the research experiments and NB and MG wrote the draft paper. ABD, YM, and MG helped in the research experiments. MG conducted Three Phase Partitioning experiments. All authors have participated in the interpretation of the results during preparation of the manuscript. All authors read and approved the final manuscript.

Author details
[1] Laboratory of Applied Microbiology, Faculty of Nature Science and Life, University of Bejaia, 06000 Bejaia, Algeria. [2] INATAA, Université des Frères Mentouri Constantine 1, Route de Ain El-Bey, 25000 Constantine, Algeria. [3] UMR1213 Herbivores, INRA, VetAgro Sup, Clermont Université, Université de Lyon, 63122 Saint-Genès-Champanelle, France. [4] Laboratory of Cellular and Molecular Biology, Microbiology Team, Faculty of Biological Sciences, University of Sciences, Technology of Houari Boumediene (USTHB), PO Box 32, El Alia, Bab Ezzouar, 16111 Algiers, Algeria.

Acknowledgements
This work was funded by the Algerian Ministry of Higher Education and Scientific Research under the National Research Program Project (Grant Number F00620110001).

Competing interests
The authors declare that they have no competing interests.

References
Amore A, Parameswaran B, Kumar R, Birolo L, Vinciguerra R, Marcolongo L, Ionata E, La Cara F, Pandey A, Faraco V (2015) Application of a new xylanase activity from *Bacillus amyloliquefaciens* XR44A in brewer's spent grain saccharification. J Chem Technol Biotechnol 90:573–581

Badhan AK, Chadha BS, Kaur J, Saini HS, Bhat MK (2007) Production of multiple xylanolytic and cellulolytic enzymes by thermophilic fungus *Myceliophthora* sp. IMI 387099. Bioresour Technol 98:504–510

Baek CU, Lee SG, Chung YR, Cho I, Kim JH (2012) Cloning of a family 11 xylanase gene from *Bacillus amyloliquefaciens* CH51 isolated from Cheonggukjang. Indian J Microbiol 52:695–700

Bastawde KB (1992) Xylan structure, microbial xylanases, and their mode of action. World J Microbiol Biotechnol 8:353–368

Bataillon M, Nunes Cardinali A, Castillon N, Duchiron F (2000) Purification and characterization of a moderately thermostable xylanase from *Bacillus* sp. strain SPS-0. Enzyme Microb Technol 26:187–192

Blanco A, Vidal T, Colom JF, Pastor FI (1995) Purification and properties of xylanase A from alkali-tolerant *Bacillus* sp. strain BP-23. Appl Environ Microbiol 61:4468–4470

Bouacem K, Bouanane-Darenfed A, Boucherba N, Joseph M, Gagaoua M, Ben Hania W, Kecha M, Benallaoua S, Hacene H, Ollivier B, Fardeau ML (2014) Partial characterization of xylanase produced by *Caldicoprobacter algeriensis*, a new thermophilic anaerobic bacterium isolated from an Algerian hot spring. Appl Biochem Biotechnol 174:1969–1981

Bouanane-Darenfed A, Boucherba N, Bouacem K, Gagaoua M, Joseph M, Kebbouche-Gana S, Nateche F, Hacene H, Ollivier B, Cayol J-L, Fardeau M-L (2016) Characterization of a purified thermostable xylanase from *Caldicoprobacter algeriensis* sp. nov. strain TH7C1T. Carbohyd Res 419:60–68

Boucherba N, Gagaoua M, Copinet E, Bettache A, Duchiron F, Benallaoua S (2014) Purification and characterization of the xylanase produced by *Jonesia denitrificans* BN-13. Appl Biochem Biotechnol 172:2694–2705

Chang S, Guo Y, Wu B, He B (2017) Extracellular expression of alkali tolerant xylanase from *Bacillus subtilis* Lucky9 in *E. coli* and application for xylooligosaccharides production from agro-industrial waste. Int J Biol Macromol 96:249–256

Collins T, Gerday C, Feller G (2005) Xylanases, xylanase families and extremophilic xylanases. FEMS Microbiol Rev 29:3–23

Dheeran P, Nandhagopal N, Kumar S, Jaiswal YK, Adhikari DK (2012) A novel thermostable xylanase of *Paenibacillus macerans* IIPSP3 isolated from the termite gut. J Ind Microbiol Biotechnol 39(6):851–860

Elgharbi F, Hlima HB, Farhat-Khemakhem A, Ayadi-Zouari D, Bejar S, Hmida-Sayari A (2015a) Expression of *A. niger* US368 xylanase in *E. coli*: purification, characterization and copper activation. Int J Biol Macromol 74:263–270

Elgharbi F, Hmida-Sayari A, Zaafouri Y, Bejar S (2015b) Expression of an *Aspergillus niger* xylanase in yeast: application in breadmaking and in vitro digestion. Int J Biol Macromol 79:103–109

Ellis JT, Magnuson TS (2012) Thermostable and alkalistable xylanases produced by the thermophilic bacterium *Anoxybacillus flavithermus* TWXYL3. ISRN microbiology

Fujimoto H, Ooi T, Wang S-L, Takizawa T, Hidaka H, Murao S, Arai M (1995) Purification and properties of three xylanases from *Aspergillus aculeatus*. Biosci Biotechnol Biochem 59:538–540

Gagaoua M, Hafid K (2016) Three phase partitioning system, an emerging non-chromatographic tool for proteolytic enzymes recovery and purification. Biosens J 5(1):100134

Gagaoua M, Boucherba N, Bouanane-Darenfed A, Ziane F, Nait-Rabah S, Hafid K, Boudechicha HR (2014) Three-phase partitioning as an efficient method for the purification and recovery of ficin from Mediterranean fig (*Ficus carica* L.) latex. Sep Purif Technol 132:461–467

Gagaoua M, Hoggas N, Hafid K (2015) Three phase partitioning of zingibain, a milk-clotting enzyme from *Zingiber officinale* Roscoe rhizomes. Int J Biol Macromol 73:245–252

Gagaoua M, Hafid K, Hoggas N (2016) Data in support of three phase partitioning of zingibain, a milk-clotting enzyme from *Zingiber officinale* Roscoe rhizomes. Data in brief 6:634–639

Gagaoua M, Ziane F, Nait Rabah S, Boucherba N, El-Hadef El-Okki Ait Kaki A, Bouanane-Darenfed A, Hafid K (2017) Three phase partitioning, a scalable method for the purification and recovery of cucumisin, a milk-clotting enzyme, from the juice of *Cucumis melo* var. *reticulatus*. Int J Biol Macromol 102:515–525

Gang GJ, Zbijewski W, Webster Stayman J, Siewerdsen JH (2012) Cascaded systems analysis of noise and detectability in dual-energy cone-beam CT. Med Phys 39:5145–5156

Garg G, Dhiman SS, Mahajan R, Kaur A, Sharma J (2011) Bleach-boosting effect of crude xylanase from *Bacillus stearothermophilus* SDX on wheat straw pulp. N Biotechnol 28:58–64

Gaur R, Tiwari S (2015) Isolation, production, purification and characterization of an organic-solvent-thermostable alkalophilic cellulase from *Bacillus vallismortis* RG-07. BMC Biotechnol 15:19

Gaur R, Tiwari S, Rai P, Srivastava V (2015) Isolation, production, and characterization of thermotolerant xylanase from solvent tolerant *Bacillus vallismortis* RSPP-15. Int J Polym Sci 2015:10

Goswami GK, Pathak RR, Krishnamohan M, Ramesh B (2013) Production, partial purification and biochemical characterization of thermostable xylanase from *Bacillus brevis*. Biomed Pharmacol J 6:435–440

Goswami GK, Krishnamohan M, Nain V, Aggarwal C, Ramesh B (2014) Cloning and heterologous expression of cellulose free thermostable xylanase from *Bacillus brevis*. SpringerPlus 3(1):20

Gowdhaman D, Manaswini VS, Jayanthi V, Dhanasri M, Jeyalakshmi G, Gunasekar V, Sugumaran KR, Ponnusami V (2014) Xylanase production from *Bacillus aerophilus* KGJ2 and its application in xylooligosaccharides preparation. Int J Biol Macromol 64:90–98

Guo G, Liu Z, Xu J, Liu J, Dai X, Xie D, Peng K, Feng X, Duan S, Zheng K, Cheng L, Fu Y (2012) Purification and characterization of a xylanase from *Bacillus subtilis* isolated from the degumming line. J Basic Microbiol 52:419–428

Gurtler V, Stanisich VA (1996) New approaches to typing and identification of bacteria using the 16S-23S rDNA spacer region. Micobiology 142:3–16

Hakulinen N, Turunen O, Janis J, Leisola M, Rouvinen J (2003) Three-dimensional structures of thermophilic beta-1,4-xylanases from *Chaetomium thermophilum* and *Nonomuraea flexuosa*. Comparison of twelve xylanases in relation to their thermal stability. Eur J Biochem 270:1399–1412

Hmida¯Sayari A, Taktek S, Elgharbi F, Bejar S (2012) Biochemical characterization, cloning and molecular modeling of a detergent and organic solvent-stable family 11 xylanase from the newly isolated Aspergillus niger US368 strain. Process Biochem 47:1839–1847

Irfan M, Nadeem M, Syed Q, Baig S (2012) Effect of medium composition on xylanase production by *Bacillus subtilis* using various agricultural wastes. Am Eurasian J Agric Environ Sci 12:561–565

Jain A, Krishnan KP (2017) A glimpse of the diversity of complex polysaccharide-degrading culturable bacteria from Kongsfjorden, Arctic Ocean. Ann Microbiol 2:203–214

Kim OS, Cho YJ, Lee K, Yoon SH, Kim M, Na H, Park SC, Jeon YS, Lee JH, Yi H, Won S, Chun J (2012) Introducing EzTaxon-e: a prokaryotic 16S rRNA gene sequence database with phylotypes that represent uncultured species. Int J Syst Evol Microbiol 62:716–721

Kumar V, Satyanarayana T (2013) Biochemical and thermodynamic characteristics of thermo-alkali-stable xylanase from a novel polyextremophilic *Bacillus halodurans* TSEV1. Extremophiles 17:797–808

Kumar V, Satyanarayana T (2014) Production of thermo-alkali-stable xylanase by a novel polyextremophilic *Bacillus halodurans* TSEV1 in cane molasses medium and its applicability in making whole wheat bread. Bioprocess Biosyst Eng 37:1043–1053

Kumar D, Verma R, Sharma P, Rana A, Sharma R, Prakash C, Bhalla TC (2010) Production and partial purification of xylanase from a new thermophilic isolate. In Biol Forum Int J 2:83–87

Kumar L, Nagar S, Mittal A, Garg N, Gupta VK (2014) Immobilization of xylanase purified from *Bacillus pumilus* VLK-1 and its application in enrichment of orange and grape juices. J Food Sci Technol 51:1737–1749

Lindner C, Stulke J, Hecker M (1994) Regulation of xylanolytic enzymes in *Bacillus subtilis*. Microbiology 140(Pt 4):753–757

Logan NA, Berkeley RC (1984) Identification of *Bacillus* strains using the API system. J Gen Microbiol 130:1871–1882

Lv Z, Yang J, Yuan H (2008) Production, purification and characterization of an alkaliphilic endo-β-1,4-xylanase from a microbial community EMSD5. Enzyme Microb Technol 43:343–348

Maalej I, Belhaj I, Masmoudi NF, Belghith H (2009) Highly thermostable xylanase of the thermophilic fungus *Talaromyces thermophilus*: purification and characterization. Appl Biochem Biotechnol 158:200–212

Mamo G, Hatti-Kaul R, Mattiasson B (2006) A thermostable alkaline active endo-β-1-4-xylanase from *Bacillus halodurans* S7: purification and characterization. Enzyme Microb Technol 39:1492–1498

Miller GL (1959) Use of dinitrosalycilic acid reagent for determination of reducing sugars. Anal Chem 31:426–428

Nagar S, Gupta VK, Kumar D, Kumar L, Kuhad RC (2010) Production and optimization of cellulase-free, alkali-stable xylanase by *Bacillus pumilus* SV-85S in submerged fermentation. J Ind Microbiol Biotechnol 37:71–83

Nagar S, Mittal A, Kumar D, Gupta VK (2012) Production of alkali tolerant cellulase free xylanase in high levels by *Bacillus pumilus* SV-205. Int J Biol Macromol 50(2):414–420

Nakamura S, Nakai R, Namba K, Kubo T, Wakabayashi K, Aono R, Horikoshi K (1995) Structure-function relationship of the xylanase from alkaliphilic *Bacillus* sp. strain 41M-1. Nucleic Acids Symp Ser 34:99–100

Ozcan BD, Coskun A, Ozcan N, Baylan M (2011) Some properties of a new thermostable xylanase from alkaliphilic and thermophilic *Bacillus* sp.isolate DM-15. J Anim Vet Adv 10:138–143

Perez-Rodriguez N, Oliveira F, Perez-Bibbins B, Belo I, Torrado Agrasar A, Dominguez JM (2014) Optimization of xylanase production by filamentous fungi in solid-state fermentation and scale-up to horizontal tube bioreactor. Appl Biochem Biotechnol 173:803–825

Poosarla VG, Chandra TS (2014) Purification and characterization of novel halo-acid-alkali-thermo-stable xylanase from *Gracilibacillus* sp. TSCPVG. Appl Biochem Biotechnol 173(6):1375–1390

Prakash P, Jayalakshmi SK, Prakash B, Rubul M, Sreeramulu K (2012) Production of alkaliphilic, halotolerent, thermostable cellulase free xylanase by *Bacillus halodurans* PPKS-2 using agro waste: single step purification and characterization. World J Microbiol Biotechnol 28:183–192

Pratumteep A, Sansernsuk J, Nitisinprasert S, Apiraksakorn J (2010) Production, characterization and hydrolysation products of xylanase from *Bacillus subtilis* GN156. KKU Res J 15:343–350

Roy I, Sharma A, Gupta MN (2004) Three phase partitioning for simultaneous renaturation and partial purification of *Aspergillus niger* xylanase. BBA Proteins Proteom 1698:107–110

Roy I, Sharma A, Gupta MN (2005) Recovery of biological activity in reversibly inactivated proteins by three phase partitioning. Enzyme Microb Technol 37:113–120

Sambrook J, Fritsch E, Maniatis T (1989) Molecular cloning: a laboratory manual, 2nd edn. Cold Spring Harbor Laboratory Press, Cold Spring Harbor

Sanghi A, Garg N, Kuhar K, Kuhad RC, Gupta VK (2009) Enhanced production of cellulase-free xylanase by alkalophilic Bacillus subtilis ASH and its application in biobleaching of kraft pulp. BioResources 4:1109–1129

Sanghi A, Garg N, Gupta VK, Mittal A, Kuhad RC (2010) One-step purification and characterization of cellulase-free xylanase produced by alkalophilic Bacillus subtilis ash. Braz J Microbiol 41:467–476

Sardar M, Sharma A, Gupta MN (2007) Refolding of a denatured α-chymotrypsin and its smart bioconjugate by three-phase partitioning. Biocatal Biotransform 25:92–97

Seo JK, Park TS, Kwon IH, Piao MY, Lee CH, Ha JK (2013) Characterization of cellulolytic and xylanolytic enzymes of Bacillus licheniformis JK7 isolated from the rumen of a native Korean goat. Asian Australas J Anim Sci 26:50–58

Shameer S (2016) Haloalkaliphilic Bacillus species from solar salterns: an ideal prokaryote for bioprospecting studies. Ann Microbiol 66:1315–1327

Sharma A, Gupta MN (2002) Macroaffinity ligand-facilitated three-phase partitioning (MLFTPP) for purification of xylanase. Biotechnol Bioeng 80:228–232

Sharma M, Chadha BS, Kaur M, Ghatora SK, Saini HS (2008) Molecular characterization of multiple xylanase producing thermophilic/thermotolerant fungi isolated from composting materials. Lett Appl Microbiol 46:526–535

Subramaniyan S (2012) Isolation, purification and characterisation of low molecular weight xylanase from Bacillus pumilus SSP-34. Appl Biochem Biotechnol 166(7):1831–1842

Subramaniyan S, Prema P (2000) Cellulase-free xylanases from Bacillus and other microorganisms. FEMS Microbiol Lett 183:1–7

Tarayre C, Brognaux A, Brasseur C, Bauwens J, Millet C, Matteotti C, Destain J, Vandenbol M, Portetelle D, De Pauw E, Haubruge E, Francis F, Thonart P (2013) Isolation and cultivation of a xylanolytic Bacillus subtilis extracted from the gut of the termite Reticulitermes santonensis. Appl Biochem Biotechnol 171:225–245

Techapun C, Poosaran N, Watanabe M, Sasaki K (2003) Optimization of aeration and agitation rates to improve cellulase-free xylanase production by thermotolerant Streptomyces sp. Ab106 and repeated fed-batch cultivation using agricultural waste. J Biosci Bioeng 95:298–301

Terrasan CR, Temer B, Duarte MC, Carmona EC (2010) Production of xylanolytic enzymes by Penicillium janczewskii. Bioresour Technol 101:4139–4143

Trajano HL, Pattathil S, Tomkins BA, Tschaplinski TJ, Hahn MG, Van Berkel GJ, Wyman CE (2014) Xylan hydrolysis in Populus trichocarpa × P. deltoides and model substrates during hydrothermal pretreatment. Bioresour Technol 179C:202–210

Viet DN, Kamio Y, Abe N, Kaneko J, Izaki K (1991) Purification and properties of beta-1, 4-xylanase from Aeromonas caviae W-61. Appl Environ Microbiol 57:445–449

Wahyuntari B, Mubarik NR, Setyahadi S (2009) Effect of pH, temperature and medium composition on xylanase production by Bacillus sp. AQ-1 and partial characterization of the crude enzyme. Microbiology 3:17–22

Wang W, Wang Z, Cheng B, Zhang J, Li C, Liu X, Yang C (2014) High secretory production of an alkaliphilic actinomycete xylanase and functional roles of some important residues. World J Microbiol Biotechnol 30:2053–2062

Wong KK, Martin LA, Gama FM, Saddler JN, de Jong E (1997) Bleach boosting and direct brightening by multiple xylanase treatments during peroxide bleaching of kraft pulps. Biotechnol Bioeng 54:312–318

Xie Z, Lin W, Luo J (2015) Genome sequence of Cellvibrio pealriver PR1, a xylanolytic and agarolytic bacterium isolated from freshwater. J Biotechnol 214:57–58

Yang YL, Zhang W, Huang JD, Lin L, Lian HX, Lu YP, Wu JD, Wang SH (2010) Purification and characterization of an extracellular xylanase from Aspergillus niger C3486. Afr J Microbiol Res 4:2249–2256

Zhang J, Viikari L (2014) Impact of xylan on synergistic effects of xylanases and cellulases in enzymatic hydrolysis of lignocelluloses. Appl Biochem Biotechnol 174:1393–1402

Zhang J, Wang J, Fang C, Song F, Xin Y, Qu L, Ding K (2010) Bacillus oceanisediminis sp. nov., isolated from marine sediment. Int J Syst Evol Microbiol 60:2924–2929

Zheng HC, Sun MZ, Meng LC, Pei HS, Zhang XQ, Yan Z, Sun JS (2014) Purification and characterization of a thermostable xylanase from Paenibacillus sp. NF1 and its application in xylooligosaccharides production. J Microbiol Biotechnol 24:489–496

Zouari Ayadi D, Hmida Sayari A, Ben Hlima H, Ben Mabrouk S, Mezghani M, Bejar S (2015) Improvement of Trichoderma reesei xylanase II thermal stability by serine to threonine surface mutations. Int J Biol Macromol 72:163–170

14

Development and validation of a stochastic molecular model of cellulose hydrolysis by action of multiple cellulose enzymes

Deepak Kumar[1,2] and Ganti S. Murthy[1*]

Abstract

Background: Cellulose is hydrolyzed to sugar monomers by the synergistic action of multiple cellulase enzymes: endo-β-1,4-glucanase, exo-β-1,4 cellobiohydrolase, and β-glucosidase. Realistic modeling of this process for various substrates, enzyme combinations, and operating conditions poses severe challenges. A mechanistic hydrolysis model was developed using stochastic molecular modeling approach. Cellulose structure was modeled as a cluster of microfibrils, where each microfibril consisted of several elementary fibrils, and each elementary fibril was represented as three-dimensional matrices of glucose molecules. Using this in-silico model of cellulose substrate, multiple enzyme actions represented by discrete hydrolysis events were modeled using Monte Carlo simulation technique. In this work, the previous model was modified, mainly to incorporate simultaneous action enzymes from multiple classes at any instant of time to account for the enzyme crowding effect, a critical phenomenon during hydrolysis process. Some other modifications were made to capture more realistic expected interactions during hydrolysis. The results were validated with experimental data of pure cellulose (Avicel, filter paper, and cotton) hydrolysis using purified enzymes from *Trichoderma reesei* for various hydrolysis conditions.

Results: Hydrolysis results predicted by model simulations showed a good fit with the experimental data under all hydrolysis conditions. Current model resulted in more accurate predictions of sugar concentrations compared to previous version of the model. Model results also successfully simulated experimentally observed trends, such as product inhibition, low cellobiohydrolase activity on high DP substrates, low endoglucanases activity on a crystalline substrate, and inverse relationship between the degree of synergism and substrate degree of polymerization emerged naturally from the model.

Conclusions: Model simulations were in qualitative and quantitative agreement with experimental data from hydrolysis of various pure cellulose substrates by action of individual as well as multiple cellulases.

Keywords: Hydrolysis, Cellulase, Bioethanol, Modeling, Cellulase purification, Synergism

Background

During bioethanol production from lignocellulosic biomass, cellulose hydrolysis can be achieved using chemicals or biological catalysts (enzymes). Although acid hydrolysis is a relatively fast process, it suffers from some

major limitations such as high operational cost, by-product formation, corrosion of equipment, neutralization requirement, high disposal cost (Bansal et al. 2009; Wang et al. 2012). Therefore, enzymatic hydrolysis is considered more feasible option during bioethanol production and has been the focus of research in last several decades. However, due to extensive hydrogen bonding, cellulose chains form a recalcitrant crystalline structure, which is difficult to degrade and require a much higher amount of

*Correspondence: Ganti.Murthy@oregonstate.edu
[1] Biological and Ecological Engineering, Oregon State University, Corvallis, OR, USA
Full list of author information is available at the end of the article

enzymes (40–100 times) for hydrolysis compared to that of starch (Merino and Cherry 2007; Wang et al. 2012).

Cellulose is hydrolyzed to glucose by synergetic action of multiple cellulase enzymes, such as endoglucanases (EG) (EC3.2.1.4), exoglucanases [also known as cellobiohydrolases (CBH)] (EC3.2.1.91) and β-glucosidase (BG) (EC3.2.1.21) (Bansal et al. 2009; Kadam et al. 2004; Zhang and Lynd 2004). Although all these enzymes have a different mode of action, they act in highly cooperative ("synergism") action for efficient degradation cellulose. Exoglucanases adsorb only from the chain ends (CBH I from reducing end and CBH II from the non-reducing end) and act in a processive manner to produce mainly cellobiose units. Processive enzymes remain bound to the glucose chain after cleaving a cellobiose molecule and will continue to cleave cellobiose units until a minimum chain length is reached. On the other side, endoglucanases are non-processive enzymes that act randomly on the surface glucose chains, hydrolyze one/few accessible internal bonds in the glucose chains and produce new chain ends. β-glucosidases hydrolyze the cellobiose and short soluble oligomers to glucose and complete the hydrolysis process (Fig. 1).

Due to high cost of cellulase enzymes (up to 30% of ethanol cost) and low sugar yields, the hydrolysis process is one of the major obstacles in the commercialization of cellulosic ethanol production (Bansal et al. 2009; Kadam et al. 2004; Kumar and Murthy 2011). There is potential for cost reduction by improving the understanding of the process, by testing a wide array of enzymes and various substrates under different conditions to determine optimum hydrolysis conditions. Designing highly efficient cellulase mixtures ("optimized enzyme cocktails") that can yield high hydrolysis rate at minimal enzyme dosage, is one such approach. Cellulase extracted from various microorganisms contain different amount of each enzyme and many commercial preparations consist of mixes from a different organism. For example, cellulase from *Trichoderma reesei* contains low fractions of β-glucosidase enzyme and this enzyme is added to the cellulase preparation to increase hydrolysis rates. It has been reported that synthetic enzyme mixture (designer combinations) of cellulase enzymes can give relatively higher hydrolysis yields (Ballesteros 2010; Banerjee et al. 2010a, b; Besselink et al. 2008). Currently, the only reliable method for designing optimal cellulase mixtures involves extensive experimentation using statistically designed combinations of various enzyme levels (Baker et al. 1998; Banerjee et al. 2010a, b; Berlin et al. 2007; Gao et al. 2010). Since conducting such large number of hydrolysis experiments is expensive, time-consuming and labor intensive, a comprehensive hydrolysis model that can that can capture process dynamics and predict hydrolysis profile under various scenarios could be an alternate feasible approach. However, due to multiple variables such as use of several enzymes acting synergistically, complex cellulose structure, and dynamic enzyme–substrate interactions make it difficult to develop mathematical models that can predict accurate hydrolysis profile under different operating conditions. Using a novel stochastic molecular modeling approach, in which each hydrolysis event is translated into a discrete event, we

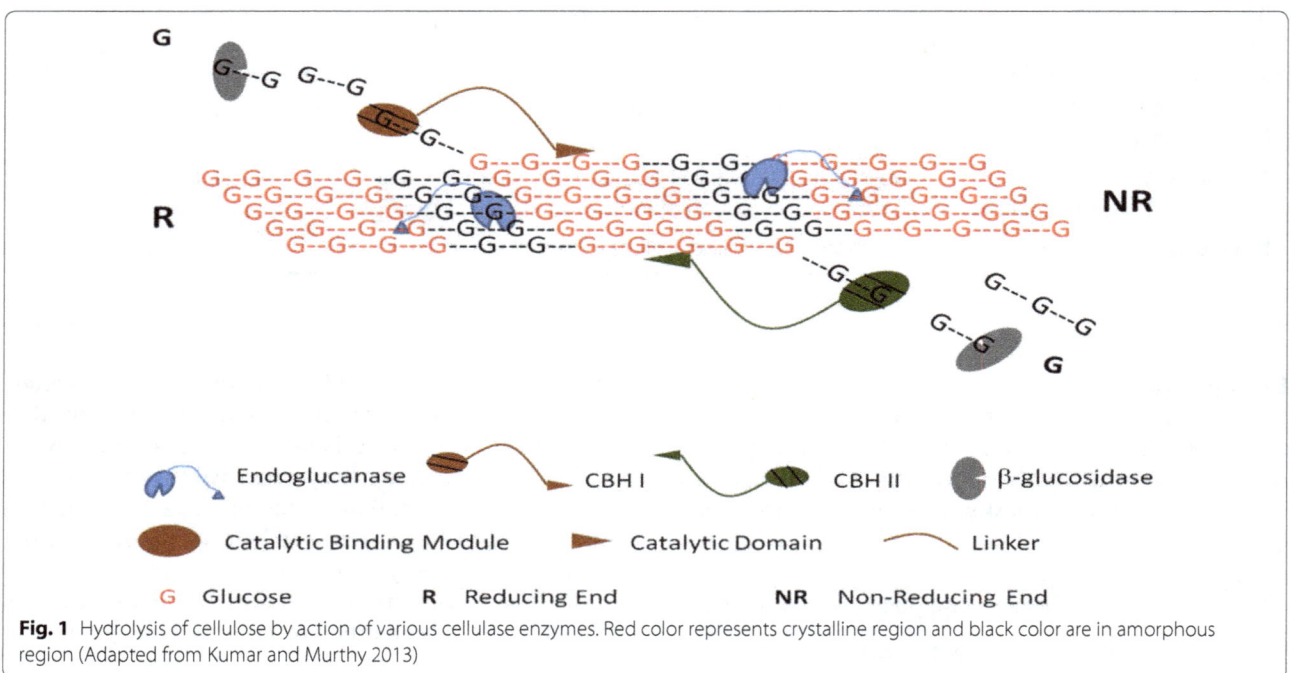

Fig. 1 Hydrolysis of cellulose by action of various cellulase enzymes. Red color represents crystalline region and black color are in amorphous region (Adapted from Kumar and Murthy 2013)

developed a first three-dimensional mechanistic cellulose hydrolysis model. The model captured the structural properties of cellulose, enzyme properties, the effect of reaction conditions, and most importantly dynamic changes in these properties (Kumar and Murthy 2013). Other than accurate predictions of hydrolysis profile, this modeling approach incorporates detailed structural features of cellulose and provides unique advantages compared to mathematical models, such as tracking of multiple oligomers as well as chain distribution, tracking of morphological changes in cellulose, elimination of the need for parameter changes with a change in experimental data set. Please refer to our earlier paper (Kumar and Murthy 2013) for more detailed comparison of this modeling approach and comparisons to other modeling approaches.

Although the previous model incorporated significant cellulose structural details and complex enzyme–substrate interactions, it did not include the simultaneous action of multiple classes of enzymes at any instant of time. During each iteration, only one class of enzymes (e.g., EG, CBH I or CBH II) was acting, which ignores the enzyme crowding/jamming effect, a critical phenomenon at high enzyme concentrations (Hall et al. 2010; Igarashi et al. 2011). This work presents the updated model with incorporation of enzyme jamming phenomenon by modeling simultaneous action of multiple enzymes, and also including other practical considerations, such as oligomer solubility and glucose production by cellobiohydrolase enzymes. Our earlier model was validated with very limited experimental data from the literature. In this work, the model simulations were validated with comprehensive experimental data sets obtained from hydrolysis of pure cellulose (Avicel, filter paper, and cotton) using purified *T. reesei* enzymes. Experiments were performed with purified CBH I and CBH II under various hydrolysis conditions, to cover the effect of enzyme loadings, substrate properties, and product inhibition.

Methods

Materials
Celluclast, a commercial cellulase from *T. reesei* (Lot # CCN03141), was donated by Novozymes (Novo, Bagsvaerd, Denmark). P-Aminophenyl β-D-cellobioside (sc-222106, Lot #K213), used as an affinity ligand for cellobiohydrolase, was purchased from Santa Cruz Biotechnology Inc. (Santa Cruz, CA, USA). All other chemicals required for protein purification and hydrolysis experiments were purchased from Sigma-Aldrich (Milwaukee, WI). Whatman No. 1 filter paper (Whatman, Inc., Florham Park, NJ) and cotton balls (Kroger Co., Cincinnati, Ohio) were used as the pure cellulose samples for the hydrolysis experiments. The commercial β-glucosidase (Novozyme 188) from *Aspergillus niger* was purchased from Sigma-Aldrich (Milwaukee, WI).

Model development
Stochastic hydrolysis model
Development of this comprehensive model consisting cellulose structural details and complex enzyme–substrate interactions consisted of in silico representative cellulose model, enzyme characterization, and developing algorithms for modeling the enzyme actions. In this model, cellulose was modeled based on the structure of cellulose Iβ, the most abundant cellulose form in higher plants. The structure was modeled as a group of microfibrils (MF) (2–20 nm diameter), and each microfibril contains multiple elementary fibrils (EF), the basic building block of cellulose with about 3.5 nm diameter and containing 36 glucose chains (Chinga-Carrasco 2011; Fan and Lee 1983; Lynd et al. 2002). The number of EF in an MF, glucose molecules in one chain of glucose (i.e., degree of polymerization, DP), was assumed to be constant during each simulation. These parameters were dynamically determined at the beginning of the cellulose structure simulation, based on the type of cellulose simulated. The degree of crystallinity in cellulose (50–90%) is a critical factor affecting the cellulose hydrolysis, as amorphous regions are believed to be relatively more susceptible to enzyme action and determine initial hydrolysis rates. To capture this important property in this model, glucose chain in each EF were assumed to pass through multiple crystalline regions (200 glucose molecules long regions) separated by amorphous regions. The concept of modeled cellulose structure and its resemblance with actual cellulose structure is illustrated in Fig. 2.

Each glucose molecule in the modeled microfibril was given a unique serial number as its identity, and a big data set containing other parameters (e.g., reducing/non-reducing end, EF surface, MF surface, crystalline or amorphous, soluble, non-soluble, distance from chain end, etc.) that describe structural properties of that bond. During developing algorithms for cellulase actions, enzyme accessibility was determined based on these parameters (data set with each glucose molecule) and action pattern of enzymes. For additional details of cellulose model please refer to earlier publications (Kumar 2014; Kumar and Murthy 2013).

Cellulase enzymes vary in mode of actions, and for this model, the enzymes were classified into eight classes depending upon their structure and mode of action (e.g., non-processive endocellulase with cellulose binding molecule (CBM), processive CBH I with CBM, processive CBH II with CBM, etc.). Please refer to our earlier paper (Kumar and Murthy 2013) for more details on enzyme classifications, their characteristics, and mode of actions. Cellulose hydrolysis is dependent on biomass-dependent extrinsic factors (crystallinity, accessibility, and DP) and enzyme action is dependent on intrinsic factors (enzyme activity, stability with pH and temperature, etc.). The extrinsic factors were modeled in the simulated cellulose structure described above. Enzyme activity (depends on enzyme origin and level of purification) and enzyme loading (amount of enzyme/g substrate; based on experimental conditions) information was transformed into theoretical maximum turnover number (maximum possible number of bonds hydrolyzed per unit time for each enzyme) (N_{hi_max}) (Eq. 1) for each class of enzyme.

$$N_{hi\_max} = E_i * U_i * 6.023 * 10^{17} * \frac{G_{Sim}}{6.023 * 10^{23}} * 162 * S_i,$$

(1)

where 'E_i' is amount of 'ith' enzyme used (mg cellulose); 'U_i' is activity of 'ith' enzyme (IU/mg enzyme); 'G_{sim}' is the number of glucose molecules simulated in the model; "162" is the average molecular weight of anhydrous glucose; 'S_i' is stability of 'ith' enzyme under experimental conditions (temperature and pH). Value of "S_i" could be calculated for any enzyme using empirical equations developed, such as Arrhenius rate relationship for temperature. Value of 'S_i' is a real number between 0 and 1.

These numbers were further transformed to numbers of hydrolyzed bonds per microfibril based on the total number of microfibrils simulated and mode of action of enzymes. For example, for endoglucanase enzymes,

these numbers were proportional to relative glucose molecules on the surface of microfibril. On the other hand, for CBH I and CBH II, these numbers were proportional to a relative number of chain ends available in one microfibril. Please refer to Kumar (2014) for more details.

The hydrolysis process was modeled using Monte Carlo simulation technique, which has been used successfully earlier for modeling the starch hydrolysis (Marchal et al. 2001, 2003; Murthy et al. 2011; Wojciechowski et al. 2001). The overall schematic for simulating the enzymatic hydrolysis for each enzyme is shown in Fig. 3 and detailed description is provided in Kumar and Murthy (2013). All the required substrate–enzyme interactions, such as binding of CBH only on chain ends, the higher binding probability of binding EG on MF surface than at EF surface, were incorporated into the model using algorithms. It was also made sure that sufficient glucose molecules (based on the size of enzyme) are available to allow binding.

Only one class of enzymes was modeled working at a time, so the model did not account for the enzyme crowding effect (locations occupied by other class of enzymes at the same time). These effects were incorporated in the modified model discussed in next section. Other than cellulose structural restrictions, some probabilities were

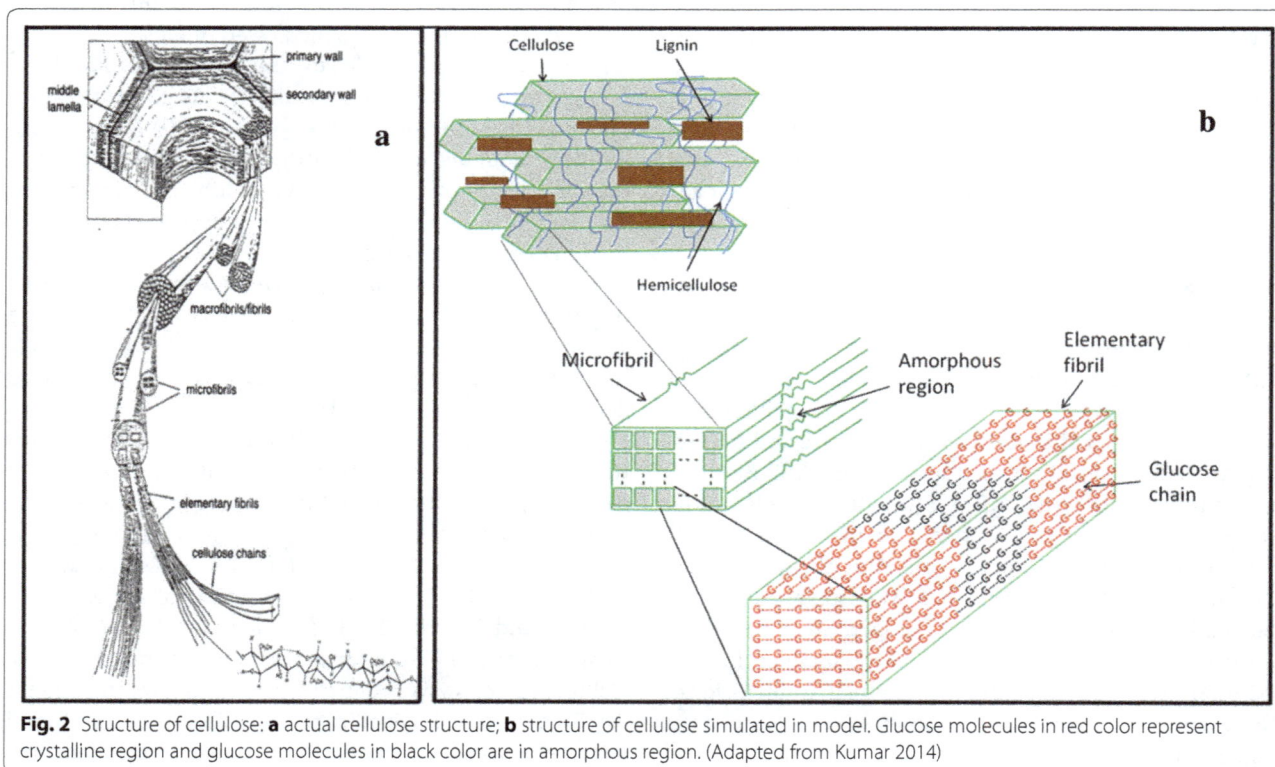

Fig. 2 Structure of cellulose: **a** actual cellulose structure; **b** structure of cellulose simulated in model. Glucose molecules in red color represent crystalline region and glucose molecules in black color are in amorphous region. (Adapted from Kumar 2014)

defined corresponding to enzyme action. For example, the probability of hydrolysis of a β-1,4 bond hydrolysis located in amorphous regions was more than that of in crystalline region by an endoglucanase enzyme. Choice was made by generating a random number at each decision point and comparing it with the defined probability. The hydrolysis event would happen only in the case when the random number was greater than the probability of hydrolysis. Number of iterations were restricted using a counter (Fig. 3). If all conditions for hydrolysis were met for that bond, it was converted to broken bond and the counter was incremented. Similarly, the counter was given an increment corresponding to unsuccessful events also (in case binding or hydrolysis does not occur). After each broken bond, it was made sure to change properties of other glucose molecules in that chain (e.g., chain length, distance from chain end, solubility, etc.). If a glucose chain becomes soluble, part of the chain just beneath the soluble chain is exposed and becomes accessible to enzymes. The concept is described in detail elsewhere (Kumar 2014).

Modifications in the model

The model described in above section was the first report of a comprehensive stochastic model for cellulose hydrolysis that successfully captured the cellulose structural features (three dimensional), enzyme characteristics, and dynamic enzyme–substrate interactions. In this work, the model was further modified to capture more realistic expected interactions during hydrolysis by incorporating the (1) simultaneous action of enzymes from multiple classes at any instant of time to account for the enzyme crowding; (2) partial solubility of cello-oligomers with DP 6–13, and (3) production of glucose by exocellulase. In the previous version of the model, the model was simulated based on the iterative concept only; however, in real conditions multiple enzyme molecules act simultaneously and block the hydrolysis sites for each other (Igarashi et al. 2011). Enzyme crowding and simultaneous action of enzymes were incorporated in the current model by calculating the number of enzyme molecules based on the enzyme loading, their molecular weight, and number of glucose molecules simulated. The iterations are performed for every minute of hydrolysis and properties of substrate are changed after that at the end of the 1-min time step. For processive enzymes, once an enzyme molecule bound to chain end, it remains bound at the end of 1-min time step and continues further down the chain until it reaches the end of the chain or desorbs from the molecule as per its probability. Exocellulase enzyme binds to multiple cellulase chains (three chains in the model) (Asztalos et al. 2012; Levine et al. 2010), so it is essential that all three chains must be accessible to the

enzyme (on surface and not blocked by other enzyme) for binding of the enzyme. In the previous version of model, it was assumed that glucose molecules equal to size of CBM only are required on surface and unblocked for binding, however in the current model whole length of enzyme was considered (except linker, because it is flexible and is compressed during movement) (Wang et al. 2012). The detailed schematics explaining algorithms developed to model CBH I and EG actions have been provided in the in Additional files 1 and 2, respectively. Cellodextrins with DP < 6 are considered to be completely soluble, DP 6–13 partially soluble and above 13 are insoluble in water (Lynd et al. 2002; Zhang and Lynd 2004). In the previous version of the model, all oligomers with DP > 6 were considered insoluble. While the CBM of the enzymes cannot bind to these chains due to its large size, the catalytic domains of the enzymes will still act on the oligomers in solution. In the absence of reliable literature data, the soluble fraction of the oligomers was set as a function of DP in the range of DP 7–13. The oligomers with DP 7–9, 10–11, and 12–13 were assumed 75, 50, and 25% soluble, respectively. Oligomers with DP < 6 were assigned a 100% solubility, and while oligomers with DP more than 13 were set to 0% solubility. In the previous model, the CBH action could only produce cellobiose during cellulose hydrolysis. However, glucose formation during cellulose hydrolysis by CBH action has been observed by some researchers (Eriksson et al. 2002; Medve et al. 1998), and was also observed in our experiments (discussed later in the "Results and discussion" section). Therefore, the model was modified to include glucose formation in addition to the cellobiose. A probability of glucose formation was included in the model, and glucose/cellobiose formation was decided by generating a random number and comparing with that probability. The probabilities and increments associated with productive various events (productive binding, no binding, non-productive binding, etc.) are listed in Additional file 3: Table S1.

Enzyme crowding/jamming phenomenon might not be critical at low enzyme dosages and during action of individual enzymes. Also, the other details incorporated into this model might be ignored if the final goal of the model is to simulate the sugar concentration only during the hydrolysis process. However, to simulate and optimize the composition/cocktail of enzymes, it is necessary to simulate the effects of each enzyme class carefully.

Model implementation and simulations

The algorithms of the hydrolysis model were written in C++ language. Random number generators were used in simulation of cellulose structure and hydrolysis process (Matsumoto and Nishimura 1998).

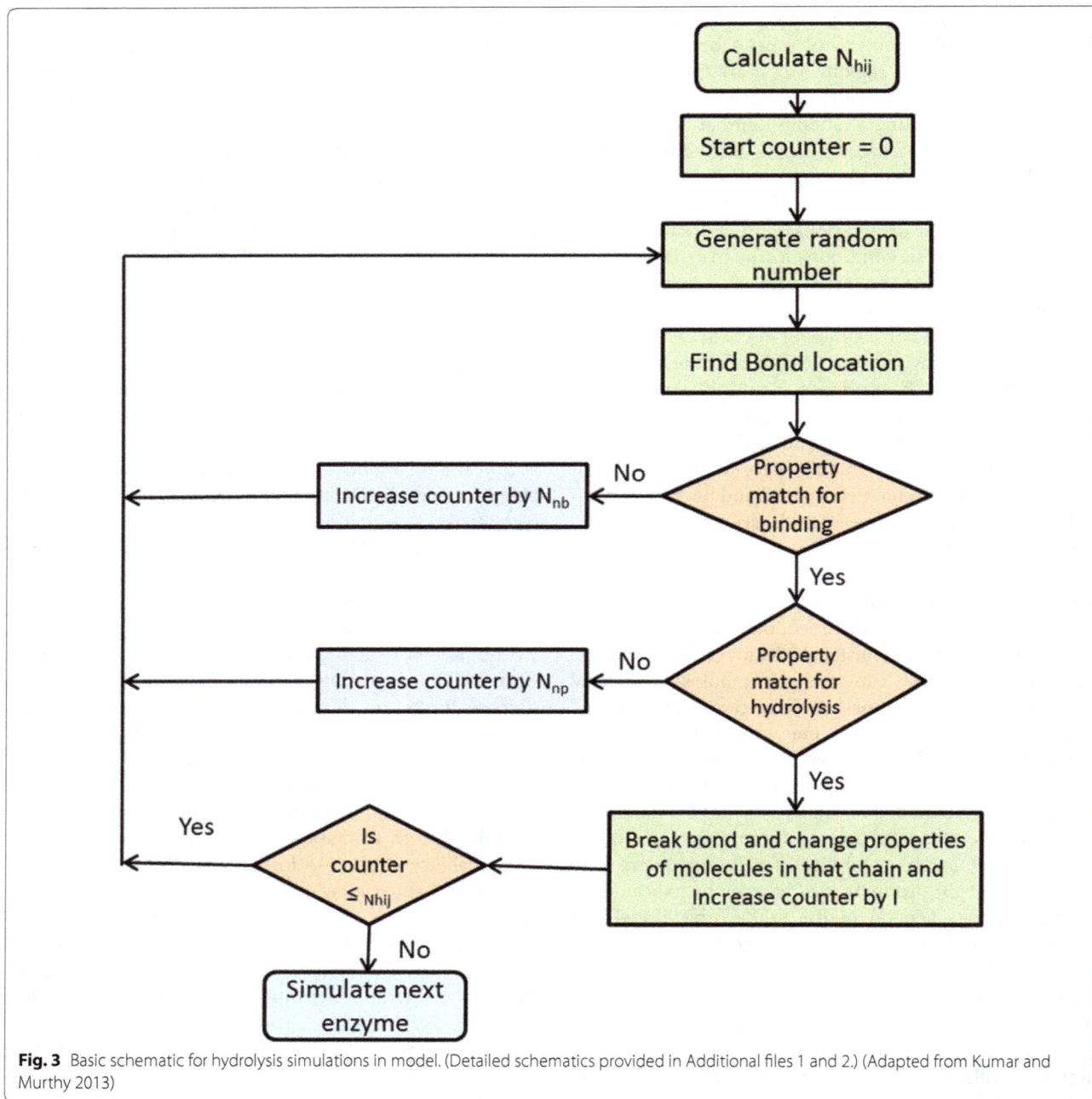

Fig. 3 Basic schematic for hydrolysis simulations in model. (Detailed schematics provided in Additional files 1 and 2.) (Adapted from Kumar and Murthy 2013)

The cellulose structure was simulated for three model cellulose substrates Avicel, filter paper and cotton, to cover the range of substrates with different structural properties (DP and degree of crystallinity). Avicel is low-DP cellulose, with DP only about 300 and crystallinity index 0.5–0.6; whereas, cotton has relatively very high DP (about 2000–2500) and crystallinity index of 0.85–0.95 (Zhang and Lynd 2004). Hydrolysis simulations were performed based on the experimental conditions: weight of solution (scale of hydrolysis), solid loading, cellulose content, total enzyme loading (mg protein/g cellulose), ratio of enzymes present (EG:CBH I:CBH II:BG), temperature, pH and hydrolysis duration. Enzyme activities can be determined from supplier, literature, or can be determined using standard protocols (Ghose 1987). Unless determined in the lab, specific activities of enzymes from *T. reesei* were assumed as 0.4, 0.08, and 0.16 IU/mg of EG I, CBH I and CBH II, respectively (Zhang and Lynd 2006) for model simulations. The output from model included several data files containing glucose concentrations, oligosaccharide concentrations, chain distribution profile (number of chains of various lengths), crystallinity index profile (ratio of crystallinity at various time intervals), solubility profile and data sheets for each microfibril (illustrating major properties associated with glucose molecules) at various times during hydrolysis.

Model validation

The data from model simulations were compared with various sets of experimental results from cellulose hydrolysis in our lab and from literature (Bezerra and Dias 2004; Bezerra et al. 2011) to validate the model under various hydrolysis conditions.

Validation with experimental data

The model was validated with the results obtained from hydrolysis of pure cellulosic substrates (filter paper and cotton) using purified CBH I and CBH II. The cellulases CBH I and CBH II were purified from Celluclast (Novozymes, Denmark) using a series of chromatography steps in BioLogic LP system (Bio-Rad Laboratories, Hercules, CA, USA). The purification experiments were performed at room temperature and the collected enzymes were transferred and stored in the refrigerator at 4 °C.

Enzyme purification

The flow diagram of steps followed in the CBH I and CBH II purification is shown in Fig. 4. In the first step of purification, the Celluclast enzyme mixture was desalted using Sephadex G-25 Fine (dimensions: 2.5 cm × 10 cm) gel filtration column. The protein was rebuffered in 50 mM Tris–HCl buffer (pH 7.0) at 5 mL/min. The desalted protein was fractionated by anion-exchange chromatography using DEAE-Sepharose column (dimensions: 2.5 cm × 10 cm). The sample was loaded using 50 mM Tris–HCl buffer (pH 7.0) at 5 mL/min flow rate and was eluted stepwise: 1st elution at 35%, and 2nd elution at 100% of 0.2 M sodium chloride in 0.05 M Tris–HCl buffer (pH 7) (Jäger et al. 2010). The flow-through from DEAE column (rich in CBH II enzymes) was concentrated and rebuffered in 50 mM sodium acetate buffer (pH 5.0) using Pellicon XL 50 Ultrafiltration Cassette, with biomax 10 (Millipore, USA). The rebuffered protein was spiked with gluconolactone (final concentration of 1 mM) and loaded on the p-aminophenyl cellobioside (pAPC) affinity column (dimensions: 1.5 cm × 10 cm) with 0.1 M sodium acetate, containing 1 mM gluconolactone and 0.2 M glucose (pH 5.0) at flow rate of 1.5 mL/min (Jeoh et al. 2007; Sangseethong and Penner 1998). The function of gluconolactone in the buffer is to suppress β-glucosidase activity, which otherwise can cleave the ligand (Sangseethong and Penner 1998). The bound CBH II protein was eluted using the running buffer

containing 0.01 M cellobiose [100 mM sodium acetate buffer containing 1 mM gluconolactone, 0.2 M glucose, and 0.1 M cellobiose (pH 5.0)]. The purified CBH II from affinity column was concentrated and loaded on the phenyl Sepharose column (dimensions: 1.0 cm × 10 cm) for hydrophobic interaction chromatography to separate core and intact proteins (Sangseethong and Penner 1998). The sample was loaded in high salt (0.35 M ammonium sulfate in 25 mM sodium acetate buffer, pH 5.0) and eluted with linear gradient from running buffer to elution buffer [25 mM acetate buffer containing 20% ethylene glycol (v/v), pH 5.0]. Hydrophobic interaction chromatography was performed on the second elution (CBH I rich) from the anion-exchange column, after concentrating and rebuffering with 25 mM sodium acetate buffer. The enzyme was loaded in very high salt (0.75 M ammonium sulfate in 25 mM sodium acetate buffer, pH 5.0) and eluted with linear gradient from running buffer to elution buffer [25 mM acetate buffer containing 5% ethylene glycol (v/v), pH 5.0]. The purified CBH II and CBH I fractions from hydrophobic interaction column were concentrated and rebuffered in 50 mM sodium acetate buffer, pH 5.0. Protein containing fractions were determined by measuring absorbance at 280 nm.

The fractions collected from the chromatographic purifications steps shown in Fig. 4 were analyzed by SDS-polyacrylamide gel electrophoresis to check for their purity. Based on the molecular weight comparison with marker, and literature data, the single bands in the numbered lanes 1 and 2 of Fig. 5 correspond to CBH II (MW 54 kDa) and CBH I (MW 61–64 kDa), respectively (Jäger et al. 2010; Medve et al. 1998; Sangseethong and Penner 1998). The activities of CBH I and CBH II on Avicel were determined as 0.478 and 0.379 IU/mg of protein, respectively.

During protein purification, the protein concentrations in the samples were determined based on Bradford assay using Quick Start™ Bradford Protein Assay Kit (Bio-Rad, USA) and bovine serum albumin (BSA) as standard. The activities of purified CBH I and CBH II were determined on Avicel in 50 mM sodium acetate buffer, pH 5.0. 1 mL of Avicel solution (10 g/L) with final enzyme concentration of 0.1 mg/mL was incubated (mixed end to end) at 45 °C in 2 mL Eppendorf centrifuge tubes for 2 h (Jäger et al. 2010). After 2 h of incubation, the samples were heated at 95 °C for 5 min to stop the hydrolysis. The samples were centrifuged at 15,000 rpm for 5 min to separate the supernatant. The reducing sugar concentration in the supernatant was determined using dinitrosalicylic acid (DNS) assay and using glucose as standard.

Enzymatic hydrolysis

The hydrolysis experiments were conducted at 25 g/L cellulose (filter paper and cotton balls) concentration

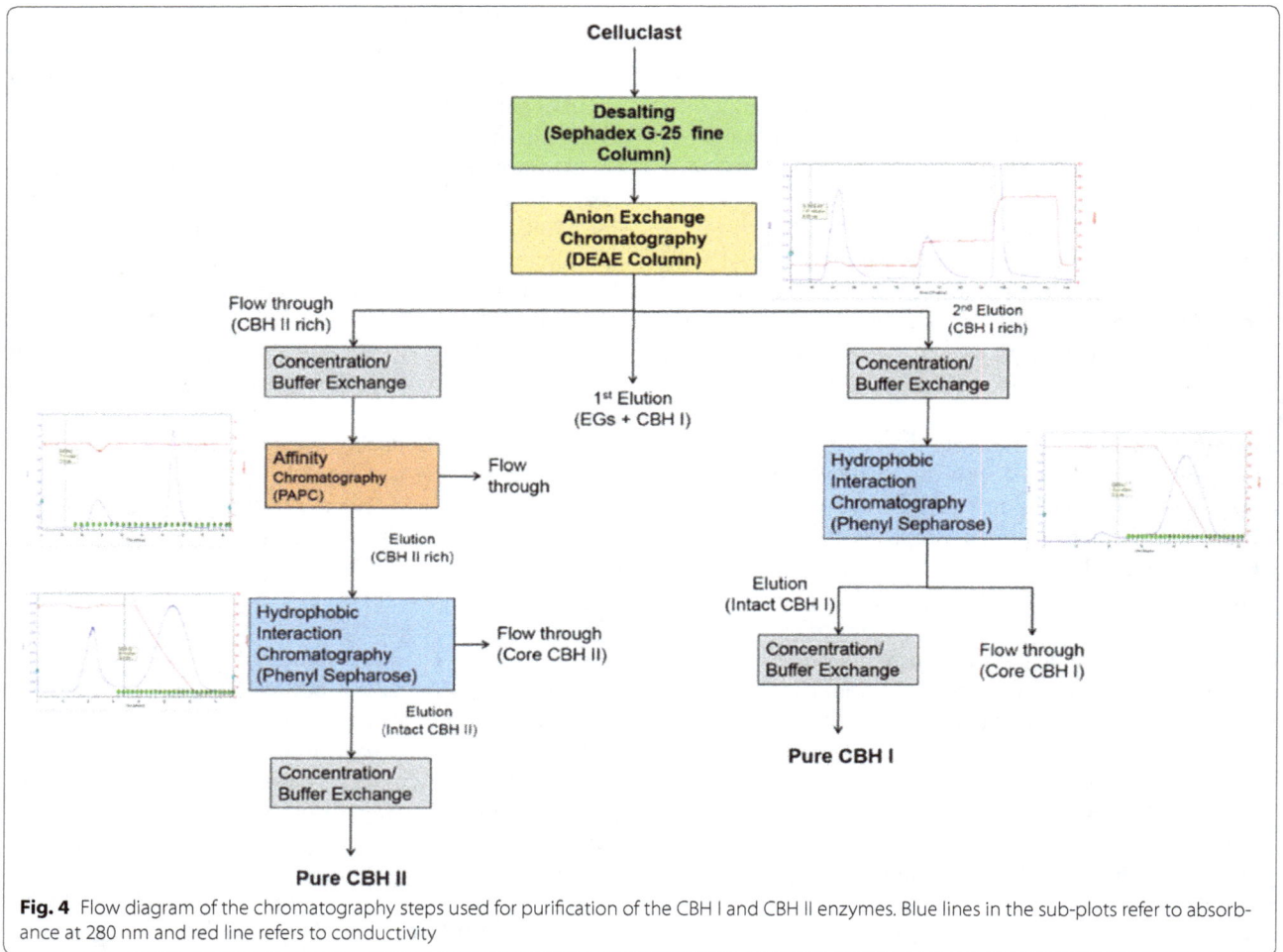

Fig. 4 Flow diagram of the chromatography steps used for purification of the CBH I and CBH II enzymes. Blue lines in the sub-plots refer to absorbance at 280 nm and red line refers to conductivity

and various enzyme loadings (5, 10, and 15 mg/g cellulose) in 50 mM sodium acetate, pH 5.0, 10 mL total volume in 25 mL Erlenmeyer flasks closed with rubber stopper. 100 µL of 2% sodium azide was added in each flask to avoid microbial contamination. The experiments were carried out in controlled environment incubator shaker set at 45 °C and 125 rpm. 200 µL of sample was withdrawn at 3, 6, 9, 12, 18, 24, 36, 48, and 72 h to determine sugar concentrations and the hydrolysis profile. The samples were heated at 95 °C for 5 min to stop the reaction and were prepared for high-performance liquid chromatography (HPLC) analysis. All experiments were performed in triplicate.

Results and discussion
Validation with literature data
Model simulations were performed for Avicel hydrolysis by CBH I using experimental conditions mentioned in Bezerra and Dias (2004). Figures 6 and 7 illustrate the comparison of model simulations and experimental data

from hydrolysis of cellulose at 5 and 2.5% solid loadings, respectively. The data from simulation of previous version of model (Kumar and Murthy 2013) were also plotted in these figures to demonstrate the differences in hydrolysis profiles. Results from experimental data and model simulation data were in qualitative and quantitative agreement at both 25 and 50 g/L Avicel loadings. Coefficient of determination (R^2) was found 0.97 (at 5% Avicel loading) and 0.94 (at 2.5% Avicel loading) and was higher than that obtained from previous model results: 0.70 (at 5% Avicel loading) and 0.89 (at 2.5% Avicel loading). Coefficient of determination value was low for previous version of model because, at such high enzyme loadings, enzyme crowding effect become predominant, which was not captured in the previous version of model. Current model captures the crowding effect and therefore results in more accurate cellobiose concentrations. Quantitative match of model simulations with experimental data for various substrate–enzyme ratios also indicated that this model successfully captured the cellobiose inhibition effect. Comparison of model simulations with additional experimental data from literature (lower enzyme loadings) is illustrated in Additional file 3: Figures S1a and S1b.

Validation with experimental data from current study
Hydrolysis of filter paper by CBH I and CBH II

The results from model simulations and actual experiments of hydrolysis of filter paper at various loadings of CBH I and CBH II are shown in Figs. 8, 9, 10 and 11, respectively.

In all of the cases, the model fitted filter paper hydrolysis data and predicted the sugar profiles and hydrolysis rates. It is very important to note that except enzyme activities (determined experimentally in this case), no other model parameter was changed during simulations under these hydrolysis conditions vs. earlier literature-based experimental conditions. Excellent fitting of model with data with studies from two different lab groups demonstrates the robustness and potential usability (scope of using for any hydrolysis conditions with parametrization issues) of this model. As expected, cellobiose, followed by glucose was the major product during hydrolysis by CBH I or CBH II. The model predictions were accurate in determining both cellobiose and glucose concentrations during hydrolysis. The previous version of model did not account for the glucose formation during cellulose hydrolysis by cellobiohydrolases, hence, did not fit with the experimental data (Additional file 3: Figure S2). Small amounts of cellotriose were also observed

both in experimental data and model simulations (data not shown). The cellobiose production rate is high at the beginning of hydrolysis and decreases significantly due to cellobiose inhibition on the CBH I and II enzymes. The inhibition effect was captured by the model and was also further validated when the effect disappears on removal of cellobiose by converting it to glucose through β-glucosidase action (discussed later in the manuscript).

The R^2 values between experimental and model data for cellobiose production during filter paper hydrolysis by CBHI and CBH II were in the range of 0.65–0.90 and 0.77–0.81, respectively. It was observed from the results that increase in enzyme loading did not result in the significant increase in the final sugar yields. The enzymes used in the hydrolysis experiments had very high activity, and possibly increasing the enzyme loading resulted in the enzyme crowding effect due to limited availability of chain ends. It can be observed from the results, that the phenomenon was well captured by the model, as the simultaneous action of multiple enzyme, their blockage by each other was considered in the model.

Effect of beta-glucosidase addition (exo-BG synergism)

Cooperative action of different enzymes, known as synergism, is one of the most important phenomenon observed in cellulose degradation (Andersen et al. 2008; Bansal et al. 2009; Wang et al. 2012; Zhang and Lynd 2004). Synergism between CBH I and/or CBH II and β-glucosidase enzymes is very important for the conversion of cellulose (Zhang and Lynd 2004). This synergism

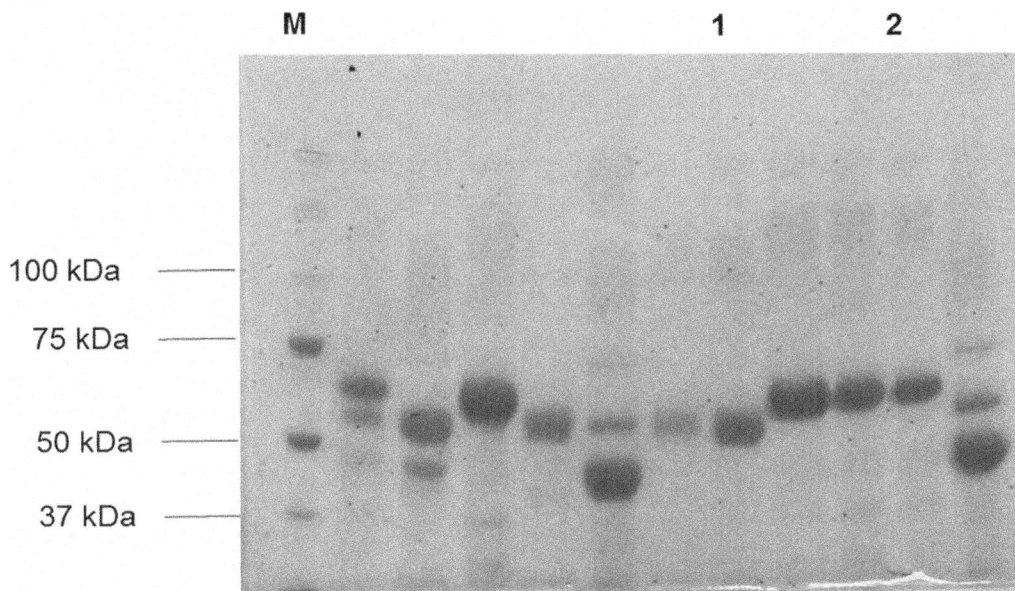

Fig. 5 SDS-polyacrylamide gel electrophoresis of the purified cellulase enzymes. *M* molecular mass marker, (1) CBH II (2) CBH I

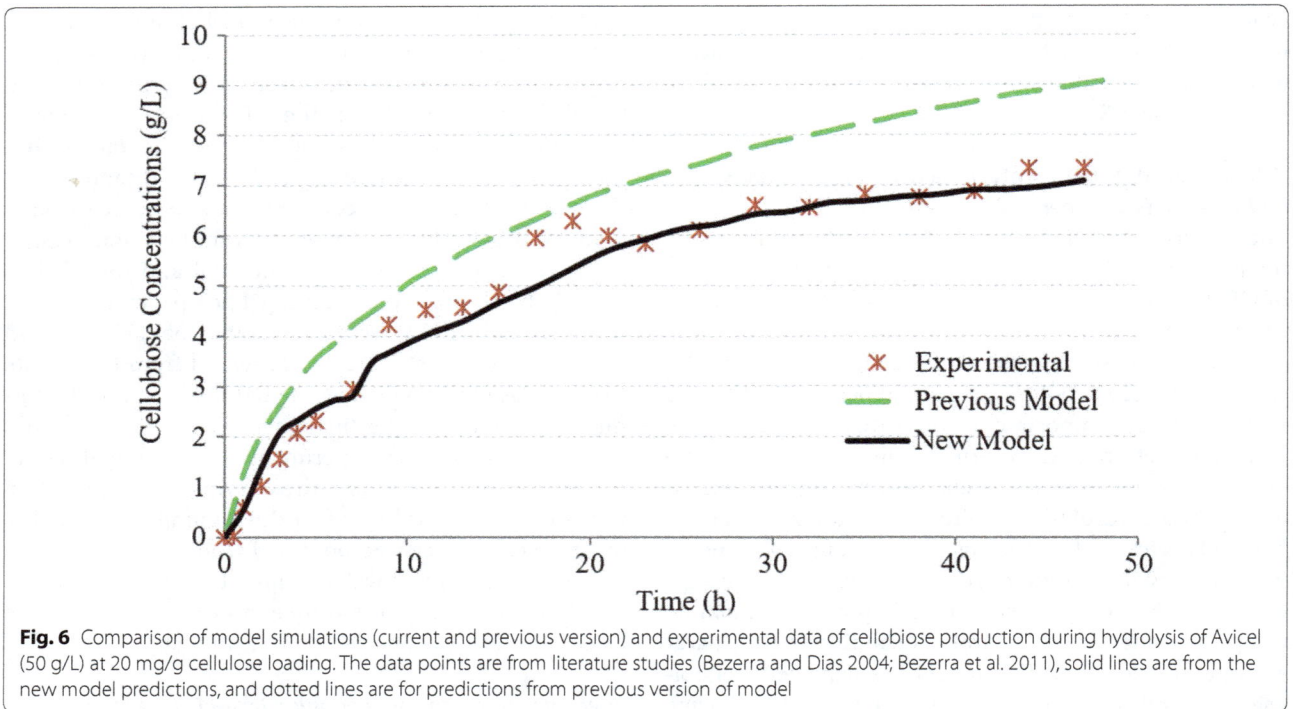

Fig. 6 Comparison of model simulations (current and previous version) and experimental data of cellobiose production during hydrolysis of Avicel (50 g/L) at 20 mg/g cellulose loading. The data points are from literature studies (Bezerra and Dias 2004; Bezerra et al. 2011), solid lines are from the new model predictions, and dotted lines are for predictions from previous version of model

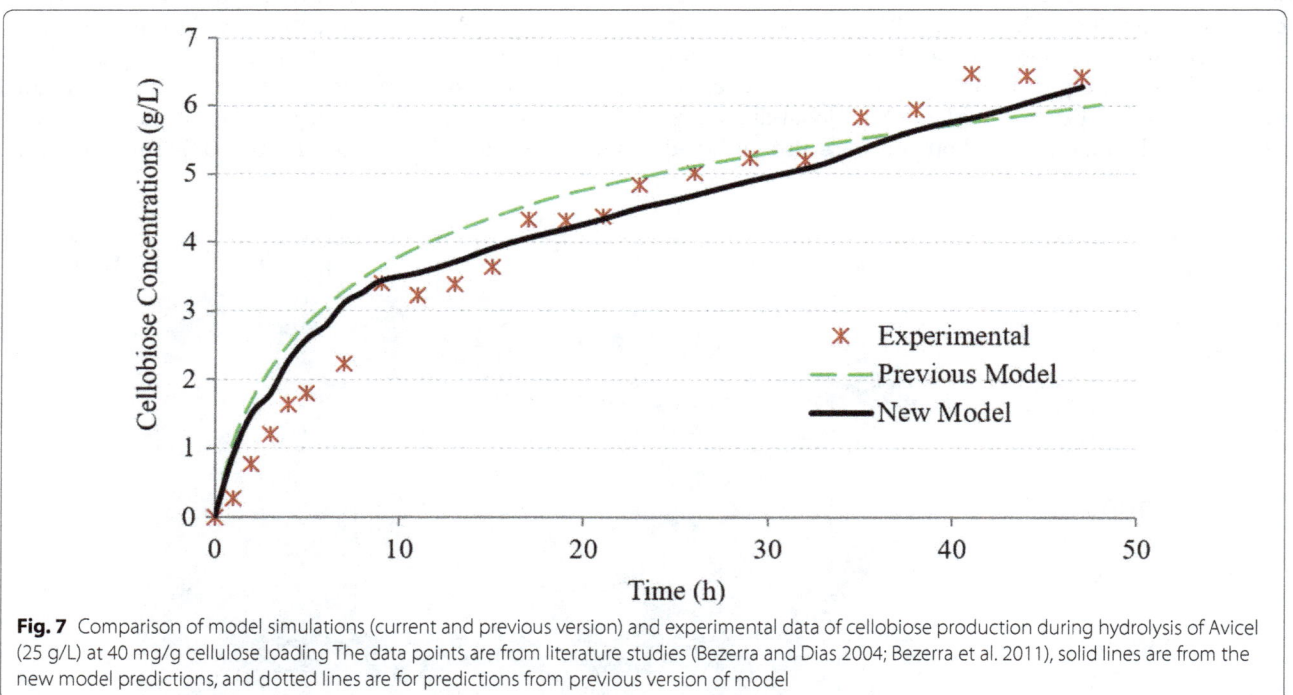

Fig. 7 Comparison of model simulations (current and previous version) and experimental data of cellobiose production during hydrolysis of Avicel (25 g/L) at 40 mg/g cellulose loading The data points are from literature studies (Bezerra and Dias 2004; Bezerra et al. 2011), solid lines are from the new model predictions, and dotted lines are for predictions from previous version of model

occurs mainly because of strong inhibition effect of cellobiose on the CBH performance. Primary product of CBH action on cellulose is cellobiose, a strong inhibitor to CBH activity (Andersen 2007; Ballesteros 2010; Fan et al. 1987; Mosier et al. 1999; Zhang and Lynd 2004). Cellobiose buildup is prevented by the action of β-glucosidase which further hydrolyzes cellobiose to glucose and results in CBH and β-glucosidase synergism. The synergistic effect of β-glucosidase addition was observed during filter paper hydrolysis by CBH I and CBH II and is illustrated in Figs. 12 and 13, respectively.

Cellulose conversions after 72 h of hydrolysis were observed about 82.7 and 15.1% higher for CBH I (10 mg/g glucans) and CBH II (10 mg/g glucans), respectively, in presence of excess β-glucosidase than those in absence of β-glucosidase. To determine the model accuracy in predicting this trend, action of CBH I and CBH II was simulated in absence and presence of β-glucosidase. It can be observed from Figs. 11 and 12 that model simulations capture this synergism successfully both for CBH I and CBH II enzymes.

The synergism was lower for CBH II compared to CBH I, possibly due to relatively less inhibitory effect of cellobiose on CBH II. The observation of relatively lower cellobiose inhibition towards CBH II was also reported in a comprehensive study on cellobiose inhibition using ^{14}C-labeled cellulose substrates, conducted by Teugjas and Väljamäe (2013). In that study, it was reported that enzymes from glycoside hydrolase (GH) family 7 were most sensitive to cellobiose inhibition followed by family 6 CBHs and endoglucanases (EGs). The model simulations successfully followed the trend observed experimentally. Other than cellulose conversion, hydrolysis rate of filter paper by CBH I and CBH II with excess

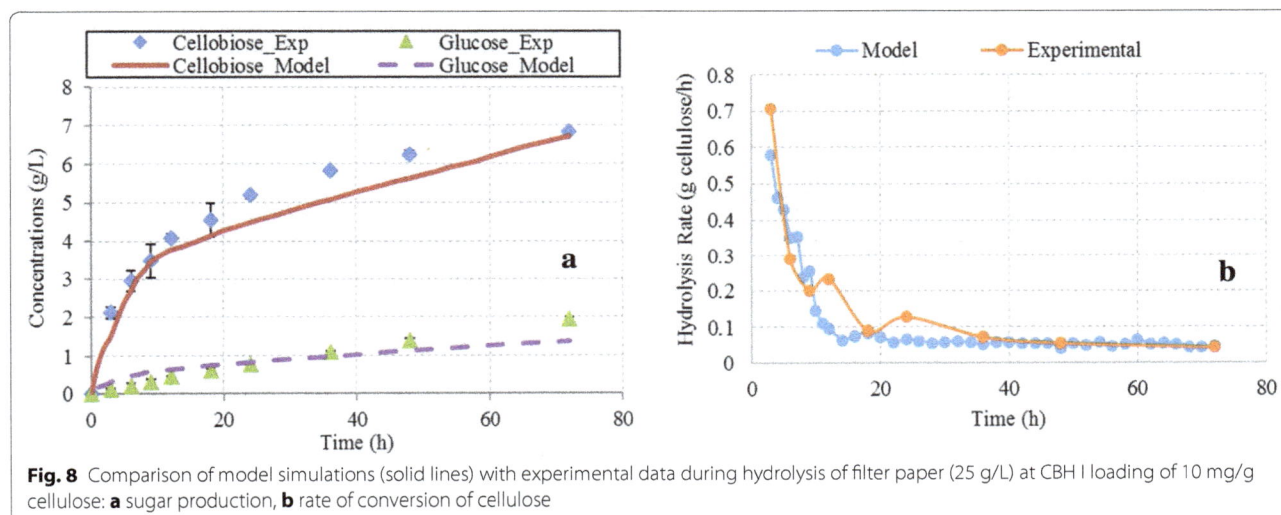

Fig. 8 Comparison of model simulations (solid lines) with experimental data during hydrolysis of filter paper (25 g/L) at CBH I loading of 10 mg/g cellulose: **a** sugar production, **b** rate of conversion of cellulose

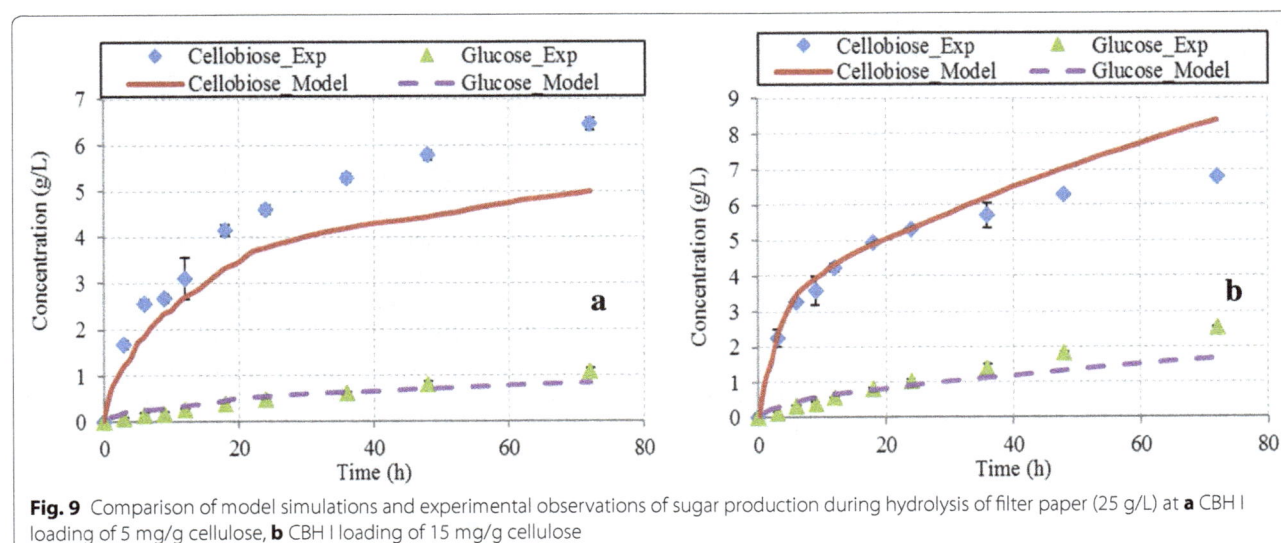

Fig. 9 Comparison of model simulations and experimental observations of sugar production during hydrolysis of filter paper (25 g/L) at **a** CBH I loading of 5 mg/g cellulose, **b** CBH I loading of 15 mg/g cellulose

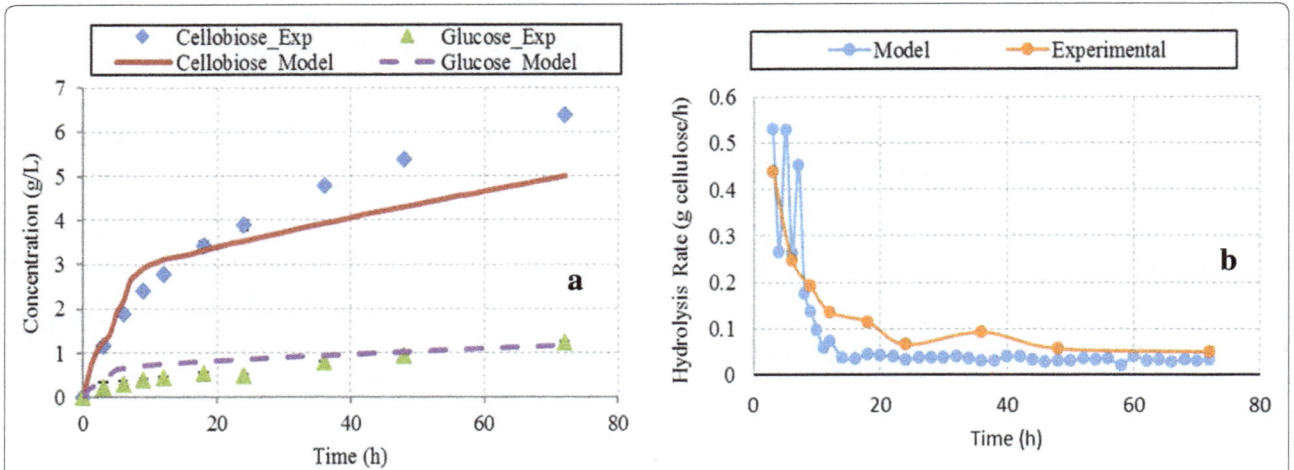

Fig. 10 Comparison of model simulations (solid lines) with experimental data during hydrolysis of filter paper (25 g/L) at CBH II loading of 10 mg/g cellulose: **a** sugar production, **b** rate of conversion of cellulose

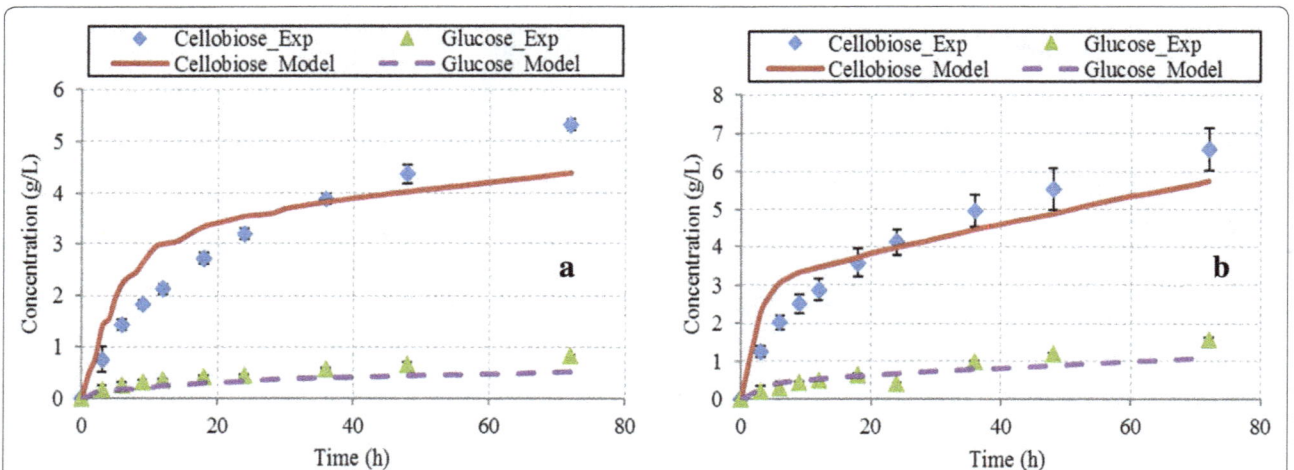

Fig. 11 Comparison of model simulations and experimental observations of sugar production during hydrolysis of filter paper (25 g/L) at **a** CBH II loading of 5 mg/g cellulose, **b** CBH II loading of 15 mg/g cellulose

of β-glucosidase was markedly higher than that of CBH enzymes acting alone (data not reported).

Effect of structural properties of cellulose

Cellulose hydrolysis is highly affected by the structural properties of cellulose. In case of cellobiohydrolases (CBH I and CBH II) action, where the enzymes act on chain ends only, fractions of reducing/non-reducing ends relative to total glucose molecules would be critical factor affecting hydrolysis. For example, the percentage of chain ends for filter paper with average chain DP of 700 is 0.13% compared to 0.05% for bacterial cellulose with average DP of 2000 and 0.033% for cotton with DP of

3000 (Zhang and Lynd 2004, 2006). Therefore, it would be expected that CBH I and CBH II would hydrolyze filter paper more efficiently compared to cotton or bacterial cellulose, as there are relatively many more reducing/non-reducing ends. The expected trends were observed in both experimental and model simulations of CBH I and CBH II action on filter paper and cotton (Fig. 14a, b).

The cellulose conversion after 72 h of cotton hydrolysis was observed 77.0 and 92.6% less than those of filter paper hydrolysis by action of CBH I and CBH II, respectively. The model had a good fit with experimental data for cotton hydrolysis in case of CBH I. For CBH II, although the absolute values of predicted cellulose conversion were higher than the actual values, the expected trend was observed. So, the model simulations successfully captured the inverse relationship between substrate DP and hydrolysis of cellulose and predicted 73 and 81.1% reduction in cellulose conversion for cotton compared to filter paper for CBH I and CBH II, respectively. Similar results have been reported in literature from both experimental as well as modeling studies (Wood 1974; Zhang and Lynd 2006).

As endoglucanases act on the surface chains, their activity is not severely affected by fraction of chain ends; however, degree of crystallinity plays an important role in deciding their performance. Bonds in the amorphous region are more susceptible to hydrolysis compared to those in crystallinity regions because of higher accessibility of enzymes in amorphous regions (Chang and Holtzapple 2000). This behavior was also successfully captured by model simulations as cellulose conversion by action of endoglucanases on cotton (highly crystalline cellulose, CrI 0.85–0.90) was found 57.2% lower than that of filter paper (semi-crystalline cellulose, CrI 0.4–0.5) after 48 h of hydrolysis (Additional file 3: Figure S3).

Other model simulations
Enzymatic hydrolysis of cellulose by individual enzymes
As discussed in sections above, model was simulated for filter paper hydrolysis by action of individual cellulase enzymes. Hydrolysis rates during action of individual enzymes (EGI, CBH I and CBH II) on filter paper (25 g/L) are presented in Fig. 15. For all enzyme classes, there was

significant drop in hydrolysis rates after few initial hours of hydrolysis and then the rate became nearly constant. This decrease in rate after few hours of initial hydrolysis is widely an observed phenomenon and is believed to occur due to morphological changes in the cellulose structure (e.g., decrease in glucose chains on the surface, increased percentage of crystallinity regions). These changes affect the enzyme–substrate interactions by limiting the accessibility of cellulase enzymes to glucose chains and results in rapid decline in hydrolysis rate (Zhang and Lynd 2004; Zhou et al. 2009). As observed in experimental results also, cellobiose is the major product formed during cellulose hydrolysis by CBH I and CBH II, which also acts as a strong inhibitor to these enzymes and negatively affects the hydrolysis rate.

After 48 h hydrolysis of filter paper by EG I, it was observed (from model simulations) that concentrations of oligomers with DP 2–4 and glucose were higher compared to cellopentaose and cellohexaose concentrations (Additional file 3: Figure S4). There was increase in concentrations of cellopentaose and cellohexaose during initial few hours (3–4 h), and after that their concentrations started decreasing. This trend was also expected because of change in availability of glucose molecules on surface. Surface glucose chains are easily accessible during initial phase of hydrolysis, where endoglucanases act randomly to producing short chains. As the hydrolysis progress, availability of these glucose chains decreases, and enzymes start acting on soluble sugars. Concentration of sugars with DP 2–4 did not decrease as EG I

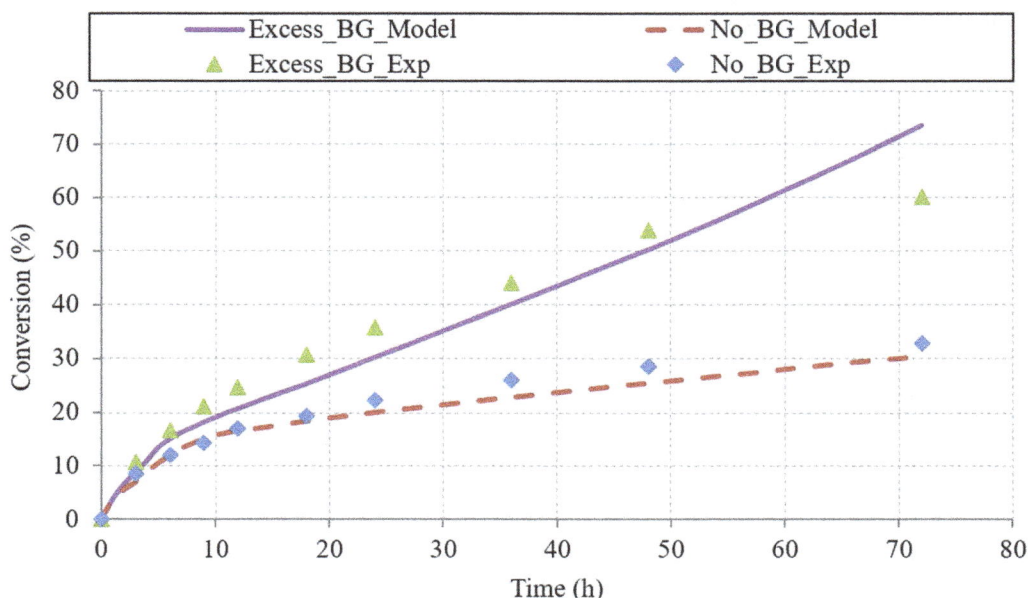

Fig. 12 Effect of β-glucosidase addition on cellulose hydrolysis by CBH I (25 g/L filter paper; CBH I: 10 mg/g cellulose)

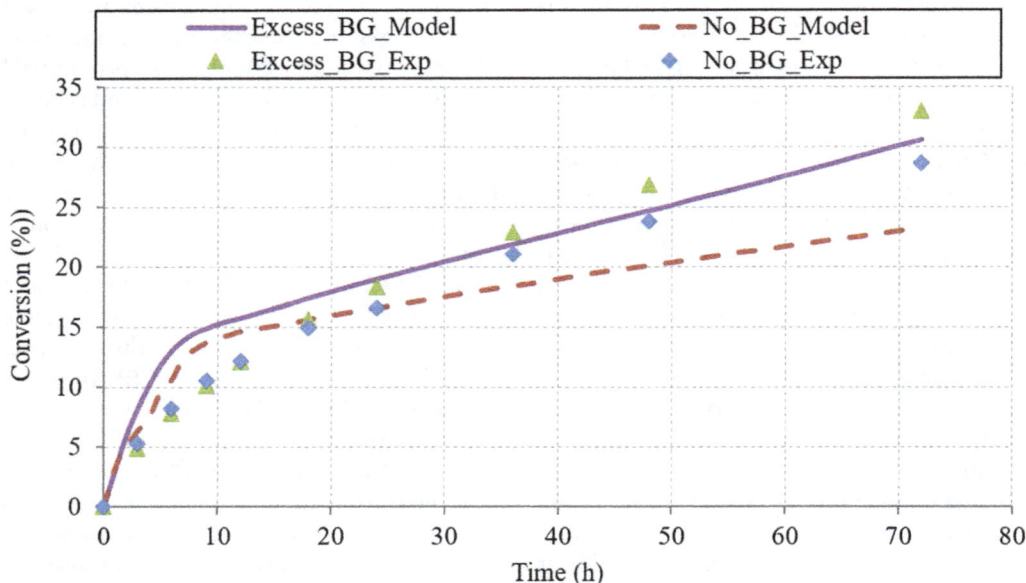

Fig. 13 Effect of β-glucosidase addition on cellulose hydrolysis by CBH II (25 g/L filter paper; CBH II: 10 mg/g cellulose)

was assumed to act only on oligomers with DP > 4. In case of hydrolysis by CBH I and CBH II, all soluble sugars except cellobiose, glucose, and cellotriose were produced in negligible amounts (less than 0.01 mg/L, data not reported).

Endo–exo synergism

The endo–exo synergism is a highly effective synergism that has been reported in many studies and plays critical role in the hydrolysis rates and yields (Andersen 2007; Medve et al. 1998; Väljamäe et al. 1999; Zhang and Lynd 2004). The model was simulated for hydrolysis of filter paper and cotton for individual and combined EG and CBH I. Simulations were performed for 48 h assuming 25 g/L substrate concentration at enzyme loadings of 10 mg/g glucans (individually and total of 20 mg/g glucans in mixture, with EG to CBH I ratio of 1:1). Figure 16 illustrates the comparison between theoretical conversion (addition of cellulose conversions during hydrolysis by individual enzymes) and actual conversion (cellulose conversion during hydrolysis by enzymes acting simultaneously).

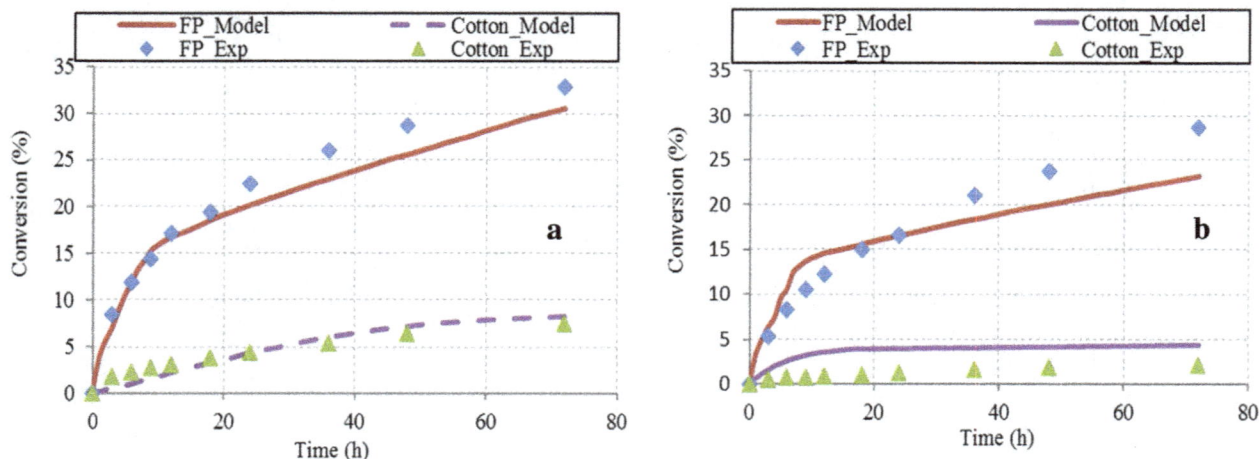

Fig. 14 Effect of cellulose structural properties on hydrolysis: cellulose conversion during hydrolysis of filter paper and cotton (25 g/L) by action of **a** CBH I at loading of 10 mg/g cellulose **b** CBH II at loading of 10 mg/g cellulose

The common measure of synergism is "Degree of synergism (DS)", which is defined as follows (Eq. 2):

$$\text{Degree of synergism} = \frac{\Delta C_{\text{mixed}}}{\sum_{i=1}^{n} \Delta C_i}, \tag{2}$$

where ΔC_{mixed} is cellulose conversion obtained from mixture of 'n' enzymes; ΔC_i is cellulose conversion obtained from an individual action of 'ith' enzyme.

It can be seen from Fig. 16 that the expected synergism was observed in the model simulations. The degree of synergism increased initially and then decreased towards the end of hydrolysis. Similar trends have been observed by other researchers: Kleman-Leyer et al. (1996) for hydrolysis of cotton and Medve et al. (1998) for hydrolysis of Avicel. The phenomenon can be explained by the fact that at the beginning of hydrolysis, the surface molecules are accessible for EG action, and chain ends were sufficient for CBH action. As the hydrolysis progresses, endoglucanases create additional chain ends and increase their availability for exoglucanases, which results in high hydrolysis rate and synergism. However, with further progress in hydrolysis, product (cellobiose and glucose mainly) inhibition becomes very dominant and total yields are not significantly higher than the case where the enzymes work individually. The highest values of degree of synergism were 1.33 and 4.35 for filter paper and cotton, respectively. The values of DS obtained from model simulations are consistent with the reported values in literature (Medve et al. 1998; Väljamäe et al. 1999; Zhang and Lynd 2004; Zhou et al. 2010). A large variation

in the DS values can be observed in literature studies, possibly because several factors such as total time of hydrolysis, purity of enzymes, activity of enzymes, and enzyme loadings can affect the synergism. The synergism was observed higher for cotton hydrolysis than that of in case of filter paper hydrolysis. The inverse relationship between DS and substrate DP was expected and has been reported in literature (Andersen 2007; Srisodsuk et al. 1998; Zhang and Lynd 2004). A comprehensive review on hydrolysis from Zhang and Lynd (2004) compiled DS values from various studies and reported low DS values (1.3–2.2) for Avicel and high DS values (4.1–10) for cotton and bacterial cellulose from synergism of *T. reeesi* enzymes. During cellulose hydrolysis by only CBH I, its accessibility to chain ends is very limited and cellulose conversion is very less. The accessibility is further reduced for substrates like cotton, with very high degree of polymerization. During combined action of EG and CBH I, creation of additional chain ends by random action of EG increases the substrate availability for action of CBH I, which results in more effective hydrolysis.

Conclusions

A novel approach of stochastic molecular modeling based on basic sciences and computer algorithms was used to model complex cellulose hydrolysis process. In this work, the model was further improved by incorporating some critical phenomenon, especially the enzyme crowding effect, and the model was validated with actual hydrolysis experiments using purified enzymes. Model was

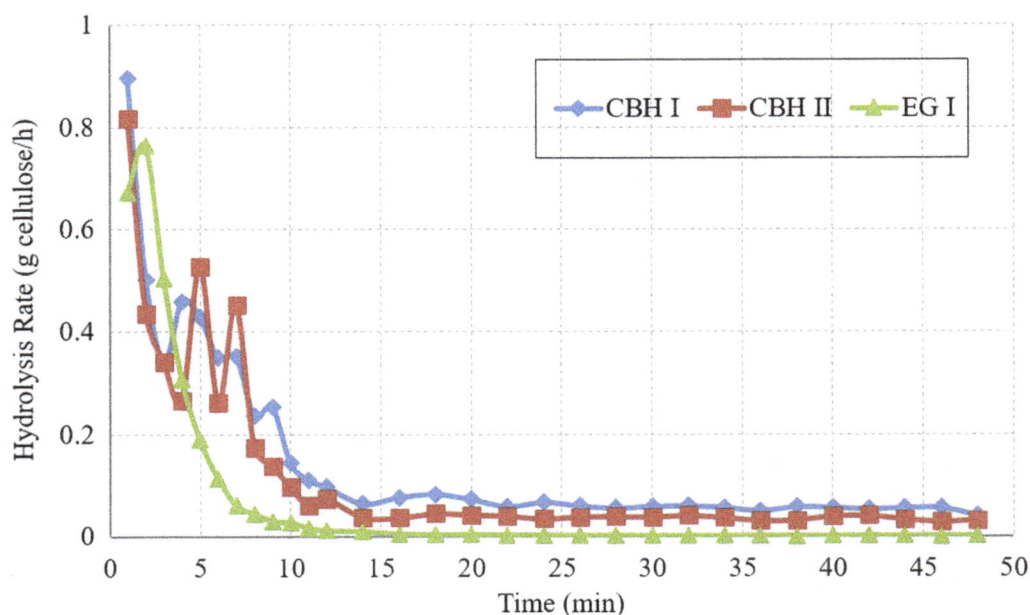

Fig. 15 Model predictions of hydrolysis rate by action of individual enzymes on filter paper (filter paper, 25 g/L; all enzymes, 10 mg/g glucans)

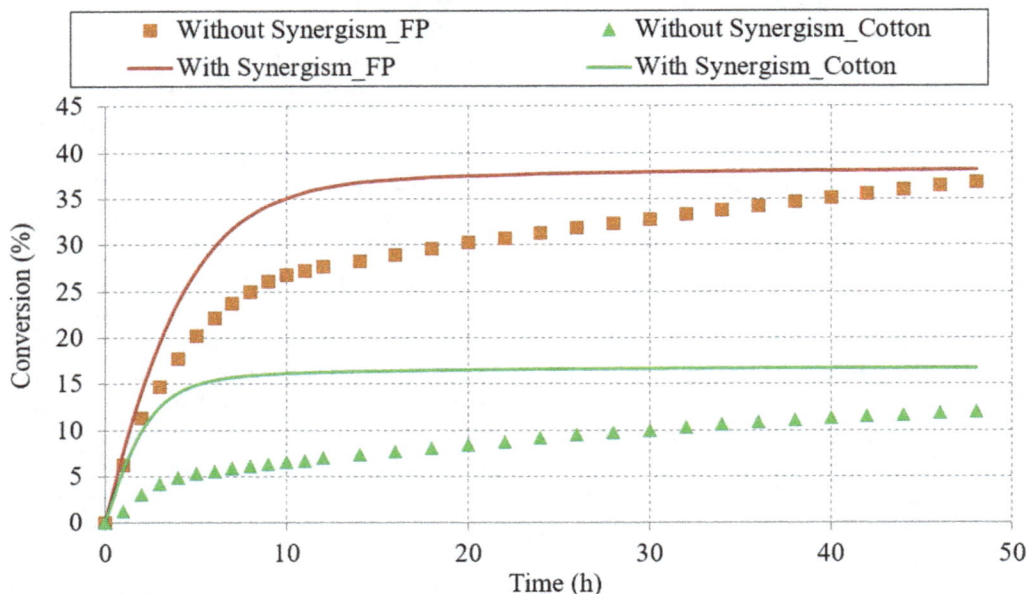

Fig. 16 Endo–exo synergism during hydrolysis of filter paper and cotton cellulose 25 g/L, EG I 10 mg/g glucans, CBHI 10 mg/g glucans (total 20 mg enzymes when acting in mixture 1:1). Solid lines are results from combined action of enzymes and lines with points (theoretical) are sum of conversions from action of individual enzymes

accurate in predicting the cellulose hydrolysis profiles obtained from experimental studies from both literature as well as this work. The model captured the dynamics of cellulose hydrolysis during action of individual as well as multiple cellulase enzymes. Model results successfully followed all important trends, such as product inhibition, low cellobiohydrolase activity on high DP substrates, low endoglucanases activity on crystalline substrate, inverse relationship between degree of synergism and substrate DP, observed experimentally and reported in literature studies. The model was robust and has high potential usability as could be observed from the fact that model simulations fitted well with the experimental data from both literature as well as current work, without changes to any model parameters (except enzyme activity).

Additional files

Additional file 1. Schematic of algorithm for CBH action for cellulose hydrolysis.

Additional file 2. Schematic of algorithm for EG action for cellulose hydrolysis.

Additional file 3: Table S1. Values of parameters used for EG I, CBH I and CBH II action; **Figure S1.** Comparison of model simulations (previous and new version) with experimental data from literature: S1a for Avicel (50 g/L) and S1b for Avicel (25 g/L); **Figure S2.** Comparison of model simulations (old version and current model) with experimental data during hydrolysis of filter paper by CBH I; **Figure S3.** Endoglucanases action on substrates with different crystallinity; **Figure S4.** Glucose production profile during action of endoglucanases on filter paper.

Authors' contributions
DK, and GM developed the model and designed experiments. DK conducted experiments, analyzed data and prepared the manuscript. GM reviewed the results, helped in data analysis and edited the manuscript. All authors read and approved the final manuscript.

Author details
[1] Biological and Ecological Engineering, Oregon State University, Corvallis, OR, USA. [2] Agricultural and Biological Engineering, University of Illinois at Urbana-Champaign, Urbana, IL, USA.

Acknowledgements
Authors gratefully acknowledge the support by National Science Foundation through NSF Grant No. 1236349 from Energy for Sustainability program, CBET Division.

Competing interests
The authors declare that they have no competing interests.

References
Andersen N (2007) Enzymatic hydrolysis of cellulose—experimental and modeling studies. Technical University of Denmark, Copenhagen
Andersen N, Johansen KS, Michelsen M, Stenby EH, Krogh KBRM, Olsson L (2008) Hydrolysis of cellulose using mono-component enzymes shows synergy during hydrolysis of phosphoric acid swollen cellulose (PASC), but competition on Avicel. Enzym Microb Technol 42:362–370
Asztalos A, Daniels M, Sethi A, Shen T, Langan P, Redondo A, Gnanakaran S _ (2012) A coarse-grained model for synergistic action of multiple enzymes on cellulose. Biotechnol Biofuels 5:55
Baker JO, Ehrman CI, Adney WS, Thomas SR, Himmel ME (1998) Hydrolysis of cellulose using ternary mixtures of purified celluloses. Appl Biochem Biotechnol 70:395–403
Ballesteros M (2010) Enzymatic hydrolysis of lignocellulosic biomass. In: Waldron K (ed) Bioalcohol production: biochemical conversion of lignocellulosic biomass. CRC Press, Boca Raton

Banerjee G, Car S, Scott-Craig JS, Borrusch MS, Bongers M, Walton JD (2010a) Synthetic multi-component enzyme mixtures for deconstruction of lignocellulosic biomass. Bioresour Technol 101:9097–9105

Banerjee G, Car S, Scott-Craig JS, Borrusch MS, Walton JD (2010b) Rapid optimization of enzyme mixtures for deconstruction of diverse pretreatment/biomass feedstock combinations. Biotechnol Biofuels 3:22

Bansal P, Hall M, Realff MJ, Lee JH, Bommarius AS (2009) Modeling cellulase kinetics on lignocellulosic substrates. Biotechnol Adv 27:833–848

Berlin A, Maximenko V, Gilkes N, Saddler J (2007) Optimization of enzyme complexes for lignocellulose hydrolysis. Biotechnol Bioeng 97:287–296

Besselink T, Baks T, Janssen AEM, Boom RM (2008) A stochastic model for predicting dextrose equivalent and saccharide composition during hydrolysis of starch by α-amylase. Biotechnol Bioeng 100:684–697

Bezerra RMF, Dias AA (2004) Discrimination among eight modified Michaelis-Menten models of cellulose hydrolysis with a large range of substrate/enzyme ratios. Appl Biochem Biotechnol 112:173–184

Bezerra RMF, Dias AA, Fraga I, Pereira AN (2011) Cellulose hydrolysis by cellobiohydrolase Cel7A Shows mixed hyperbolic product inhibition. Appl Biochem Biotechnol 165:178–189

Chang VS, Holtzapple MT (2000) Fundamental factors affecting biomass enzymatic reactivity. Appl Biochem Biotechnol 84:5–37

Chinga-Carrasco G (2011) Cellulose fibres, nanofibrils and microfibrils: the morphological sequence of MFC components from a plant physiology and fibre technology point of view. Nanoscale Res Lett 6:417

Eriksson T, Karlsson J, Tjerneld F (2002) A model explaining declining rate in hydrolysis of lignocellulose substrates with cellobiohydrolase I (Cel7A) and endoglucanase I (Cel7B) of Trichoderma reesei. Appl Biochem Biotechnol 101:41–60

Fan L, Lee Y (1983) Kinetic studies of enzymatic hydrolysis of insoluble cellulose: derivation of a mechanistic kinetic model. Biotechnol Bioeng 25:2707–2733

Fan L, Gharpuray MM, Lee YH (1987) Cellulose hydrolysis. Biotechnology monographs, vol 3. Springer, Berlin

Gao D, Chundawat SPS, Krishnan C, Balan V, Dale BE (2010) Mixture optimization of six core glycosyl hydrolases for maximizing saccharification of ammonia fiber expansion (AFEX) pretreated corn stover. Bioresour Technol 101:2770–2781

Ghose T (1987) Measurement of cellulase activities. Pure Appl Chem 59:257–268

Hall M, Bansal P, Lee JH, Realff MJ, Bommarius AS (2010) Cellulose crystallinity—a key predictor of the enzymatic hydrolysis rate. FEBS J 277:1571–1582

Igarashi K et al (2011) Traffic jams reduce hydrolytic efficiency of cellulase on cellulose surface. Science 333:1279–1282

Jäger G et al (2010) Practical screening of purified cellobiohydrolases and endoglucanases with α-cellulose and specification of hydrodynamics. Biotechnol Biofuels 3:18

Jeoh T, Ishizawa CI, Davis MF, Himmel ME, Adney WS, Johnson DK (2007) Cellulase digestibility of pretreated biomass is limited by cellulose accessibility. Biotechnol Bioeng 98:112–122

Kadam KL, Rydholm EC, McMillan JD (2004) Development and validation of a kinetic model for enzymatic saccharification of lignocellulosic biomass. Biotechnol Prog 20:698–705

Kleman-Leyer KM, Siika-Aho M, Teeri TT, Kirk TK (1996) The cellulases endoglucanase I and cellobiohydrolase II of Trichoderma reesei act synergistically to solubilize native cotton cellulose but not to decrease Its molecular size. Appl Environ Microbiol 62:2883–2887

Kumar D (2014) Biochemical conversion of lignocellulosic biomass to ethanol: experimental, enzymatic hydrolysis modeling, techno-economic and life cycle assessment studies. Oregon State University, Corvallis

Kumar D, Murthy GS (2011) Impact of pretreatment and downstream processing technologies on economics and energy in cellulosic ethanol production. Biotechnol Biofuels 4:27

Kumar D, Murthy GS (2013) Stochastic molecular model of enzymatic hydrolysis of cellulose for ethanol production. Biotechnol Biofuels 6:63

Levine SE, Fox JM, Blanch HW, Clark DS (2010) A mechanistic model of the enzymatic hydrolysis of cellulose. Biotechnol Bioeng 107:37–51

Lynd LR, Weimer PJ, Van Zyl WH, Pretorius IS (2002) Microbial cellulose utilization: fundamentals and biotechnology. Microbiol Mol Biol Rev 66:506–577

Marchal L, Zondervan J, Bergsma J, Beeftink H, Tramper J (2001) Monte Carlo simulation of the α-amylolysis of amylopectin potato starch. Bioprocess Biosyst Eng 24:163–170

Marchal L, Ulijn R, Gooijer CD, Franke G, Tramper J (2003) Monte Carlo simulation of the α-amylolysis of amylopectin potato starch. 2. α-amylolysis of amylopectin. Bioprocess Biosyst Eng 26:123–132

Matsumoto M, Nishimura T (1998) Mersenne twister: a 623-dimensionally equidistributed uniform pseudo-random number generator. ACM Trans Model Comput Simul 8:3–30

Medve J, Karlsson J, Lee D, Tjerneld F (1998) Hydrolysis of microcrystalline cellulose by cellobiohydrolase I and endoglucanase II from Trichoderma reesei: adsorption, sugar production pattern, and synergism of the enzymes. Biotechnol Bioeng 59:621–634

Merino S, Cherry J (2007) Progress and challenges in enzyme development for biomass utilization. Biofuels 108:95–120

Mosier N, Hall P, Ladisch C, Ladisch M (1999) Reaction kinetics, molecular action, and mechanisms of cellulolytic proteins. Recent Progress Bioconversion Lignocellul 65:23–40

Murthy GS, Johnston DB, Rausch KD, Tumbleson M, Singh V (2011) Starch hydrolysis modeling: application to fuel ethanol production. Bioprocess Biosyst Eng 34:879–890

Sangseethong K, Penner MH (1998) p-Aminophenyl β-cellobioside as an affinity ligand for exo-type cellulases. Carbohydr Res 314:245–250

Srisodsuk M, Kleman-Leyer K, Keränen S, Kirk TK, Teeri TT (1998) Modes of action on cotton and bacterial cellulose of a homologous endoglucanase–exoglucanase pair from Trichoderma reesei. Eur J Biochem 251:885–892

Teugjas H, Väljamäe P (2013) Product inhibition of cellulases studied with ^{14}C-labeled cellulose substrates. Biotechnol Biofuels 6:104

Väljamäe P, Sild V, Nutt A, Pettersson G, Johansson G (1999) Acid hydrolysis of bacterial cellulose reveals different modes of synergistic action between cellobiohydrolase I and endoglucanase I. Eur J Biochem 266:327–334

Wang M, Li Z, Fang X, Wang L, Qu Y (2012) Cellulolytic enzyme production and enzymatic hydrolysis for second-generation bioethanol production. Adv Biochem Eng Biotechnol. 128:1–24. https://doi.org/10.1007/10_2011_131

Wojciechowski PM, Koziol A, Noworyta A (2001) Iteration model of starch hydrolysis by amylolytic enzymes. Biotechnol Bioeng 75:530–539

Wood T (1974) Properties and mode of action of cellulases. In: Biotechnology and bioengineering symposium, vol 5, pp 111–133

Zhang YHP, Lynd LR (2004) Toward an aggregated understanding of enzymatic hydrolysis of cellulose: noncomplexed cellulase systems. Biotechnol Bioeng 88:797–824

Zhang YHP, Lynd LR (2006) A functionally based model for hydrolysis of cellulose by fungal cellulase. Biotechnol Bioeng 94:888–898

Zhou W, Hao Z, Xu Y, Schüttler HB (2009) Cellulose hydrolysis in evolving substrate morphologies II: numerical results and analysis. Biotechnol Bioeng 104:275–289

Zhou W, Xu Y, Schüttler HB (2010) Cellulose hydrolysis in evolving substrate morphologies III: time-scale analysis. Biotechnol Bioeng 107:224–234

15

Changes of membrane fatty acids and proteins of *Shewanella putrefaciens* treated with cinnamon oil and gamma irradiation

Fei Lyu*, Fei Gao, Qianqian Wei and Lin Liu*

Abstract

Background: In order to detect the antimicrobial mechanism of combined treatment of cinnamon oil and gamma irradiation (GI), the membrane fatty acids and proteins characteristics of *Shewanella putrefaciens* (*S. putrefaciens*) treated with cinnamon oil and GI, and the distribution of cinnamon oil in *S. putrefaciens* were observed in this study.

Results: The membrane lipid profile of *S. putrefaciens* was notably damaged by treatments of cinnamon oil and the combination of cinnamon oil and GI, with significantly fatty acids decrease in C14:0, C16:0, C16:1, C17:1, C18:1 ($p < 0.05$). The SDS-PAGE result showed that GI did not have obvious effect on membrane proteins (MP), but GI combined with cinnamon oil changed the MP subunits. Cinnamaldehyde, the main component of cinnamon oil, can not transport into *S. putrefaciens* obviously. It was transformed into cinnamyl alcohol in the nutrient broth with the action of *S. putrefaciens*. This indicated that the antimicrobial action of cinnamon oil mainly happened on the membrane of *S. putrefaciens*.

Conclusion: Cinnamon oil could act on the membrane of *S. putrefaciens* with the damage of fatty acids and proteins, and GI would increase the destructive capability of cinnamon oil on the membrane fatty acids and proteins of *S. putrefaciens*.

Keywords: Gamma irradiation, *Shewanella putrefaciens*, Cinnamon oil, Fatty acids

Background

The interest of the food industry and consumers in natural antimicrobials to prevent spoilage and pathogenic microorganisms has increased significantly. *Shewanella putrefaciens* has been proved to be one of the main specific spoilage organisms (SSO) presented in refrigerated meat and fish products (Borch et al. 1996; Gram and Huss 1996; Gram et al. 1990; Leblanc et al. 2001). It can generate hydrogen sulfide (H_2S) from cysteine, reduce trimethylamine oxide (TMAO) to trimethylamine (TMA), participate in the proteolytic and lipolytic degradations, thus producing unpleasant off-odors leading to organoleptic alterations of food products (Gennari et al. 1999; Leblanc et al. 2001; López-Caballero et al. 2001; Stenström and Molin 1990). Nowadays, many studies

have focused on the antimicrobial techniques against *S. putrefaciens* in different foods (Cai et al. 2015; Jasour et al. 2015; Shokri et al. 2015; Zhang et al. 2015b).

Essential oils (EOs) are characterized by a wide range of volatile compounds, some of which are important to food flavor quality, and they are generally recognized as safe (GRAS) (Belletti et al. 2004). Cinnamon oil has a strong antimicrobial activity against Gram-positive and Gram-negative bacteria (Almariri and Safi 2014; Urbaniak et al. 2014). It has been proved that cinnamon oil used in fish and meat products could extend their microbial shelf life (Van Haute et al. 2016). Cinnamaldehyde, the main component of cinnamon oil, has been shown to be effective against a broad spectrum of food-borne pathogens (Burt 2004; Holley and Patel 2005). It is common for reviewers of spice oils to ascribe the interactions of spice oils with the cell membrane (Brul and Coote 1999; Roller and Board 2003). Gill and Holley (2004) observed that

*Correspondence: lvfei_zju@163.com; 834943625@qq.com
Department of Food Science, Ocean College, Zhejiang University of Technology, 18 Chaowang Road, Hangzhou 310014, China

there was a rapid decline in cellular adenosine triphosphate (ATP) in *Listeria monocytogenes* treated with cinnamaldehyde. It was hypothesized that cinnamaldehyde acted as an ion transporter and interacted with the cell membrane causes disruption sufficient to disperse the proton motive force by leakage of small ions and inhibition of energy generation (Gill and Holley 2004). Hammer and Heel (2012) demonstrated that cinnamaldehyde could decrease the membrane polarity before increasing the membrane permeability. It was also reported that cell membrane integrity of *Escherichia coli* and *Staphylococcus aureus* was damaged by cinnamaldehyde (Shen et al. 2015). Mousavi et al. (2016) successfully demonstrated that cinnamaldehyde could change *E. coli* metabolism through interactions with different biochemical families such as proteins, nucleic acids, lipids, and carbohydrates.

Irradiation technology has been used for decontamination and/or sterilization of dehydrated vegetables, fruits, meats, poultry, fish, and seafood in order to improve product safety and shelf life (Arvanitoyannis et al. 2009; Lacroix and Ouattara 2000). The action of gamma irradiation (GI) on DNA molecules and cell division inhibition is now well understood (Bonura et al. 1975; Le-Tien et al. 2007). Various reactive oxygen species (ROS) are produced during the irradiation treatment of foodstuff which contributes to cellular damage (Bonura et al. 1975). Although much literature has reported the mechanism of cinnamon oil and GI on bacteria alone against different bacteria, the combined antimicrobial mechanism of cinnamon oil and GI on *S. putrefaciens* has not been reported. The aim of the experiments was to evaluate the membrane damage capacity of the combination treatments of cinnamon oil and GI on *S. putrefaciens* by analyzing the membrane protein and fatty acid profiles as well as the distribution of cinnamaldehyde in *S. putrefaciens*, thus to analyze the antimicrobial mechanism of the combination treatment against *S. putrefaciens*.

Methods

Antimicrobial compounds

Cinnamon oil was extracted from *Cinnamomum zeylanicum* leaves by steam distillation method. It was purchased from Erin Limited Company, Australia. Cinnamon oil stock solution was prepared by emulsifying cinnamon oil in deionized water with 1% Tween-80 by stirring 30 min to get a colloidal suspension for use within 24 h with final cinnamon oil concentrations of 207 and 414 mg/mL, respectively.

Chemicals and reagents

Cinnamaldehyde [99.5%, chromatographic pure (GCP)] and cinnamyl alcohol (99%, GCP) were purchased from Aladdin, Shanghai, China. HPLC-grade methylene

dichloride was purchased from Tianjin Shield Specialty Chemical Co., Ltd., Tianjin, China. HPLC-grade acetonitrile and methanol were purchased from Tedia Company, Inc., Ohio, USA. Other solvents and chemicals were purchased from Dingguo biological technology Co., Ltd., Shanghai, China. Ultrapure water was purified on a Milli-Q system (Millipore, Bedford, USA). Millipore syringe filters (Millex-GP, 0.22 mm pore size) were purchased from Nihon Millipore, Tokyo, Japan.

Shewanella putrefaciens preparation

Shewanella putrefaciens was isolated from spoiled fish and identified by China Center of Industrial Culture Collection. When shipped to our laboratory, the strain was cultured twice in nutrient broth (NB) at 30 °C for 24 h, then streaked on nutrient agar (NA) slants and cultured under the same conditions. The slants were stored at 4 °C and sub-cultured monthly until use. Before each experiment, stock cultures were propagated through two consecutive 24-h growth cycles in NB at 30 °C and then cultivated to the exponential phase (5 h). The working cultures contained approximately 10^8 CFU/mL *S. putrefaciens* were obtained by diluting the exponential phase cells in nutrient broth.

Treatments of *S. putrefaciens*

Each 50 mL working culture containing approximately 10^8 colony-forming unit (CFU)/mL *S. putrefaciens* was transferred into 100-mL test tube. These test tubes were treated as follows: group one without adding cinnamon oil was used as control (CK); group two was added cinnamon oil with the final concentration of 207 µg/mL (C1); group three contained 207 µg/mL cinnamon oil and irradiated by GI (C1+G); group four only was irradiated (G); group five was added cinnamon oil with the final concentration of 414 µg/mL (C2). G and C1+G were irradiated 0.080 kGy GI as soon as possible after cinnamon oil treatment at the Institute of Crops and Nuclear Technology Utilization, Zhejiang Academy of Agricultural Sciences, China. The data of 207 and 414 µg/mL were the ½ MIC (minimal inhibitory concentration) and MIC values of cinnamon oil for 10^8 CFU/mL *S. putrefaciens*. The doses were conducted by using 55-cm distance to the irradiation source with 30 min. The self-contained GI source was ^{137}Cs with an approximate dose rate of 0.10 kGy/min. The dose rate was established by using National Physical Laboratory (Middlesex, United Kingdom) dosimeters. After irradiation, all samples were transported to our laboratory within 180 min. Moreover, the nutrient broths containing 207 and 414 µg/mL cinnamon oil without *S. putrefaciens* were also used as controls marked as CC1 and CC2, respectively. Each treatment group contained 5 test tubes. The assays were tested in triplicate, and values are presented as mean ± standard deviation of replicated measurements.

Cinnamon oil analysis

The cinnamon oil was analyzed using a GC–MS system (Agilent 7890A/5975C) on a HP-5MS capillary column (30.0 m × 250 μm × 0.25 μm) using helium as the carrier gas with a split flow was 1.5 mL/min and a 100:1 split ratio. The initial oven temperature of GC was 50 °C, and programmed to 250 °C at a rate of 30 °C/min and then kept constant at 250 °C for 10 min. The inject volume was 0.5 μL and the source temperature was 230 °C. MS was taken at 70 eV and a mass range of 29–450 amu. A library search was carried out using NIST98.L database. Relative percentage amount was calculated from total ions chromatograms (TIC) by the computer (Ooi et al. 2006).

Analysis of membrane fatty acids in *S. putrefaciens*

The working cultures of *S. putrefaciens* with different treatments at 180 min were centrifuged for 10 min at 5000g, and the cell pellet was harvested and re-suspended in phosphate-buffered saline (0.1 M, pH 7.0), afterwards frozen at −80 °C and freeze-dried using a freeze-dryer (FD-1-50, Bo Yikang Co. Ltd., Beijing, China). The samples were submitted for membrane fatty acid extraction. Extraction of fatty acid from cellular materials was carried out as described by Evans et al. (1998).

Lipid samples were trans-methylated for analysis of their acyl groups as fatty acid methyl esters (FAME). The samples of total lipid extract were evaporated to dryness in a round-bottom flask using a boiling water bath. To the dried samples, heated under a reflux condenser (20–30 cm), 10 mL of KOH in methanol (0.2 M) and 1 mL of heptane were added. After 10 min, 5 mL of boron trifluoride (BF$_3$) was added and followed, after 2 min, by 4 mL of hexane. After waiting for 1 min, the samples were cooled. A saturated solution of Na$_2$SO$_4$ was added to the samples, and after settling into a two-phase system, the upper layer was taken and transferred into a vial. The samples were stored at −30 °C until further analysis (Dussault et al. 2009).

GC-MS analysis was performed using a Thermo Trace 1300 Gas Chromatograph equipped with a Thermo TG-WAXMS capillary column (dimensions: 30 m × 0.25 mm × 0.25 μm; Thermo Scientific, USA), coupled to a Thermo ISQ LT Single Quadrupole Mass Spectrometer (MS) through a heated transfer line (220 °C). Helium (99.999%) was used as the carrier gas with a constant flow rate of 1 mL/min and a 1:50 split ratio. The GC inlet temperature is 220 °C. 1 mL aliquots were injected using an AI 1310 autosampler, and the GC oven was programmed to hold 120 °C for 5 min, then raise the temperature by 6 °C/min to 210 °C, which was

held for 5 min, then raise the temperature by 1 °C/min to 230 °C, which was held for 20 min. The MS was operated with the ion source at 220 °C. The solvent delay is 1.46 min. FAME peaks were identified by comparison of their retention times with those of a standard solution (GLC NESTLE 37 Component FAME MIX) and quantified with the internal standard. In all cases, the mass spectrometer was operated in the electron ionization mode (EI) at 70 eV. The retention times and the characteristic fragments of the EI mass spectra were obtained from m/z 20–400 with the scan rate of 500 amu/s. The most abundant ions and/or ions with-out apparent cross-contribution and interferences were chosen as target ions for the quantification (SIM mode) (Schummer et al. 2009).

Electrophoretic analysis

All the implements must be precooled at −20 °C before the tests. The membrane protein (MP) was prepared with Bacterial Membrane Extraction Kit (BestBio, Shanghai, China) according to the instructions with minor modification. The working cultures of *S. putrefaciens* with different treatments at 180 min were centrifuged at 10,000g for 5 min at 0 °C, then washed with PBS buffer (0.1 M, pH 7.0) for two times. The extraction buffer A (500 μL, combined with 2 μL Protease inhibitor) was added to the sediments (20 mg), then put on ice for 2–3 h with shaking for 30 s with every 30 min. The mixture was then centrifuged at 12,000g for 5 min at 0 °C, the supernatant was collected, and then 10 μL. Extraction buffer B was added to the supernatant, then kept at 37 °C for 10 min. The mixture was centrifuged at 1000g for 5 min at 37 °C. The under layer samples were the membrane protein (MP). MP concentration was measured with enhanced bicinchoninic acid (BCA) Protein Assay Kit (Beyotime Biotechnology, Jiangsu, China) according to the manufacturer's instructions. The samples may need to be diluted further to same concentration with appropriate volume of membrane protein dissolution buffer in the kit. The loaded concentration of different treatments' MP was adjusted to 1 mg/mL. 20 μL of protein lysates was then mixed with 5 μL 5× SDS gel-loading buffer (60 mM Tris–HCl, pH 6.8, 14.4 mM dithiothreitol, 2% SDS, 25% glycerol, and 0.1% bromphenol blue), boiled at 100 °C for 5–10 min before use or stored at −20 °C until use. The loaded volume was 15 μL. Sodium-dodecyl-sulfate polyacrylamide gel electrophoresis (SDS-PAGE) was used to measure the molecular weight changes of MP, using 12% separating gel and a 5% stacking gel. The running buffer was 25 mM Tris-192 mM glycin-0.1% SDS, pH 8.3. The power supply was settled at 80 V during the stacking of the proteins, then at 120 V. Gels were stained with Coomassie Brilliant Blue R250 and destained with a solution of methanol and acetic acid to visualize proteins (Carraro and Catani 1983).

Changes of cinnamon oil in nutrient broth with S. putrefaciens

GC–MS was used to analyze chemical compositions of cinnamon oil in nutrient broth (NB) of C1 at 0 and 180 min. GC–MS analysis was performed using a Thermo Trace GC Ultra system equipped with a Thermo TR-5MS capillary column (dimensions: 30 m × 250 mm × 0.25 mm; Thermo Scientific, Runcorn, UK) operating with helium as a carrier gas, coupled to a Thermo ITQ 1100 mass spectrometer (MS) through a heated transfer line (230 °C). The GC injector (230 °C) was operated in a pulsed split mode (50:1); 1 μL aliquots were injected using an autosampler. The initial oven temperature of GC was 100 °C, and programmed to 130 °C at a rate of 20 °C/min, then programmed to 170 °C at a rate of 5 °C/min, after then, programmed to 230 °C at a rate of 25 °C/min and then kept constant at 230 °C for 5 min. The MS was operated with the ion source at 230 °C and a damping flow of 1 mL/min. MS was taken at 70 eV and a mass range of 35–425 amu. The solvent delay time was 3 min. A library search was carried out using NIST98.L database. Relative percentage amount was calculated from TIC by the computer (Ooi et al. 2006).

Distribution of cinnamaldehyde and cinnamyl alcohol

After the analysis of cinnamon oil changes in broth in C1, we found that in the nutrient broth, cinnamaldehyde decreased and cinnamyl alcohol emerged (data are shown in results section). Thus, the distribution of cinnamaldehyde and cinnamyl alcohol in nutrient broth and S. putrefaciens was analyzed using HPLC system (Waters 2695, Milford, USA) consisting of quaternary gradient pump, autosampler, column oven, and photodiode array detector (PDA, Waters 2996). Chromatographic data were acquired using Empower software. The HPLC column consisted of a Waters symmetry C18 column (200 mm × 4.6 mm × 5 mm) connected to Nova-Pak C18 Guard-PakTM guard column (2 mm × 4 mm × 5 mm). The gradient elution was employed using deionized water and acetonitrile at 30 °C for 20 min. The flow rate was set at 1 mL/min. A volume of 10 μL of sample was injected into HPLC system for analysis. The detection wavelengths were set at 290 and 250 nm for cinnamaldehyde and cinnamyl alcohol, respectively. To calculate their concentration, the standard curves ranging from 0.2 to 200 μg/mL were used to obtain a linear relationship between concentrations of drugs versus peak area response, which resulted in a R^2 (coefficient of determination) value of 0.9999. A re-equilibration period of 5 min was used between individual runs. Because only

treatments of C1, C1+G, and C2 were treated with cinnamon oil, so here the changes of cinnamaldehyde and cinnamyl alcohol of C1, C1+G, and C2 at 0 and 180 min were observed. In order to observe the effect of S. putrefaciens, the changes of cinnamaldehyde and cinnamyl alcohol of CC1 and CC2 at 0 and 180 min were also detected.

Statistical analysis

One-way analysis of variance and Duncan's multiple ranges tests were employed to determine the effect of the combination of cinnamon oil and GI treatments on fatty acids, cinnamaldehyde, and cinnamyl alcohol. Calculations were performed using SPSS software Base 19.0 (SPSS, Inc., Chicago, IL, USA). Differences between means were considered significant at $p \leq 0.05$.

Results and discussion

GC–MS analysis of cinnamon oil

The chemical compositions of cinnamon oil analyzed by GC–MS method. There were eight primary phytochemicals of cinnamon oil were identified. Trans-cinnamaldehyde was the major compound, according for 78.54%. It has also been reported in previous studies that cinnamaldehyde was the major compound of cinnamon oil (Cheng et al. 2006; Senanayake et al. 1978). Cinnamon oil (*Cinnamomum cassia* Blume) was identified 85% trans-cinnamaldehyde by GC–MS analysis (Ooi et al. 2006). Trans-cinnamaldehyde was also detected in *Cinnamomum osmophloeum* leaves' constituents, which accounted for 76.00% (Chang et al. 2001).

Changes of fatty acids in S. putrefaciens

Fatty acid composition of microorganisms at various points is required for understanding membrane-associated processes of cells (Špitsmeister et al. 2010). The changes of membrane fatty acid compositions of S. putrefaciens with different treatments at 180 min are shown in Fig. 1. Saturated fatty acids (SFA), C12:0, C14:0, C16:0, C18:0, and unsaturated fatty acids (UFA), C16:1(9c), C17:1(10c), C18:1(9c) were observed in S. putrefaciens in our study. Our result is in agreement with Moule and Wilkinson (1987) who reported that S. putrefaciens can biosynthesize a variety of fatty acids, such as saturated fatty acids (SFA), C12:0, C14:0, C16:0, C18:0, unsaturated fatty acids (UFA), C16:1(9c), C17:1(10c), C18:1(9c). Wang et al. (2009) reported that *Shewanella* inhabiting various environments contain SFA (16:0, 18:0) and UFA (16:1, 18:1).

Compared with CK, all of FAs of G, C1, C1+G, and C2 showed a decrease trend, especially for C14:0, C16:0, C16:1, C17:1, C18:1 ($p < 0.05$). C2 showed the largest decrease level on these fatty acids, followed by C1+G,

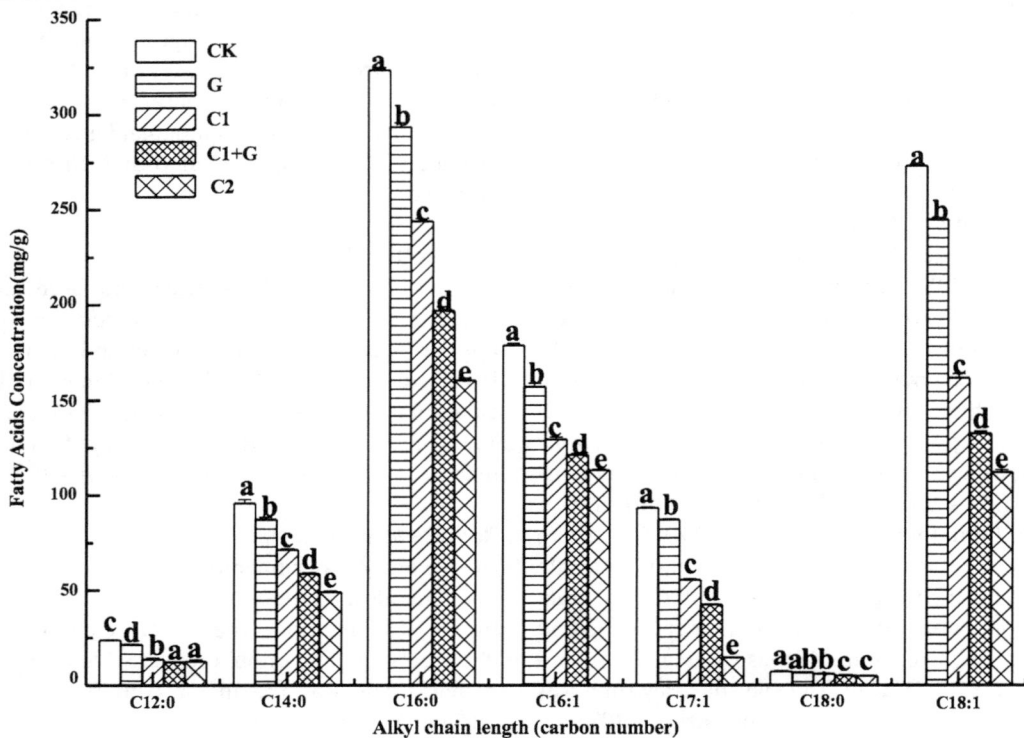

Fig. 1 Changes of membrane fatty acids of *S. putrefaciens* with different treatments. CK, *S. putrefaciens* without treatment; C1, *S. putrefaciens* treated with 207 µg/mL cinnamon oil; C2, *S. putrefaciens* treated with 414 µg/mL cinnamon oil; G, *S. putrefaciens* irradiated with 0.080 kGy gamma irradiation dose; C1+G, *S. putrefaciens* treated with 207 µg/mL cinnamon oil and then irradiated with 0.080 kGy gamma irradiation dose

C1, and G ($p < 0.05$). The higher concentration of cinnamon oil was, the stronger damage on the membrane of *S. putrefaciens*. The combination of C1+G showed a higher damage on fatty acids than C1 and G alone, and C1 was stronger than G. This indicated that the treatments of cinnamon oil and irradiation had an obvious influence on the membrane of *S. putrefaciens*. Biological membranes are essential for the cell integrity, providing a barrier between the inside and outside environments for the cell (Pedersen et al. 2006), and the membrane lipids play an essential role in microbial adaptation under different environmental changes. Irradiation, cinnamon oil, and the combination treatments all showed obviously effect on the lipid profiles, and the effect of cinnamon oil was stronger than irradiation. This is similar with the study of Di Pasqua et al. (2007) who reported that modifications of the membrane by GI were not important when compared with those treated with antimicrobial agents such as essential oils (i.e., thymol).

The UFA ratio of the untreated *S. putrefaciens* was above 0.50, showed in Table 1. *Shewanella* bacteria naturally reside/inhabit the deep-sea environment with low temperature, the mechanism for their survival lies in that the high percentage of UFA in the bacterial membrane

incorporated with phospholipid confers the better membrane fluidity, which in turn enhances its capability of cold adaptation (Zhang et al. 2015a). Di Pasqua et al. (2006) reported that UFAs are always present at a higher amount than SFAs in the total lipid profile of the microorganisms used in their research. Wang et al. (2009) also reported that *Shewanella* has appreciable ability to produce various types of low-melting-point fatty acids with monounsaturated fatty acids (MUFA) included.

The saturated/unsaturated fatty acids (SFA/UFA) ratio can effectively demonstrate the destructive effect of different treatments on the membrane fatty acids. The SFA/UFA ratios are shown in Table 1. The SFA/UFA ratios of CK, G, C1, C1+G, and C2 were 0.825, 0.836, 0.966, 0.922, and 0.946, respectively. CK and G showed a lower SFA/UFA ratio values than C1, C1+G, and C2 ($p < 0.05$). The increase of SFA/UFA ratio indicated a reduction of UFA or an increase of SFA. Significant decrease of UFA ratio and increase of SFA ratio took place in C1, C1+G, and C2 treatment, as compared with CK and G ($p < 0.05$). This result revealed that the antimicrobial mechanism of cinnamon oil against the cell membrane of *S. putrefaciens*, interacting with the membrane lipid profile and causing membrane structural alterations. It is well known

Table 1 Changes in the concentrations of principal fatty acids of *S. putrefaciens*

Fatty acid indexes	Treatments				
	CK	G	C1	C1+G	C2
UFA ratio	0.548 ± 0.001^a	0.545 ± 0.002^a	0.509 ± 0.003^d	0.520 ± 0.003^b	0.514 ± 0.003^c
SFA ratio	0.452 ± 0.001^d	0.455 ± 0.002^d	0.491 ± 0.003^a	0.480 ± 0.003^c	0.486 ± 0.003^b
SFA/UFA ratio	0.825 ± 0.008^a	0.836 ± 0.006^a	0.966 ± 0.005^d	0.922 ± 0.012^b	0.946 ± 0.009^c

UFA ratio, SFA ratio, and SFA/UFA ratio represent ratio of UFA/TFA (total fatty acids), ratio of SFA/TFA, ratio of SFA to UFA, respectively; CK, *S. putrefaciens* without treatment; C1, *S. putrefaciens* treated with 207 µg/mL cinnamon oil; C2, *S. putrefaciens* treated with 414 µg/mL cinnamon oil; G, *S. putrefaciens* irradiated with 0.080 kGy gamma irradiation dose; C1+G, *S. putrefaciens* treated with 207 µg/mL cinnamon oil and then irradiated with 0.080 kGy gamma irradiation dose; Values are the mean ± standard deviation of the weight percentages of fatty acids in total lipid isolated from at least three independent cultures. Different letters mean significantly different ($p < 0.05$) data in the same row

that UFAs give the membrane a high degree of fluidity, thus affecting the adaptive capacity of bacteria. The decrease of UFA and increase of SFA were detected in *Pseudomonas fluorescens* and *Staphylococcus aureus* treated with a sublethal concentration of some essential oil antimicrobial compounds (Di Pasqua et al. 2006).

Electrophoresis of the membrane protein (MP) of *S. putrefaciens*

SDS-PAGE analysis was performed to investigate the effect of cinnamon oil and GI on protein changes of *S. putrefaciens* (Katayama et al. 2002). Figure 2 shows the changes in the MP subunits during different treatments with the same sampling protein content and same volume. The loading concentration of proteins was adjusted according BAC Protein Assay Kit (see method section of electrophoretic analysis). More than 30 protein bands in the MP of *S. putrefaciens* were resolved ranging in size from 30 to 172 kD as determined by visual assessment of their approximate molecular masses. MP of C1, C1+G, and C2 showed more obvious brands than CK and G, especially the higher concentration cinnamon oil. It indicated that the secondary or tertiary structure of membrane proteins of C1, C1+G, and C2 were more affected with more exposure of amino and sulfhydryl groups in membrane proteins, and thus more Coomassie dye R-250 was bound to protein or protein subunit (Knauf and Rothstein 1971).

Changes of cinnamon oil in nutrient broth with *S. putrefaciens*

Figure 3 shows component changes of cinnamon oil in nutrient broths at 0 and 180 min. At 0 min, the main component in the nutrient broth was cinnamaldehyde (peak a in Fig. 3A), accounting for 82.41%. After 180 min, cinnamaldehyde reduced obviously to 1.93% (peak a in Fig. 3B), while the main component in the nutrient broth changed into cinnamyl alcohol (peak b in Fig. 3B), which increased from 2.03% at 0 min to 82.04% at 180 min

(peak b in Fig. 3). Cinnamyl alcohol can be prepared by selective hydrogenation of cinnamaldehyde though biocatalysis of enzymes and microorganisms (Hollmann et al. 2011). *Bacillus stearothermophilus* alcohol dehydrogenase can reduce cinnamaldehyde to cinnamyl alcohol with the consumption of reduced form of nicotinamide-adenine dinucleotid (NADH) (Pennacchio et al. 2013). The transformation of cinnamaldehyde to cinnamyl alcohol in *S. putrefaciens* nutrient broth would support the theory of selective hydrogenation of cinnamaldehyde catalyzed by enzymes or microorganisms.

Distributions of cinnamaldehyde and cinnamyl alcohol in nutrient broth and *S. putrefaciens*

The distributions of cinnamaldehyde and cinnamyl alcohol of all cinnamon oil treatments in nutrient broth and *S. putrefaciens* are shown in Fig. 4. As shown in Fig. 4A, after 180-min treatment of cinnamon oil, the cinnamaldehyde concentrations of C1, C1+G, and C2 in the nutrient broth decreased obviously as compared with them at 0 min. As concerned on the controls of CC1 and CC2 contained 207 and 414 µg/mL cinnamon oil without *S. putrefaciens*, almost no change of cinnamaldehyde concentration was observed in nutrient broth from 0 to 180 min. This result proved that cinnamaldehyde was transformed to cinnamyl alcohol with *S. putrefaciens*.

Distribution of cinnamaldehyde in *S. putrefaciens* was shown in Fig. 4B. Cinnamaldehyde was detected in the sediments of CC1 and CC2 at 0 and 180 min, indicating that the sediments without bacteria cells contain cinnamon oil. While cinnamaldehyde concentrations of C1, C1+G, and C2 treatments containing bacterial cells were lower than CC1 and CC2, respectively, no matter at 0 or 180 min. There were no significant differences in the treatments of C1, C1+G, and C2 both at 0 and 180 min. It may be due to that cinnamon oil cannot pass through the bacterial membrane, or even little of them pass the membrane; it may be metabolized into the other components as soon as possible.

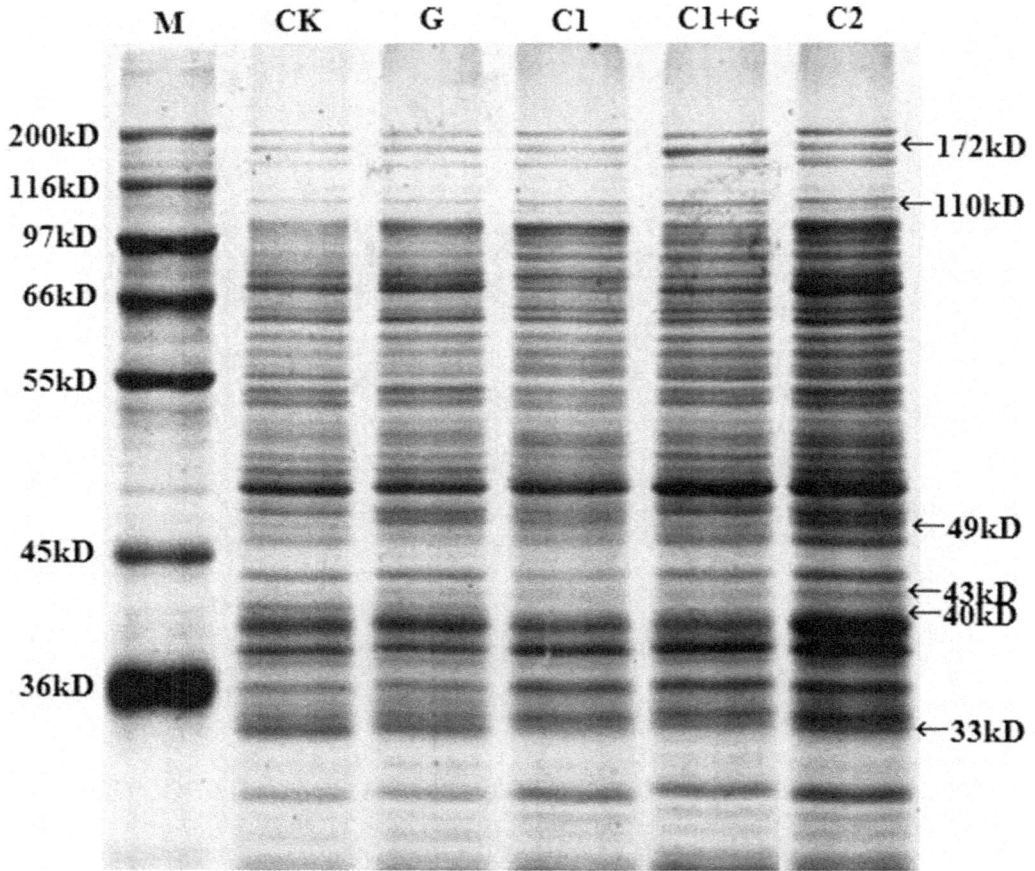

Fig. 2 SDS-PAGE fractionation of *S. putrefaciens* membrane proteins with different treatments. The sampling amounts of the SDS-PAGE for each lane contained equal protein content based on BCA Protein Assay Kit (Beyotime). M, Marker; CK, *S. putrefaciens* without treatment; C1, *S. putrefaciens* treated with 207 μg/mL cinnamon oil; C2, *S. putrefaciens* treated with 414 μg/mL cinnamon oil; G, *S. putrefaciens* irradiated with 0.080 kGy gamma irradiation dose; C1+G, *S. putrefaciens* treated with 207 μg/mL cinnamon oil and then irradiated with 0.080 kGy gamma irradiation dose. Molecular masses were estimated from commercial standards (Sigma Chemical Co.)

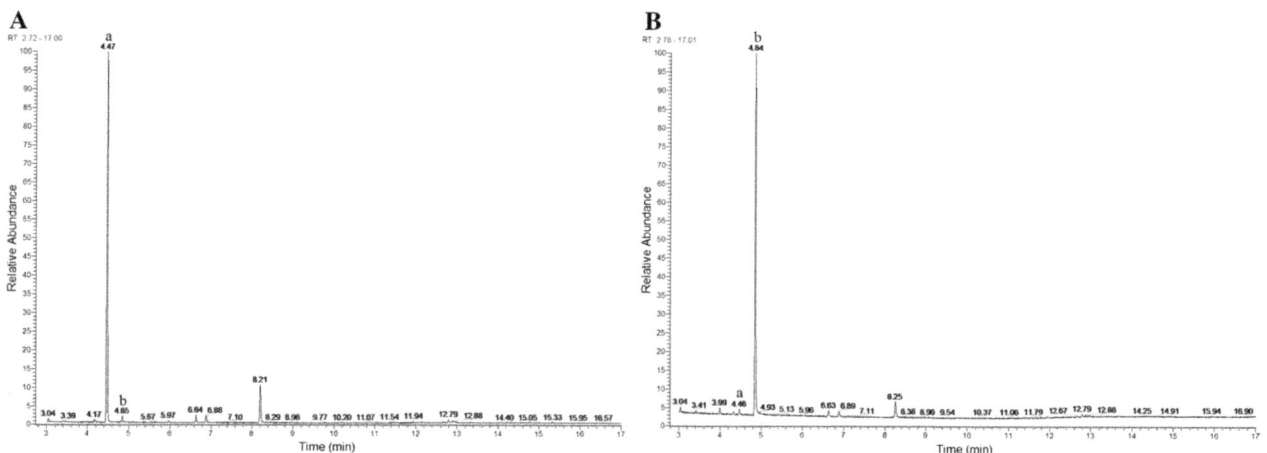

Fig. 3 Changes of chemical compositions of cinnamon oil in nutrient broth of treatment C1 by GC–MS analysis at 0 (**A**) and 180 min (**B**). *a* Trans-cinnamaldehyde, *b* cinnamyl alcohol. C1, *S. putrefaciens* treated with 207 μg/mL cinnamon oil

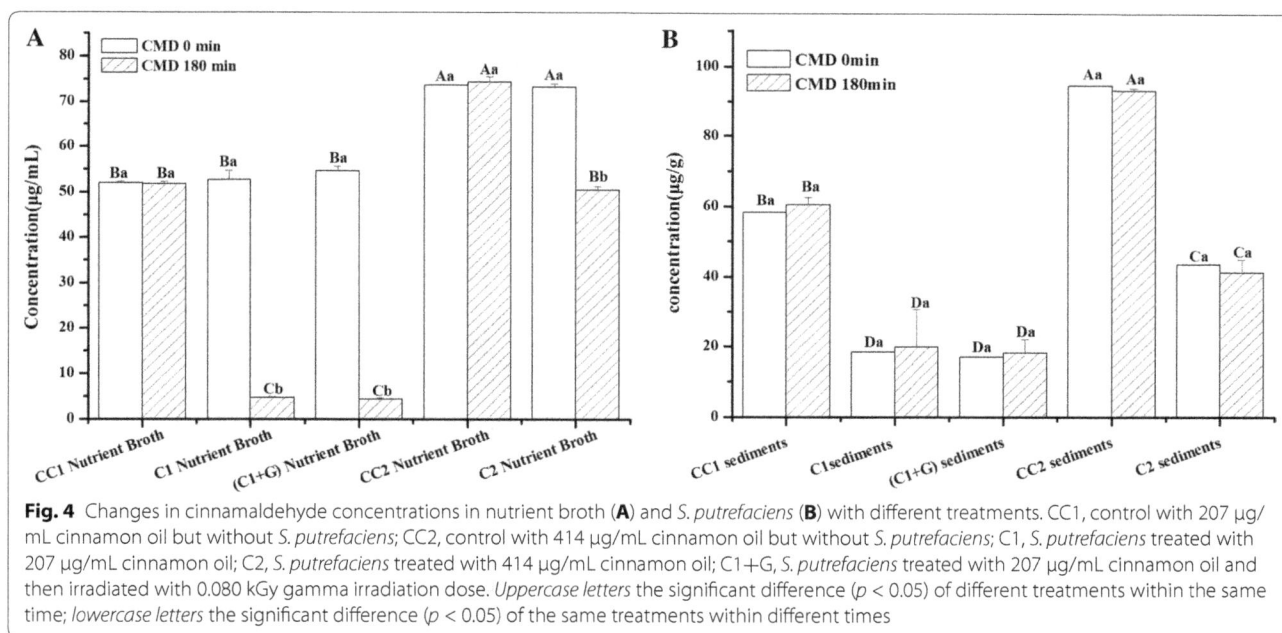

Fig. 4 Changes in cinnamaldehyde concentrations in nutrient broth (**A**) and *S. putrefaciens* (**B**) with different treatments. CC1, control with 207 µg/mL cinnamon oil but without *S. putrefaciens*; CC2, control with 414 µg/mL cinnamon oil but without *S. putrefaciens*; C1, *S. putrefaciens* treated with 207 µg/mL cinnamon oil; C2, *S. putrefaciens* treated with 414 µg/mL cinnamon oil; C1+G, *S. putrefaciens* treated with 207 µg/mL cinnamon oil and then irradiated with 0.080 kGy gamma irradiation dose. *Uppercase letters* the significant difference ($p < 0.05$) of different treatments within the same time; *lowercase letters* the significant difference ($p < 0.05$) of the same treatments within different times

Changes of cinnamyl alcohol of all cinnamon oil treatments in broth and in *S. putrefaciens* are shown in Fig. 5. The cinnamyl alcohol concentration of treatments of C1, C1+G, and C2 in nutrient broth increased at 0 or 180 min as compared with CC1 and CC2, especially at 180 min. As concerned for the cinnamyl alcohol concentration at 180 min, there were obviously increases in C1, C1+G, and C2 in nutrient broth (Fig. 5A). This result was consistent with the changes of cinnamaldehyde as shown in Fig. 4A. It is the first time to report that cinnamaldehyde was transformed into cinnamyl alcohol in the presence of *S. putrefaciens*. The changes of cinnamyl alcohol concentration in *S. putrefaciens* of different treatments are shown in Fig. 5B. The cinnamyl alcohol concentrations of C1, C1+G, and C2 at 0 min were obviously higher than that of CC1 and CC2, respectively. It means that with the action of *S. putrefaciens*, the cinnamaldehyde in the cell can be quickly transformed into cinnamyl alcohol. But at 180 min, cinnamyl alcohol concentrations of C1, C1+G, and C2 decreased obviously.

Because of the hydrophobic character of cinnamon oil, it was not easy to transfer the membrane of *S. putrefaciens* for cinnamaldehyde. The changes of cinnamaldehyde and cinnamyl alcohol in broth and *S. putrefaciens* indicated that the biotransformation of cinnamaldehyde to cinnamyl alcohol occurred on the membrane of *S. putrefaciens*. With transformation of cinnamaldehyde to cinnamyl alcohol, obvious damages of fatty acids of *S. putrefaciens* were also found in our study. All these illustrated that the action site of cinnamon oil appears to be on the membrane of *S. putrefaciens*. GI could help cinnamon oil to damage the membrane lipid and protein profile, but no effect on the transformation of cinnamaldehyde.

Conclusion

In order to detect the antimicrobial mechanism of the combination of cinnamon oil and GI on *S. putrefaciens*, the changes of cinnamaldehyde, membrane fatty acids, and proteins of *S. putrefaciens* were observed. The changes of membrane fatty acids and proteins of treated with C1+G were obvious than CK, C1, and G. Cinnamon oil or the combination with GI could affect the membrane proteins. Interestingly, after cinnamon oil treatments, the content of cinnamaldehyde in the nutrient broth decreased, and cinnamyl alcohol increased obviously. The antimicrobial mechanism might be complex, with the action of GI, the destruction ability of cinnamon oil on membrane fatty acids and proteins increased. Cinnamaldehyde was transformed into cinnamyl alcohol with *S. putrefaciens*. It might be catalyzed by alcohol dehydrogenase with consumption of NADH in *S. putrefaciens*, thus the activity of *S. putrefaciens* also might be inhibited with the transformation of cinnamaldehyde.

Abbreviations

S. putrefaciens: *Shewanella putrefaciens*; GI: gamma irradiation; MP: membrane proteins; SDS-PAGE: sodium dodecyl sulfate polyacrylamide gel electrophoresis; SSO: specific spoilage organisms; H$_2$S: hydrogen sulfide; TMAO:

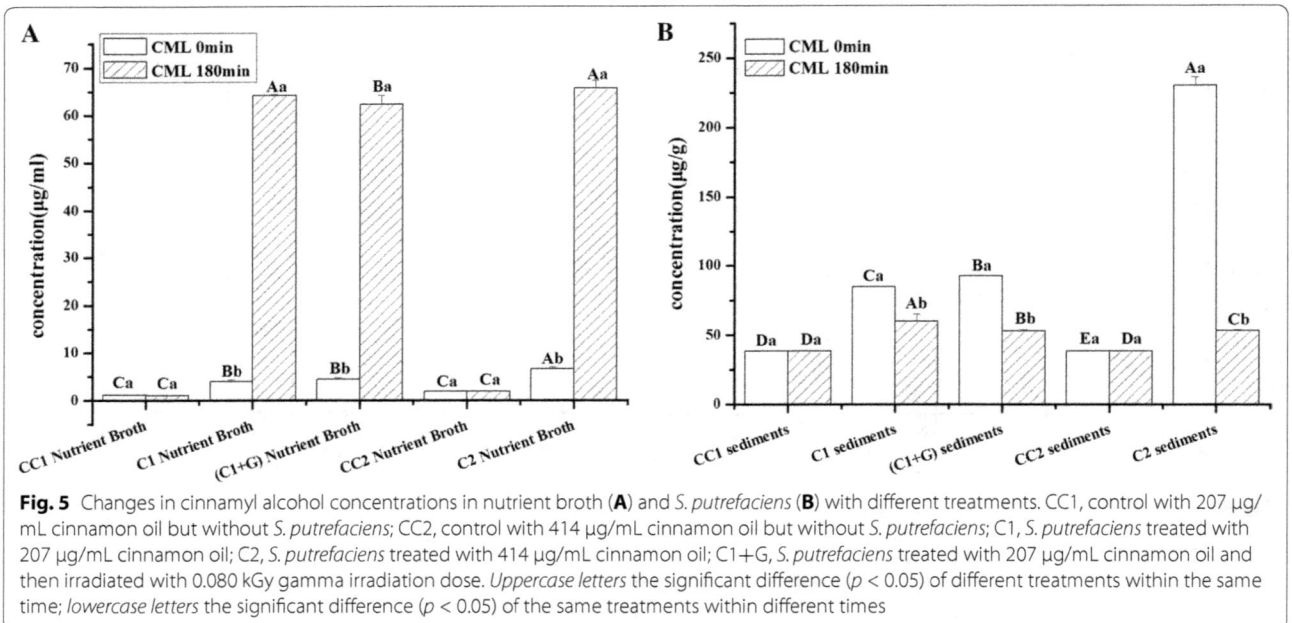

Fig. 5 Changes in cinnamyl alcohol concentrations in nutrient broth (**A**) and *S. putrefaciens* (**B**) with different treatments. CC1, control with 207 μg/mL cinnamon oil but without *S. putrefaciens*; CC2, control with 414 μg/mL cinnamon oil but without *S. putrefaciens*; C1, *S. putrefaciens* treated with 207 μg/mL cinnamon oil; C2, *S. putrefaciens* treated with 414 μg/mL cinnamon oil; C1+G, *S. putrefaciens* treated with 207 μg/mL cinnamon oil and then irradiated with 0.080 kGy gamma irradiation dose. *Uppercase letters* the significant difference ($p < 0.05$) of different treatments within the same time; *lowercase letters* the significant difference ($p < 0.05$) of the same treatments within different times

trimethylamine oxide; TMA: trimethylamine; EO: essential oil; GRAS: generally recognized as safe; ATP: adenosine triphosphate; DNA: deoxyribonucleic acid; ROS: reactive oxygen species; GCP: chromatographic pure; NB: nutrient broth; NA: nutrient agar; CFU: colony-forming unit; CK: control, *S. putrefaciens* without any treatments; C1: treatment, *S. putrefaciens* were added 207 μg/mL cinnamon oil; C1+G: treatment, *S. putrefaciens* were added 207 μg/mL cinnamon oil and then and irradiated with 0.080 kGy; G: treatment, *S. putrefaciens* were gamma irradiated with 0.080 kGy; C2: treatment, *S. putrefaciens* were added 414 μg/mL cinnamon oil; MIC: minimal inhibitory concentration; GC–MS: gas chromatography-mass spectrometer; CC1: control, nutrient broths containing 207 μg/mL cinnamon oil without *S. putrefaciens*; CC2: control, nutrient broths containing 414 μg/mL cinnamon oil without *S. putrefaciens*; HPLC: high performance liquid chromatography; TIC: total ions chromatograms; FAME: fatty acid methyl esters; GC: gas chromatography; MS: mass spectrometer; BCA: bicinchoninic acid; PDA: photodiode array detector; SFA: saturated fatty acids; UFA: unsaturated fatty acids; TFA: total fatty acids; MUFA: monounsaturated fatty acids; NADH: reduced form of nicotinamide-adenine dinucleotide.

Authors' contributions
FL and LL conceived the study. FG and QW carried out the laboratory analysis. QW and FL participated in the study design and coordination and drafting of the manuscript. All authors read and approved the final manuscript.

Acknowledgements
The authors thank Zhejiang Academy of Agricultural Sciences, Hangzhou, China for providing gamma irradiation facilities during this study. The authors also thank Editor and reviewers for their constructive suggestions that have lead to the improvement in the manuscript.

Competing interests
The authors declare that they have no competing interests. There are non-financial competing interests.

Funding
This work was supported by National Natural Science Foundation of China with Grant Number 31301579 and 31371799.

References
Almariri A, Safi M (2014) In vitro antibacterial activity of several plant extracts and oils against some Gram-negative bacteria. Iran J Med Sci 39(1):36–43
Arvanitoyannis IS, Stratakos AC, Tsarouhas P (2009) Irradiation applications in vegetables and fruits: a review. Crit Rev Food Sci 49(5):427–462
Belletti N, Ndagijimana M, Sisto C, Guerzoni ME, Lanciotti R, Gardini F (2004) Evaluation of the antimicrobial activity of citrus essences on *Saccharomyces cerevisiae*. J Agr Food Chem 52(23):6932–6938
Bonura T, Smith KC, Kaplan HS (1975) Enzymatic induction of DNA double-strand breaks in gamma-irradiated *Escherichia coli* K-12. Proc Natl Acad Sci 72(11):4265–4269
Borch E, KantMuermans ML, Blixt Y (1996) Bacterial spoilage of meat and cured meat products. Int J Food Microbiol 33(1):103–120. doi:10.1016/0168-1605(96)01135-x
Brul S, Coote P (1999) Preservative agents in foods. Mode of action and microbial resistance mechanisms. Int J Food Microbiol 50(1–2):1–17
Burt S (2004) Essential oils: their antibacterial properties and potential applications in foods—a review. Int J Food Microbiol 94(3):223–253. doi:10.1016/J.Ijfoodmicro.2004.03.022
Cai LY, Cao AL, Li TT, Wu XS, Xu YX, Li JR (2015) Effect of the fumigating with essential oils on the microbiological characteristics and quality changes of refrigerated turbot (*Scophthalmus maximus*) fillets. Food Bioprocess Technol 8(4):844–853. doi:10.1007/s11947-014-1453-0
Carraro U, Catani C (1983) A sensitive SDS-PAGE method separating myosin heavy chain isoforms of rat skeletal muscles reveals the heterogeneous nature of the embryonic myosin. Biochem Biophys Res Commun 116(3):793–802
Chang S-T, Chen P-F, Chang S-C (2001) Antibacterial activity of leaf essential oils and their constituents from *Cinnamomum osmophloeum*. J Ethnopharmacol 77(1):123–127. doi:10.1016/S0378-8741(01)00273-2
Cheng S-S, Liu J-Y, Hsui Y-R, Chang S-T (2006) Chemical polymorphism and antifungal activity of essential oils from leaves of different provenances of indigenous cinnamon (*Cinnamomum osmophloeum*). Bioresour Technol 97(2):306–312
Di Pasqua R, Hoskins N, Betts G, Mauriello G (2006) Changes in membrane fatty acids composition of microbial cells induced by addiction of thymol, carvacrol, limonene, cinnamaldehyde, and eugenol in the growing media. J Agric Food Chem 54(7):2745–2749. doi:10.1021/jf052722I
Di Pasqua R, Betts G, Hoskins N, Edwards M, Ercolini D, Mauriello G (2007) Membrane toxicity of antimicrobial compounds from essential oils. J Agric Food Chem 55(12):4863–4870. doi:10.1021/jf0636465
Dussault D, Caillet S, Le Tien C, Lacroix M (2009) Effect of γ-irradiation on membrane fatty acids and peptidoglycan's muropeptides of *Pantoea*

agglomerans, a plant pathogen. J Appl Microbiol 106(3):1033–1040

Evans R, McClure P, Gould G, Russell N (1998) The effect of growth temperature on the phospholipid and fatty acyl compositions of non-proteolytic *Clostridium botulinum*. Int J Food Microbiol 40(3):159–167

Gennari M, Tomaselli S, Cotrona V (1999) The microflora of fresh and spoiled sardines (*Sardina pilchardus*) caught in Adriatic (Mediterranean) Sea and stored in ice. Food Microbiol 16(1):15–28. doi:10.1006/fmic.1998.0210

Gill AO, Holley RA (2004) Mechanisms of bactericidal action of cinnamaldehyde against *Listeria monocytogenes* and of eugenol against *L. monocytogenes* and *Lactobacillus sakei*. Appl Environ Microbiol 70(10):5750–5755

Gram L, Huss HH (1996) Microbiological spoilage of fish and fish products. Int J Food Microbiol 33(1):121–137. doi:10.1016/0168-1605(96)01134-8

Gram L, Wedellneergaard C, Huss HH (1990) The bacteriology of fresh and spoiling Lake Victorian Nile Perch (*Lates niloticus*). Int J Food Microbiol 10(3–4):303–316. doi:10.1016/0168-1605(90)90077-i

Hammer KA, Heel KA (2012) Use of multiparameter flow cytometry to determine the effects of monoterpenoids and phenylpropanoids on membrane polarity and permeability in Staphylococci and Enterococci. Int J Antimicrob Agents 40(3):239–245. doi:10.1016/j.ijantimicag.2012.05.015

Holley RA, Patel D (2005) Improvement in shelf-life and safety of perishable foods by plant essential oils and smoke antimicrobials. Food Microbiol 22(4):273–292. doi:10.1016/j.fm.2004.08.006

Hollmann F, Arends IWCE, Holtmann D (2011) Enzymatic reductions for the chemist. Green Chem 13(9):2285–2314. doi:10.1039/C1GC15424A

Jasour MS, Ehsani A, Mehryar L, Naghibi SS (2015) Chitosan coating incorporated with the lactoperoxidase system: an active edible coating for fish preservation. J Sci Food Agr 95(6):1373–1378. doi:10.1002/jsfa.6838

Katayama S, Shima J, Saeki H (2002) Solubility improvement of shellfish muscle proteins by reaction with glucose and its soluble state in low-ionic-strength medium. J Agr Food Chem 50(15):4327–4332

Knauf PA, Rothstein A (1971) Chemical modification of membranes 1. Effects of sulfhydryl and amino reactive reagents on anion and cation permeability of the human red blood cell. J General Physiol 58(2):190–210

Lacroix M, Ouattara B (2000) Combined industrial processes with irradiation to assure innocuity and preservation of food products—a review. Food Res Int 33(9):719–724

Leblanc L, Gouffi K, Leroi F, Hartke A, Blanco C, Auffray Y, Pichereau V (2001) Uptake of choline from salmon flesh and its conversion to glycine betaine in response to salt stress in *Shewanella putrefaciens*. Int J Food Microbiol 65(1–2):93–103. doi:10.1016/S0168-1605(00)00516-X

Le-Tien C, Lafortune R, Shareck F, Lacroix M (2007) DNA analysis of a radio-tolerant bacterium *Pantoea agglomerans* by FT-IR spectroscopy. Talanta 71(5):1969–1975

López-Caballero ME, Sánchez-Feránndez JA, Moral A (2001) Growth and metabolic activity of *Shewanella putrefaciens* maintained under different CO_2 and O_2 concentrations. Int J Food Microbiol 64(3):277–287. doi:10.1016/S0168-1605(00)00473-6

Moule AL, Wilkinson SG (1987) Polar lipids, fatty acids, and isoprenoid quinones of Alteromonas putrefaciens (*Shewanella putrefaciens*). Syst Appl Microbiol 9(3):192–198

Mousavi F, Bojko B, Bessonneau V, Pawliszyn J (2016) Cinnamaldehyde characterization as an antibacterial agent toward *E. coli* metabolic profile using 96-blade solid-phase microextraction coupled to liquid chromatography-mass spectrometry. J Proteome Res 15(3):963–975. doi:10.1021/acs.jproteome.5b00992

Ooi LS, Li Y, Kam S-L, Wang H, Wong EY, Ooi VE (2006) Antimicrobial activities of cinnamon oil and cinnamaldehyde from the Chinese medicinal herb *Cinnamomum cassia* Blume. Am J Chin Med 34(03):511–522

Pedersen UR, Leidy C, Westh P, Peters GH (2006) The effect of calcium on the properties of charged phospholipid bilayers. Biochim Biophys Acta 1758(5):573–582

Pennacchio A, Rossi M, Raia CA (2013) Synthesis of cinnamyl alcohol from cinnamaldehyde with *Bacillus stearothermophilus* alcohol dehydrogenase as the isolated enzyme and in recombinant *E. coli* cells. Appl Biochem Biotechnol 170(6):1482–1490. doi:10.1007/s12010-013-0282-3

Roller S, Board RG (2003) Naturally occurring antimicrobial systems. Springer, Berlin, p 262

Schummer C, Delhomme O, Appenzeller BM, Wennig R, Millet M (2009) Comparison of MTBSTFA and BSTFA in derivatization reactions of polar compounds prior to GC/MS analysis. Talanta 77(4):1473–1482

Senanayake UM, Lee TH, Wills RB (1978) Volatile constituents of cinnamon (*Cinnamomum zeylanicum*) oils. J Agr Food Chem 26(4):822–824

Shen SX, Zhang TH, Yuan Y, Lin SY, Xu JY, Ye HQ (2015) Effects of cinnamaldehyde on *Escherichia coli* and *Staphylococcus aureus* membrane. Food Control 47:196–202. doi:10.1016/j.foodcont.2014.07.003

Shokri S, Ehsani A, Jasour MS (2015) Efficacy of lactoperoxidase system-whey protein coating on shelf-life extension of rainbow trout fillets during cold storage (4 °C). Food Bioprocess Technol 8(1):54–62. doi:10.1007/s11947-014-1378-7

Špitsmeister M, Adamberg K, Vilu R (2010) UPLC/MS based method for quantitative determination of fatty acid composition in Gram-negative and Gram-positive bacteria. J. Microbiol Meth 82(3):288–295

Stenström IM, Molin G (1990) Classification of the spoilage flora of fish, with special reference to *Shewanella putrefaciens*. J Appl Bacteriol 68(6):601–618. doi:10.1111/j.1365-2672.1990.tb05226.x

Urbaniak A, Głowacka A, Kowalczyk E, Lysakowska M, Sienkiewicz M (2014) The antibacterial activity of cinnamon oil on the selected Gram-positive and Gram-negative bacteria. Med Dosw Mikrobiol 66(2):131–141

Van Haute S, Raes K, Van der Meeren P, Sampers I (2016) The effect of cinnamon, oregano and thyme essential oils in marinade on the microbial shelf life of fish and meat products. Food Control 68:30–39. doi:10.1016/j.foodcont.2016.03.025

Wang F, Xiao X, Ou HY, Gai YB, Wang FP (2009) Role and regulation of fatty acid biosynthesis in the response of *Shewanella piezotolerans* WP3 to different temperatures and pressures. J Bacteriol 191(8):2574–2584. doi:10.1128/Jb.00498-08

Zhang HM, Zheng BW, Gao RS, Feng YJ (2015a) Binding of Shewanella FadR to the fabA fatty acid biosynthetic gene: implications for contraction of the fad regulon. Protein Cell 6(9):667–679. doi:10.1007/s13238-015-0172-2

Zhang Q, Lin H, Sui JX, Wang JX, Cao LM (2015b) Effects of Fab'fragments of specific egg yolk antibody (IgY-Fab') against *Shewanella putrefaciens* on the preservation of refrigerated turbot. J Sci Food Agr 95(1):136–140.

16

A food-grade expression system for D-psicose 3-epimerase production in *Bacillus subtilis* using an alanine racemase-encoding selection marker

Jingqi Chen[1,2], Zhaoxia Jin[3], Yuanming Gai[1], Jibin Sun[1,2] and Dawei Zhang[1,2]*

Abstract

Background: Food-grade expression systems require that the resultant strains should only contain materials from food-safe microorganisms, and no antibiotic resistance marker can be utilized. To develop a food-grade expression system for D-psicose 3-epimerase production, we use an alanine racemase-encoding gene as selection marker in *Bacillus subtilis*.

Results: In this study, the D-alanine racemase-encoding gene *dal* was deleted from the chromosome of *B. subtilis* 1A751 using Cre/*lox* system to generate the food-grade host. Subsequently, the plasmid-coded selection marker *dal* was complemented in the food-grade host, and RDPE was thus successfully expressed in *dal* deletion strain without addition of D-alanine. The selection appeared highly stringent, and the plasmid was stably maintained during culturing. The highest RDPE activity in medium reached 46 U/ml at 72 h which was comparable to RDPE production in kanamycin-based system. Finally, the capacity of the food-grade *B. subtilis* 1A751D2R was evaluated in a 7.5 l fermentor with a fed-batch fermentation.

Conclusion: The alanine racemase-encoding gene can be used as a selection marker, and the food-grade expression system was suitable for heterologous proteins production in *B. subtilis*.

Keywords: *Bacillus subtilis*, Cre/*lox* system, D-Psicose 3-epimerase, Fed-batch fermentation, Food-grade system

Background

D-Psicose is a hexoketose monosaccharide sweetener, which is a C-3 epimer of D-fructose and is rarely found in nature (Mu et al. 2012). It has 70% relative sweetness but 0.3% energy of sucrose and is suggested as an ideal sucrose substitute for food products (Matsuo et al. 2002; Oshima et al. 2006). It shows important physiological functions, such as blood glucose suppressive effect (Hayashi et al. 2010; Iida et al. 2008), reactive oxygen species scavenging activity (Matsuo et al. 2003), and neuroprotective effect (Takata et al. 2005). It also improves the gelling behavior and products good flavor during

food process (Sun et al. 2004). In virtue of its outstanding advantages, the conversion of D-fructose to D-psicose using the D-psicose 3-epimerase has been investigated for the commercial production of D-psicose.

Bacillus subtilis is a food-safe microorganism, which has been used to food fermentation for a long period of time (Song et al. 2015). Several gene expression systems have been developed for high-level production of heterologous proteins such as amylase (Chen et al. 2015a, b), lysozyme (Zhang et al. 2014), protease (Degering et al. 2010), and lipase (Lu et al. 2010). However, few reports are concerned with application of recombinant *B. subtilis* directly to food processing. One important reason is that most vectors (mainly based on antibiotics) used in this systems are not food-grade. Food-grade expression systems have been widely developed and investigated for

*Correspondence: zhang_dw@tib.cas.cn
[1] Tianjin Institute of Industrial Biotechnology, Chinese Academy of Sciences, Tianjin 300308, People's Republic of China
Full list of author information is available at the end of the article

lactic acid bacteria. Such systems require that the resultant strains should only contain materials from food-safe microorganisms, and no antibiotic resistance marker can be utilized. Usually, food-grade selection markers can be classified as dominant markers or complementation markers (de Vos 1999). Compared with dominant markers, selection markers based on complementation do not require supplements in the cultivation medium. In order to develop a food-grade complementation-based system, usually a gene on the host chromosome is mutated or deleted, and a wild type copy is inserted into the expression vector. The alanine racemase gene *dal* is involved in the conversion of D-alanine and L-alanine (Bron et al. 2002), and D-alanine is not a common ingredient of large-scale fermentation media (Nguyen et al. 2011); the *dal* gene thus has considerable potential as a food-grade selection marker in *B. subtilis*.

In the present study, we developed a food-grade expression system for the production of D-psicose 3-epimerase (RDPE) from *Ruminococcus* sp. 5_1_39BFAA in *B. subtilis*, using alanine racemase gene *dal* as the selection marker. The selection appeared highly stringent, and the plasmid was stably maintained during culturing. Moreover, the expression level of RDPE in the newly developed food-grade system was comparable to the level obtained in the conventional kanamycin-based system. This new expression system was therefore suitable for food-grade production of various heterologous proteins.

Methods

Bacterial strains, plasmids, and growth conditions

Bacterial strains and plasmids used in this study are listed in Additional file 1: Table S1. *Escherichia coli* DH5α was used as a host for cloning and plasmid preparation. *Bacillus subtilis* 1A751, which is deficient in two extracellular proteases (*nprE*, *aprE*), served as the parental strain. The plasmid pMA5 is an *E. coli/B. subtilis* shuttle vector and used to clone and express protein. The plasmids p7Z6 containing *lox*71-*zeo-lox*66 cassette and p148-cre containing *cre* expression cassette were used for the knockout of target gene. Transformants of *E. coli* and *B. subtilis* were selected on Luria–Bertani (LB) agar [1% (w/v) peptone, 0.5% (w/v) yeast extract, 1% (w/v) NaCl, and 2% (w/v) agar], supplemented with ampicillin (100 μg/ml), zeocin (20 μg/ml), or kanamycin (50 μg/ml) depending on the plasmid antibiotic marker. *E. coli* DH5α was incubated in LB medium supplemented with ampicillin (100 μg/ml) at 37 °C. *Bacillus subtilis* was cultivated in SR medium [1.5% (w/v) peptone, 2.5% (w/v) yeast extract, and 0.3% (w/v) K_2HPO_4, pH 7.2] containing additionally kanamycin (50 μg/ml) or zeocin (20 μg/ml) at 37 °C. All of the strains were incubated under a shaking condition at 200 rpm. Except the fed-batch fermentation, all of the

experiments were repeated at least 3 times, and mean values were used for comparison.

Primers and oligonucleotides

Polymerase chain reaction (PCR) primers and oligonucleotides used in this study were synthesized by GENEWIZ (Suzhou, China) and listed in Additional file 1: Table S2.

Genetic manipulation

PCRs were performed using PrimeSTAR Max DNA Polymerase (TaKaRa, Japan). DNA fragments and PCR products were excised from a 0.8% agarose gel and purified by E.Z.N.A.™ Gel Extraction Kit (200) (Omega Bio-tek, Inc., USA) according to the manufactures' instruction. E.Z.N.A.™ Plasmid Mini Kit I (Omega Bio-tek, Inc., USA) was applied for plasmid extraction according to the manufactures' instruction. Genomic DNA isolation was carried out by TIANamp Bacteria DNA Kit (TIANGEN BIOTECH (BEIJING) CO., LTD., China). All the DNA constructs were sequenced by GENEWIZ (Suzhou, China).

Construction of the *dal* deletion mutant

The deletion of the alanine racemase gene *dal* in *B. subtilis* was performed using Cre/*lox* system as described previously (Yan et al. 2008; Dong and Zhang 2014). The two flanking fragments upstream and downstream (~1 kb) of the *dal* gene were amplified using genomic DNA from *B. subtilis* 168 as template and UP-F/UP-R and DN-F/DN-R as primers, respectively. The *lox*71-*zeo-lox*66 cassette (~0.5 kb) was amplified from the plasmid p7Z6 using the primers lox-F and lox-R. Then, the flanking fragments and the *lox*71-*zeo-lox*66 fragment were fused together by splicing by overlap extension PCR (SOE-PCR) using the primers UP-F and DN-R. Subsequently, the fused fragment was transformed into *B. subtilis* 1A751. Selection of the double crossover mutant (*B. subtilis* 1A751D1), marker excision of by Cre-dependent recombination of the *lox*-sites, and selection of the *dal* deletion mutant (*B. subtilis* 1A751D2) were performed by the previous strategy (Yan et al. 2008). Finally, the entire *dal* gene was thus successfully deleted via double crossing over and marker excision in the chromosome of *B. subtilis* 1A751, which was further confirmed by PCR amplification.

Construction of plasmids

The food-grade expression plasmid was constructed based on pMA5 by replacing the zeocin resistance gene *zeo* with the alanine racemase gene *dal* from *B. subtilis* 168 using a sequence-independent method named "simple cloning" developed by Chun You (2012). Based on the nucleotide sequence of *dal*, the primers dal-F/dal-R were designed to amplify the fragment *dal* using the *B.*

subtilis 168 as the template. The linear vector backbone was amplified using the primers pMA5-F1 and pMA5-R1 as the primers and the plasmid pMA5 as the template. Dal-F/dal-R had the reverse complementary sequences of pMA5-F1/pMA5-R1, respectively. Then, the DNA multimer was generated based on these DNA templates by prolonged overlap extension PCR (POE-PCR). Eventually, the POE-PCR products (DNA multimer) were directly transformed into competent *E. coli* DH5α, yielding the recombinant plasmid pMA5-DAL. Likewise, the *rdpe* gene from pET-RDPE was inserted into the plasmid pMA5-DAL downstream of the promoter P_{HpaII}, resulting into the recombinant plasmid pMA5-DAL-RDPE.

Stability of the recombinant plasmids

The evaluation of the stability of the plasmid pMA5-DAL-RDPE was conducted using the method described by Nguyen (2005). The recombinant strains were inoculated onto Plate A (LB agar plate without supplement of D-alanine with selection pressure) and Plate B (LB agar plate with supplement of 200 μg D-alanine/ml without selection pressure). Colony numbers of the strain 1A751D2R on Plate A and Plate B were named as C_A and C_B. The value of C_A/C_B was regarded as the stability of the plasmid pMA5-DAL-RDPE at the certain generation of cultivation.

Fed-batch fermentation in 7.5 l fermentor

The food-grade RDPE production in *B. subtilis* 1A751D2R was evaluated in 7.5 l BIO FLO 310 fermentor (New Brunswick Scientific co Inc., USA) with a fed-batch strategy. The airflow rate was 6.0 l/min, and dissolved oxygen tension was maintained between 20 and 40% air saturation by automatic adjustment of speed of the stirrer. The temperature was kept at 37 °C and the pH was controlled at pH 7.2. Foam was controlled by the addition of a silicone-based anti-foaming agent. The fermentation medium was SR medium. The fermentation was performed with an initial working volume of 3.5 l. When the cell growth rate became constant, the substrate fed-batch mode was started by adding 8.0% soluble starch at a constant flow rate, until the final concentration of soluble starch was up to 4.0%. Cell growth was monitored by measuring dry cell weight of the fermentation broth. The activity of RDPE was determined by measuring the supernatant of broth.

Enzyme assays

The RDPE activity was analyzed by determining the amount of D-psicose obtained from D-fructose. One milliliter of reactions mixture contained D-fructose (20 g/l) in sodium phosphate buffer (50 mM, pH 8.0) and 200 μl crude enzyme. The reaction was incubated at 55 °C for 10 min, following by boiling at 100 °C for 10 min. The obtained D-psicose in the mixture was determined via high-performance HPLC system with a refractive index detector and a Sugar-PakTM column (6.5 mm × 300 mm; Waters), which was eluted with ultrapure water at 80 °C and 0.4 ml/min. One unit of DPEase activity is defined as the amount of enzyme that catalyzed the production of 1 μmol D-psicose per minute. For the determination of extracellular enzyme activity, the crude enzyme was the supernatant of fermentation broth. For the determination of intracellular enzyme activity, the cells need to be broken. *Bacillus subtilis* cells expressing RDPE were harvested from the culture broth by centrifugation at 6000×g for 10 min at 4 °C. The cells were then suspended in lysis buffer (25 mM Tris/HCl, 300 mM NaCl, and 40 mM imidazole, pH 8.0). The suspended cells were disrupted using a high-pressure homogenizer (APV, Denmark) at 900–1000 bar. The supernatant was obtained by centrifugation at 15,000×g for 30 min at 4 °C and filtration through a 0.45 μm filter. Then, the crude extract was applied in the enzyme assay.

SDS-PAGE analysis

Culture samples (1 ml) were harvested and the supernatant was separated from the culture medium by centrifugation (12,000g, 10 min, 4 °C). After adding 5× SDS-PAGE sample buffer, the supernatants were boiled for 10 min, and proteins were separated in SDS-PAGE using the NuPAGE 10% Bis–Tris Gel (Novex by Life Technologies, USA) in combination with MOPS SDS Running Buffer (Invitrogen Life Technologies, USA). PageRuler Prestained Protein Ladder (Invitrogen Life Technologies, USA) was used to determine the apparent molecular weight of separated proteins. Proteins were visualized with Coomassie Brilliant Blue.

Results and discussion

Construction of the food-grade host strain with deficiency of *dal*

In order to obtain the mutant strain with deficiency of *dal*, we attempted to knock out the gene *dal* from *B. subtilis* chromosome using Cre/*lox* system. The flow chart for construction of the food-grade host strain is shown in Fig. 1. The two flanking fragments upstream and downstream of *dal* and the fragment *lox*71-*zeo*-*lox*66 cassette were fused into a long DNA fragment by SOE-PCR, and the fused fragment was then directly transformed into *B. subtilis* 1A751. Because the fused fragment had two efficient homology regions with the chromosome of *B. subtilis* 1A751, the homologous recombination via double crossing over event thus occurred, and the *lox*71-*zeo*-*lox*66 cassette was integrated into the chromosome. The transformants (zeocin-resistant phenotype) were selected

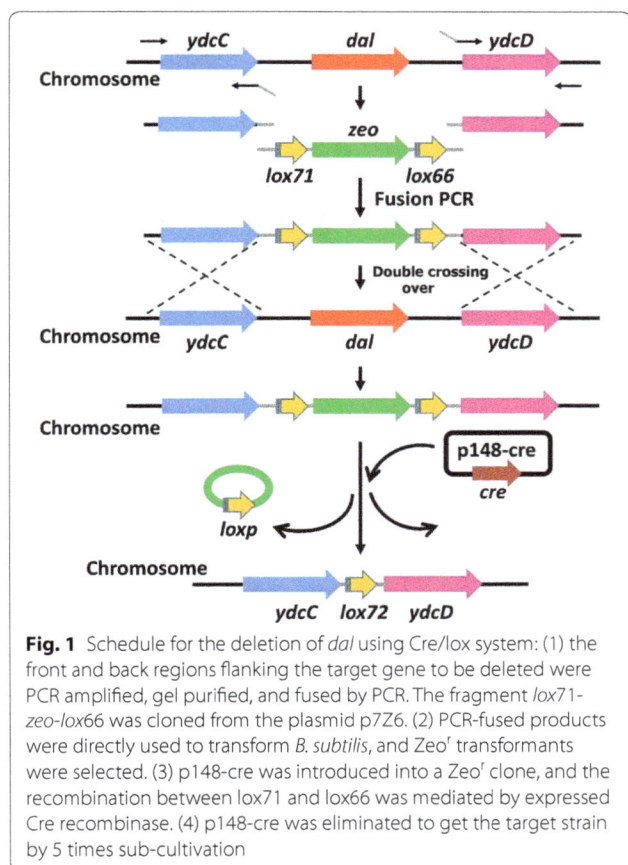

Fig. 1 Schedule for the deletion of *dal* using Cre/lox system: (1) the front and back regions flanking the target gene to be deleted were PCR amplified, gel purified, and fused by PCR. The fragment *lox*71-*zeo*-lox66 was cloned from the plasmid p7Z6. (2) PCR-fused products were directly used to transform *B. subtilis*, and Zeo^r transformants were selected. (3) p148-cre was introduced into a Zeo^r clone, and the recombination between lox71 and lox66 was mediated by expressed Cre recombinase. (4) p148-cre was eliminated to get the target strain by 5 times sub-cultivation

on LB agar plate with the addition of 20 µg zeocin/ml and 200 µg D-alanine/ml. To exclude the false positive, the transformants with zeocin-resistant phenotype were replica-plated on LB agar plate and LB agar plate with the supplement of zeocin and D-alanine. The strains with the *lox*71-*zeo*-lox66 cassette could not grow on LB agar plate but grown on LB agar plate with the supplement of zeocin and D-alanine. Of 36 tested transformants, 28 displayed the desired phenotype (Additional file 1: Figure S1a); 5 of these were confirmed by PCR, cultivated in liquid LB medium (Additional file 1: Figure S1b) and named as *B. subtilis* 1A751D1. The plasmid p148-cre with P$_{spac}$-cre cassette was transformed into 1A751D1 and selected on LB agar plate with kanamycin and D-alanine. With the induction of IPTG, the Cre recombinase was expressed and the *lox*-sequence-flanked zeocin resistance gene was then excised. 30 colonies were identified by PCR and 100% of these lost *zeo* gene, indicating high efficiency of gene deletion using Cre/lox system. The obtained strain 1A751D1C was subcultured in liquid LB medium with D-alanine five times to lose the plasmid p148-cre and then incubated on LB agar plate with D-alanine, followed by replica plating on LB agar plate with D-alanine and LB

agar plate with kanamycin and D-alanine. 23 of 25 tested colonies displayed the desired phenotype; 5 of these were further identified by PCR, with loss of the plasmid p148-cre because of the plasmid instability. At last, we successfully obtained the D-alanine-auxotrophic strain 1A751D2.

Construction of food-grade expression plasmids with auxotrophic marker

As described in "Methods" section and shown in Fig. 2, the new expression plasmids were constructed based on the conventional plasmid pMA5. First, the *neo* gene of pMA5 was replaced with the gene *dal*, and *dal* was under the control of the native promoter of the *neo* gene, yielding the plasmid pMA5-DAL. Then, the gene *rdpe* encoding D-psicose 3-epimerase was cloned and inserted into pMA5-DAL downstream of a strong and constitutive promoter P$_{HpaII}$ which is from *Staphylococcus aureus*, resulting into the food-grade expression plasmid pMA5-DAL-RDPE.

Expression of D-psicose 3-epimerase in the food-grade system

The food-grade plasmids pMA5-DAL-RDPE and pMA5-DAL were transformed into the food-grade host strain 1A751D2 with the deficiency of alanine racemase gene to generate the recombinant strain 1A751D2R and 1A751D2C, respectively. Subsequently, the strains 1A751D2R, and 1A751D2C was inoculated in 250 ml shake flask containing 30 ml SR medium at 37 °C and 200 rpm for 78 h. The strain 1A751D2 was inoculated in SR medium with the addition of 200 µg D-alanine/ml as the negative control. The activity of RDPE in the cells or medium was determined throughout all the cultivation process. As shown in Fig. 3a, the extracellular activity of RDPE was gradually increased with the fermentation process, and the highest activity reached 46 U/ml at 72 h, which was comparable to RDPE production in the conventional *neo*-based system (44 U/ml at 72 h). In our previous study, we have demonstrated that RDPE is one of non-classically secreted proteins which are secreted via non-classical secretion pathway in *B. subtilis* (Chen et al. 2016). Thus, RDPE can still be exported into the extracellular milieu without any classical signal peptide. To provide further evidence supporting the above result, the SDS-PAGE analysis was performed. Distinct bands with a molecular mass of about 34 kDa were observed which was in good agreement with the deduced value (Fig. 3d), and the result was consistent with the activity analysis. Meanwhile, we noted that there was still residual RDPE in the cell fraction over the whole fermentation (Fig. 3a, c). We also traced the growth states of the recombinant strains. As shown in

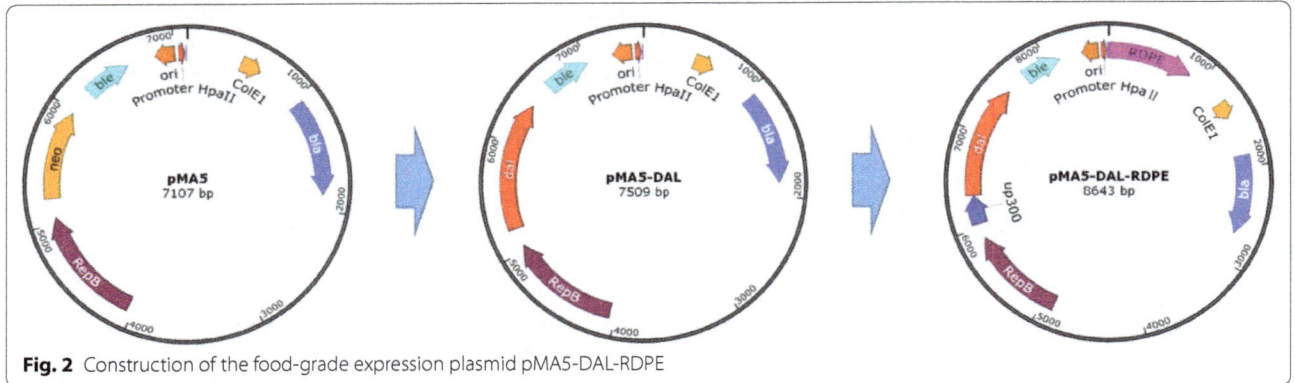

Fig. 2 Construction of the food-grade expression plasmid pMA5-DAL-RDPE

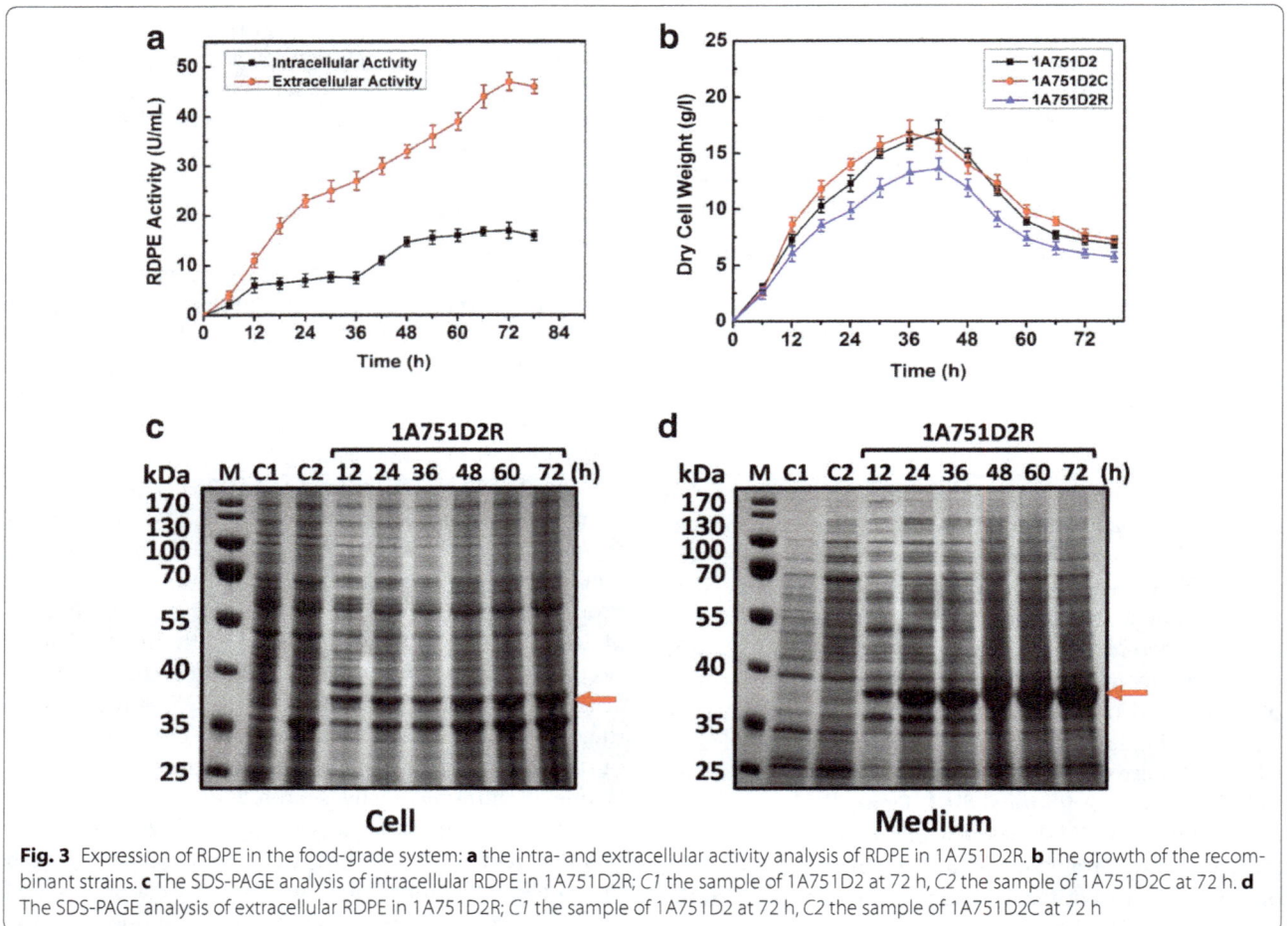

Fig. 3 Expression of RDPE in the food-grade system: **a** the intra- and extracellular activity analysis of RDPE in 1A751D2R. **b** The growth of the recombinant strains. **c** The SDS-PAGE analysis of intracellular RDPE in 1A751D2R; *C1* the sample of 1A751D2 at 72 h, *C2* the sample of 1A751D2C at 72 h. **d** The SDS-PAGE analysis of extracellular RDPE in 1A751D2R; *C1* the sample of 1A751D2 at 72 h, *C2* the sample of 1A751D2C at 72 h

Fig. 3b, there was no difference between the biomass of 1A751D2 and that of 1A751D2C, however, the biomass of 1A751D2R was slightly lower than that of 1A751D2 and 1A751D2C. This was caused by increased metabolic burden with the overexpression of RDPE, because high protein production was usually accompanied by reduced growth rates. In addition, the biomass of all the three strains sharply decreases after about 40 h, we speculated that the carbon source might have been depleted at that time.

Table 1 Stability of *dal*-based and *neo*-based plasmids in different medium

Plasmid	5 Generations (%)	15 Generations (%)	40 Generations (%)	80 Generations (%)
pMA5-RDPE/SR	56	10	3	0
pMA5-RDPE/SR + Kan	100	95	88	85
pMA5-DAL-RDPE/SR	100	100	100	100
pMA5-DAL-RDPE/SR + D-alanine	88	76	54	30

Bacillus subtilis 1A751R harboring pMA5-RDPE and *B. subtilis* 1A751D2R harboring pMA5-DAL-RDPE were cultivated in selective and nonselective medium. The kanamycin concentration was 50 μg/ml; the D-alanine concentration was 200 μg/ml. The strains were cultivated at 37 °C

Every 10 h, 5 generations passed. The stability of plasmids was calculated by dividing the number of colonies on selective medium with the number of colonies on nonselective medium

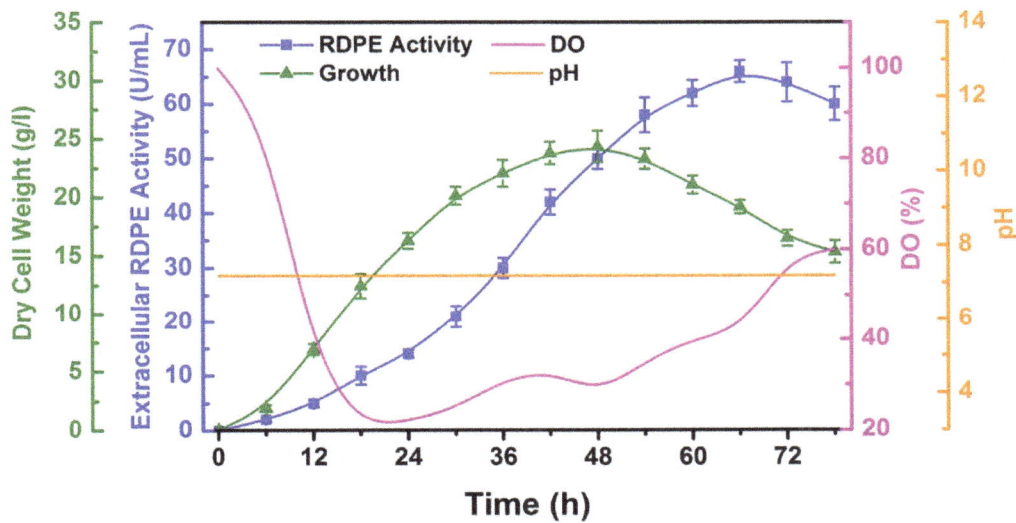

Fig. 4 Production of RDPE in recombinant strain 1A751D2R by fed-batch fermentation in 7.5 l fermentor. *Blue line* RDPE activity in medium. *Green line* biomass. *Pink line* DO concentration. *Orange line* pH

Evaluation of plasmid stability and copy numbers

The stability of the *dal*- and *neo*-based plasmids was determined using pMA5-DAL-RDPE (*dal*) in *B. subtilis* 1A751D2 and pMA5-RDPE (*neo*) in *B. subtilis* 1A751. The strains were cultivated for an estimated 80 generations (160 h) at 37 °C in selective and nonselective medium, followed by replica plating of diluted cultures in order to determine the stability of plasmids. The fraction of cells retaining the plasmid pMA5-RDPE after 80 generations in nonselective medium (SR without kanamycin) was 0%, whereas in selective medium 85% of the colonies still contained the plasmid. Interestingly, the plasmid pMA5-DAL-RDPE showed much better stability: after 80 generations, it was retained in about 30 and 100% of cells of *B. subtilis* 1A751D2R under nonselective and selective conditions, respectively (Table 1).

Food-grade production of RDPE in 7.5 l fermentor with fed-batch fermentation

The expression efficiency of *B. subtilis* 1A751D2R was further explored in 7.5 l fermentor. The fermentor was inoculated with 5% (v/v) of freshly cultured 1A751D2R grown in SR medium at 37 °C for 18 h. To maintain cell growth and RDPE production, we chose a fed-batch strategy. When the cell growth rate was constant, 8.0% (w/v) soluble starch was added at a constant flow rate until the final concentration of soluble starch was up to 4.0% (w/v). As shown in Fig. 4, during the growth phase, the maximum biomass in the fermentor reached 23.9 g/l (dry cell weight) at 48 h. Compared with the biomass decrease in shake flask, the biomass decrease after 48 h in 7.5 l fermentor was very mild, which was attributed to the supplement of soluble starch. The activity of RDPE in medium was continuously increased and reached the maximum of 65 U/ml with a high productivity of 0.9 U/ml h at 72 h. The RDPE concentration in the supernatant reached about 1.8 g/l. The high activity of RDPE indicates that *B. subtilis* is a suitable host for the industrial production of heterologous protein.

Conclusion

In this study, we developed a food-grade expression system for D-psicose 3-epimerase production in *B. subtilis*. The plasmid co-expressing *rdpe* and *dal* was introduced into *dal* mutant, selection appeared highly stringent, and plasmids were stably maintained during culturing. Moreover, the production of RDPE in this food-grade expression system was comparable to that in *neo*-based system. The results showed that this system was very suitable for food-grade expression of heterologous proteins.

Abbreviations
RDPE: D-psicose 3-epimerase from *Ruminococcus* sp. 5_1_39BFAA; *Dal*: D-alanine racemase-encoding gene; LB: Luria–Bertani; SR: super rich; PCR: polymerase chain reaction; SOE-PCR: splicing by overlap extension PCR; POE –PCR: prolonged overlap extension PCR; HPLC: high-performance liquid chromatography; SDS-PAGE: sodium dodecyl sulfate polyacrylamide gel electrophoresis; IPTG: isopropyl-β-D-thiogalactoside; g: grams; l: liter; h: hours; DO: dissolved oxygen; DCW: dry cell weight.

Authors' contributions
JC, ZJ, and DZ designed the experiments; JC and YG performed the experiments; JC, JS, and DZ wrote this manuscript; and all authors contributed to the discussion of the research. All authors read and approved the final manuscript.

Author details
[1] Tianjin Institute of Industrial Biotechnology, Chinese Academy of Sciences, Tianjin 300308, People's Republic of China. [2] Key Laboratory of Systems Microbial Biotechnology, Chinese Academy of Sciences, Tianjin 300308, People's Republic of China. [3] School of Biological Engineering, Dalian Polytechnic University, Dalian 116034, People's Republic of China.

Acknowledgements
Professor Shupeng Li from Key Laboratory for Microbiological Engineering of Agricultural Environment of Ministry of Agriculture, Nanjing Agricultural University, friendly gave us the plasmids p7Z6 and p148-cre as the gifts.

Competing interests
The authors declare that they have no competing interests.

Funding
This work was supported by National Nature Science Foundation of China (31370089, 31670604, 31570303), State Key Development 973 Program for Basic Research of China (2013CB733601), Nature Science Foundation of Tianjin City (CN) (16JCYBJC23500), the Key Projects in the Tianjin Science & Technology Pillar Program(11ZCZDSY08400), and Natural Science Foundation of Liaoning Province of China (2014026012).

References
Bron PA, Benchimol MG, Lambert J, Palumbo E, Deghorain M, Delcour J, de Vos WM, Kleerebezem M, Hols P (2002) Use of the alr gene as a food-grade selection marker in lactic acid bacteria. Appl Environ Microbiol 68:5663–5670
Chen J, Fu G, Gai Y, Zheng P, Zhang D, Wen J (2015a) Combinatorial Sec pathway analysis for improved heterologous protein secretion in *Bacillus subtilis*: identification of bottlenecks by systematic gene overexpression. Microb Cell Fact 14:92
Chen J, Gai Y, Fu G, Zhou W, Zhang D, Wen J (2015b) Enhanced extracellular production of alpha-amylase in *Bacillus subtilis* by optimization of regulatory elements and over-expression of PrsA lipoprotein. Biotechnol Lett 37:899–906

Chen J, Zhao L, Fu G, Zhou W, Sun Y, Zheng P, Sun J, Zhang D (2016) A novel strategy for protein production using non-classical secretion pathway in *Bacillus subtilis*. Microb Cell Fact 15:69
de Vos WM (1999) Safe and sustainable systems for food-grade fermentations by genetically modified lactic acid bacteria. Int Dairy J 9:3–10
Degering C, Eggert T, Puls M, Bongaerts J, Evers S, Maurer KH, Jaeger KE (2010) Optimization of protease secretion in *Bacillus subtilis* and *Bacillus licheniformis* by screening of homologous and heterologous signal peptides. Appl Environ Microbiol 76:6370–6376
Dong H, Zhang D (2014) Current development in genetic engineering strategies of Bacillus species. Microb Cell Fact 13:63
Hayashi N, Iida T, Yamada T, Okuma K, Takehara I, Yamamoto T, Yamada K, Tokuda M (2010) Study on the postprandial blood glucose suppression effect of D-psicose in borderline diabetes and the safety of long-term ingestion by normal human subjects. Biosci Biotechnol Biochem 74:510–519
Iida T, Kishimoto Y, Yoshikawa Y, Hayashi N, Okuma K, Tohi M, Yagi K, Matsuo T, Izumori K (2008) Acute D-psicose administration decreases the glycemic responses to an oral maltodextrin tolerance test in normal adults. J Nutr Sci Vitaminol 54:511–514
Lu Y, Lin Q, Wang J, Wu Y, Bao W, Lv F, Lu Z (2010) Overexpression and characterization in *Bacillus subtilis* of a positionally nonspecific lipase from *Proteus vulgaris*. J Ind Microbiol Biotechnol 37:919–925
Matsuo T, Suzuki H, Hashiguchi M, Izumori K (2002) D-Psicose is a rare sugar that provides no energy to growing rats. J Nutr Sci Vitaminol 48:77–80
Matsuo T, Tanaka T, Hashiguchi M, Izumori K, Suzuki H (2003) Metabolic effects of D-psicose in rats: studies on faecal and urinary excretion and caecal fermentation. Asia Pacific J Clin Nutr 12:225–231
Mu W, Zhang W, Feng Y, Jiang B, Zhou L (2012) Recent advances on applications and biotechnological production of D-psicose. Appl Microbiol Biotechnol 94:1461–1467
Nguyen HD, Nguyen QA, Ferreira RC, Ferreira LC, Tran LT, Schumann W (2005) Construction of plasmid-based expression vectors for *Bacillus subtilis* exhibiting full structural stability. Plasmid 54:241–248
Nguyen TT, Mathiesen G, Fredriksen L, Kittl R, Nguyen TH, Eijsink VG, Haltrich D, Peterbauer CK (2011) A food-grade system for inducible gene expression in *Lactobacillus plantarum* using an alanine racemase-encoding selection marker. J Agric Food Chem 59:5617–5624
Oshima H, Kimura I, Izumori K (2006) Psicose contents in various food products and its origin. Food Sci Technol Res 12:137–143
Song Y, Nikoloff JM, Zhang D (2015) Improving protein production on the level of regulation both of expression and secretion pathways in *Bacillus subtilis*. J Microbiol Biotechnol 25:963–977
Sun Y, Hayakawa S, Izumori K (2004) Modification of ovalbumin with a rare ketohexose through the maillard reaction: effect on protein structure and gel properties. J Agric Food Chem 52:1293–1299
Takata MK, Yamaguchi F, Nakanose K, Watanabe Y, Hatano N, Tsukamoto I, Nagata M, Izumori K, Tokuda M (2005) Neuroprotective effect of D-psicose on 6-hydroxydopamine-induced apoptosis in rat pheochromocytoma (PC12) cells. J Biosci Bioeng 100:511–516
Yan X, Yu HJ, Hong Q, Li SP (2008) Cre/lox system and PCR-based genome engineering in *Bacillus subtilis*. Appl Environ Microbiol 74:5556–5562
You C, Zhang XZ, Zhang YH (2012) Simple cloning via direct transformation of PCR product (DNA Multimer) to *Escherichia coli* and *Bacillus subtilis*. Appl Environ Microbiol 78:1593–1595
Zhang HF, Fu G, Zhang D (2014) Cloning, characterization and production of a novel lysozyme by different expression hosts. J Microbiol Biotechnol 24:1405–1412

17

An efficient multi-stage fermentation strategy for the production of microbial oil rich in arachidonic acid in *Mortierella alpina*

Wen-Jia Wu[1], Ai-Hui Zhang[1], Chao Peng[5], Lu-Jing Ren[1,4], Ping Song[1], Ya-Dong Yu[4], He Huang[2,3,4] and Xiao-Jun Ji[1,4*] (ID)

Abstract

Background: Fungal morphology and aeration play a significant role in the growth process of *Mortierella alpina*. The production of microbial oil rich in arachidonic acid (ARA) in *M. alpina* was enhanced by using a multi-stage fermentation strategy which combined fed-batch culture with precise control of aeration and agitation rates at proper times.

Results: The fermentation period was divided into four stages according to the cultivation characteristics of *M. alpina*. The dissolved oxygen concentration was well suited for ARA biosynthesis. Moreover, the ultimate dry cell weight (DCW), lipid, and ARA yields obtained using this strategy reached 41.4, 22.2, 13.5 g/L, respectively. The respective values represent 14.8, 25.8, and 7.8% improvements over traditional fed-batch fermentation processes.

Conclusions: This strategy provides promising control insights for the mass production of ARA-rich oil on an industrial scale. Pellet-like fungal morphology was transformed into rice-shaped particles which were beneficial for oxygen transfer and thus highly suitable for biomass accumulation.

Keywords: Arachidonic acid, *Mortierella alpina*, Multi-stage fermentation, Aeration, Agitation, Morphology

Background

Arachidonic acid, (5, 8, 11, 14-cis-eicosatetraenoic acid, ARA), a representative of the omega-6 group of essential polyunsaturated fatty acids (PUFAs), acts as a precursor for eicosanoid hormones such as prostaglandins, leukotrienes, and thromboxanes (Ji et al. 2014a). The application of ARA as the active ingredient in drugs and food additives thus has great potential. Owing to its unique physiological functions, it has been widely applied in the food industry as well as cosmetics, medicine, and many other fields (Ward and Singh 2005). For a long time, egg yolk, animal liver, and adrenal glands were the main sources of ARA. However, their low intrinsic ARA content (Higashiyama et al. 2002) restricts their application, and it is not possible to source sufficient material for

use of ARA in infant formula. On the other hand, ARA-rich oil derived from the oleaginous fungus *Mortierella alpina* has received GRAS status from the US FDA in 2001 (Ryan et al. 2010), and *M. alpina* is regarded as one of the most promising candidates for the mass production of ARA-rich oil (Ji et al. 2014a). The ARA biosynthesis pathway in *M. alpina* proceeds via the formation of C16 or C18 saturated fatty acids, which are further modified through a series of elongation and desaturation steps, culminating in the formation of ARA. It is known that these reactions require NADPH, an electron transport system, a terminal desaturase, and molecular oxygen (Ward and Singh 2005).

In general, fungal mycelia are brittle and physically weak. Therefore, the agitation rate in mechanically stirred bioreactors, which are normally used for the production of ARA-rich oil, has to be controlled within a very precise range. High agitation rates increase the shear forces, which can break mycelial integrity and influence the broth characteristics. Low agitation rates, on the other hand, lead to

*Correspondence: xiaojunji@njtech.edu.cn
[4] Jiangsu National Synergetic Innovation Center for Advanced Materials (SICAM), No. 5 Xinmofan Road, Nanjing 210009, People's Republic of China
Full list of author information is available at the end of the article

a low dissolved oxygen concentration insufficient for ARA biosynthesis. Overall, mycelial morphology has a strong effect on the physical properties of the broth and often leads to a number of different problems in large bioreactors with respect to gas dispersion, as well as mass and heat transfer (Higashiyama et al. 2002). There are many reports that discuss the size and shape of fungal mycelial pellets (Xu et al. 2010; Tai et al. 2010), but little is known about the true features of the internal pellet structure, including geometry and mycelial viability (Hamanaka et al. 2001). Interestingly, pellets with a moderate compactness are the more productive morphological form for the production of ARA-rich oil, compared to free filamentous mycelia. Therefore, controlling proper aeration and agitation rates in the whole process to balance the contradiction between these two factors is vitally important for the fermentation of fungal producers of ARA-rich oils. There have been some attempts to fulfill this objective by controlling the aeration (Higashiyama et al. 1999; Nie et al. 2014) and agitation rates (Higashiyama et al. 1999; Peng et al. 2010), respectively. ARA yields in these reports reached 4.7 g/L by strictly monitoring the mycelial morphology and employing a two-stage control strategy for the aeration rate, which represents an increase of 38.2% (Gao et al. 2016). However, until now, no efforts have been made to simultaneously evaluate the aeration and agitation rate in relation to the proper mycelial morphology for increasing the biomass yield of the filamentous fungus *M. alpina*.

In this study, an innovative multi-stage strategy was investigated to optimize ARA productivity in bioreactors. We thereby aimed at balancing the contradiction between the aeration and agitation controls required for optimal dissolved oxygen concentration and fungal morphology, respectively. The strategy was further assessed regarding its effectiveness in improving the biomass yield, which reached >40 g/L. This approach gives a detailed insight into the mycelial morphology control of oil-producing filamentous fungi and will provide guidance for the large-scale production of ARA and similar polyunsaturated fatty acids.

Methods

Microorganism

Mortierella alpina R807 (CCTCC M 2012118), preserved in the China Center for Type Culture Collection, was used in the present study. It was maintained on potato dextrose agar (PDA) slants by culturing for 10 days at 25 °C, and transferred every 3 months.

Culture medium

Slant medium: Potato dextrose agar (PDA). The PDA medium contained (g/L): potatoes 200; glucose 25; agar 20. Inoculation medium (g/L): yeast extract

(Angel Yeast Co., Ltd, China) 6; glucose 30; KH_2PO_4 3; $NaNO_3$ 3; $MgSO_4 \cdot 7H_2O$ 3. Fermentation medium (g/L): yeast extract 10; glucose 80; KH_2PO_4 4; $NaNO_3$ 3; $MgSO_4 \cdot 7H_2O$ 0.6; initial pH 6.0.

Fermentation methods

Pellets of *M. alpina* were used to inoculate the PDA slants which were cultivated at 25 °C. After 12–15 days of incubation, a loop was used to transfer mycelial material into deep 250 mL baffled flasks containing 50 mL inoculation medium, and the cultures were subsequently incubated for 2 days at 25 °C under constant orbital shaking at 125 rpm. Fed-batch fermentations were carried out in a 7.5 L bioreactor (New Brunswick Scientific, USA) containing 5 L of fermentation medium.

The multi-stage process was carried out according to our proposed stepwise aeration and agitation control strategy. The aeration rate was set at 6 L/min to achieve an aeration rate of 1.2 volumes of air per volume of liquid per minute (vvm), without agitation in stage I (0–48 h). The agitation rate was increased stepwise from 50 to 150 revolutions per minute (rpm) in stage II and subsequently kept constant at 200 rpm until the end of the fermentation. The aeration rate was set at 1.0 vvm from step II to step IV. Glucose (500 g/L stock solution) was fed into the fermentation broth during the entire fermentation process to maintain the glucose concentration at 5–20 g/L. Samples comprising 100 mL of the fermentation broth were taken periodically for further examination.

Determination of dry cell weight (DCW) and glucose concentration

Aliquots comprising 100 mL of the fermentation broth were used to determine the DCW using the filtration method. The broth samples were transferred to a suction filter under 0.1 MPa negative pressure. The cell pellet was washed twice with distilled water and dried at 60 °C to constant weight (12 h). An aliquot comprising 1 mL of fermentation broth was transferred to a centrifuge tube, centrifuged at $3000 \times g$ for 3 min, and the resulting supernatant was used to measure the glucose concentration, which was determined enzymatically using a simultaneous Bioanalyzer (SBA-40C, Institute of Biology, Shandong Academy of Sciences, China).

Total lipids (TLs)

The dry cell material was ground into a fine powder for lipid extraction and fatty acid determination. A 2 g aliquot of the resulting powder was loaded onto a Soxhlet extractor with 150 mL chloroform/methanol (2:1, v/v) and extracted for 8 h at 75 °C. Finally, the solvent was removed on a rotary evaporator and recycled, with TLs remaining as evaporation residue (Ji et al. 2014b).

Fatty acid methyl esters (FAMEs) were prepared according to the established method (Ji et al. 2014b; Ren et al. 2009) as follows: 1.5 mL n-hexane and 0.2 mL 0.5 M KOH– methanol were added to a centrifuge tube containing 0.1 g of the powdered dry cells, and mixed thoroughly by vortexing for 3 min, followed by stewing for 15 min. A 0.3 mL aliquot of the resulting upper phase was combined with 0.5 mL distilled water in another centrifuge tube and the tube was centrifuged at $5000 \times g$ for 3 min. The upper phase containing FAMEs was applied to a Thermo Finnigan trace GC2000 DSQ gas chromatograph equipped with a 30 m × 0.25 mm × 0.25 μm DB-23MS capillary column (Agilent Technologies). The column temperature was increased from 80 to 200 °C at 40 °C/min, and subsequently to 300 °C at 10 °C/min. The temperature of both the injector and detector was set to 250 °C. Nitrogen was used as the carrier gas at 1 mL/min. Peaks were identified using authentic standards of the corresponding fatty acid methyl esters (Sigma-Aldrich). Fatty acids were quantified based on their corresponding peak areas relative to the peak areas of the standards.

Results and discussion

Fermentation disparities between batch and fed-batch protocols

The effects of batch and fed-batch fermentation on DCW, lipid, ARA contents, and ARA production were investigated using a 7.5 L bioreactor (Fig. 1). Though batch fermentation was found to be optimal for growth and total lipid production, ARA was synthesized more rapidly in fed-batch cultures. The lipid concentration reached a maximum value of 17.6 g/L at 6.5 days. However, higher ARA contents (42.8%) and ARA yield (10.0 g/L) were obtained with fed-batch fermentation. These results suggest that the optimal culture conditions for lipid accumulation and ARA biosynthesis are different. A higher C/N ratio was achieved in batch fermentations, which stimulated lipid accumulation. However, it also led to lower ARA biosynthesis. Furthermore, the C/N ratio is an important fermentation parameter which can affect mycelial morphology (Koike et al. 2001; Park et al. 2001), and it has been demonstrated previously that the morphology of M. alpina mycelia has a strong effect on physical properties of the broth, which in turn might lead to poor mass transfer performance. Due to the difficulty of controlling the mycelial morphology of M. alpina under constant aeration and agitation rates, the dry cell weight obtained in fed-batch fermentation (36.1 g/L) was not much higher than what was obtained in batch fermentation (31.1 g/L). However, the ARA contents and ARA yield (42.8% and 10.0 g/L, respectively) were clearly enhanced over the batch fermentation. A low initial

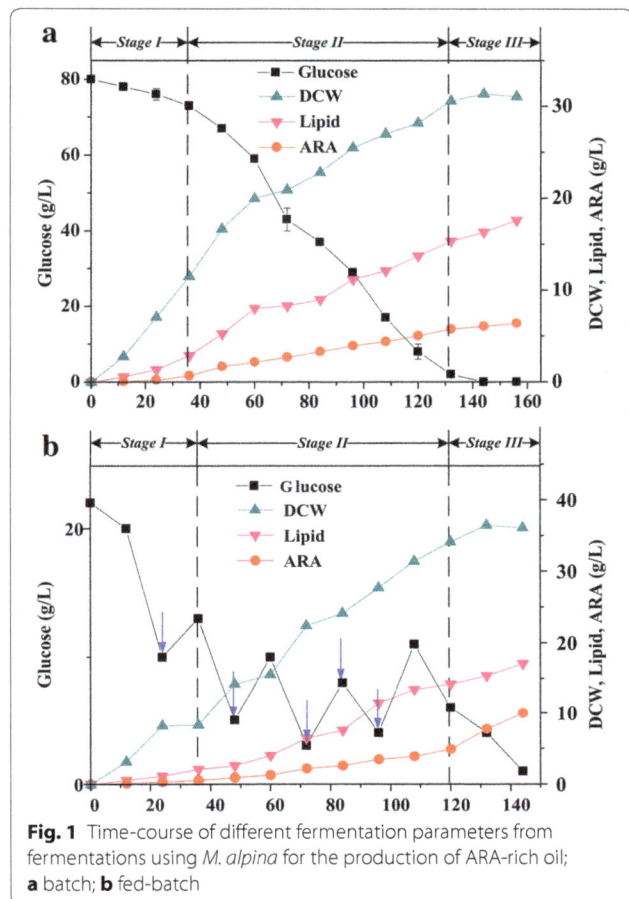

Fig. 1 Time-course of different fermentation parameters from fermentations using M. alpina for the production of ARA-rich oil; **a** batch; **b** fed-batch

glucose concentration is usually used to shorten the lag phase of fungal growth (Zhu et al. 2006). The glucose consumption during this stage was rapid, and the glucose was used up almost completely by day 6. The ARA productivity achieved by this method was 1.67 g/L day^{-1}, which is 1.70-fold higher than what was obtained in batch fermentations.

Controlling the morphology of Mortierella alpina using a multi-stage fermentation strategy

Seed culture morphology was found to be a significant factor in fermentations producing ARA-rich oil. This was directly due to the effects of mycelial morphology on the physical properties of the fermentation broth (Higashiyama et al. 2002). Thus, mycelial morphology is considered to be a key parameter in ARA fermentation, and the fungus must consequently be cultivated in the correct morphological form in order to obtain maximal ARA concentration (Ji et al. 2014b). Although feather-like hyphal filaments (Fig. 2a) were observed to be optimal for ARA production at low densities (Park et al. 1999),

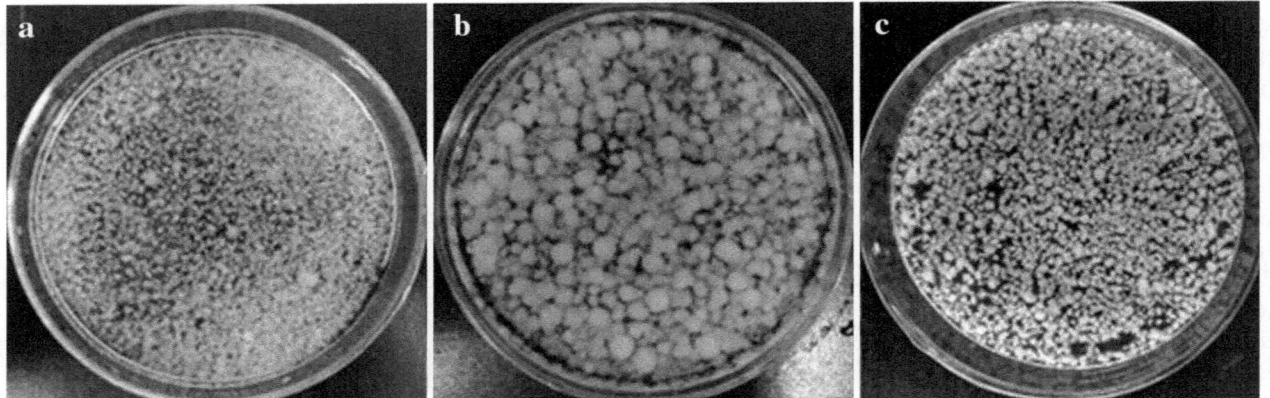

Fig. 2 Morphology of *M. alpina* mycelia cultured using different fermentation strategies. **a** At the end of the fed-batch fermentation; **b** at 48 h of the multi-stage strategy; **c** at the end of the multi-stage fermentation strategy

this morphology is disadvantageous at high cell densities because viscosity of the ferrmentation broth may be increased to an extent that oxygen transmission becomes limited (Wynn and Ratledge 2005).

In stage I (from 0 to 48 h), pellet-like mycelia of *M. alpina* were formed using baffled shake flasks and transferred into the nutrient-rich fermentation medium. Even though this pellet-like morphology allowed easier mixing and better mass transfer, pellet-like mycelia were highly sensitive to shear stress (Fig. 2b). No agitation combined with 1.2 vvm aeration was used to maintain this pellet-like morphology during the 2-day lag phase. During the fermentation period, the pellet-like particles became fragmented into both small pellets and filamentous mycelia. As the cultivation processed, the pellet-like cores became smaller, and more cell material displaying the pellet-like morphology was formed. After 132 h, the cells started to autolyse, and particles with a rice-shaped morphology were formed (Fig. 2c). This rice-shaped morphology proved to be optimal during the stage at which a high agitation rate was used to maintain the DO level.

Development of a multi-stage strategy for ARA fermentation

In this paper, the effects of aeration and agitation on cell morphology, lipid accumulation, and ARA production were investigated systemically, and a multi-stage strategy was developed aimed at achieving a high cell density, high accumulation rate, and high ARA yield. The final dry cell weight, total lipids, ARA contents, and ARA yield reached 41.41, 22.17 g/L, 61.05%, and 13.53 g/L,

respectively. The highest ARA productivity obtained in this study, which stood at 1.81 g/L day^{-1}, was achieved using the stepwise aeration and agitation control developed here. This was the same as the highest value published for *M. alpina* ME-1 (Jin et al. 2008) and was much higher than the 1.50 g/L day^{-1} reported for *M. alpina* DSA-12 using conventional protocols (Hwang et al. 2005).

Compared to the standard batch and fed-batch fermentation protocols, this multi-stage culture method prolongs the fermentation period by nearly 24 h, and while the total consumption glucose also increased sharply from 80 to 100 g/L, an obvious increase of dry cell weight was also noticed (41.4 g/L). ARA productivity consequently increased to 1.81 g/L day^{-1}, which was 1.08-fold higher than in the fed-batch fermentation.

To optimally analyze the process of cell growth and ARA accumulation, as well as to understand the effects of aeration and agitation on mycelial morphology, the fermentation process was divided into four stages according to cell growth characteristics (Ren et al. 2010). Stage I represents the beginning of the process until the morphological adaptation period; stage II was the phase of high cell density fermentation; stage III encompasses the lipid biosynthesis period; and stage IV comprises the period of most efficient ARA accumulation.

High cell density is the first precondition for high production of intracellular products, and it was obvious that high aeration was beneficial to cell growth. Our study also showed this positive effect, but with a reduction of $Y_{x/s}$ at stage II and a slight increase at stage III (Table 1).

Table 1 Comparison of fermentation parameters at different stages of ARA fermentation via a multi-stage fermentation strategy

	Stage			
	I	II	III	IV
Glucose consumption rate (g/L h^{-1})	0.479 ± 0.28	1.104 ± 0.19	0.694 ± 0.29	None
ARA increase (%)	None	10.676 ± 0.81	2.492 ± 0.13	18.862 ± 0.23
$Y_{x/s}$	0.614 ± 0.21	0.373 ± 0.32	0.419 ± 0.18	None
$Y_{l/s}$	0.165 ± 0.33	0.208 ± 0.38	0.341 ± 0.21	None
$Y_{ARA/s}$	0.048 ± 0.33	0.098 ± 0.33	0.159 ± 0.18	None

$Y_{x/s}$: conversion of glucose to biomass

$Y_{l/s}$: conversion of glucose to lipids

$Y_{ARA/s}$: conversion of glucose to ARA

This might be explained by the fact that cell respiration would be intensified and additional carbon flux channeled towards the tricarboxylic acid cycle under a high aeration rate. Consequently, dissolved oxygen can be controlled by using an appropriate aeration and agitation rate. To maintain the dissolved oxygen concentration, usually either an oxygen-enrichment method or a pressurization method is used (Higashiyama et al. 2002). In this research, on the other hand, the DO concentration was maintained via a combined stepwise aeration and agitation control strategy.

During stage I (from 0 to 48 h), glucose was consumed to below 10 g/L after a single feeding. Nitrogen was considered to be exhausted at 48 h (Lu et al. 2011; Ling et al. 2016). With ample carbon and nitrogen source, the dry cell weight increased slightly to 13 g/L, albeit with only 3 g/L of lipids and nearly 10 g/L of non-lipid dry cell weight. At the same time, pellet-like morphology could be preserved better. In stage II (from 48 to 96 h), the agitation rate was increased stepwise from 50 to 150 rpm, while the DO concentration was maintained between 10 and 20%. The glucose consumption rate reached its maximum, which might be explained by increased consumption for cell maintenance. At this stage, the glucose consumption rate was so high that glucose needed to be fed every 12 h, and a sharp increase of biomass from 13 to 28 g/L was also noticed. In stage III (from 96 to 132 h), the dry cell weight reached its maximum value of 44.2 g/L, whereas the non-lipid dry cell weight increased slightly and remained at a constant level. At the same time, the lipid contents increased from 11 to 25 g/L (Fig. 3), which indicated that cell metabolism had shifted away from cell growth towards lipid accumulation. After

the seventh batch of glucose feed, the consumption rate began to decline at 96 h. Moreover, the ARA contents slightly decreased, and other fatty acids (such as C18:0 and C18:1) increased, which can likely be explained by the high accumulation of lipid droplets.

High ARA content in total fatty acids is the prerequisite for high-quality ARA-rich oil, and a high aeration rate was favorable for efficient ARA biosynthesis. It is reported that the pathway of ARA biosynthesis was most widespread in oleaginous yeasts, and fungi encompasses both desaturation and elongation steps (Ji et al. 2014a). Palmitic acid (C16:0) is the main saturated fatty acid formed by fatty acid synthase (FAS), and ARA is subsequently produced from it via desaturase and elongase reactions. Our study showed that adequate oxygen was needed to increase the levels of unsaturated fatty acids, especially C18:3 and C20:4 (ARA). In stage IV (from 132 to 216 h), an improved mycelium-aging protocol was used to enhance ARA production (Zhang et al. 2015). When cells were cultivated under high aeration and agitation rates (Fig. 3), glucose was exhausted at 132 h, with dry cell weight reaching 44.2 g/L. Overall, the cells consumed 100 g/L of glucose, which was 20 g/L more than in fed-batch fermentation. The main increase in ARA contents was observed during stage IV, at which point the ARA contents reached 61.1%, with total yield also increasing sharply.

Changes in PUFA distribution in response to different dissolved oxygen conditions

PUFAs are produced via desaturation and elongation reactions, which involve aerobic oxygenation. Therefore, dissolved oxygen (DO) is a very significant factor

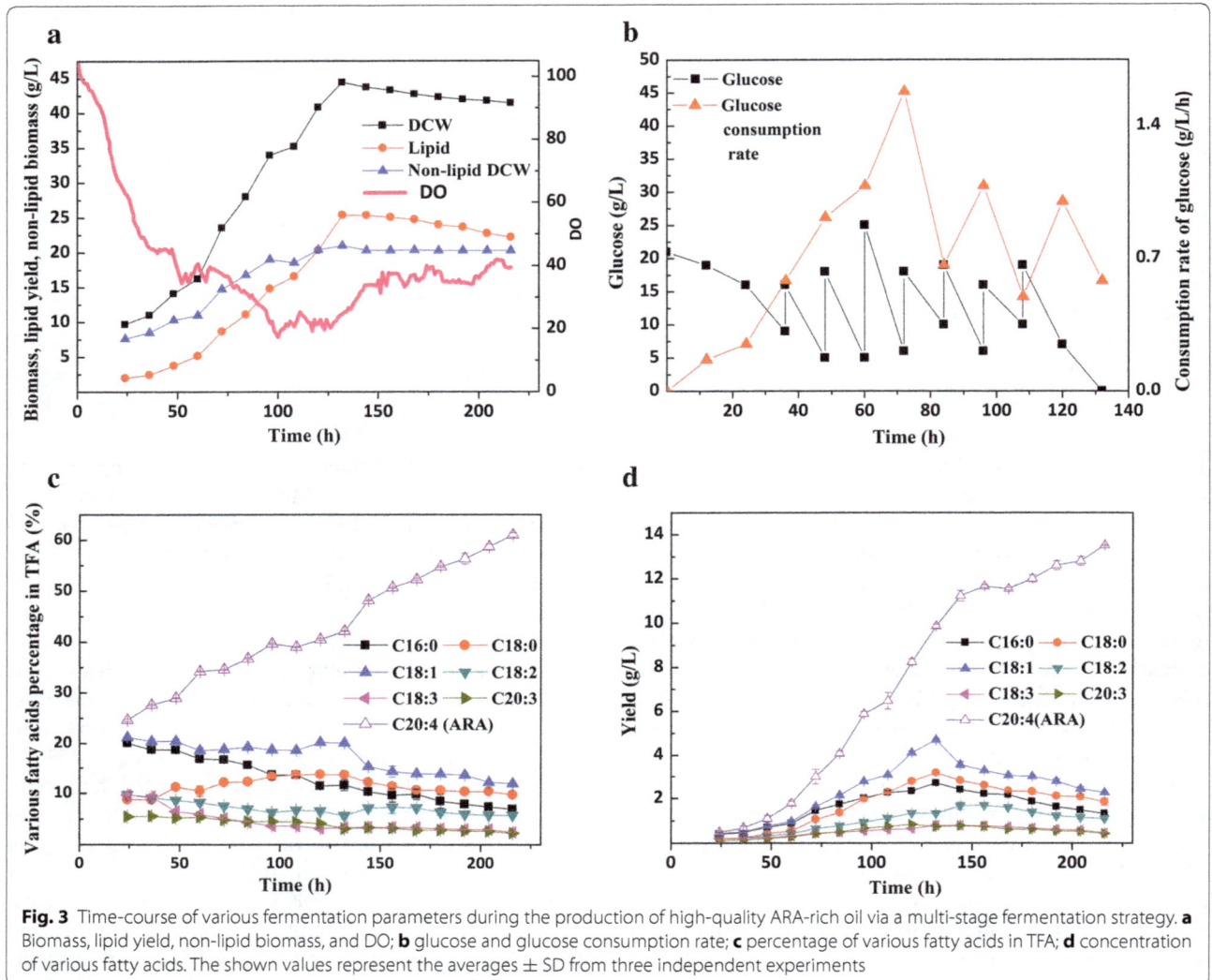

Fig. 3 Time-course of various fermentation parameters during the production of high-quality ARA-rich oil via a multi-stage fermentation strategy. **a** Biomass, lipid yield, non-lipid biomass, and DO; **b** glucose and glucose consumption rate; **c** percentage of various fatty acids in TFA; **d** concentration of various fatty acids. The shown values represent the averages ± SD from three independent experiments

for PUFA production, as reported by previous studies (Higashiyama et al. 2002; Su et al. 2016). There have also been some attempts to monitor and control the DO concentration in order to prevent DO limitation during ARA production (Higashiyama et al. 1999; Totani et al. 1992). Cultivations were carried out at different DO levels in the range of 30–40, 10–20, and 0–5%, respectively, and the optimum DO concentration range was found to be 30–40%, as shown in Fig. 4. No significant differences were observed in the contents of C18:2, C18:3, and C20:3 among the 30–40,

10–20, and 0–5% DO groups. The contents of C16:0 and C18:0 were increased slightly at 30–40 and 10–20% DO, respectively, compared to 0–5%. In this optimal DO concentration range, the ARA yield was enhanced about 1.2-fold compared to that obtained at 10–20%, and ARA contents decreased drastically from 39.4 to 22.8% at 0–5% DO. This decrease was likely due to stress caused by the very limited DO concentration. This observation underscores that DO concentration is indeed one of the most important factors influencing ARA productivity.

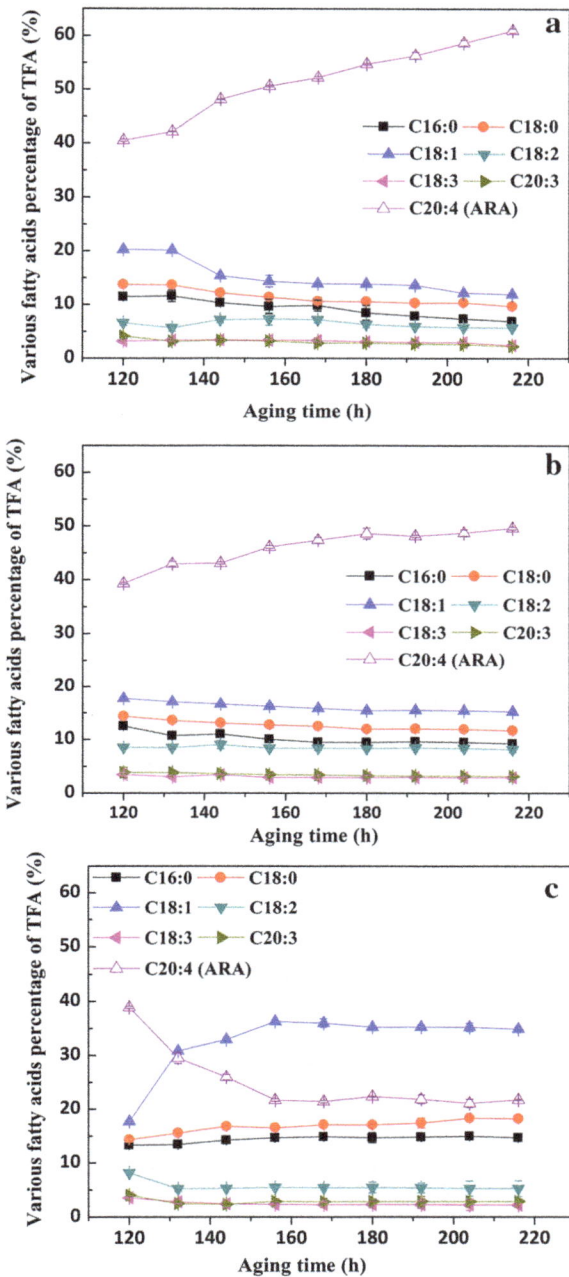

Fig. 4 Effect of different DO concentrations on the accumulation of ARA-rich oil during aging. **a** DO kept between 30 and 40%; **b** 10–20%; **c** 0–5%

Kinetic parameters of the multi-stage strategy

To analyze the kinetic characteristics of the multi-stage fermentation process, five parameters, including glucose consumption rate, ARA increase rate, $Y_{x/s}$, $Y_{l/s}$, and $Y_{ARA/s}$, were compared at different stages. The corresponding data are summarized in Table 1. The values of these kinetic parameters all fluctuated with time and were especially influenced by glucose feeding. At the early fermentation stage (stage I), high aeration without mechanical (Table 1) agitation was able to ensure a higher $Y_{x/s}$ (conversion of glucose to biomass) than was observed at the other stages. This indicates that high aeration could not only improve cell growth and glucose consumption, but could also accelerate the conversion of glucose to biomass. What is more, omitting the mechanical agitation was beneficial for maintaining a pellet-like morphology. At stage II, the value of $Y_{x/s}$ showed a decrease, whereas the values of ARA and $Y_{ARA/s}$ (conversion of glucose to ARA) increased sharply, indicating that a high lipid accumulation rate could be maintained under the high aeration and low agitation conditions found in stage II. After 96 h, in stage III, the value of lipid had increased significantly, indicating that the ARA contents were slightly decreased, whereas the other fatty acids, especially C18:0 and C18:1, increased sharply. After 132 h, the increasing ARA concentration reached the value of 18.9%, while glucose was exhausted. Although the ARA increase at the end of the multi-stage process (18.9%) was higher than during stages I to III, beginning cell autolysis resulted in less glucose consumption, and led to a slight decrease of overall dry cell weight. This further confirmed the importance of the multi-stage strategy combined with an efficient control of mycelial morphology.

To sum up, this study systematically examined the effects of aeration and agitation rates on ARA production by *M. alpina* and proposed a stepwise aeration and agitation rate control strategy to achieve a high cell growth rate and optimal overall productivity (Table 2).

Conclusions

This paper compared the experimental results of fermentations using *M. alpina* to produce oils rich in ARA via different culture strategies. The morphology of the fungal mycelia could be maintained in an optimal state throughout the fermentation. It could be shown that the multi-stage strategy provides a favorable gas–liquid mixture,

Table 2 Comparison of parameters from different ARA fermentation strategies

	Fermentation strategy			Rate of increase[a]
	Batch	Fed-batch	Multi-stage	
DCW (g/L)	31.06	36.08	41.41	14.77%
Mycelial specific growth rate (g/L d^{-1})	4.78	6.01	5.92	−1.50%
Lipids (g/L)	17.63	17.09	22.17	25.75%
Fermentation duration (d)	6.5	6	7.5	7.69%
Total glucose (g/L)	80	80	100	25%
Glucose consumption rate (g/L h^{-1})	0.51	0.56	0.76	35.71%
ARA contents (%)	36.30	42.81	61.05	42.61%
ARA yield (g/L)	6.40	10.01	13.53	35.16%
ARA productivity (g/L d^{-1})	0.98	1.67	1.81	7.78

[a] This value represents the corresponding data from the multi-stage strategy divided by the highest respective value from either the batch or fed-batch fermentation

and consequently increase biomass accumulation significantly. This work offers insights into the control of aeration and agitation and provides a reference for the fermentation of filamentous fungi at a mass industrial scale.

Authors' contributions
WJW and AHZ carried out the main experiments. CP, LJR, PS, and YDY helped in the cultivation of the strain and fatty acids assay. WJW, HH, and XJJ are involved in the drafting and revision of the manuscript. XJJ has given final approval of the version to be published. All authors read and approved the final manuscript.

Author details
[1] College of Biotechnology and Pharmaceutical Engineering, Nanjing Tech University, No. 30 South Puzhu Road, Nanjing 211816, People's Republic of China. [2] School of Pharmaceutical Sciences, Nanjing Tech University, No. 30 South Puzhu Road, Nanjing 211816, People's Republic of China. [3] State Key Laboratory of Materials-Oriented Chemical Engineering, Nanjing Tech University, No. 5 Xinmofan Road, Nanjing 210009, People's Republic of China. [4] Jiangsu National Synergetic Innovation Center for Advanced Materials (SICAM), No. 5 Xinmofan Road, Nanjing 210009, People's Republic of China. [5] Beijing Key Laboratory of Nutrition Health and Food Safety, COFCO Nutrition and Health Research Institute, Beijing 102209, People's Republic of China.

Acknowledgements
The authors wish to acknowledge the financial support from the National Science Foundation for Distinguished Young Scholars of China (No. 21225626), the National Natural Science Foundation of China (Nos. 21376002, and 21476111), the Natural Science Foundation of Jiangsu Province (No. BK20131405), the National High-Tech R&D Program of China (No. 2014AA021703), and the Priority Academic Program Development of Jiangsu Higher Education Institutions.

Competing interests
The authors declare that they have no competing interests.

References
Gao MJ, Wang C, Zheng ZY, Zhu L, Zhan XB, Lin CC (2016) Improving arachidonic acid fermentation by *Mortierella alpina* through multistage temperature and aeration rate control in bioreactor. Prep Biochem Biotechnol 46:360–367

Hamanaka T, Higashiyama K, Fujikawa S, Park EY (2001) Mycelial pellet intrastructure and visualization of mycelia and intracellular lipid in a culture of *Mortierella alpina*. Appl Microbiol Biotechnol 56:233–238

Higashiyama K, Murakami K, Tsujimura H, Matsumoto N, Fujikawa S (1999) Effects of dissolved oxygen on the morphology of an arachidonic acid production by *Mortierella alpina* 1S-4. Biotechnol Bioeng 63:442–448

Higashiyama K, Fujikawa S, Park EY, Shimizu S (2002) Production of arachidonic acid by *Mortierella* fungi. Biotechnol Bioprocess Eng 7:252–262

Hwang BH, Kim JW, Park CY, Park CS, Kim YS, Ryu YW (2005) High-level production of arachidonic acid by fed-batch culture of *Mortierella alpina* using NH4OH as a nitrogen source and pH control. Biotechnol Lett 27:731–735

Ji XJ, Ren LJ, Nie ZK, Huang H, Ouyang PK (2014a) Fungal arachidonic acid-rich oil: research, development and industrialization. Crit Rev Biotechnol 34:197–214

Ji XJ, Zhang AH, Nie ZK, Wu WJ, Ren LJ, Huang H (2014b) Efficient arachidonic acid-rich oil production by *Mortierella alpina* through a repeated fed-batch fermentation strategy. Bioresour Technol 170:356–360

Jin MJ, Huang H, Xiao AH, Zhang K, Liu X, Li S, Peng C (2008) A novel two-step fermentation process for improved arachidonic acid production by *Mortierella alpina*. Biotechnol Lett 30:1087–1091

Koike Y, Cai HJ, Higashiyama K, Fujikawa S, Park EY (2001) Effect of consumed carbon to nitrogen ratio of mycelial morphology and arachidonic acid production in cultures of *Mortierella alpina*. J Biosci Bioeng 91:382–389

Ling XP, Zeng SY, Chen CX, Liu XT, Lu YH (2016) Enhanced arachidonic acid production using a bioreactor culture of *Mortierella alpina* with a combined organic nitrogen source. Bioresour Bioprocess 3:43

Lu JM, Peng C, Ji XJ, You JY, Cong LL, Ouyang PK, Huang H (2011) Fermentation characteristics of *Mortierella alpina* in response to different nitrogen sources. Appl Biochem Biotechnol 164:979–990

Nie ZK, Ji XJ, Shang JS, Zhang AH, Ren LJ, Huang H (2014) Arachidonic acid-rich oil production by *Mortierella alpina* with different gas distributors. Bioprocess Biosyst Eng 37:1127–1132

Park EY, Koike Y, Higashiyama K, Fujikawa S, Okabe M (1999) Effect of nitrogen source on mycelial morphology and arachidonic acid production in cultures of *Mortierella alpina*. J Biosci Bioeng 88:61–67

Park EY, Koike Y, Cai HJ, Higashiyama K, Fujikawa S (2001) Morphological diversity of *Mortierella alpina*: effect of consumed carbon to nitrogen ratio in flask culture. Bioprocess Biosyst Eng 6:161–166

Peng C, Huang H, Ji X, Liu X, Ren LJ, Yu W, You JY, Lu JM (2010) Effects of n-hexadecane concentration and a two-stage oxygen supply control strategy on arachidonic acid production by *Mortierella alpina* ME-1. Chem Eng Technol 33:692–697

Ren LJ, Huang H, Xiao AH, Lian M, Jin LJ, Ji XJ (2009) Enhanced docosahexaenoic acid production by reinforcing acetyl-CoA and NADPH supply in *Schizochytrium* sp. HX-308. Bioprocess Biosyst Eng 32:837–843

Ren LJ, Ji XJ, Huang H, Qu L, Feng Y, Tong QQ, Ouyang PK (2010) Development of a stepwise aeration control strategy for efficient docosahexaenoic acid production by *Schizochytrium* sp. Appl Microbiol Biotechnol 87:1649–1656

Ryan AS, Zeller S, Nelson EB, Cohen Z, Ratledge C (2010) Safety evaluation of single cell oils and the regulatory requirements for use as food ingredients. In: Cohen Z, Ratledge C (eds) Single cell oils: microbial and algal oils, 2nd edn. AOCS Press, Urbana

Su GM, Jiao KL, Chang JY, Li Z, Guo XY, Sun Y, Zeng XH, Lu YH, Lin L (2016) Enhancing total fatty acids and arachidonic acid production by the red microalgae *Porphyridium purpureum*. Bioresour Bioprocess 3:33

Tai C, Li S, Xu Q, Ying H, Huang H, Ouyang PK (2010) Chitosan production from hemicellulose hydrolysate of corn straw: impact of degradation products on *Rhizopus oryzae* growth and chitosan fermentation. Lett Appl Microbiol 51:278–284

Totani N, Someya K, Oba K (1992) Industrial production of arachidonic acid by *Mortierella*. In: Kyle DJ, Ratledge C (eds) Industrial applications of single cell oils. AOCS Press, Urbana

Ward OP, Singh A (2005) Omega-3/6 fatty acids: alternative sources of production. Process Biochem 40:3627–3652

Wynn JP, Ratledge C (2005) Oils from microorganisms. In: Shahidi F (ed) Bailey's industrial oil and fat products, 6th edn. Wiley, New York, pp 121–153

Xu Q, Li S, Fu Y, Tai C, Huang H (2010) Two-stage utilization of corn straw by *Rhizopus oryzae* for fumaric acid production. Bioresour Technol 101:6262–6264

Zhang AH, Ji XJ, Wu WJ, Ren LJ, Yu YD, Huang H (2015) Lipid fraction and intracellular metabolite analysis reveal the mechanism of arachidonic acid-rich oil accumulation in the aging process of *Mortierella alpina*. J Agric Food Chem 63:9812–9819

Zhu M, Yu LJ, Li W, Zhou PP, Li CY (2006) Optimization of arachidonic acid production by fed-batch culture of *Mortierella alpina* based on dynamic analysis. Enzyme Microb Technol 38:735–740

PERMISSIONS

All chapters in this book were first published in BB, by Springer International Publishing AG.; hereby published with permission under the Creative Commons Attribution License or equivalent. Every chapter published in this book has been scrutinized by our experts. Their significance has been extensively debated. The topics covered herein carry significant findings which will fuel the growth of the discipline. They may even be implemented as practical applications or may be referred to as a beginning point for another development.

The contributors of this book come from diverse backgrounds, making this book a truly international effort. This book will bring forth new frontiers with its revolutionizing research information and detailed analysis of the nascent developments around the world.

We would like to thank all the contributing authors for lending their expertise to make the book truly unique. They have played a crucial role in the development of this book. Without their invaluable contributions this book wouldn't have been possible. They have made vital efforts to compile up to date information on the varied aspects of this subject to make this book a valuable addition to the collection of many professionals and students.

This book was conceptualized with the vision of imparting up-to-date information and advanced data in this field. To ensure the same, a matchless editorial board was set up. Every individual on the board went through rigorous rounds of assessment to prove their worth. After which they invested a large part of their time researching and compiling the most relevant data for our readers.

The editorial board has been involved in producing this book since its inception. They have spent rigorous hours researching and exploring the diverse topics which have resulted in the successful publishing of this book. They have passed on their knowledge of decades through this book. To expedite this challenging task, the publisher supported the team at every step. A small team of assistant editors was also appointed to further simplify the editing procedure and attain best results for the readers.

Apart from the editorial board, the designing team has also invested a significant amount of their time in understanding the subject and creating the most relevant covers. They scrutinized every image to scout for the most suitable representation of the subject and create an appropriate cover for the book.

The publishing team has been an ardent support to the editorial, designing and production team. Their endless efforts to recruit the best for this project, has resulted in the accomplishment of this book. They are a veteran in the field of academics and their pool of knowledge is as vast as their experience in printing. Their expertise and guidance has proved useful at every step. Their uncompromising quality standards have made this book an exceptional effort. Their encouragement from time to time has been an inspiration for everyone.

The publisher and the editorial board hope that this book will prove to be a valuable piece of knowledge for researchers, students, practitioners and scholars across the globe.

LIST OF CONTRIBUTORS

Adepu Kiran Kumar and Shaishav Sharma
Bioconversion Technology Division, Sardar Patel Renewable Energy Research Institute, Vallabh Vidyanagar, Anand 388 120, Gujarat, India

Li-Li Jiang, Jin-Jie Zhou and Zhi-Long Xiu
School of Life Science and Biotechnology, Dalian University of Technology, Linggong Road 2, Dalian 116024, Liaoning Province, China.

Chun-Shan Quan
Key Laboratory of Biotechnology and Bioresources Utilization, College of Life Science, Dalian Minzu University, Liaohe West Road 18, Jinzhou New District, Dalian 116600, Liaoning Province, China.

Víctor Hugo Grisales Díaz
School of Chemical Engineering and Advanced Materials, Newcastle University, Newcastle upon Tyne NE1 7RU, UK.

Gerard Olivar Tost
Control y Percepción Inteligente, Departamento de Ingeniería Eléctrica, Electrónica y Computación, Universidad Nacional de Colombia, Cra. 27 No. 64-60, Manizales, Colombia.

Puttaswamy Manjula, Govindan Srinikethan and K. Vidya Shetty
Department of Chemical Engineering, National Institute of Technology Karnataka, Surathkal, India

Vinay Mohan Pathak and Navneet
Department of Botany and Microbiology, Gurukul Kangri University, Haridwar, Uttarakhand 249-404, India

Eugene M. Obeng, Siti Nurul Nadzirah Adam and Cahyo Budiman
Biotechnology Research Institute, Universiti Malaysia Sabah, 88400 Kota Kinabalu, Sabah, Malaysia.

Clarence M. Ongkudon
Biotechnology Research Institute, Universiti Malaysia Sabah, 88400 Kota Kinabalu, Sabah, Malaysia.
Energy Research Institute, Universiti Malaysia Sabah, 88400 Kota Kinabalu, Sabah, Malaysia.

Ruth Maas
Autodisplay Biotech GmbH, Lifescience Center, Merowinger Platz 1a, 40225 Dusseldorf, Germany.

Joachim Jose
Autodisplay Biotech GmbH, Lifescience Center, Merowinger Platz 1a, 40225 Dusseldorf, Germany. Institute of Pharmaceutical and Medicinal Chemistry, PharmaCampus, Westphalian Wilhelms-University of Münster, Corrensstraße 48, 48149 Münster, Germany.

Qiang Zhang and Jie Bao
State Key Laboratory of Bioreactor Engineering, East China University of Science and Technology, 130 Meilong Road, Shanghai 200237, China

Krishan Kumar
Department of Food Technology, Akal College of Agriculture, Eternal University, Sirmour 173001, India.

Ajar Nath Yadav, Vinod Kumar, Pritesh Vyas and Harcharan Singh Dhaliwal
Department of Biotechnology, Akal College of Agriculture, Eternal University, Sirmour 173001, India.

Urooj Javed and Afsheen Aman
The Karachi Institute of Biotechnology and Genetic Engineering, University of Karachi, Karachi, Pakistan.

Shah Ali Ul Qader
Department of Biochemistry, University of Karachi, Karachi, Pakistan.

Shih-I Tan and You-Jin Yu
Department of Chemical Engineering, National Cheng Kung University, Tainan 70101, Taiwan.

I-Son Ng
Department of Chemical Engineering, National Cheng Kung University, Tainan 70101, Taiwan. Research Center for Energy Technology and Strategy, National Cheng Kung University, Tainan 70101, Taiwan.

Naseer Hussain and Shahid Abbas Abbasi
Centre for Pollution Control & Environmental Engineering, Pondicherry University, Chinakalapet, Puducherry 605 014, India.

Tasneem Abbasi
Centre for Pollution Control & Environmental Engineering, Pondicherry University, Chinakalapet, Puducherry 605 014, India.
Department of Fire Protection Engineering, Worcester Polytechnic Institute, Worcester, MA 01609, USA.

Andrew R. Barber and Wei Zhang
Centre for Marine Bioproducts Development, Flinders University, Adelaide, Australia.
Department of Medical Biotechnology, School of Medicine, Flinders University, Adelaide, Australia.

Trung T. Nguyen
Centre for Marine Bioproducts Development, Flinders University, Adelaide, Australia.
Department of Medical Biotechnology, School of Medicine, Flinders University, Adelaide, Australia.
Department of Food Science and Technology, Agricultural and Natural Resources Faculty, An Giang University, Long Xuyen, Vietnam.

Kendall Corbin
Centre for Marine Bioproducts Development, Flinders University, Adelaide, Australia.
Department of Medical Biotechnology, School of Medicine, Flinders University, Adelaide, Australia.
Centre for NanoScale Science Technology (CNST), Chemical and Physical Sciences, Flinders University, Adelaide, Australia.

Nawel Boucherba, Cilia Bouiche, Mohamed Yacine Kerbous, Yacine Maafa and Said Benallaoua
Laboratory of Applied Microbiology, Faculty of Nature Science and Life, University of Bejaia, 06000 Bejaia, Algeria.

Mohammed Gagaoua
NATAA, Université des Frères Mentouri Constantine 1, Route de Ain El-Bey, 25000 Constantine, Algeria.
UMR1213 Herbivores, INRA, VetAgro Sup, Clermont Université, Université de Lyon, 63122 Saint-Genès-Champanelle, France.

Amel Bouanane-Darenfed and Khelifa Bouacem
Laboratory of Cellular and Molecular Biology, Microbiology Team, Faculty of Biological Sciences, University of Sciences, Technology of Houari Boumediene (USTHB), PO Box 32, El Alia, Bab Ezzouar, 16111 Algiers, Algeria.

Ganti S. Murthy
Biological and Ecological Engineering, Oregon State University, Corvallis, OR, USA.

Deepak Kumar
Biological and Ecological Engineering, Oregon State University, Corvallis, OR, USA.
Agricultural and Biological Engineering, University of Illinois at Urbana-Champaign, Urbana, IL, USA.

Fei Lyu, Fei Gao, Qianqian Wei and Lin Liu
Department of Food Science, Ocean College, Zhejiang University of Technology, 18 Chaowang Road, Hangzhou 310014, China

Yuanming Gai
Tianjin Institute of Industrial Biotechnology, Chinese Academy of Sciences, Tianjin 300308, People's Republic of China.

Jingqi Chen, Jibin Sun and Dawei Zhang
Tianjin Institute of Industrial Biotechnology, Chinese Academy of Sciences, Tianjin 300308, People's Republic of China.
Key Laboratory of Systems Microbial Biotechnology, Chinese Academy of Sciences, Tianjin 300308, People's Republic of China.

Zhaoxia Jin
School of Biological Engineering, Dalian Polytechnic University, Dalian 116034, People's Republic of China.

Wen-Jia Wu, Ai-Hui Zhang and Ping Song
College of Biotechnology and Pharmaceutical Engineering, Nanjing Tech University, No. 30 South Puzhu Road, Nanjing 211816, People's Republic of China.

Lu-Jing Ren and Xiao-Jun Ji
College of Biotechnology and Pharmaceutical Engineering, Nanjing Tech University, No. 30 South Puzhu Road, Nanjing 211816, People's Republic of China.
Jiangsu National Synergetic Innovation Center for Advanced Materials (SICAM), No. 5 Xinmofan Road, Nanjing 210009, People's Republic of China.

Ya-Dong Yu and He Huang
School of Pharmaceutical Sciences, Nanjing Tech University, No. 30 South Puzhu Road, Nanjing 211816, People's Republic of China.
State Key Laboratory of Materials-Oriented Chemical Engineering, Nanjing Tech University, No. 5 Xinmofan Road, Nanjing 210009, People's Republic of China.

Jiangsu National Synergetic Innovation Center for Advanced Materials (SICAM), No. 5 Xinmofan Road, Nanjing 210009, People's Republic of China.

Chao Peng
Beijing Key Laboratory of Nutrition Health and Food Safety, COFCO Nutrition and Health Research Institute, Beijing 102209, People's Republic of China.

Index

www.ingramcontent.com/pod-product-compliance
Lightning Source LLC
Chambersburg PA
CBHW061245190326
41458CB00011B/3582